Sixth Edition

A First Course in Probability

Sheldon Ross

University of California, Berkeley

PEARSON
Education

ISBN 81-7758-361-1

First Impression, 2006
Second Impression, 2007

Published by Dorling Kindersley (India) Pvt. Ltd., licensees of Pearson Education in South Asia.

Head Office: 482, F.I.E., Pataparganj, Delhi 110 092, India.
Registered Office: 14 Local Shopping Centre, Panchsheel Park, New Delhi 110 017, India.

Printed in India by Anand Sons.

Contents

(optional sections are noted with an asterisk)

Preface

"We see that the theory of probability is at bottom only common sense reduced to calculation; it makes us appreciate with exactitude what reasonable minds feel by a sort of instinct, often without being able to account for it .... It is remarkable that this science, which originated in the consideration of games of chance, should have become the most important object of human knowledge .... The most important questions of life are, for the most part, really only problems of probability." So said the famous French mathematician and astronomer (the "Newton of France") Pierre Simon, Marquis de Laplace. Although many people might feel that the famous marquis, who was also one of the great contributors to the development of probability, might have exaggerated somewhat, it is nevertheless true that probability theory has become a tool of fundamental importance to nearly all scientists, engineers, medical practitioners, jurists, and industrialists. In fact, the enlightened individual had learned to ask not "Is it so?" but rather "What is the probability that it is so?"

This book is intended as an elementary introduction to the theory of probability for students in mathematics, statistics, engineering, and the sciences (including computer science, the social sciences and management science) who possess the prerequisite knowledge of elementary calculus. It attempts to present not only the mathematics of probability theory, but also, through numerous examples, the many diverse possible applications of this subject.

In Chapter 1 we present the basic principles of combinatorial analysis, which are most useful in computing probabilities.

In Chapter 2 we consider the axioms of probability theory and show how they can be applied to compute various probabilities of interest.

Chapter 3 deals with the extremely important subjects of conditional probability and independence of events. By a series of examples we illustrate how conditional probabilities come into play not only when some partial information is available, but also as a tool to enable us to compute probabilities more easily, even when no partial information is present. This extremely important technique of obtaining probabilities by "conditioning" reappears in Chapter 7, where we use it to obtain expectations.

In Chapters 4, 5, and 6 we introduce the concept of random variables. Discrete random variables are dealt with in Chapter 4, continuous random variables in Chapter 5, and jointly distributed random variables in Chapter 6. The important concepts of the expected value and the variance of a random variable are introduced in Chapters 4 and 5: These quantities are then determined for many of the common types of random variables.

Additional properties of the expected value are considered in Chapter 7. Many examples illustrating the usefulness of the result that the expected value of a sum of random variables is equal to the sum of their expected values are presented. Sections on conditional expectation, including its use in prediction, and moment generating functions are contained in this chapter. In addition, the final section introduces the multivariate normal distribution and presents a simple proof concerning the joint distribution of the sample mean and sample variance of a sample from a normal distribution.

In Chapter 8 we present the major theoretical results of probability theory. In particular, we prove the strong law of large numbers and the central limit theorem. Our proof of the strong law is a relatively simple one which assumes that the random variables have a finite fourth moment, and our proof of the central limit theorem assumes Levy's continuity theorem. Also in this chapter we present such probability inequalities as Markov's inequality, Chebyshev's inequality, and Chernoff bounds. The final section of Chapter 8 gives a bound on the error involved when a probability concerning a sum of independent Bernoulli random variables is approximated by the corresponding probability for a Poisson random variable having the same expected value.

Chapter 9 presents some additional topics, such as Markov chains, the Poisson process, and an introduction to information and coding theory, and Chapter 10 considers simulation.

The sixth edition continues the evolution and fine tuning of the text. There are many new exercises and examples. Among the latter are examples on utility (Example 4c of Chapter 4), on normal approximations (Example 4i of Chapter 5), on applying the lognormal distribution to finance (Example 3d of Chapter 6), and on coupon collecting with general collection probabilities (Example 2v of Chapter 7). There are also new optional subsections in Chapter 7 dealing with the probabilistic method (Subsection 7.2.1), and with the maximum-minimums identity (Subsection 7.2.2).

As in the previous edition, three sets of exercises are given at the end of each chapter. They are designated as **Problems, Theoretical Exercises**, and **Self-Test Problems and Exercises**. This last set of exercises, for which complete solutions appear in Appendix B, is designed to help students test their comprehension and study for exams.

All materials included on the Probability Models diskette from previous editions can now be downloaded from the Ross companion website at http://www.prenhall.com/Ross. Using the website students will be able to perform calculations and simulations quickly and easily in six key areas:

- Three of the modules derive probabilities for, respectively, binomial, Poisson, and normal random variables.
- Another module illustrates the central limit theorem. It considers random variables that take on one of the values $0, 1, 2, 3, 4$ and allows the user to enter the probabilities for these values along with a number n. The module then plots the probability mass function of the sum of n independent random variables of this type. By increasing n one can "see" the mass function converge to the shape of a normal density function.
- The other two modules illustrate the strong law of large numbers. Again the user enters probabilities for the five possible values of the random variable along with an integer n. The program then uses random numbers to simulate n random variables having the prescribed distribution. The modules graph the number of times each outcome occurs along with the average of all outcomes. The modules differ in how they graph the results of the trials.

We would like to thank the following reviewers whose helpful comments and suggestions contributed to recent editions of this book: Anastasia Ivanova, University of North Carolina; Richard Bass, University of Connecticut; Ed Wheeler, University of Tennessee; Jean Cadet, State University of New York at SUNY, Stony Brook; Jim Propp, University of Wisconsin; Mike Hardy, Massachusetts Institute of Technology; Anant Godbole, Michigan Technical University; Zakkula Govindarajulu, University of Kentucky; Richard Groeneveld. Iowa State University; Bernard Harris, University of Wisconsin; Stephen Herschkorn, Rutgers University; Robert Keener, University of Michigan; Thomas Liggett, University of California, Los Angeles; Bill McCormick, University of Georgia; and Kathryn Prewitt, Arizona State University. Special thanks go to Hossein Hamedani, Marquette University, and Ben Perles for their hard work in accuracy checking this manuscript.

We also express gratitude to the reviewers on earlier editions: Thomas R. Fischer, Texas A & M University; Jay DeVore, California Polytechnic University, San Luis Obispo; Robb J. Muirhead, University of Michigan; David Heath, Cornell University; Myra Samuels, Purdue University; I. R. Savage, Yale University; R. Miller, Stanford University; K. B. Athreya, Iowa State University; Phillip Beckwith, Michigan Tech; Howard Bird, St. Cloud State University; Steven Chiappari, Santa Clara University; James Clay, University of Arizona at Tucson; Francis Conlan, University of Santa Clara; Fred Leysieffer, Florida State University; Ian McKeague, Florida State University; Helmut Mayer, University of Georgia; N. U. Prabhu, Cornell University; Art Schwartz, University of Michigan at Ann Arbor; Therese Shelton, Southwestern University; and Allen Webster, Bradley University.

S. R.
smross@newton.me.berkeley.edu

DEDICATION

For Rebecca

CHAPTER 1

Combinatorial Analysis

1.1 INTRODUCTION

Here is a typical problem of interest involving probability. A communication system is to consist of n seemingly identical antennas that are to be lined up in a linear order. The resulting system will then be able to receive all incoming signals—and will be called *functional*—as long as no two consecutive antennas are defective. If it turns out that exactly m of the n antennas are defective, what is the probability that the resulting system will be functional? For instance, in the special case where $n = 4$ and $m = 2$ there are 6 possible system configurations—namely,

$$\begin{matrix} 0 & 1 & 1 & 0 \\ 0 & 1 & 0 & 1 \\ 1 & 0 & 1 & 0 \\ 0 & 0 & 1 & 1 \\ 1 & 0 & 0 & 1 \\ 1 & 1 & 0 & 0 \end{matrix}$$

where 1 means that the antenna is working and 0 that it is defective. As the resulting system will be functional in the first 3 arrangements and not functional in the remaining 3, it seems reasonable to take $\frac{3}{6} = \frac{1}{2}$ as the desired probability. In the case of general n and m, we could compute the probability that the system is functional in a similar fashion. That is, we could count the number of configurations that result in the system being functional and then divide by the total number of all possible configurations.

From the preceding we see that it would be useful to have an effective method for counting the number of ways that things can occur. In fact, many problems in probability theory can be solved simply by counting the number of different ways that a certain event can occur. The mathematical theory of counting is formally known as *combinatorial analysis*.

1.2 THE BASIC PRINCIPLE OF COUNTING

The following principle of counting will be basic to all our work. Loosely put, it states that if one experiment can result in any of m possible outcomes and if another experiment can result in any of n possible outcomes, then there are mn possible outcomes of the two experiments.

> ***The basic principle of counting***
>
> Suppose that two experiments are to be performed. Then if experiment 1 can result in any one of m possible outcomes and if for each outcome of experiment 1 there are n possible outcomes of experiment 2, then together there are mn possible outcomes of the two experiments.

Proof of the Basic Principle: The basic principle may be proved by enumerating all the possible outcomes of the two experiments as follows:

$$\begin{array}{c}(1,1),(1,2),\ldots,(1,n)\\(2,1),(2,2),\ldots,(2,n)\\\vdots\\(m,1),(m,2),\ldots,(m,n)\end{array}$$

where we say that the outcome is (i, j) if experiment 1 results in its ith possible outcome and experiment 2 then results in the jth of its possible outcomes. Hence the set of possible outcomes consists of m rows, each row containing n elements, which proves the result.

Example 2a. A small community consists of 10 women, each of whom has 3 children. If one woman and one of her children are to be chosen as mother and child of the year, how many different choices are possible?

Solution By regarding the choice of the woman as the outcome of the first experiment and the subsequent choice of one of her children as the outcome of the second experiment, we see from the basic principle that there are $10 \times 3 = 30$ possible choices. ■

When there are more than two experiments to be performed, the basic principle can be generalized as follows.

> ***The generalized basic principle of counting***
>
> If r experiments that are to be performed are such that the first one may result in any of n_1 possible outcomes, and if for each of these n_1 possible outcomes there are n_2 possible outcomes of the second experiment, and

if for each of the possible outcomes of the first two experiments there are n_3 possible outcomes of the third experiment, and if ..., then there is a total of $n_1 \cdot n_2 \cdots n_r$ possible outcomes of the r experiments.

Example 2b. A college planning committee consists of 3 freshmen, 4 sophomores, 5 juniors, and 2 seniors. A subcommittee of 4, consisting of 1 person from each class, is to be chosen. How many different subcommittees are possible?

Solution We may regard the choice of a subcommittee as the combined outcome of the four separate experiments of choosing a single representative from each of the classes. Hence it follows from the generalized version of the basic principle that there are $3 \times 4 \times 5 \times 2 = 120$ possible subcommittees. ■

Example 2c. How many different 7-place license plates are possible if the first 3 places are to be occupied by letters and the final 4 by numbers?

Solution By the generalized version of the basic principle the answer is $26 \cdot 26 \cdot 26 \cdot 10 \cdot 10 \cdot 10 \cdot 10 = 175{,}760{,}000$. ■

Example 2d. How many functions defined on n points are possible if each functional value is either 0 or 1?

Solution Let the points be $1, 2, \ldots, n$. Since $f(i)$ must be either 0 or 1 for each $i = 1, 2, \ldots, n$, it follows that there are 2^n possible functions. ■

Example 2e. In Example 2c, how many license plates would be possible if repetition among letters or numbers were prohibited?

Solution In this case there would be $26 \cdot 25 \cdot 24 \cdot 10 \cdot 9 \cdot 8 \cdot 7 = 78{,}624{,}000$ possible license plates. ■

1.3 PERMUTATIONS

How many different ordered arrangements of the letters a, b, and c are possible? By direct enumeration we see that there are 6: namely, *abc*, *acb*, *bac*, *bca*, *cab*, and *cba*. Each arrangement is known as a *permutation*. Thus there are 6 possible permutations of a set of 3 objects. This result could also have been obtained from the basic principle, since the first object in the permutation can be any of the 3, the second object in the permutation can then be chosen from any of the remaining 2, and the third object in the permutation is then chosen from the remaining 1. Thus there are $3 \cdot 2 \cdot 1 = 6$ possible permutations.

Suppose now that we have n objects. Reasoning similar to that we have just used for the 3 letters shows that there are

$$n(n-1)(n-2)\cdots 3 \cdot 2 \cdot 1 = n!$$

different permutations of the n objects.

Example 3a. How many different batting orders are possible for a baseball team consisting of 9 players?

Solution There are $9! = 362{,}880$ possible batting orders. ■

Example 3b. A class in probability theory consists of 6 men and 4 women. An examination is given, and the students are ranked according to their performance. Assume that no two students obtain the same score.

(a) How many different rankings are possible?
(b) If the men are ranked just among themselves and the women among themselves, how many different rankings are possible?

Solution (a) As each ranking corresponds to a particular ordered arrangement of the 10 people, we see that the answer to this part is $10! = 3{,}628{,}800$.

(b) As there are $6!$ possible rankings of the men among themselves and $4!$ possible rankings of the women among themselves, it follows from the basic principle that there are $(6!)(4!) = (720)(24) = 17{,}280$ possible rankings in this case. ■

Example 3c. Ms. Jones has 10 books that he is going to put on his bookshelf. Of these, 4 are mathematics books, 3 are chemistry books, 2 are history books, and 1 is a language book. Jones wants to arrange his books so that all the books dealing with the same subject are together on the shelf. How many different arrangements are possible?

Solution There are $4!\ 3!\ 2!\ 1!$ arrangements such that the mathematics books are first in line, then the chemistry books, then the history books, and then the language book. Similarly, for each possible ordering of the subjects, there are $4!\ 3!\ 2!\ 1!$ possible arrangements. Hence, as there are $4!$ possible orderings of the subjects, the desired answer is $4!\ 4!\ 3!\ 2!\ 1! = 6912$. ■

We shall now determine the number of permutations of a set of n objects when certain of the objects are indistinguishable from each other. To set this straight in our minds, consider the following example.

Example 3d. How many different letter arrangements can be formed using the letters $P\,E\,P\,P\,E\,R$?

Solution We first note that there are $6!$ permutations of the letters $P_1\,E_1\,P_2\,P_3\,E_2R$ when the $3P$'s and the $2E$'s are distinguished from each other. However, consider any one of these permutations—for instance, $P_1\,P_2\,E_1\,P_3\,E_2\,R$. If we now permute the P's among themselves and the E's among themselves, then the resultant arrangement would still be of the form $P\,P\,E\,P\,E\,R$. That is, all $3!\ 2!$ permutations

$$\begin{array}{ll} P_1\,P_2\,E_1\,P_3\,E_2\,R & P_1\,P_2\,E_2\,P_3\,E_1\,R \\ P_1\,P_3\,E_1\,P_2\,E_2\,R & P_1\,P_3\,E_2\,P_2\,E_1\,R \\ P_2\,P_1\,E_1\,P_3\,E_2\,R & P_2\,P_1\,E_2\,P_3\,E_1\,R \\ P_2\,P_3\,E_1\,P_1\,E_2\,R & P_2\,P_3\,E_2\,P_1\,E_1\,R \\ P_3\,P_1\,E_1\,P_2\,E_2\,R & P_3\,P_1\,E_2\,P_2\,E_1\,R \\ P_3\,P_2\,E_1\,P_1\,E_2\,R & P_3\,P_2\,E_2\,P_1\,E_1\,R \end{array}$$

are of the form $P\,P\,E\,P\,E\,R$. Hence there are $6!/(3!\,2!) = 60$ possible letter arrangements of the letters $P\,E\,P\,P\,E\,R$. ■

In general, the same reasoning as that used in Example 3d shows that there are

$$\frac{n!}{n_1!n_2!\cdots n_r!}$$

different permutations of n objects, of which n_1 are alike, n_2 are alike, ..., n_r are alike.

Example 3e. A chess tournament has 10 competitors of which 4 are Russian, 3 are from the United States, 2 from Great Britain, and 1 from Brazil. If the tournament result lists just the nationalities of the players in the order in which they placed, how many outcomes are possible?

Solution There are

$$\frac{10!}{4!\,3!\,2!\,1!} = 12{,}600$$

possible outcomes. ■

Example 3f. How many different signals, each consisting of 9 flags hung in a line, can be made from a set of 4 white flags, 3 red flags, and 2 blue flags if all flags of the same color are identical?

Solution There are

$$\frac{9!}{4!\,3!\,2!} = 1260$$

different signals. ■

1.4 COMBINATIONS

We are often interested in determining the number of different groups of r objects that could be formed from a total of n objects. For instance, how many different groups of 3 could be selected from the 5 items A, B, C, D, and E? To answer this, reason as follows: Since there are 5 ways to select the initial item, 4 ways to then select the next item, and 3 ways to select the final item, there are thus $5 \cdot 4 \cdot 3$ ways of selecting the group of 3 when the order in which the items are selected is relevant. However, since every group of 3, say, the group consisting of items A, B, and C, will be counted 6 times (that is, all of the permutations ABC, ACB, BAC, BCA, CAB, and CBA will be counted when the order of selection is relevant), it follows that the total number of groups that can be formed is

$$\frac{5 \cdot 4 \cdot 3}{3 \cdot 2 \cdot 1} = 10$$

In general, as $n(n-1)\cdots(n-r+1)$ represents the number of different ways that a group of r items could be selected from n items when the order of selection is relevant, and as each group of r items will be counted $r!$ times in this count, it follows that the number of different groups of r items that could be formed from a set of n items is

$$\frac{n(n-1)\cdots(n-r+1)}{r!} = \frac{n!}{(n-r)!\,r!}$$

Notation and terminology

We define $\binom{n}{r}$, for $r \le n$, by

$$\binom{n}{r} = \frac{n!}{(n-r)!\,r!}$$

and say that $\binom{n}{r}$ represents the number of possible combinations of n objects taken r at a time.†

Thus $\binom{n}{r}$ represents the number of different groups of size r that could be selected from a set of n objects when the order of selection is not considered relevant.

Example 4a. A committee of 3 is to be formed from a group of 20 people. How many different committees are possible?

Solution There are $\binom{20}{3} = \frac{20 \cdot 19 \cdot 18}{3 \cdot 2 \cdot 1} = 1140$ possible committees. ■

†By convention, 0! is defined to be 1. Thus $\binom{n}{0} = \binom{n}{n} = 1$. We also take $\binom{n}{i}$ to be equal to 0 when either $i < 0$ or $i > n$.

Example 4b. From a group of 5 women and 7 men, how many different committees consisting of 2 women and 3 men can be formed? What if 2 of the men are feuding and refuse to serve on the committee together?

Solution As there are $\binom{5}{2}$ possible groups of 2 women, and $\binom{7}{3}$ possible groups of 3 men, it follows from the basic principle that there are $\binom{5}{2}\binom{7}{3}$ $= \left(\frac{5 \cdot 4}{2 \cdot 1}\right)\frac{7 \cdot 6 \cdot 5}{3 \cdot 2 \cdot 1} = 350$ possible committees consisting of 2 women and 3 men.

On the other hand, if 2 of the men refuse to serve on the committee together, then, as there are $\binom{2}{0}\binom{5}{3}$ possible groups of 3 men not containing either of the 2 feuding men and $\binom{2}{1}\binom{5}{2}$ groups of 3 men containing exactly 1 of the feuding men, it follows that there are $\binom{2}{0}\binom{5}{3} + \binom{2}{1}\binom{5}{2} = 30$ groups of 3 men not containing both of the feuding men. Since there are $\binom{5}{2}$ ways to choose the 2 women, it follows that in this case there are $30\binom{5}{2} = 300$ possible committees. ■

Example 4c. Consider a set of n antennas of which m are defective and $n - m$ are functional and assume that all of the defectives and all of the functionals are considered indistinguishable. How many linear orderings are there in which no two defectives are consecutive?

Solution Imagine that the $n - m$ functional antennas are lined up among themselves. Now, if no two defectives are to be consecutive, then the spaces between the functional antennas must each contain at most one defective antenna. That is, in the $n - m + 1$ possible positions—represented in Figure 1.1 by carets—between the $n - m$ functional antennas, we must select m of these in which to put the defective antennas. Hence there are $\binom{n - m + 1}{m}$ possible orderings in which there is at least one functional antenna between any two defective ones. ■

$$\wedge 1 \wedge 1 \wedge 1 \ldots \wedge 1 \wedge 1 \wedge$$

1 = functional

$\wedge$ = place for at most one defective

Figure 1.1

A useful combinatorial identity is

$$\binom{n}{r} = \binom{n-1}{r-1} + \binom{n-1}{r} \qquad 1 \leq r \leq n \tag{4.1}$$

Equation (4.1) may be proved analytically or by the following combinatorial argument. Consider a group of n objects and fix attention on some particular one of these objects—call it object 1. Now, there are $\binom{n-1}{r-1}$ groups of size r that contain object 1 (since each such group is formed by selecting $r - 1$ from the remaining $n - 1$ objects). Also, there are $\binom{n-1}{r}$ groups of size r that do not contain object 1. As there is a total of $\binom{n}{r}$ groups of size r, Equation (4.1) follows.

The values $\binom{n}{r}$ are often referred to as *binomial coefficients.* This is so because of their prominence in the binomial theorem.

The binomial theorem

$$(x + y)^n = \sum_{k=0}^{n} \binom{n}{k} x^k y^{n-k} \tag{4.2}$$

We shall present two proofs of the binomial theorem. The first is a proof by mathematical induction, and the second is a proof based on combinatorial considerations.

Proof of the Binomial Theorem by Induction: When $n = 1$, Equation (4.2) reduces to

$$x + y = \binom{1}{0}x^0y^1 + \binom{1}{1}x^1y^0 = y + x$$

Assume Equation (4.2) for $n - 1$. Now,

$$\begin{aligned}(x + y)^n &= (x + y)(x + y)^{n-1} \\ &= (x + y)\sum_{k=0}^{n-1}\binom{n-1}{k}x^k y^{n-1-k} \\ &= \sum_{k=0}^{n-1}\binom{n-1}{k}x^{k+1}y^{n-1-k} + \sum_{k=0}^{n-1}\binom{n-1}{k}x^k y^{n-k}\end{aligned}$$

Letting $i = k + 1$ in the first sum and $i = k$ in the second sum, we find that

$$
\begin{aligned}
(x+y)^n &= \sum_{i=1}^{n}\binom{n-1}{i-1}x^i y^{n-i} + \sum_{i=0}^{n-1}\binom{n-1}{i}x^i y^{n-i} \\
&= x^n + \sum_{i=1}^{n-1}\left[\binom{n-1}{i-1}+\binom{n-1}{i}\right]x^i y^{n-i} + y^n \\
&= x^n + \sum_{i=1}^{n-1}\binom{n}{i}x^i y^{n-i} + y^n \\
&= \sum_{i=0}^{n}\binom{n}{i}x^i y^{n-i}
\end{aligned}
$$

where the next-to-last equality follows by Equation (4.1). By induction the theorem is now proved.

Combinatorial Proof of the Binomial Theorem: Consider the product

$$(x_1 + y_1)(x_2 + y_2)\cdots(x_n + y_n)$$

Its expansion consists of the sum of 2^n terms, each term being the product of n factors. Furthermore, each of the 2^n terms in the sum will contain as a factor either x_i or y_i for each $i = 1, 2, \ldots, n$. For example,

$$(x_1 + y_1)(x_2 + y_2) = x_1x_2 + x_1y_2 + y_1x_2 + y_1y_2$$

Now, how many of the 2^n terms in the sum will have as factors k of the x_i's and $(n - k)$ of the y_i's? As each term consisting of k of the x_i's and $(n - k)$ of the y_i's corresponds to a choice of a group of k from the n values $x_1, x_2, \ldots, x_n$, there are $\binom{n}{k}$ such terms. Thus, letting $x_i = x$, $y_i = y$, $i = 1, \ldots, n$, we see that

$$(x + y)^n = \sum_{k=0}^{n}\binom{n}{k}x^k y^{n-k}$$

Example 4d. Expand $(x + y)^3$.

Solution

$$
\begin{aligned}
(x+y)^3 &= \binom{3}{0}x^0y^3 + \binom{3}{1}x^1y^2 + \binom{3}{2}x^2y + \binom{3}{3}x^3y^0 \\
&= y^3 + 3xy^2 + 3x^2y + x^3
\end{aligned}
$$

■

Example 4e. How many subsets are there of a set consisting of n elements?

Solution Since there are $\binom{n}{k}$ subsets of size k, the desired answer is

$$\sum_{k=0}^{n}\binom{n}{k} = (1 + 1)^n = 2^n$$

This result could also have been obtained by assigning to each element in the set either the number 0 or the number 1. To each assignment of numbers there corresponds, in a one-to-one fashion, a subset, namely, that subset consisting of all elements that were assigned the value 1. As there are 2^n possible assignments, the result follows.

Note that we have included as a subset the set consisting of 0 elements (that is, the null set). Hence the number of subsets that contain at least one element is $2^n - 1$. ■

1.5 MULTINOMIAL COEFFICIENTS

In this section we consider the following problem: A set of n distinct items is to be divided into r distinct groups of respective sizes $n_1, n_2, \ldots, n_r$, where $\sum_{i=1}^{r} n_i = n$. How many different divisions are possible? To answer this, we note that there are $\binom{n}{n_1}$ possible choices for the first group; for each choice of the first group there are $\binom{n - n_1}{n_2}$ possible choices for the second group; for each choice of the first two groups there are $\binom{n - n_1 - n_2}{n_3}$ possible choices for the third group; and so on. Hence it follows from the generalized version of the basic counting principle that there are

$$\binom{n}{n_1}\binom{n - n_1}{n_2}\cdots\binom{n - n_1 - n_2 - \cdots - n_{r-1}}{n_r}$$
$$= \frac{n!}{(n - n_1)!\,n_1!}\,\frac{(n - n_1)!}{(n - n_1 - n_2)!\,n_2!}\cdots\frac{(n - n_1 - n_2 - \cdots - n_{r-1})!}{0!\,n_r!}$$
$$= \frac{n!}{n_1!\,n_2!\cdots n_r!}$$

possible divisions.

Notation

If $n_1 + n_2 + \cdots + n_r = n$, we define $\binom{n}{n_1, n_2, \ldots, n_r}$ by

$$\binom{n}{n_1, n_2, \ldots, n_r} = \frac{n!}{n_1!\,n_2!\cdots n_r!}$$

Thus $\binom{n}{n_1, n_2, \ldots, n_r}$ represents the number of possible divisions of n distinct objects into r distinct groups of respective sizes $n_1, n_2, \ldots, n_r$.

Example 5a. A police department in a small city consists of 10 officers. If the department policy is to have 5 of the officers patrolling the streets, 2 of the officers working full time at the station, and 3 of the officers on reserve at the station, how many different divisions of the 10 officers into the 3 groups are possible?

Solution There are $\frac{10!}{5!\,2!\,3!} = 2520$ possible divisions. ■

Example 5b. Ten children are to be divided into an A team and a B team of 5 each. The A team will play in one league and the B team in another. How many different divisions are possible?

Solution There are $\frac{10!}{5!\,5!} = 252$ possible divisions. ■

Example 5c. In order to play a game of basketball, 10 children at a playground divide themselves into two teams of 5 each. How many different divisions are possible?

Solution Note that this example is different from Example 5b because now the order of the two teams is irrelevant. That is, there is no A and B team but just a division consisting of 2 groups of 5 each. Hence the desired answer is

$$\frac{10!/(5!\,5!)}{2!} = 126$$ ■

The proof of the following theorem, which generalizes the binomial theorem, is left as an exercise.

The multinomial theorem

$$(x_1 + x_2 + \cdots + x_r)^n = \sum_{\substack{(n_1, \ldots, n_r): \\ n_1 + \cdots + n_r = n}} \binom{n}{n_1, n_2, \ldots, n_r} x_1^{n_1} x_2^{n_2} \cdots x_r^{n_r}$$

That is, the sum is over all nonnegative integer-valued vectors $(n_1, n_2, \ldots, n_r)$ such that $n_1 + n_2 + \cdots + n_r = n$.

The numbers $\binom{n}{n_1, n_2, \ldots, n_r}$ are known as *multinomial coefficients.*

Example 5d.

$$\begin{aligned}(x_1 + x_2 + x_3)^2 &= \binom{2}{2,0,0}x_1^2x_2^0x_3^0 + \binom{2}{0,2,0}x_1^0x_2^2x_3^0 \\ &\quad + \binom{2}{0,0,2}x_1^0x_2^0x_3^2 + \binom{2}{1,1,0}x_1^1x_2^1x_3^0 \\ &\quad + \binom{2}{1,0,1}x_1^1x_2^0x_3^1 + \binom{2}{0,1,1}x_1^0x_2^1x_3^1 \\ &= x_1^2 + x_2^2 + x_3^2 + 2x_1x_2 + 2x_1x_3 + 2x_2x_3\end{aligned}$$

*1.6 THE NUMBER OF INTEGER SOLUTIONS OF EQUATIONS

There are r^n possible outcomes when n distinguishable balls are to be distributed into r distinguishable urns. This follows because each ball may be distributed into any of r possible urns. Let us now, however, suppose that the n balls are indistinguishable from each other. In this case, how many different outcomes are possible? As the balls are indistinguishable, it follows that the outcome of the experiment of distributing the n balls into r urns can be described by a vector $(x_1, x_2, \dots x_r)$, where x_i denotes the number of balls that are distributed into the ith urn. Hence the problem reduces to finding the number of distinct nonnegative integer-valued vectors $(x_1, x_2, \dots, x_r)$ such that

$$x_1 + x_2 + \cdots + x_r = n$$

To compute this, let us start by considering the number of positive integer-valued solutions. Toward this end, imagine that we have n indistinguishable objects lined up and that we want to divide them into r nonempty groups. To do so, we can select $r - 1$ of the $n - 1$ spaces between adjacent objects as our dividing points (see Figure 1.2). For instance, if we have $n = 8$ and $r = 3$ and choose the 2 divisors as shown

ooo|ooo|oo

then the vector obtained is $x_1 = 3, x_2 = 3, x_3 = 2$. As there are $\binom{n-1}{r-1}$ possible selections, we obtain the following proposition.

$0 \wedge 0 \wedge 0 \wedge \dots \wedge 0 \wedge 0$

n objects 0

Choose $r - 1$ of the spaces $\wedge$.

Figure 1.2

* Note that asterisks denote material that is optional.

Proposition 6.1

There are $\binom{n-1}{r-1}$ distinct positive integer-valued vectors $(x_1, x_2, \ldots, x_r)$ satisfying.

$$x_1 + x_2 + \cdots + x_r = n \qquad x_i > 0, i = 1, \ldots, r$$

To obtain the number of nonnegative (as opposed to positive) solutions, note that the number of nonnegative solutions of $x_1 + x_2 + \cdots + x_r = n$ is the same as the number of positive solutions of $y_1 + \cdots + y_r = n + r$ (seen by letting $y_i = x_i + 1, i = 1, \ldots, r$). Hence, from Proposition 6.1, we obtain the following proposition.

Proposition 6.2

There are $\binom{n+r-1}{r-1}$ distinct nonnegative integer-valued vectors $(x_1, x_2, \ldots, x_r)$ satisfying

$$x_1 + x_2 + \cdots + x_r = n \tag{6.1}$$

Example 6a. How many distinct nonnegative integer-valued solutions of $x_1 + x_2 = 3$ are possible?

Solution There are $\binom{3+2-1}{2-1} = 4$ such solutions: $(0, 3)$, $(1, 2)$, $(2, 1)$, $(3, 0)$. ■

Example 6b. An investor has 20 thousand dollars to invest among 4 possible investments. Each investment must be in units of a thousand dollars. If the total 20 thousand is to be invested, how many different investment strategies are possible? What if not all the money need be invested?

Solution If we let $x_i, i = 1, 2, 3, 4$, denote the number of thousands invested in investment number i, then, when all is to be invested, x_1, x_2, x_3, x_4 are integers satisfying.

$$x_1 + x_2 + x_3 + x_4 = 20 \qquad x_i \geq 0$$

Hence, by Proposition 6.2, there are $\binom{23}{3} = 1771$ possible investment strategies. If not all of the money need be invested, then if we let x_5 denote the amount kept in reserve, a strategy is a nonnegative integer-valued vector $(x_1, x_2, x_3, x_4, x_5)$ satisfying

$$x_1 + x_2 + x_3 + x_4 + x_5 = 20$$

Hence, by Proposition 6.2, there are now $\binom{24}{4} = 10{,}626$ possible strategies. ■

Example 6c. How many terms are there in the multinomial expansion of $(x_1 + x_2 + \cdots + x_r)^n$?

Solution

$$(x_1 + x_2 + \cdots + x_r)^n = \sum \binom{n}{n_1, \ldots, n_r} x_1^{n_1} \cdots x_r^{n_r}$$

where the sum is over all nonnegative integer-valued $(n_1, \ldots, n_r)$ such that $n_1 + \cdots + n_r = n$. Hence, by Proposition 6.2, there are $\binom{n + r - 1}{r - 1}$ such terms. ■

Example 6d. Let us reconsider Example 4c, in which we have a set of n items, of which m are (indistinguishable and) defective and the remaining $n - m$ are (also indistinguishable and) functional. Our objective is to determine the number of linear orderings in which no two defectives are next to each other. To determine this quantity, let us imagine that the defective items are lined up among themselves and the functional ones are now to be put in position. Let us denote x_1 as the number of functional items to the left of the first defective, x_2 as the number of functional items between the first two defectives, and so on. That is, schematically we have

$$x_1 \ 0 \ x_2 \ 0 \cdots x_m \ 0 \ x_{m+1}$$

Now there will be at least one functional item between any pair of defectives as long as $x_i > 0, i = 2, \ldots, m$. Hence the number of outcomes satisfying the condition is the number of vectors $x_1, \ldots, x_{m+1}$ that satisfy

$$x_1 + \cdots + x_{m+1} = n - m \qquad x_1 \geq 0, x_{m+1} \geq 0, x_i > 0, i = 2, \ldots, m$$

But on letting $y_1 = x_1 + 1, y_i = x_i, i = 2, \ldots, m, y_{m+1} = x_{m+1} + 1$, we see that this is equal to the number of positive vectors $(y_1, \ldots, y_{m+1})$ that satisfy

$$y_1 + y_2 + \cdots + y_{m+1} = n - m + 2$$

Hence, by Proposition 6.1, there are $\binom{n - m + 1}{m}$ such outcomes, which is in agreement with the results of Example 4c.

Suppose now that we are interested in the number of outcomes in which each pair of defective items is separated by at least 2 functional ones. By the same reasoning as that applied previously, this would equal the number of vectors satisfying

$$x_1 + \cdots + x_{m+1} = n - m \qquad x_1 \geq 0, x_{m+1} \geq 0, x_i \geq 2, i = 2, \ldots, m$$

Upon letting $y_1 = x_1 + 1$, $y_i = x_i - 1$, $i = 2, \ldots, m$, $y_{m+1} = x_{m+1} + 1$, we see that this is the same as the number of positive solutions of

$$y_1 + \cdots + y_{m+1} = n - 2m + 3$$

Hence, from Proposition 6.1, there are $\binom{n-2m+2}{m}$ such outcomes. ■

SUMMARY

The basic principle of counting states that if an experiment consisting of two phases is such that there are n possible outcomes of phase 1, and for each of these n outcomes there are m possible outcomes of phase 2, then there are nm possible outcomes of the experiment.

There are $n! = n(n-1)\cdots 3 \cdot 2 \cdot 1$ possible linear orderings of n items. The quantity 0! is defined to equal 1.

Let

$$\binom{n}{i} = \frac{n!}{(n-i)!\,i!}$$

when $0 \leq i \leq n$, and let it equal 0 otherwise. This quantity represents the number of different subgroups of size i that can be chosen from a set of size n. It is often called a *binomial coefficient* because of its prominence in the binomial theorem, which states that

$$(x+y)^n = \sum_{i=0}^{n} \binom{n}{i} x^i y^{n-i}$$

For nonnegative integers $n_1, \ldots, n_r$ summing to n,

$$\binom{n}{n_1, n_2, \ldots, n_r} = \frac{n!}{n_1!n_2!\cdots n_r!}$$

is the number of divisions of n items into r distinct nonoverlapping subgroups of sizes $n_1, n_2, \ldots, n_r$.

PROBLEMS

1. **(a)** How many different 7-place license plates are possible if the first 2 places are for letters and the other 5 for numbers?
   (b) Repeat part (a) under the assumption that no letter or number can be repeated in a single license plate.
2. How many outcome sequences are possible when a die is rolled four times, where we say, for instance, that the outcome is 3, 4, 3, 1 if the first roll landed on 3, the second on 4, the third on 3, and the fourth on 1?
3. Twenty workers are to be assigned to 20 different jobs, one to each job. How many different assignments are possible?

4. John, Jim, Jay, and Jack have formed a band consisting of 4 instruments. If each of the boys can play all 4 instruments, how many different arrangements are possible? What if John and Jim can play all 4 instruments, but Jay and Jack can each play only piano and drums?

5. For years, telephone area codes in the United States and Canada consisted of a sequence of three digits. The first digit was an integer between 2 and 9; the second digit was either 0 or 1; the third digit was any integer between 1 and 9. How many area codes were possible? How many area codes starting with a 4 were possible?

6. A well-known nursery rhyme starts as follows:

As I was going to St. Ives
I met a man with 7 wives.
Each wife had 7 sacks.
Each sack had 7 cats.
Each cat had 7 kittens.
How many kittens did the traveler meet?

7. **(a)** In how many ways can 3 boys and 3 girls sit in a row?
(b) In how many ways can 3 boys and 3 girls sit in a row if the boys and the girls are each to sit together?
(c) In how many ways if only the boys must sit together?
(d) In how many ways if no two people of the same sex are allowed to sit together?

8. How many different letter arrangements can be made from the letters
(a) FLUKE;
(b) PROPOSE;
(c) MISSISSIPPI;
(d) ARRANGE?

9. A child has 12 blocks, of which 6 are black, 4 are red, 1 is white, and 1 is blue. If the child puts the blocks in a line, how many arrangements are possible?

10. In how many ways can 8 people be seated in a row if
(a) there are no restrictions on the seating arrangement;
(b) persons A and B must sit next to each other;
(c) there are 4 men and 4 women and no 2 men or 2 women can sit next to each other;
(d) there are 5 men and they must sit next to each other;
(e) there are 4 married couples and each couple must sit together?

11. In how many ways can 3 novels, 2 mathematics books, and 1 chemistry book be arranged on a bookshelf if
(a) the books can be arranged in any order;
(b) the mathematics books must be together and the novels must be together;
(c) the novels must be together but the other books can be arranged in any order?

12. Five separate awards (best scholarship, best leadership qualities, and so on) are to be presented to selected students from a class of 30. How many different outcomes are possible if
(a) a student can receive any number of awards;
(b) each student can receive at most 1 award?

13. Consider a group of 20 people. If everyone shakes hands with everyone else, how many handshakes take place?

14. How many 5-card poker hands are there?

15. A dance class consists of 22 students, 10 women and 12 men. If 5 men and 5 women are to be chosen and then paired off, how many results are possible?

16. A student has to sell 2 books from a collection of 6 math, 7 science, and 4 economics books. How many choices are possible if

(a) both books are to be on the same subject;
(b) the books are to be on different subjects?

17. A total of 7 different gifts are to be distributed among 10 children. How many distinct results are possible if no child is to receive more than one gift?

18. A committee of 7, consisting of 2 Republicans, 2 Democrats, and 3 Independents, is to be chosen from a group of 5 Republicans, 6 Democrats, and 4 Independents. How many committees are possible?

19. From a group of 8 women and 6 men a committee consisting of 3 men and 3 women is to be formed. How many different committees are possible if
(a) 2 of the men refuse to serve together;
(b) 2 of the women refuse to serve together;
(c) 1 man and 1 woman refuse to serve together?

20. A person has 8 friends, of whom 5 will be invited to a party.
(a) How many choices are there if 2 of the friends are feuding and will not attend together?
(b) How many choices if 2 of the friends will only attend together?

21. Consider the grid of points shown below. Suppose that starting at the point labeled A you can go one step up or one step to the right at each move. This is continued until the point labeled B is reached. How many different paths from A to B are possible?

HINT: Note that to reach B from A you must take 4 steps to the right and 3 steps upward.

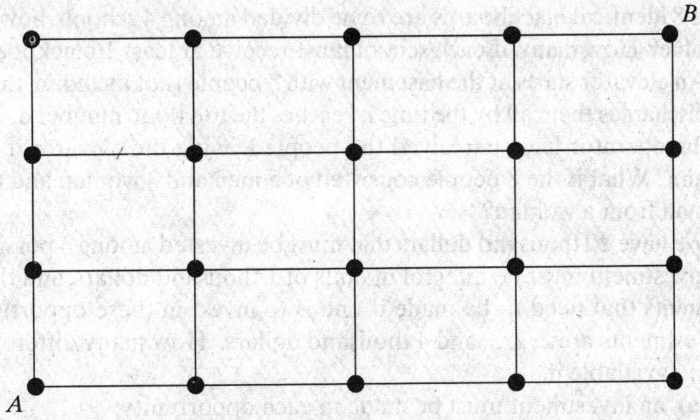

22. In Problem 21, how many different paths are there from A to B that go through the point circled below?

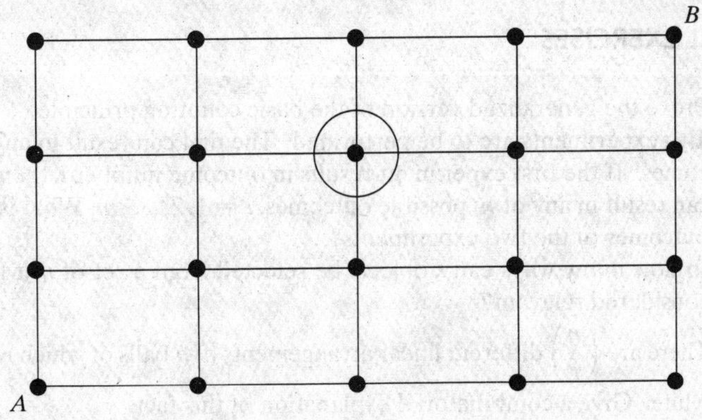

23. A psychology laboratory conducting dream research contains 3 rooms, with 2 beds in each room. If 3 sets of identical twins are to be assigned to these 6 beds so that each set of twins sleeps in different beds in the same room, how many assignments are possible?

24. Expand $(3x^2 + y)^5$.

25. The game of bridge is played by 4 players, each of whom is dealt 13 cards. How many bridge deals are possible?

26. Expand $(x_1 + 2x_2 + 3x_3)^4$.

27. If 12 people are to be divided into 3 committees of respective sizes 3, 4, and 5, how many divisions are possible?

28. If 8 new teachers are to be divided among 4 schools, how many divisions are possible? What if each school must receive 2 teachers?

29. Ten weight lifters are competing in a team weight-lifting contest. Of the lifters, 3 are from the United States, 4 are from Russia, 2 are from China, and 1 is from Canada. If the scoring takes account of the countries that the lifters represent but not their individual identities, how many different outcomes are possible from the point of view of scores? How many different outcomes correspond to results in which the United States has 1 competitor in the top three and 2 in the bottom three?

30. Delegates from 10 countries, including Russia, France, England, and the United States, are to be seated in a row. How many different seating arrangements are possible if the French and English delegates are to be seated next to each other, and the Russian and U.S. delegates are not to be next to each other?

***31.** If 8 identical blackboards are to be divided among 4 schools, how many divisions are possible? How many, if each school must receive at least 1 blackboard?

***32.** An elevator starts at the basement with 8 people (not including the elevator operator) and discharges them all by the time it reaches the top floor, number 6. In how many ways could the operator have perceived the people leaving the elevator if all people look alike to him? What if the 8 people consisted of 5 men and 3 women and the operator could tell a man from a woman?

***33.** We have 20 thousand dollars that must be invested among 4 possible opportunities. Each investment must be integral in units of 1 thousand dollars, and there are minimal investments that need to be made if one is to invest in these opportunities. The minimal investments are 2, 2, 3, and 4 thousand dollars. How many different investment strategies are available if

(a) an investment must be made in each opportunity;

(b) investments must be made in at least 3 of the 4 opportunities?

THEORETICAL EXERCISES

1. Prove the generalized version of the basic counting principle.

2. Two experiments are to be performed. The first can result in any one of m possible outcomes. If the first experiment results in outcome number i, then the second experiment can result in any of n_i possible outcomes, $i = 1, 2, \ldots, m$. What is the number of possible outcomes of the two experiments?

3. In how many ways can r objects be selected from a set of n if the order of selection is considered relevant?

4. There are $\binom{n}{r}$ different linear arrangements of n balls of which r are black and $n - r$ are white. Give a combinatorial explanation of this fact.

5. Determine the number of vectors $(x_1, \ldots, x_n)$, such that each x_i is either 0 or 1 and

$$\sum_{i=1}^{n} x_i \geq k$$

6. How many vectors $x_1, \ldots, x_k$ are there for which each x_i is a positive integer such that $1 \leq x_i \leq n$ and $x_1 < x_2 < \cdots < x_k$?

7. Give an analytic proof of Equation (4.1).

8. Prove that

$$\binom{n+m}{r} = \binom{n}{0}\binom{m}{r} + \binom{n}{1}\binom{m}{r-1} + \cdots + \binom{n}{r}\binom{m}{0}$$

HINT: Consider a group of n men and m women. How many groups of size r are possible?

9. Use Theoretical Exercise 8 to prove that

$$\binom{2n}{n} = \sum_{k=0}^{n} \binom{n}{k}^2$$

10. From a group of n people, suppose that we want to choose a committee of $k, k \leq n$, one of whom is to be designated as chairperson.

(a) By focusing first on the choice of the committee and then on the choice of the chair, argue that there are $\binom{n}{k}k$ possible choices.

(b) By focusing first on the choice of the nonchair committee members and then on the choice of the chair, argue that there are $\binom{n}{k-1}(n-k+1)$ possible choices.

(c) By focusing first on the choice of the chair and then on the choice of the other committee members, argue that there are $n\binom{n-1}{k-1}$ possible choices.

(d) Conclude from parts (a), (b), and (c) that

$$k\binom{n}{k} = (n-k+1)\binom{n}{k-1} = n\binom{n-1}{k-1}$$

(e) Use the factorial definition of $\binom{m}{r}$ to verify the identity in part (d).

11. The following identity is known as Fermat's combinatorial identity.

$$\binom{n}{k} = \sum_{i=k}^{n} \binom{i-1}{k-1} \qquad n \geq k$$

Give a combinatorial argument (no computations are needed) to establish this identity.

HINT: Consider the set of numbers 1 through n. How many subsets of size k have i as their highest-numbered member?

12. Consider the following combinatorial identity:

$$\sum_{k=1}^{n} k\binom{n}{k} = n \cdot 2^{n-1}$$

(a) Present a combinatorial argument for the above by considering a set of n people and determining, in two ways, the number of possible selections of a committee of any size and a chairperson for the committee.

HINT: (i) How many possible selections are there of a committee of size k and its chairperson?
(ii) How many possible selections are there of a chairperson and the other committee members?

(b) Verify the following identity for $n = 1, 2, 3, 4, 5$:

$$\sum_{k=1}^{n} \binom{n}{k} k^2 = 2^{n-2} n(n+1)$$

For a combinatorial proof of the preceding, consider a set of n people, and argue that both sides of the identity represent the number of different selections of a committee, its chairperson, and its secretary (possibly the same as the chairperson).

HINT: (i) How many different selections result in the committee containing exactly k people?
(ii) How many different selections are there in which the chairperson and the secretary are the same?
(ANSWER: $n2^{n-1}$.)
(iii) How many different selections result in the chairperson and the secretary being different?

(c) Now argue that

$$\sum_{k=1}^{n} \binom{n}{k} k^3 = 2^{n-3} n^2 (n+3)$$

13. Show that for $n > 0$,

$$\sum_{i=0}^{n} (-1)^i \binom{n}{i} = 0$$

HINT: Use the binomial theorem.

14. From a set of n people a committee of size j is to be chosen, and from this committee a subcommittee of size i, $i \leq j$, is also to be chosen.

(a) Derive a combinatorial identity by computing, in two ways, the number of possible choices of the committee and subcommittee—first by supposing that the committee is chosen first and then the subcommittee, and second by supposing that the subcommittee is chosen first and then the remaining members of the committee are chosen.

(b) Use part (a) to prove the following combinatorial identity:

$$\sum_{j=1}^{n} \binom{n}{j} \binom{j}{i} = \binom{n}{i} 2^{n-i} \qquad i \leq n$$

(c) Use part (a) and Theoretical Exercise 13 to show that

$$\sum_{j=i}^{n} \binom{n}{j} \binom{j}{i} (-1)^{n-j} = 0 \qquad i \leq n$$

15. Let $H_k(n)$ be the number of vectors $x_1, \ldots, x_k$ for which each x_i is a positive integer satisfying $1 \le x_i \le n$ and $x_1 \le x_2 \le \cdots \le x_k$.

(a) Without any computations, argue that

$$H_1(n) = n$$

$$H_k(n) = \sum_{j=1}^{n} H_{k-1}(j) \qquad k > 1$$

HINT: How many vectors are there in which $x_k = j$?

(b) Use the preceding recursion to compute $H_3(5)$.

HINT: First compute $H_2(n)$ for $n = 1, 2, 3, 4, 5$.

16. Consider a tournament of n contestants in which the outcome is an ordering of these contestants, with ties allowed. That is, the outcome partitions the players into groups, with the first group consisting of the players that tied for first place, the next group being those that tied for the next best position, and so on. Let $N(n)$ denote the number of different possible outcomes. For instance, $N(2) = 3$ since in a tournament with 2 contestants, player 1 could be uniquely first, player 2 could be uniquely first, or they could tie for first.

(a) List all the possible outcomes when $n = 3$.

(b) With $N(0)$ defined to equal 1, argue, without any computations, that

$$N(n) = \sum_{i=1}^{n} \binom{n}{i} N(n-i)$$

HINT: How many outcomes are there in which i players tie for last place?

(c) Show that the formula of part (b) is equivalent to the following:

$$N(n) = \sum_{i=0}^{n-1} \binom{n}{i} N(i)$$

(d) Use the recursion to find $N(3)$ and $N(4)$.

17. Present a combinatorial explanation of why $\binom{n}{r} = \binom{n}{r, n-r}$.

18. Argue that

$$\binom{n}{n_1, n_2, \ldots, n_r} = \binom{n-1}{n_1 - 1, n_2, \ldots, n_r} + \binom{n-1}{n_1, n_2 - 1, \ldots, n_r} + \cdots + \binom{n-1}{n_1, n_2, \ldots, n_r - 1}$$

HINT: Use an argument similar to the one used to establish Equation (4.1).

19. Prove the multinomial theorem.

***20.** In how many ways can n identical balls be distributed into r urns so that the ith urn contains at least m_i balls, for each $i = 1, \ldots, r$? Assume that $n \ge \sum_{i=1}^{r} m_i$.

***21.** Argue that there are exactly $\binom{r}{k}\binom{n-1}{n-r+k}$ solutions of

$$x_1 + x_2 + \cdots + x_r = n$$

for which exactly k of the x_i are equal to 0.

*22. Consider a function $f(x_1, \ldots, x_n)$ of n variables. How many different partial derivatives of order r does it possess?

*23. Determine the number of vectors $(x_1, \ldots, x_n)$, such that each x_i is a nonnegative integer and

$$\sum_{i=1}^{n} x_i \leq k$$

SELF-TEST PROBLEMS AND EXERCISES

1. How many different linear arrangements are there of the letters A, B, C, D, E, F for which
   (a) A and B are next to each other;
   (b) A is before B;
   (c) A is before B and B is before C;
   (d) A is before B and C is before D;
   (e) A and B are next to each other and C and D are also next to each other;
   (f) E is not last in line?
2. If 4 Americans, 3 Frenchmen, and 3 Englishmen are to be seated in a row, how many seating arrangements are possible when people of the same nationality must sit next to each other?
3. A president, treasurer, and secretary, all different, are to be chosen from a club consisting of 10 people. How many different choices of officers are possible if
   (a) there are no restrictions;
   (b) A and B will not serve together;
   (c) C and D will serve together or not at all;
   (d) E must be an officer;
   (e) F will serve only if he is president?
4. A student is to answer 7 out of 10 questions in an examination. How many choices has she? How many if she must answer at least 3 of the first 5 questions?
5. In how many ways can a man divide 7 gifts among his 3 children if the eldest is to receive 3 gifts and the others 2 each?
6. How many different 7-place license plates are possible when 3 of the entries are letters and 4 are digits? Assume that repetition of letters and numbers is allowed and that there is no restriction on where the letters or numbers can be placed.
7. Give a combinatorial explanation of the identity

$$\binom{n}{r} = \binom{n}{n-r}$$

8. Consider n-digit numbers where each digit is one of the 10 integers $0, 1, \ldots, 9$. How many such numbers are there for which
   (a) no two consecutive digits are equal;
   (b) 0 appears as a digit a total of i times, $i = 0, \ldots, n$?
9. Consider three classes, each consisting of n students. From this group of $3n$ students, a group of 3 students is to be chosen.
   (a) How many choices are possible?
   (b) How many choices are there in which all 3 students are in the same class?

(c) How many choices are there in which 2 of the 3 students are in the same class and the other student is in a different class?

(d) How many choices are there in which all 3 students are in different classes?

(e) Using the results of parts (a) through (d), write a combinatorial identity.

10. How many 5-digit numbers can be formed from the integers $1, 2, \ldots, 9$ if no digit can appear more than twice? (Thus, for instance, 41434 is not allowed.)

11. A tennis tournament has $2n$ contestants. In the first round, the players are partitioned into n pairs and each pair plays a game. How many different first round results are possible, where a result specifies the n pairs along with the resulting n winners?

12. A committee of 6 people is to be chosen from a group consisting of 7 men and 8 women. If the committee must consist of at least 3 women and at least 2 men, how many different committees are possible?

***13.** An art collection on auction consisted of 4 Dalis, 5 van Goghs, and 6 Picassos. At the auction were 5 art collectors. If a reporter noted only the number of Dalis, van Goghs, and Picassos acquired by each collector, how many different results could have been recorded if all works were sold?

***14.** Determine the number of vectors $(x_1, \ldots, x_n)$ such that each x_i is a positive integer and

$$\sum_{i=1}^{n} x_i \leq k$$

where $k \geq n$.

15. A total of n students are enrolled in a review course for the actuarial examination in probability. The posted results of the examination will list the names of those who passed, in decreasing order of their scores. For instance, the posted result will be "Brown, Cho" if Brown and Cho are the only ones to pass, with Brown receiving the higher score. Assuming that all scores are distinct (no ties), how many posted results are possible?

CHAPTER 2

Axioms of Probability

2.1 INTRODUCTION

In this chapter we introduce the concept of the probability of an event and then show how these probabilities can be computed in certain situations. As a preliminary, however, we need the concept of the sample space and the events of an experiment.

2.2 SAMPLE SPACE AND EVENTS

Consider an experiment whose outcome is not predictable with certainty. However, although the outcome of the experiment will not be known in advance, let us suppose that the set of all possible outcomes is known. This set of all possible outcomes of an experiment is known as the *sample space* of the experiment and is denoted by S. Some examples follow.

1. If the outcome of an experiment consists in the determination of the sex of a newborn child, then

$$S = \{g, b\}$$

where the outcome g means that the child is a girl and b that it is a boy.

2. If the outcome of an experiment is the order of finish in a race among the 7 horses having post positions 1, 2, 3, 4, 5, 6, 7, then

$$S = \{\text{all } 7! \text{ permutations of } (1, 2, 3, 4, 5, 6, 7)\}$$

The outcome $(2, 3, 1, 6, 5, 4, 7)$ means, for instance, that the number 2 horse comes in first, then the number 3 horse, then the number 1 horse, and so on.

3. If the experiment consists of flipping two coins, then the sample space consists of the following four points:

$$S = \{(H, H), (H, T), (T, H), (T, T)\}$$

The outcome will be (H, H) if both coins are heads, (H, T) if the first coin is heads and the second tails, (T, H) if the first is tails and the second heads, and (T, T) if both coins are tails.

4. If the experiment consists of tossing two dice, then the sample space consists of the 36 points

$$S = \{(i, j): i, j = 1, 2, 3, 4, 5, 6\}$$

where the outcome (i, j) is said to occur if i appears on the leftmost die and j on the other die.

5. If the experiment consists of measuring (in hours) the lifetime of a transistor, then the sample space consists of all nonnegative real numbers. That is,

$$S = \{x: 0 \leq x < \infty\}$$

Any subset E of the sample space is known as an *event*. That is, an event is a set consisting of possible outcomes of the experiment. If the outcome of the experiment is contained in E, then we say that E has occurred. Some examples of events are the following.

In the preceding example 1, if $E = \{g\}$, then E is the event that the child is a girl. Similarly, if $F = \{b\}$, then F is the event that the child is a boy.

In example 2, if

$$E = \{\text{all outcomes in } S \text{ starting with a } 3\}$$

then E is the event that horse 3 wins the race.

In example 3, if $E = \{(H, H), (H, T)\}$, then E is the event that a head appears on the first coin.

In example 4, if $E = \{(1, 6), (2, 5), (3, 4), (4, 3), (5, 2), (6, 1)\}$, then E is the event that the sum of the dice equals 7.

In example 5, if $E = \{x: 0 \leq x \leq 5\}$, then E is the event that the transistor does not last longer than 5 hours.

For any two events E and F of a sample space S, we define the new event $E \cup F$ to consist of all outcomes that are either in E or in F or in both E and F. That is, the event $E \cup F$ will occur if *either* E or F occurs. For instance, in example 1 if event $E = \{g\}$ and $F = \{b\}$, then

$$E \cup F = \{g, b\}$$

That is, $E \cup F$ would be the whole sample space S. In example 3, if $E = \{(H, H), (H, T)\}$ and $F = \{(T, H)\}$, then

$$E \cup F = \{(H, H), (H, T), (T, H)\}$$

Thus $E \cup F$ would occur if a head appeared on either coin.

The event $E \cup F$ is called the *union* of the event E and the event F.

Similarly, for any two events E and F we may also define the new event EF, called the *intersection* of E and F, to consist of all outcomes that are both in E and in F. That is, the event EF (sometimes written $E \cap F$) will occur only if both E and F occur. For instance, in example 3 if $E = \{(H, H), (H, T), (T, H)\}$ is the event that at least 1 head occurs, and $F = \{(H, T), (T, H), (T, T)\}$ is the event that at least 1 tail occurs, then

$$EF = \{(H, T), (T, H)\}$$

is the event that exactly 1 head and 1 tail appear. In example 4, if $E = \{(1,6),(2,5),(3,4),(4,3),(5,2),(6,1)\}$ is the event that the sum of the dice is 7 and $F = \{(1,5),(2,4),(3,3),(4,2),(5,1)\}$ is the event that the sum is 6, then the event EF does not contain any outcomes and hence could not occur. To give such an event a name, we shall refer to it as the null event and denote it by $\emptyset$ (that is, $\emptyset$ refers to the event consisting of no outcomes). If $EF = \emptyset$, then E and F are said to be *mutually exclusive*.

We also define unions and intersections of more than two events in a similar manner. If $E_1, E_2, \ldots$ are events, the union of these events, denoted by $\bigcup_{n=1}^{\infty} E_n$, is defined to be that event which consists of all outcomes that are in E_n for at least one value of $n = 1, 2, \ldots$. Similarly, the intersection of the events E_n, denoted by $\bigcap_{n=1}^{\infty} E_n$, is defined to be the event consisting of those outcomes that are in all of the events $E_n, n = 1, 2, \ldots$.

Finally, for any event E we define the new event E^c, referred to as the complement of E, to consist of all outcomes in the sample space S that are not in E. That is, E^c will occur if and only if E does not occur. In example 4, if event $E = \{(1,6),(2,5),(3,4),(4,3),(5,2),(6,1)\}$, then E^c will occur when the sum of the dice does not equal 7. Also note that because the experiment must result in some outcome, it follows that $S^c = \emptyset$.

For any two events E and F, if all of the outcomes in E are also in F, then we say that E is contained in F and write $E \subset F$ (or equivalently, $F \supset E$). Thus, if $E \subset F$, the occurrence of E necessarily implies the occurrence of F. If $E \subset F$ and $F \subset E$, we say that E and F are equal and write $E = F$.

A graphical representation that is very useful for illustrating logical relations among events is the Venn diagram. The sample space S is represented as consisting of all the outcomes in a large rectangle, and the events $E, F, G, \ldots$ are represented as consisting of all the outcomes in given circles within the rectangle. Events of interest can then be indicated by shading appropriate regions of the diagram. For instance, in the three Venn diagrams shown in Figure 2.1, the shaded areas represent, respectively, the events $E \cup F$, EF, and E^c. The Venn diagram in Figure 2.2 indicates that $E \subset F$.

The operations of forming unions, intersections, and complements of events obey certain rules not dissimilar to the rules of algebra. We list a few of these rules.

Commutative laws	$E \cup F = F \cup E$	$EF = FE$
Associative laws	$(E \cup F) \cup G = E \cup (F \cup G)$	$(EF)G = E(FG)$
Distributive laws	$(E \cup F)G = EG \cup FG$	$EF \cup G = (E \cup G)(F \cup G)$

These relations are verified by showing that any outcome that is contained in the event on the left side of the equality sign is also contained in the event on the right side, and vice versa. One way of showing this is by means of Venn diagrams. For instance, the distributive law may be verified by the sequence of diagrams in Figure 2.3.

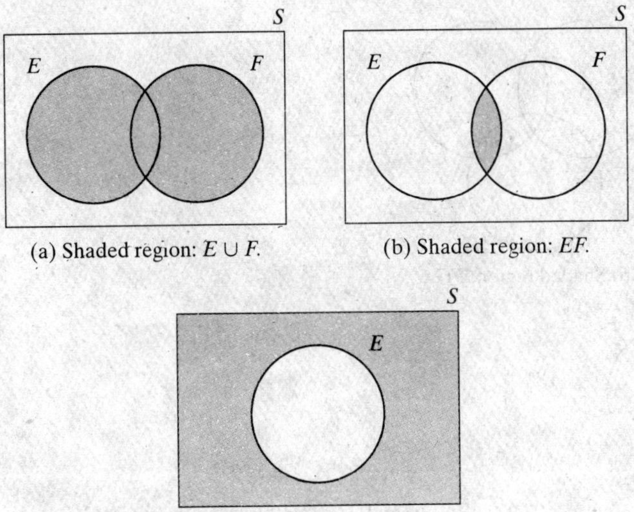

(a) Shaded region: $E \cup F$. (b) Shaded region: EF.

(c) Shaded region: E^c.

Figure 2.1

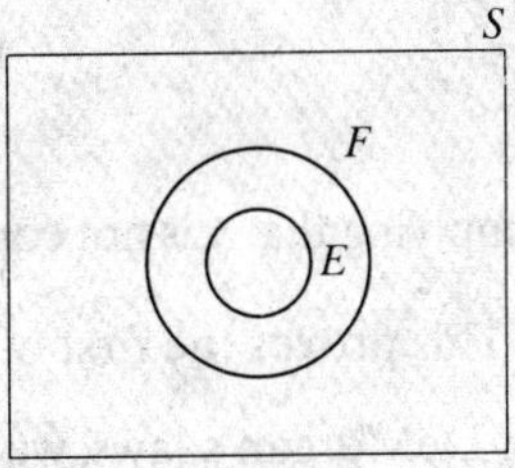

$E \subset F$

Figure 2.2

The following useful relationships between the three basic operations of forming unions, intersections, and complements are known as *DeMorgan's laws*:

$$\left(\bigcup_{i=1}^{n} E_i\right)^c = \bigcap_{i=1}^{n} E_i^c$$

$$\left(\bigcap_{i=1}^{n} E_i\right)^c = \bigcup_{i=1}^{n} E_i^c$$

To prove DeMorgan's laws, suppose first that x is an outcome of $\left(\bigcup_{i=1}^{n} E_i\right)^c$. Then x is not contained in $\bigcup_{i=1}^{n} E_i$, which means that x is not contained in any of the events $E_i, i = 1, 2, \ldots, n$, implying that x is contained in E_i^c for all $i = 1, 2, \ldots, n$ and thus is contained in $\bigcap_{i=1}^{n} E_i^c$. To go the other way, suppose that x is an outcome of $\bigcap_{i=1}^{n} E_i^c$. Then x is contained in E_i^c for all $i = 1, 2, \ldots, n$, which means that x is not contained

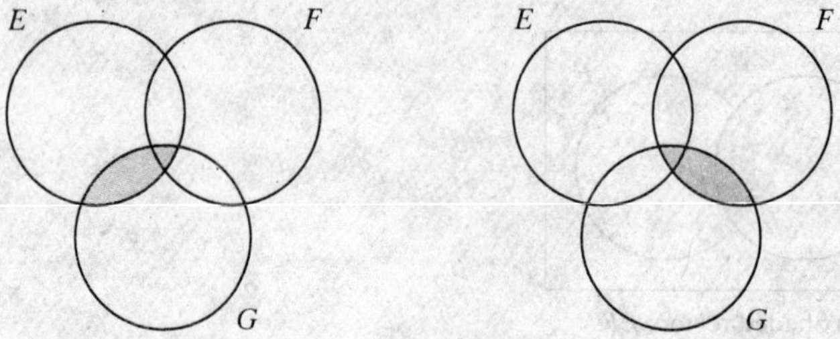

(a) Shaded region: EG. (b) Shaded region: FG.

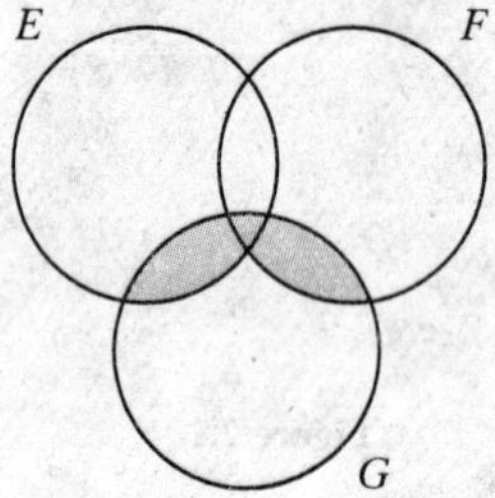

(c) Shaded region: $(E \cup F)G$.
$(E \cup F)G = EG \cup FG$

Figure 2.3

in E_i for any $i = 1, 2, \ldots, n$, implying that x is not contained in $\bigcup_i^n E_i$, which yields that x is contained in $\left(\bigcup_1^n E_i\right)^c$. This proves the first of DeMorgan's laws.

To prove the second of DeMorgan's laws, we use the first law to obtain

$$\left(\bigcup_{i=1}^n E_i^c\right)^c = \bigcap_{i=1}^n (E_i^c)^c$$

which, since $(E^c)^c = E$, is equivalent to

$$\left(\bigcup_1^n E_i^c\right)^c = \bigcap_1^n E_i$$

Taking complements of both sides of the preceding yields the result, namely,

$$\bigcup_1^n E_i^c = \left(\bigcap_1^n E_i\right)^c$$

2.3 AXIOMS OF PROBABILITY

One way of defining the probability of an event is in terms of its relative frequency. Such a definition usually goes as follows: We suppose that an experiment, whose sample space is S, is repeatedly performed under exactly the same conditions. For each event E of the sample space S, we define $n(E)$ to be the number of times in the first n repetitions of the experiment that the event E occurs. Then $P(E)$, the probability of the event E, is defined by

$$P(E) = \lim_{n \to \infty} \frac{n(E)}{n}$$

That is, $P(E)$ is defined as the (limiting) proportion of time that E occurs. It is thus the limiting frequency of E.

Although the preceding definition is certainly intuitively pleasing and should always be kept in mind by the reader, it possesses a serious drawback: How do we know that $n(E)/n$ will converge to some constant limiting value that will be the same for each possible sequence of repetitions of the experiment? For example, suppose that the experiment to be repeatedly performed consists of flipping a coin. How do we know that the proportion of heads obtained in the first n flips will converge to some value as n gets large? Also, even if it does converge to some value, how do we know that, if the experiment is repeatedly performed a second time, we shall again obtain the same limiting proportion of heads?

Proponents of the relative frequency definition of probability usually answer this objection by stating that the convergence of $n(E)/n$ to a constant limiting value is an assumption, or an *axiom*, of the system. However, to assume that $n(E)/n$ will necessarily converge to some constant value seems to be a very complex assumption. For, although we might indeed hope that such a constant limiting frequency exists, it does not at all seem to be a priori evident that this need be the case. In fact, would it not be more reasonable to assume a set of simpler and more self-evident axioms about probability and then attempt to prove that such a constant limiting frequency does in some sense exist? This latter approach is the modern axiomatic approach to probability theory that we shall adopt in this text. In particular, we shall assume that for each event E in the sample space S there exists a value $P(E)$, referred to as the probability of E. We shall then assume that the probabilities satisfy a certain set of axioms, which, we hope the reader will agree, is in accordance with our intuitive notion of probability.

Consider an experiment whose sample space is S. For each event E of the sample space S we assume that a number $P(E)$ is defined and satisfies the following three axioms.

Axiom 1

$$0 \leq P(E) \leq 1$$

Axiom 2

$$P(S) = 1$$

Axiom 3

For any sequence of mutually exclusive events $E_1, E_2, \ldots$ (that is, events for which $E_iE_j = \emptyset$ when $i \neq j$),

$$P\left(\bigcup_{i=1}^{\infty} E_i\right) = \sum_{i=1}^{\infty} P(E_i)$$

We refer to $P(E)$ as the probability of the event E.

Thus Axiom 1 states that the probability that the outcome of the experiment is an outcome in E is some number between 0 and 1. Axiom 2 states that, with probability 1, the outcome will be a point in the sample space S. Axiom 3 states that for any sequence of mutually exclusive events the probability of at least one of these events occurring is just the sum of their respective probabilities.

If we consider a sequence of events $E_1, E_2, \ldots$, where $E_1 = S$, $E_i = \emptyset$ for $i > 1$, then, as the events are mutually exclusive and as $S = \bigcup_{i=1}^{\infty} E_i$, we have from Axiom 3 that

$$P(S) = \sum_{i=1}^{\infty} P(E_i) = P(S) + \sum_{i=2}^{\infty} P(\emptyset)$$

implying that

$$P(\emptyset) = 0$$

That is, the null event has probability 0 of occurring.

It should also be noted that it follows that for any finite sequence of mutually exclusive events $E_1, E_2, \ldots, E_n$,

$$P\left(\bigcup_{1}^{n} E_i\right) = \sum_{i=1}^{n} P(E_i) \tag{3.1}$$

This follows from Axiom 3 by defining E_i to be the null event for all values of i greater than n. Axiom 3 is equivalent to Equation (3.1) when the sample space is finite (why?). However, the added generality of Axiom 3 is necessary when the sample space consists of an infinite number of points.

Example 3a. If our experiment consists of tossing a coin and if we assume that a head is as likely to appear as a tail, then we would have

$$P(\{H\}) = P(\{T\}) = \tfrac{1}{2}$$

On the other hand, if the coin were biased and we felt that a head were twice as likely to appear as a tail, then we would have

$$P(\{H\}) = \tfrac{2}{3} \qquad P(\{T\}) = \tfrac{1}{3}$$

Example 3b. If a die is rolled and we suppose that all six sides are equally likely to appear, then we would have $P(\{1\}) = P(\{2\}) = P(\{3\}) = P(\{4\}) = P(\{5\}) = P(\{6\}) = \frac{1}{6}$. From Axiom 3 it would thus follow that the probability of rolling an even number would equal

$$P(\{2,4,6\}) = P(\{2\}) + P(\{4\}) + P(\{6\}) = \frac{1}{2}$$

The assumption of the existence of a set function P, defined on the events of a sample space S, and satisfying Axioms 1, 2, and 3, constitutes the modern mathematical approach to probability theory. Hopefully, the reader will agree that the axioms are natural and in accordance with our intuitive concept of probability as related to chance and randomness. Furthermore, using these axioms we shall be able to prove that if an experiment is repeated over and over again, then, with probability 1, the proportion of time during which any specific event E occurs will equal $P(E)$. This result, known as the strong law of large numbers, is presented in Chapter 8. In addition, we present another possible interpretation of probability—as being a measure of belief—in Section 2.7.

Technical Remark. We have supposed that $P(E)$ is defined for all the events E of the sample space. Actually, when the sample space is an uncountably infinite set, $P(E)$ is defined only for a class of events called measurable. However, this restriction need not concern us as all events of any practical interest are measurable.

2.4 SOME SIMPLE PROPOSITIONS

In this section we prove some simple propositions regarding probabilities. We first note that as E and E^c are always mutually exclusive and since $E \cup E^c = S$, we have by Axioms 2 and 3 that

$$1 = P(S) = P(E \cup E^c) = P(E) + P(E^c)$$

Or, equivalently, we have the statement given in Proposition 4.1.

Proposition 4.1

$$P(E^c) = 1 - P(E)$$

In words, Proposition 4.1 states that the probability that an event does not occur is 1 minus the probability that it does occur. For instance, if the probability of obtaining a head on the toss of a coin is $\frac{3}{8}$, the probability of obtaining a tail must be $\frac{5}{8}$.

Our second proposition states that if the event E is contained in the event F, then the probability of E is no greater than the probability of F.

Proposition 4.2

If $E \subset F$, then $P(E) \leq P(F)$.

Proof: Since $E \subset F$, it follows that we can express F as

$$F = E \cup E^cF$$

Hence, as E and E^cF are mutually exclusive, we obtain from Axiom 3 that

$$P(F) = P(E) + P(E^cF)$$

which proves the result, since $P(E^cF) \geq 0$.

Proposition 4.2 tells us, for instance, that the probability of rolling a 1 with a die is less than or equal to the probability of rolling an odd value with the die.

The next proposition gives the relationship between the probability of the union of two events in terms of the individual probabilities and the probability of the intersection.

Proposition 4.3

$$P(E \cup F) = P(E) + P(F) - P(EF)$$

Proof: To derive a formula for $P(E \cup F)$, we first note that $E \cup F$ can be written as the union of the two disjoint events E and E^cF. Thus from Axiom 3 we obtain that

$$P(E \cup F) = P(E \cup E^cF)$$
$$= P(E) + P(E^cF)$$

Furthermore, since $F = EF \cup E^cF$, we again obtain from Axiom 3 that

$$P(F) = P(EF) + P(E^cF)$$

or, equivalently,

$$P(E^cF) = P(F) - P(EF)$$

thus completing the proof.

Proposition 4.3 could also have been proved by making use of the Venn diagram in Figure 2.4.

Let us divide $E \cup F$ into three mutually exclusive sections, as shown in Figure 2.5. In words, section I represents all the points in E that are not in F (that is, EF^c);

Figure 2.4 Venn diagram.

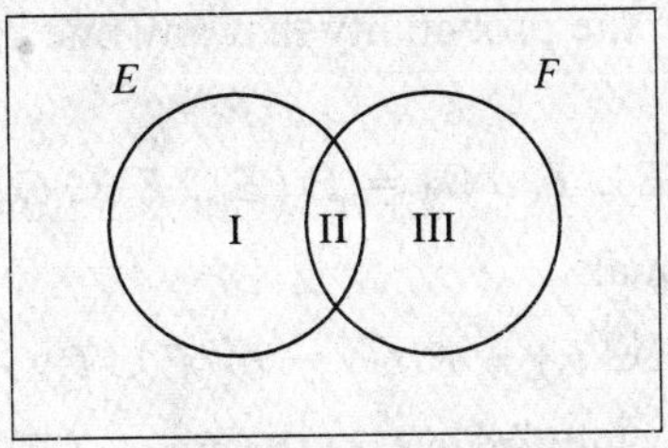

Figure 2.5 Venn diagram in sections.

section II represents all points both in E and in F (that is, EF): and section III represents all points in F that are not in E (that is, E^cF).

From Figure 2.5 we see that

$$E \cup F = \text{I} \cup \text{II} \cup \text{III}$$
$$E = \text{I} \cup \text{II}$$
$$F = \text{II} \cup \text{III}$$

As I, II, and III are mutually exclusive, it follows from Axiom 3 that

$$P(E \cup F) = P(\text{I}) + P(\text{II}) + P(\text{III})$$
$$P(E) = P(\text{I}) + P(\text{II})$$
$$P(F) = P(\text{II}) + P(\text{III})$$

which shows that

$$P(E \cup F) = P(E) + P(F) - P(\text{II})$$

and Proposition 4.3 is proved, since II $= EF$.

Example 4a. Suppose that we toss two coins and suppose that each of the four points in the sample space $S = \{(H, H), (H, T), (T, H), (T, T)\}$ is equally likely and hence has probability $\frac{1}{4}$. Let

$$E = \{(H, H), (H, T)\} \qquad \text{and} \qquad F = \{(H, H), (T, H)\}$$

That is, E is the event that the first coin falls heads, and F is the event that the second coin falls heads.

By Proposition 4.3 we have that $P(E \cup F)$, the probability that either the first or second coin falls heads, is given by

$$\begin{aligned} P(E \cup F) &= P(E) + P(F) - P(EF) \\ &= \tfrac{1}{2} + \tfrac{1}{2} - P(\{(H, H)\}) \\ &= 1 - \tfrac{1}{4} \\ &= \tfrac{3}{4} \end{aligned}$$

This probability could, of course, have been computed directly because

$$P(E \cup F) = P(\{(H, H), (H, T), (T, H)\}) = \tfrac{3}{4}$$

■

We may also calculate the probability that any one of the three events E or F or G occurs:

$$P(E \cup F \cup G) = P[(E \cup F) \cup G]$$

which by Proposition 4.3 equals

$$P(E \cup F) + P(G) - P[(E \cup F)G]$$

Now, it follows from the distributive law that the events $(E \cup F)G$ and $EG \cup FG$ are equivalent, and hence we obtain from the preceding equations that

$$\begin{aligned}
&P(E \cup F \cup G) \\
&\quad = P(E) + P(F) - P(EF) + P(G) - P(EG \cup FG) \\
&\quad = P(E) + P(F) - P(EF) + P(G) - P(EG) - P(FG) + P(EGFG) \\
&\quad = P(E) + P(F) + P(G) - P(EF) - P(EG) - P(FG) + P(EFG)
\end{aligned}$$

In fact, the following proposition can be proved by induction.

Proposition 4.4

$$\begin{aligned}
P(E_1 \cup E_2 \cup \cdots \cup E_n) = \sum_{i=1}^{n} P(E_i) - \sum_{i_1<i_2} P(E_{i_1}E_{i_2}) + \cdots \\
+ (-1)^{r+1} \sum_{i_1<i_2<\cdots<i_r} P(E_{i_1}E_{i_2}\cdots E_{i_r}) \\
+ \cdots + (-1)^{n+1}P(E_1E_2\cdots E_n)
\end{aligned}$$

The summation $\sum_{i_1<i_2<\cdots<i_r} P(E_{i_1}E_{i_2}\cdots E_{i_r})$ is taken over all of the $\binom{n}{r}$ possible subsets of size r of the set $\{1, 2, \ldots, n\}$.

In words Proposition 4.4 states that the probability of the union of n events equals the sum of the probabilities of these events taken one at a time, minus the sum of the probabilities of these events taken two at a time, plus the sum of the probabilities of these events taken three at a time, and so on.

REMARK. For a noninductive argument for Proposition 4.4, note first that if an outcome of the sample space is not a member of any of the sets E_i, then its probability does not contribute anything to either side of the equality. On the other hand, suppose that an outcome is in exactly m of the events E_i, where $m > 0$. Then since it is in $\bigcup_i E_i$ its probability is counted once in $P\left(\bigcup_i E_i\right)$; also as this outcome is contained in $\binom{m}{k}$ subsets of the type $E_{i_1}E_{i_2}\cdots E_{i_k}$, its probability is counted

$$\binom{m}{1} - \binom{m}{2} + \binom{m}{3} - \cdots \pm \binom{m}{m}$$

times on the right of the equality sign in Proposition 4.4. Thus, for $m > 0$, we must show that

$$1 = \binom{m}{1} - \binom{m}{2} + \binom{m}{3} - \cdots \pm \binom{m}{m}$$

However, since $1 = \binom{m}{0}$, the preceding is equivalent to

$$\sum_{i=0}^{m} \binom{m}{i} (-1)^i = 0$$

and the latter equation follows from the binomial theorem since

$$0 = (-1 + 1)^m = \sum_{i=0}^{m} \binom{m}{i} (-1)^i (1)^{m-i}$$

2.5 SAMPLE SPACES HAVING EQUALLY LIKELY OUTCOMES

For many experiments it is natural to assume that all outcomes in the sample space are equally likely to occur. That is, consider an experiment whose sample space S is a finite set, say $S = \{1, 2, \ldots, N\}$. Then it is often natural to assume that

$$P(\{1\}) = P(\{2\}) = \cdots = P(\{N\})$$

which implies from Axioms 2 and 3 (why?) that

$$P(\{i\}) = \frac{1}{N} \qquad i = 1, 2, \ldots, N$$

From this it follows from Axiom 3 that for any event E

$$P(E) = \frac{\text{number of points in } E}{\text{number of points in } S}$$

In words, if we assume that all outcomes of an experiment are equally likely to occur, then the probability of any event E equals the proportion of outcomes in the sample space that are contained in E.

Example 5a. If two dice are rolled, what is the probability that the sum of the upturned faces will equal 7?

Solution We shall solve this problem under the assumption that all of the 36 possible outcomes are equally likely. Since there are 6 possible outcomes—namely (1, 6), (2, 5), (3, 4), (4, 3), (5, 2), (6, 1)—that result in the sum of the dice being equal to 7, the desired probability is $\frac{6}{36} = \frac{1}{6}$. ■

Example 5b. If 3 balls are "randomly drawn" from a bowl containing 6 white and 5 black balls, what is the probability that one of the drawn balls is white and the other two black?

Solution If we regard the order in which the balls are selected as being relevant, then the sample space consists of $11 \cdot 10 \cdot 9 = 990$ outcomes. Furthermore, there are $6 \cdot 5 \cdot 4 = 120$ outcomes in which the first ball selected is white and the other two black; $5 \cdot 6 \cdot 4 = 120$ outcomes in which the first is black, the second white, and the third black; and $5 \cdot 4 \cdot 6 = 120$ in which the first two are black and the third white. Hence, assuming that "randomly drawn" means that each outcome in the sample space is equally likely to occur, we see that the desired probability is

$$\frac{120 + 120 + 120}{990} = \frac{4}{11}$$

This problem could also have been solved by regarding the outcome of the experiment as the unordered set of drawn balls. From this point of view, there are $\binom{11}{3} = 165$ outcomes in the sample space. Now, each set of 3 balls corresponds to 3! outcomes when the order of selection is noted. As a result, if all outcomes are assumed equally likely when the order of selection is noted, then it follows that they remain equally likely when the outcome is taken to be the unordered set of selected balls. Hence, using the latter representation of the experiment, we see that the desired probability is

$$\frac{\binom{6}{1}\binom{5}{2}}{\binom{11}{3}} = \frac{4}{11}$$

which, of course, agrees with the answer obtained previously. ∎

Example 5c. A committee of 5 is to be selected from a group of 6 men and 9 women. If the selection is made randomly, what is the probability that the committee consists of 3 men and 2 women?

Solution Let us assume that *randomly selected* means that each of the $\binom{15}{5}$ possible combinations is equally likely to be selected. Hence the desired probability equals

$$\frac{\binom{6}{3}\binom{9}{2}}{\binom{15}{5}} = \frac{240}{1001}$$ ∎

Example 5d. An urn contains n balls, of which one is special. If k of these balls are withdrawn one at a time, with each selection being equally likely to be any of the balls that remain at the time, what is the probability that the special ball is chosen?

Solution Since all of the balls are treated in an identical manner, it follows that the set of k balls selected is equally likely to be any of the $\binom{n}{k}$ sets of k balls. Therefore,

$$P\{\text{special ball is selected}\} = \frac{\binom{1}{1}\binom{n-1}{k-1}}{\binom{n}{k}} = \frac{k}{n}$$

We could also have obtained the preceding result by letting A_i denote the event that the special ball is the ith ball to be chosen, $i = 1, \ldots, k$. Then, since each one of the n balls is equally likely to be the ith ball chosen, it follows that $P(A_i) = 1/n$. Hence, since these events are clearly mutually exclusive, we have that

$$P\{\text{special ball is selected}\} = P\left(\bigcup_{i=1}^{k} A_i\right) = \sum_{i=1}^{k} P(A_i) = \frac{k}{n}$$

We could have argued that $P(A_i) = 1/n$, by noting that there are $n(n-1)\cdots(n-k+1) = n!/(n-k)!$ equally likely outcomes of the experiment, of which $(n-1)(n-2)\cdots(n-i+1)(1)(n-i)\cdots(n-k+1) = (n-1)!/(n-k)!$ result in the special ball being the ith one chosen. From this it follows that

$$P(A_i) = \frac{(n-1)!}{n!} = \frac{1}{n}$$ ■

Example 5e. Suppose that $n + m$ balls, of which n are red and m are blue, are arranged in a linear order in such a way that all $(n + m)!$ possible orderings are equally likely. If we record the result of this experiment by only listing the colors of the successive balls, show that all the possible results remain equally likely.

Solution Consider any one of the $(n + m)!$ possible orderings and note that any permutation of the red balls among themselves and of the blue balls among themselves does not change the sequence of colors. As a result, every ordering of colorings corresponds to $n!\,m!$ different orderings of the $n + m$ balls, so every ordering of the colors has probability $\frac{n!\,m!}{(n+m)!}$ of occurring.

For example, suppose that there are 2 red balls, numbered r_1, r_2, and 2 blue balls, numbered b_1, b_2. Then, of the 4! possible orderings, there will be 2! 2! orderings that result in any specified color combination. For instance, the following orderings result in the successive balls alternating in color with a red ball first:

$$r_1, b_1, r_2, b_2 \qquad r_1, b_2, r_2, b_1 \qquad r_2, b_1, r_1, b_2 \qquad r_2, b_2, r_1, b_1$$

Hence each of the possible orderings of the colors has probability $\frac{4}{24} = \frac{1}{6}$ of occurring. ■

Example 5f. A poker hand consists of 5 cards. If the cards have distinct consecutive values and are not all of the same suit, we say that the hand is a straight. For instance, a hand consisting of the five of spades, six of spades, seven of spades, eight of spades, and nine of hearts is a straight. What is the probability that one is dealt a straight?

Solution We start by assuming that all $\binom{52}{5}$ possible poker hands are equally likely. To determine the number of outcomes that are straights, let us first determine the number of possible outcomes for which the poker hand consists of an ace, two, three, four, and five (the suits being irrelevant). Since the ace can be any 1 of the 4 possible aces, and similarly for the two, three, four, and five, it follows that there are 4^5 outcomes leading to exactly one ace, two, three, four, and five. Hence, since in 4 of these outcomes all the cards will be of the same suit (such a hand is called a straight flush), it follows that there are $4^5 - 4$ hands that make up a straight of the form ace, two, three, four, and five. Similarly, there are $4^5 - 4$ hands that make up a straight of the form ten, jack, queen, king, and ace. Hence there are $10(4^5 - 4)$ hands that are straights. Thus the desired probability is

$$\frac{10(4^5 - 4)}{\binom{52}{5}} \approx .0039$$

■

Example 5g. A 5-card poker hand is said to be a full house if it consists of 3 cards of the same denomination and 2 cards of the same denomination. (That is, a full house is three of a kind plus a pair.) What is the probability that one is dealt a full house?

Solution Again we assume that all $\binom{52}{5}$ possible hands are equally likely. To determine the number of possible full houses, we first note that there are $\binom{4}{2}\binom{4}{3}$ different combinations of, say, 2 tens and 3 jacks. Because there are 13 different choices for the kind of pair and, after a pair has been chosen, there are 12 other choices for the denomination of the remaining 3 cards, it follows that the probability of a full house is

$$\frac{13 \cdot 12 \cdot \binom{4}{2}\binom{4}{3}}{\binom{52}{5}} \approx .0014$$

■

Example 5h. In the game of bridge the entire deck of 52 cards is dealt out to 4 players. What is the probability that

(a) one of the players receives all 13 spades;
(b) each player receives 1 ace?

Solution (a) There are $\binom{52}{13, 13, 13, 13}$ possible divisions of the cards among the 4 distinct players. As there are $\binom{39}{13, 13, 13}$ possible divisions of the cards leading to a fixed player having all 13 spades, it follows that the desired probability is given by

$$\frac{4\binom{39}{13, 13, 13}}{\binom{52}{13, 13, 13, 13}} \approx 6.3 \times 10^{-12}$$

(b) To determine the number of outcomes in which each of the distinct players receives exactly 1 ace, put aside the aces and note that there are $\binom{48}{12, 12, 12, 12}$ possible divisions of the other 48 cards when each player is to receive 12. As there are 4! ways of dividing the 4 aces so that each player receives 1, we see that the number of possible outcomes in which each player receives exactly 1 ace is $4!\binom{48}{12, 12, 12, 12}$. Hence the desired probability is

$$\frac{4!\binom{48}{12, 12, 12, 12}}{\binom{52}{13, 13, 13, 13}} \approx .105$$ ■

Some results in probability are quite surprising when initially encountered. Our next two examples illustrate this phenomenon.

Example 5i. If n people are present in a room, what is the probability that no two of them celebrate their birthday on the same day of the year? How large need n be so that this probability is less than $\frac{1}{2}$?

Solution As each person can celebrate his or her birthday on any one of 365 days, there is a total of $(365)^n$ possible outcomes. (We are ignoring the possibility of someone's having been born on February 29.) Assuming that each outcome is equally likely, we see that the desired probability is $(365)(364)(363)\cdots(365 - n + 1)/(365)^n$. It is a rather surprising fact that when $n \geq 23$, this probability is less than $\frac{1}{2}$. That is, if there are 23 or more people in a room, then the probability that at least two of them have the same birthday exceeds $\frac{1}{2}$. Many people are initially surprised by this result, since 23 seems so small in relation to 365, the number of days of the year. However, every pair of individuals has probability $\frac{365}{(365)^2} = \frac{1}{365}$ of having the same birthday, and in a group of 23 people there are $\binom{23}{2} = 253$ different pairs of individuals. Looked at this way, the result no longer seems so surprising.

When there are 50 people in the room, the probability that at least two share the same birthday is approximately .970. And with 100 persons in the room, the odds are better than 3,000,000:1 (that is, the probability is greater than $\frac{3 \times 10^6}{3 \times 10^6 + 1}$) that at least two people have the same birthday. ■

Example 5j. A deck of 52 playing cards is shuffled and the cards turned up one at a time until the first ace appears. Is the next card—that is, the card following the first ace—more likely to be the ace of spades or the two of clubs?

Solution To determine the probability that the card following the first ace is the ace of spades, we need to calculate how many of the $(52)!$ possible orderings of the cards have the ace of spades immediately following the first ace. To begin, note that each ordering of the 52 cards can be obtained by first ordering the 51 cards different from the ace of spades and then inserting the ace of spades into that ordering. Furthermore, for each of the $(51)!$ orderings of the other cards, there is only one place where the ace of spades can be placed so that it follows the first ace. For instance, if the ordering of the other 51 cards is

$$4c, 6h, Jd, 5s, Ac, 7d, \ldots, Kh$$

then the only insertion of the ace of spades into this ordering that results in its following the first ace is

$$4c, 6h, Jd, 5s, Ac, As, 7d, \ldots, Kh$$

Therefore, we see that there are $(51)!$ orderings that result in the ace of spades following the first ace, so

$$P\{\text{the ace of spades follows the first ace}\} = \frac{(51)!}{(52)!} = \frac{1}{52}$$

In fact, by exactly the same argument, it follows that the probability that the two of clubs (or any other specified card) follows the first ace is also $\frac{1}{52}$. In other words, each of the 52 cards of the deck is equally likely to be the one that follows the first ace!

Many people find this result rather surprising. Indeed, a common reaction is to suppose initially that it is more likely that the two of clubs (rather than the ace of spades) follows the first ace, since that first ace might itself be the ace of spades. This reaction is often followed by the realization that the two of clubs might itself appear before the first ace, thus negating its chance of immediately following the first ace. However, as there is one chance in four that the ace of spades will be the first ace (because all 4 aces are equally likely to be first) and only one chance in five that the two of clubs will appear before the first ace (because each of the set of 5 cards consisting of the two of clubs and the

4 aces is equally likely to be the first of this set to appear), it again appears that the two of clubs is more likely. However, this is not the case and a more complete analysis shows that they are equally likely. ■

Example 5k. A football team consists of 20 offensive and 20 defensive players. The players are to be paired in groups of 2 for the purpose of determining roommates. If the pairing is done at random, what is the probability that there are no offensive–defensive roommate pairs? What is the probability that there are $2i$ offensive–defensive roommate pairs, $i = 1, 2, \ldots, 10$?

Solution There are

$$\binom{40}{2, 2, \ldots, 2} = \frac{(40)!}{(2!)^{20}}$$

ways of dividing the 40 players into 20 *ordered* pairs of two each. [That is, there are $(40)!/2^{20}$ ways of dividing the players into a *first* pair, a *second* pair, and so on.] Hence there are $(40)!/2^{20}(20)!$ ways of dividing the players into (unordered) pairs of 2 each. Furthermore, since a division will result in no offensive–defensive pairs if the offensive (and defensive) players are paired among themselves, it follows that there are $[(20)!/2^{10}(10)!]^2$ such divisions. Hence the probability of no offensive–defensive roommate pairs, call it P_0, is given by

$$P_0 = \frac{\left(\dfrac{(20)!}{2^{10}(10)!}\right)^2}{\dfrac{(40)!}{2^{20}(20)!}} = \frac{[(20)!]^3}{[(10)!]^2(40)!}$$

To determine P_{2i}, the probability that there are $2i$ offensive–defensive pairs, we first note that there are $\binom{20}{2i}^2$ ways of selecting the $2i$ offensive players and the $2i$ defensive players who are to be in the offensive–defensive pairs. These $4i$ players can then be paired up into $(2i)!$ possible offensive–defensive pairs. (This is so because the first offensive can be paired with any of the $2i$ defensives, the second offensive with any of the remaining $2i - 1$ defensives, and so on.) As the remaining $20 - 2i$ offensives (and defensives) must be paired among themselves, it follows that there are

$$\binom{20}{2i}^2 (2i)! \left[\frac{(20 - 2i)!}{2^{10-i}(10 - i)!}\right]^2$$

divisions which lead to $2i$ offensive–defensive pairs. Hence

$$P_{2i} = \frac{\binom{20}{2i}^2 (2i)! \left[\dfrac{(20 - 2i)!}{2^{10-i}(10 - i)!}\right]^2}{\dfrac{(40)!}{2^{20}(20)!}} \qquad i = 0, 1, \ldots, 10$$

The $P_{2i}, i = 0, 1, \ldots, 10$, can now be computed or they can be approximated by making use of a result of Stirling which shows that $n!$ can be approximated by $n^{n+1/2} + e^{-n}\sqrt{2\pi}$. For instance, we obtain that

$$P_0 \approx 1.3403 \times 10^{-6}$$

$$P_{10} \approx .345861$$

$$P_{20} \approx 7.6068 \times 10^{-6}$$ ■

Our next three examples illustrate the usefulness of Proposition 4.4. In Example 5l, the introduction of probability enables us to obtain a quick solution to a counting problem.

Example 5l. A total of 36 members of a club play tennis, 28 play squash, and 18 play badminton. Furthermore, 22 of the members play both tennis and squash, 12 play both tennis and badminton, 9 play both squash and badminton, and 4 play all three sports. How many members of this club play at least one of these sports?

Solution Let N denote the number of members of the club, and introduce probability by assuming that a member of the club is randomly selected. If for any subset C of members of the club, we let $P(C)$ denote the probability that the selected member is contained in C, then

$$P(C) = \frac{\text{number of members in } C}{N}$$

Now, with T being the set of members that plays tennis, S being the set that plays squash, and B being the set that plays badminton, we have from Proposition 4.4 that

$$P(T \cup S \cup B) = P(T) + P(S) + P(B) - P(TS) - P(TB) - P(SB) + P(TSB)$$

$$= \frac{36 + 28 + 18 - 22 - 12 - 9 + 4}{N}$$

$$= \frac{43}{N}$$

Hence we can conclude that 43 members play at least one of the sports. ■

The next example in this section not only possesses the virtue of giving rise to a somewhat surprising answer but is also of theoretical interest.

Example 5m. ***The matching problem.*** Suppose that each of N men at a party throws his hat into the center of the room. The hats are first mixed up, and then each man randomly selects a hat. What is the probability that

(a) none of the men selects his own hat;
(b) exactly k of the men select their own hats?

Solution (a) We first calculate the complementary probability of at least one man's selecting his own hat. Let us denote by $E_i, i = 1, 2, \ldots, N$ the event that the ith man selects his own hat. Now, by Proposition 4.4 $P\left(\bigcup_{i=1}^{N} E_i\right)$, the probability that at least one of the men selects his own hat, is given by

$$P\left(\bigcup_{i=1}^{N} E_i\right) = \sum_{i=1}^{N} P(E_i) - \sum_{i_1<i_2} P(E_{i_1}E_{i_2}) + \cdots$$
$$+ (-1)^{n+1} \sum_{i_1<i_2\cdots<i_n} P(E_{i_1}E_{i_2}\cdots E_{i_n})$$
$$+ \cdots + (-1)^{N+1}P(E_1E_2\cdots E_N)$$

If we regard the outcome of this experiment as a vector of N numbers, where the ith element is the number of the hat drawn by the ith man, then there are $N!$ possible outcomes. [The outcome $(1, 2, 3, \ldots, N)$ means, for example, that each man selects his own hat.] Furthermore, $E_{i_1}E_{i_2}\ldots E_{i_n}$, the event that each of the n men $i_1, i_2, \ldots, i_n$ selects his own hat, can occur in any of $(N - n)(N - n - 1)\cdots 3 \cdot 2 \cdot 1 = (N - n)!$ possible ways; for, of the remaining $N - n$ men, the first can select any of $N - n$ hats, the second can then select any of $N - n - 1$ hats, and so on. Hence, assuming that all $N!$ possible outcomes are equally likely, we see that

$$P(E_{i_1}E_{i_2}\cdots E_{i_n}) = \frac{(N - n)!}{N!}$$

Also, as there are $\binom{N}{n}$ terms in $\sum_{i_1<i_2\cdots<i_n} P(E_{i_1}E_{i_2}\cdots E_{i_n})$, we see that

$$\sum_{i_1<i_2\cdots<i_n} P(E_{i_1}E_{i_2}\cdots E_{i_n}) = \frac{N!(N - n)!}{(N - n)!\,n!\,N!} = \frac{1}{n!}$$

and thus

$$P\left(\bigcup_{i=1}^{N} E_i\right) = 1 - \frac{1}{2!} + \frac{1}{3!} - \cdots + (-1)^{N+1}\frac{1}{N!}$$

Hence the probability that none of the men selects his own hat is

$$1 - 1 + \frac{1}{2!} - \frac{1}{3!} + \cdots + \frac{(-1)^N}{N!}$$

which for N large is approximately equal to $e^{-1} \approx .36788$. In other words, for N large, the probability that none of the men selects his own hat is approximately .37. (How many readers would have incorrectly thought that this probability would go to 1 as $N \to \infty$?)

(b) To obtain the probability that exactly k of the N men select their own hats, we first fix attention on a particular set of k men. The number of ways in which these and only these k men can select their own hats is equal to the number of ways in which the other $N - k$ men can select among their hats in such a way that none of them selects his own hat. But, as

$$1 - 1 + \frac{1}{2!} - \frac{1}{3!} + \cdots + \frac{(-1)^{N-k}}{(N-k)!}$$

is the probability that not one of $N - k$ men, selecting among their hats, selects his own, it follows that the number of ways in which the set of men selecting their own hats corresponds to the set of k men under consideration is

$$(N-k)!\left[1 - 1 + \frac{1}{2!} - \frac{1}{3!} + \cdots + \frac{(-1)^{N-k}}{(N-k)!}\right]$$

Hence, as there are $\binom{N}{k}$ possible selections of a group of k men, it follows that there are

$$\binom{N}{k}(N-k)!\left[1 - 1 + \frac{1}{2!} - \frac{1}{3!} + \cdots + \frac{(-1)^{N-k}}{(N-k)!}\right]$$

ways in which exactly k of the men select their own hats. The desired probability is thus

$$\frac{\binom{N}{k}(N-k)!\left[1 - 1 + \frac{1}{2!} - \frac{1}{3!} + \cdots + \frac{(-1)^{N-k}}{(N-k)!}\right]}{N!}$$

$$= \frac{1 - 1 + \frac{1}{2!} - \frac{1}{3!} + \cdots + \frac{(-1)^{N-k}}{(N-k)!}}{k!}$$

which for N large is approximately $e^{-1}/k!$. The values $e^{-1}/k!$, $k = 0, 1, \ldots$, are of some theoretical importance as they represent the values associated with the Poisson distribution. This is elaborated upon in Chapter 4.† ■

For another illustration of the usefulness of Proposition 4.4, consider the following example.

Example 5n. If 10 married couples are seated at random at a round table, compute the probability that no wife sits next to her husband.

Solution If we let E_i, $i = 1, 2, \ldots, 10$ denote the event that the ith couple sits next to each other, it follows that the desired probability is $1 - P\left(\bigcup_{i=1}^{10} E_i\right)$.

† See Example 5c of Chapter 3 for another approach to this problem.

Now, from Proposition 4.4,

$$P\left(\bigcup_{1}^{10} E_i\right) = \sum_{1}^{10} P(E_i) - \cdots + (-1)^{n+1} \sum_{i_1 < i_2 < \cdots < i_n} P(E_{i_1} E_{i_2} \cdots E_{i_n}) + \cdots - P(E_1 E_2 \cdots E_{10})$$

To compute $P(E_{i_1} E_{i_2} \cdots E_{i_n})$, we first note that there are 19! ways of arranging 20 people around a round table (why?). The number of arrangements that result in a specified set of n men sitting next to their wives can most easily be obtained by first thinking of each of the n married couples as being single entities. If this were the case, then we would need to arrange $20 - n$ entities around a round table, and there are clearly $(20 - n - 1)!$ such arrangements. Finally, since each of the n married couples can be arranged next to each other in one of two possible ways, it follows that there are $2^n(20 - n - 1)!$ arrangements that result in a specified set of n men each sitting next to their wives. Therefore,

$$P(E_{i_1} E_{i_2} \cdots E_{i_n}) = \frac{2^n(19 - n)!}{(19)!}$$

Thus, from Proposition 4.4, we obtain that the probability that at least one married couple sits together equals

$$\binom{10}{1} 2^1 \frac{(18)!}{(19)!} - \binom{10}{2} 2^2 \frac{(17)!}{(19)!} + \binom{10}{3} 2^3 \frac{(16)!}{(19)!} - \cdots - \binom{10}{10} 2^{10} \frac{9!}{(19)!} \approx .6605$$

and the desired probability is approximately .3395. ■

Example 5o.** ***Runs. Consider an athletic team that had just finished its season with a final record of n wins and m losses. By examining the sequence of wins and losses, we are hoping to determine whether the team had stretches of games in which it was more likely to win than at other times. One way to gain some insight into this question is to count the number of runs of wins and then see how likely that result would be when all $(n + m)!/(n!\, m!)$ orderings of the n wins and m losses are assumed equally likely. By a run of wins we mean a consecutive sequence of wins. For instance, if $n = 10$, $m = 6$ and the sequence of outcomes was $WWLLWWWLWLLLWWWW$, then there would be 4 runs of wins—the first run being of size 2, the second of size 3, the third of size 1, and the fourth of size 4.

Suppose now that a team has n wins and m losses. Assuming that all $(n + m)!/(n!\, m!) = \binom{n+m}{n}$ orderings are equally likely, let us determine the probability that there will be exactly r runs of wins. To do so, consider first any vector of positive integers $x_1, x_2, \ldots, x_r$ with $x_1 + \cdots + x_r = n$, and let us see how many outcomes result in r runs of wins in which the ith run is of size x_i, $i = 1, \ldots, r$. For any such outcome, if we let y_1 denote the number of losses before the first run of wins, y_2 the number of losses between the first 2 runs of wins, ..., y_{r+1} the number of losses after the last run of wins, then the y_i satisfy

$$y_1 + y_2 + \cdots + y_{r+1} = m \qquad y_1 \geq 0, y_{r+1} \geq 0, y_i > 0, i = 2, \ldots, r$$

and the outcome can be represented schematically as

$$\underbrace{LL\ldots L}_{y_1}\ \underbrace{WW\ldots W}_{x_1}\ \underbrace{L\ldots L}_{y_2}\ \underbrace{WW\ldots W}_{x_2}\ \ldots\ \underbrace{WW}_{x_r}\ \underbrace{L\ldots L}_{y_{r+1}}$$

Hence the number of outcomes that result in r runs of wins—the ith of size x_i, $i = 1, \ldots r$—is equal to the number of integers $y_1, \ldots, y_{r+1}$ that satisfy the foregoing, or, equivalently, to the number of positive integers

$$\bar{y}_1 = y_1 + 1 \quad \bar{y}_i = y_i, i = 2, \ldots, r, \quad \bar{y}_{r+1} = y_{r+1} + 1$$

that satisfy

$$\bar{y}_1 + \bar{y}_2 + \cdots + \bar{y}_{r+1} = m + 2$$

By Proposition 6.1 in Chapter 1 there are $\binom{m+1}{r}$ such outcomes. Hence the total number of outcomes that result in r runs of wins is $\binom{m+1}{r}$ multiplied by the number of positive integral solutions of $x_1 + \cdots + x_r = n$. Hence, again from Proposition 6.1, there are thus $\binom{m+1}{r}\binom{n-1}{r-1}$ outcomes resulting in r runs of wins. As there are $\binom{n+m}{n}$ equally likely outcomes, we thus see that

$$P(\{r \text{ runs of wins}\}) = \frac{\binom{m+1}{r}\binom{n-1}{r-1}}{\binom{m+n}{n}} \qquad r \geq 1$$

For example, if $n = 8$, $m = 6$, then the probability of 7 runs is $\binom{7}{7}\binom{7}{6}\Big/\binom{14}{8} = 1/429$ if all $\binom{14}{8}$ outcomes are equally likely. Hence, if the outcome was *WLWLWLWLWWLWLW*, then we might suspect that the team's win probability was changing over time. (In particular, the probability that the team wins seems to be quite high when it lost its last game and quite low when it won its last game.) On the other extreme, if the outcome were *WWWWWWWWLLLLLL*, then there would have been only 1 run, and as $P(\{1 \text{ run}\}) = \binom{7}{1}\binom{7}{0}\Big/\binom{14}{8} = 1/429$, it would thus again seem unlikely that the team's win probability remained unchanged over its 14 games. ■

*2.6 PROBABILITY AS A CONTINUOUS SET FUNCTION

A sequence of events $\{E_n, n \geq 1\}$ is said to be an increasing sequence if

$$E_1 \subset E_2 \subset \cdots \subset E_n \subset E_{n+1} \subset \cdots$$

whereas it is said to be a decreasing sequence if

$$E_1 \supset E_2 \supset \cdots \supset E_n \supset E_{n+1} \supset \cdots$$

If $\{E_n, n \geq 1\}$ is an increasing sequence of events, then we define a new event, denoted by $\lim_{n\to\infty} E_n$, by

$$\lim_{n\to\infty} E_n = \bigcup_{i=1}^{\infty} E_i$$

Similarly, if $\{E_n, n \geq 1\}$ is a decreasing sequence of events, we define $\lim_n E_n$ by

$$\lim_{n\to\infty} E_n = \bigcap_{i=1}^{\infty} E_i$$

We now prove Proposition 6.1.

Proposition 6.1

If $\{E_n, n \geq 1\}$ is either an increasing or a decreasing sequence of events, then

$$\lim_{n\to\infty} P(E_n) = P\left(\lim_{n\to\infty} E_n\right)$$

Proof: Suppose, first, that $\{E_n, n \geq 1\}$ is an increasing sequence and define the events $F_n, n \geq 1$ by

$$F_1 = E_1$$

$$F_n = E_n\left(\bigcup_{1}^{n-1} E_i\right)^c = E_n E_{n-1}^c \qquad n > 1$$

where we have used the fact that $\bigcup_{1}^{n-1} E_i = E_{n-1}$, since the events are increasing. In words, F_n consists of those outcomes in E_n that are not in any of the earlier $E_i, i < n$. It is easy to verify that the F_n are mutually exclusive events such that

$$\bigcup_{i=1}^{\infty} F_i = \bigcup_{i=1}^{\infty} E_i \quad \text{and} \quad \bigcup_{i=1}^{n} F_i = \bigcup_{i=1}^{n} E_i \quad \text{for all } n \geq 1$$

Thus

$$
\begin{aligned}
P\left(\bigcup_1^\infty E_i\right) &= P\left(\bigcup_1^\infty F_i\right) \\
&= \sum_1^\infty P(F_i) \qquad \text{(by Axiom 3)} \\
&= \lim_{n\to\infty} \sum_1^n P(F_i) \\
&= \lim_{n\to\infty} P\left(\bigcup_1^n F_i\right) \\
&= \lim_{n\to\infty} P\left(\bigcup_1^n E_i\right) \\
&= \lim_{n\to\infty} P(E_n)
\end{aligned}
$$

which proves the result when $\{E_n, n \geq 1\}$ is increasing.

If $\{E_n, n \geq 1\}$ is a decreasing sequence, then $\{E_n^c, n \geq 1\}$ is an increasing sequence; hence, from the preceding equations,

$$P\left(\bigcup_1^\infty E_i^c\right) = \lim_{n\to\infty} P(E_n^c)$$

But as $\bigcup_1^\infty E_i^c = \left(\bigcap_1^\infty E_i\right)^c$, we see that

$$P\left(\left(\bigcap_1^\infty E_i\right)^c\right) = \lim_{n\to\infty} P(E_n^c)$$

or, equivalently,

$$1 - P\left(\bigcap_1^\infty E_i\right) = \lim_{n\to\infty} [1 - P(E_n)] = 1 - \lim_{n\to\infty} P(E_n)$$

or

$$P\left(\bigcap_1^\infty E_i\right) = \lim_{n\to\infty} P(E_n)$$

which proves the result.

Example 6a. ***Probability and a paradox.*** Suppose that we possess an infinitely large urn and an infinite collection of balls labeled ball number 1, number 2, number 3, and so on. Consider an experiment performed as follows. At 1 minute to 12 P.M., balls numbered 1 through 10 are placed in the urn, and ball number 10 is withdrawn.

(Assume the withdrawal takes no time.) At $\frac{1}{2}$ minute to 12 P.M., balls numbered 11 through 20 are placed in the urn, and ball number 20 is withdrawn. At $\frac{1}{4}$ minute to 12 P.M., balls numbered 21 through 30 are placed in the urn, and ball number 30 is withdrawn. At $\frac{1}{8}$ minute to 12 P.M., and so on. The question of interest is, how many balls are in the urn at 12 P.M.?

The answer to this question is clearly that there is an infinite number of balls in the urn at 12 P.M., since any ball whose number is not of the form $10n$, $n \geq 1$, will have been placed in the urn and will not have been withdrawn before 12 P.M. Hence the problem is solved when the experiment is performed as described.

However, let us now change the experiment and suppose that at 1 minute to 12 P.M. balls numbered 1 through 10 are placed in the urn, and ball number 1 is withdrawn; at $\frac{1}{2}$ minute to 12 P.M., balls numbered 11 through 20 are placed in the urn, and ball number 2 is withdrawn; at $\frac{1}{4}$ minute to 12 P.M., balls numbered 21 through 30 are placed in the urn, and ball number 3 is withdrawn; at $\frac{1}{8}$ minute to 12 P.M., balls numbered 31 through 40 are placed in the urn, and ball number 4 is withdrawn, and so on. For this new experiment how many balls are in the urn at 12 P.M.?

Surprisingly enough, the answer now is that the urn is *empty* at 12 P.M. For, consider any ball—say, ball number n. At some time prior to 12 P.M. [in particular, at $\left(\frac{1}{2}\right)^{n-1}$ minutes to 12 P.M.], this ball would have been withdrawn from the urn. Hence for each n, ball number n is not in the urn at 12 P.M.; therefore, the urn must be empty at this time.

Thus we see from the preceding discussion that the manner in which the withdrawn balls are selected makes a difference. For, in the first case only balls numbered $10n$, $n \geq 1$, are ever withdrawn; whereas in the second case all of the balls are eventually withdrawn. Let us now suppose that whenever a ball is to be withdrawn that ball is randomly selected from among those present. That is, suppose that at 1 minute to 12 P.M. balls numbered 1 through 10 are placed in the urn, and a ball is randomly selected and withdrawn, and so on. In this case how many balls are in the urn at 12 P.M.?

Solution We shall show that, with probability 1, the urn is empty at 12 P.M. Let us first consider ball number 1. Define E_n to be the event that ball number 1 is still in the urn after the first n withdrawals have been made. Clearly,

$$P(E_n) = \frac{9 \cdot 18 \cdot 27 \cdots (9n)}{10 \cdot 19 \cdot 28 \cdots (9n + 1)}$$

[To understand this equation, just note that if ball number 1 is still to be in the urn after the first n withdrawals, the first ball withdrawn can be any one of 9, the second any one of 18 (there are 19 balls in the urn at the time of the second withdrawal, one of which must be ball number 1), and so on. The denominator is similarly obtained.]

Now, the event that ball number 1 is in the urn at 12 P.M. is just the event $\bigcap_{n=1}^{\infty} E_n$. As the events $E_n, n \geq 1$, are decreasing events, it follows from Proposition 6.1 that

$$\begin{aligned}
P\{\text{ball number 1 is in the urn at 12 P.M.}\} \\
&= P\left(\bigcap_{n=1}^{\infty} E_n\right) \\
&= \lim_{n\to\infty} P(E_n) \\
&= \prod_{n=1}^{\infty}\left(\frac{9n}{9n+1}\right)
\end{aligned}$$

We now show that

$$\prod_{n=1}^{\infty}\frac{9n}{9n+1} = 0$$

Since

$$\prod_{n=1}^{\infty}\left(\frac{9n}{9n+1}\right) = \left[\prod_{n=1}^{\infty}\left(\frac{9n+1}{9n}\right)\right]^{-1}$$

this is equivalent to showing that

$$\prod_{n=1}^{\infty}\left(1+\frac{1}{9n}\right) = \infty$$

Now, for all $m \geq 1$,

$$\begin{aligned}
\prod_{n=1}^{\infty}\left(1+\frac{1}{9n}\right) &\geq \prod_{n=1}^{m}\left(1+\frac{1}{9n}\right) \\
&= \left(1+\frac{1}{9}\right)\left(1+\frac{1}{18}\right)\left(1+\frac{1}{27}\right)\cdots\left(1+\frac{1}{9m}\right) \\
&> \frac{1}{9}+\frac{1}{18}+\frac{1}{27}+\cdots+\frac{1}{9m} \\
&= \frac{1}{9}\sum_{i=1}^{m}\frac{1}{i}
\end{aligned}$$

Hence, letting $m \to \infty$ and using the fact that $\sum_{i=1}^{\infty} 1/i = \infty$ yields

$$\prod_{n=1}^{\infty}\left(1+\frac{1}{9n}\right) = \infty$$

Hence, letting F_i denote the event that ball number i is in the urn at 12 P.M., we have shown that $P(F_1) = 0$. Similarly, we can show that $P(F_i) = 0$ for all i.

(For instance, the same reasoning shows that $P(F_i) = \prod_{n=2}^{\infty} [9n/(9n+1)]$ for $i = 11, 12, \ldots, 20$.) Therefore, the probability that the urn is not empty at 12 P.M., $P\left(\bigcup_{1}^{\infty} F_i\right)$, satisfies

$$P\left(\bigcup_{1}^{\infty} F_i\right) \leq \sum_{1}^{\infty} P(F_i) = 0$$

by Boole's inequality (see Self-Test Exercise 14).

Thus, with probability 1, the urn will be empty at 12 P.M. ■

2.7 PROBABILITY AS A MEASURE OF BELIEF

Thus far we have interpreted the probability of an event of a given experiment as being a measure of how frequently the event will occur when the experiment is continually repeated. However, there are also other uses of the term *probability*. For instance, we have all heard such statements as, "it is 90 percent probable that Shakespeare actually wrote *Hamlet*," or "the probability that Oswald acted alone in assassinating Kennedy is .8." How are we to interpret these statements?

The most simple and natural interpretation is that the probabilities referred to are measures of the individual's belief in the statements that he or she is making. In other words, the individual making the foregoing statements is quite certain that Oswald acted alone and is even more certain that Shakespeare wrote *Hamlet*. This interpretation of probability as being a measure of one's belief is often referred to as the *personal* or *subjective* view of probability.

It seems logical to suppose that a "measure of belief" should satisfy all of the axioms of probability. For example, if we are 70 percent certain that Shakespeare wrote *Julius Caesar* and 10 percent certain that it was actually Marlowe, then it is logical to suppose that we are 80 percent certain that it was either Shakespeare or Marlowe. Hence, whether we interpret probability as a measure of belief or as a long-run frequency of occurrence, its mathematical properties remain unchanged.

Example 7a. Suppose that in a 7-horse race you feel that each of the first 2 horses has a 20 percent chance of winning, horses 3 and 4 each has a 15 percent chance, and the remaining 3 horses, a 10 percent chance each. Would it be better for you to wager at even money, that the winner will be one of the first three horses, or to wager, again at even money, that the winner will be one of the horses 1, 5, 6, 7?

Solution Based on your personal probabilities concerning the outcome of the race, your probability of winning the first bet is $.2 + .2 + .15 = .55$, whereas it is $.2 + .1 + .1 + .1 = .5$ for the second. Hence the first wager is more attractive. ■

It should be noted that in supposing that person's subjective probabilities are always consistent with the axioms of probability, we are dealing with an idealized rather than an actual person. For instance, if we were to ask someone what he or she thought the chances were of

(a) rain today,
(b) rain tomorrow,
(c) rain both today and tomorrow,
(d) rain either today or tomorrow,

it is quite possible that after some deliberation that this person might give 30 percent, 40 percent, 20 percent, and 60 percent as answers. Unfortunately, however, such answers (or such subjective probabilities) are not consistent with the axioms of probability (why not?). We would of course hope that after this was pointed out to the respondent, he or she would change the answers. (One possibility we could accept is 30 percent, 40 percent, 10 percent, and 60 percent.)

SUMMARY

Let S denote the set of all possible outcomes of an experiment. S is called the *sample space* of the experiment. An event is a subset of S. If $A_i, i = 1, \ldots, n$, are events, then $\bigcup_{i=1}^{n} A_i$, called the *union* of these events, consists of all outcomes that are in at least one of the events $A_i, i = 1, \ldots, n$. Similarly, $\bigcap_{i=1}^{n} A_i$, sometimes written as $A_1 \cdots A_n$, is called the *intersection* of the events A_i, and consists of all outcomes that are in all of the events $A_i, i = 1, \ldots, n$.

For any event A, we define A^c to consist of all outcomes in the sample space that are not in A. We call A^c the *complement* of the event A. The event S^c, which is empty of outcomes, is designated by $\emptyset$ and is called the *null* set. If $AB = \emptyset$, then we say that A and B are mutually exclusive.

For each event A of the sample space S we suppose that a number $P(A)$, called the probability of A, is defined and is such that

(i) $0 \leq P(A) \leq 1$
(ii) $P(S) = 1$
(iii) For mutually exclusive events $A_i, i \geq 1$,

$$P\left(\bigcup_{i=1}^{\infty} A_i\right) = \sum_{i=1}^{\infty} P(A_i)$$

$P(A)$ represents the probability that the outcome of the experiment is in A.

It can be shown that

$$P(A^c) = 1 - P(A)$$

A useful result is that

$$P(A \cup B) = P(A) + P(B) - P(AB)$$

which can be generalized to give

$$P\left(\bigcup_{i=1}^{n} A_i\right) = \sum_{i=1}^{n} P(A_i) - \sum\sum_{i<j} P(A_iA_j) + \sum\sum\sum_{i<j<k} P(A_iA_jA_k)$$
$$+ \cdots + (-1)^{n+1}P(A_1 \cdots A_n)$$

If S is finite and each one point set is assumed to have equal probability, then

$$P(A) = \frac{|A|}{|S|}$$

where $|E|$ denotes the number of outcomes in the event E.

$P(A)$ can be interpreted either as a long-run relative frequency or as a measure of one's degree of belief.

PROBLEMS

1. A box contains 3 marbles, 1 red, 1 green, and 1 blue. Consider an experiment that consists of taking 1 marble from the box, then replacing it in the box and drawing a second marble from the box. Describe the sample space. Repeat when the second marble is drawn without first replacing the first marble.

2. A die is rolled continually until a 6 appears, at which point the experiment stops. What is the sample space of this experiment? Let E_n denote the event that n rolls are necessary to complete the experiment. What points of the sample space are contained in E_n? What is $\left(\bigcup_{1}^{\infty} E_n\right)^c$?

3. Two dice are thrown. Let E be the event that the sum of the dice is odd; let F be the event that at least one of the dice lands on 1; and let G be the event that the sum is 5. Describe the events EF, $E \cup F$, FG, EF^c, and EFG.

4. A, B, and C take turns in flipping a coin. The first one to get a head wins. The sample space of this experiment can be defined by

$$S = \begin{cases} 1, 01, 001, 0001, \ldots, \\ 0000 \cdots \end{cases}$$

(a) Interpret the sample space.

(b) Define the following events in terms of S:

(i) A wins $= A$.

(ii) B wins $= B$.

(iii) $(A \cup B)^c$.

Assume that A flips first, then B, then C, then A, and so on.

5. A system is composed of 5 components, each of which is either working or failed. Consider an experiment that consists of observing the status of each component, and let the outcome of the experiment be given by the vector $(x_1, x_2, x_3, x_4, x_5)$, where x_i is equal to 1 if component i is working and is equal to 0 if component i is failed.

(a) How many outcomes are in the sample space of this experiment?

(b) Suppose that the system will work if components 1 and 2 are both working, or if components 3 and 4 are both working, or if components 1, 3, and 5 are all working. Let W be the event that the system will work. Specify all the outcomes in W.

(c) Let A be the event that components 4 and 5 are both failed. How many outcomes are contained in the event A?

(d) Write out all the outcomes in the event AW.

6. A hospital administrator codes incoming patients suffering gunshot wounds according to whether they have insurance (coding 1 if they do and 0 if they do not) and according to their condition, which is rated as good (g), fair (f), or serious (s). Consider an experiment that consists of the coding of such a patient.

(a) Give the sample space of this experiment.

(b) Let A be the event that the patient is in serious condition. Specify the outcomes in A.

(c) Let B be the event that the patient is uninsured. Specify the outcomes in B.

(d) Give all the outcomes in the event $B^c \cup A$.

7. Consider an experiment that consists of determining the type of job—either blue-collar or white-collar—and the political affiliation—Republican, Democratic, or Independent—of the 15 members of an adult soccer team. How many outcomes are

(a) in the sample space;

(b) in the event that at least one of the team members is a blue-collar worker;

(c) in the event that none of the team members considers himself or herself an Independent?

8. Suppose that A and B are mutually exclusive events for which $P(A) = .3$ and $P(B) = .5$. What is the probability that

(a) either A or B occurs;

(b) A occurs but B does not;

(c) both A and B occur?

9. A retail establishment accepts either the American Express or the VISA credit card. A total of 24 percent of its customers carry an American Express card, 61 percent carry a VISA card, and 11 percent carry both. What percentage of its customers carry a credit card that the establishment will accept?

10. Sixty percent of the students at a certain school wear neither a ring nor a necklace. Twenty percent wear a ring and 30 percent wear a necklace. If one of the students is chosen randomly, what is the probability that this student is wearing

(a) a ring or a necklace;

(b) a ring and a necklace?

11. A total of 28 percent of American males smoke cigarettes, 7 percent smoke cigars, and 5 percent smoke both cigars and cigarettes.

(a) What percentage of males smoke neither cigars nor cigarettes?

(b) What percentage smoke cigars but not cigarettes?

12. An elementary school is offering 3 language classes: one in Spanish, one in French, and one in German. These classes are open to any of the 100 students in the school. There are 28 students in the Spanish class, 26 in the French class, and 16 in the German class. There are 12 students that are in both Spanish and French, 4 that are in both Spanish and German, and 6 that are in both French and German. In addition, there are 2 students taking all 3 classes.

(a) If a student is chosen randomly, what is the probability that he or she is not in any of these classes?

(b) If a student is chosen randomly, what is the probability that he or she is taking exactly one language class?

(c) If 2 students are chosen randomly, what is the probability that at least 1 is taking a language class?

13. A certain town of population size 100,000 has 3 newspapers: I, II, and III. The proportions of townspeople who read these papers are as follows:

I:10 percent	I and II: 8 percent	I and II and III: 1 percent
II:30 percent	I and III: 2 percent	
III:5 percent	II and III: 4 percent	

(The list tells us, for instance, that 8000 people read newspapers I and II.)

(a) Find the number of people who read only one newspaper.

(b) How many people read at least two newspapers?

(c) If I and III are morning papers and II is an evening paper, how many people read at least one morning paper plus an evening paper?

(d) How many people do not read any newspapers?

(e) How many people read only one morning paper and one evening paper?

14. The following data were given in a study of a group of 1000 subscribers to a certain magazine: In reference to job, marital status, and education, there were 312 professionals, 470 married persons, 525 college graduates, 42 professional college graduates, 147 married college graduates, 86 married professionals, and 25 married professional college graduates. Show that the numbers reported in the study must be incorrect.

HINT: Let M, W, and G denote, respectively, the set of professionals, married persons, and college graduates. Assume that one of the 1000 persons is chosen at random and use Proposition 4.4 to show that if the numbers above are correct, then $P(M \cup W \cup G) > 1$.

15. If it is assumed that all $\binom{52}{5}$ poker hands are equally likely, what is the probability of being dealt

(a) a flush? (A hand is said to be a flush if all 5 cards are of the same suit.)

(b) one pair? (This occurs when the cards have denominations a, a, b, c, d, where a, b, c, and d are all distinct.)

(c) two pairs? (This occurs when the cards have denominations a, a, b, b, c, where a, b, and c are all distinct.)

(d) three of a kind? (This occurs when the cards have denominations a, a, a, b, c, where a, b, and c are all distinct.)

(e) four of a kind? (This occurs when the cards have denominations a, a, a, a, b.)

16. Poker dice is played by simultaneously rolling 5 dice. Show that

(a) $P\{\text{no two alike}\} = .0926$; **(b)** $P\{\text{one pair}\} = .4630$;

(c) $P\{\text{two pair}\} = .2315$; **(d)** $P\{\text{three alike}\} = .1543$;

(e) $P\{\text{full house}\} = .0386$; **(f)** $P\{\text{four alike}\} = .0193$;

(g) $P\{\text{five alike}\} = .0008$.

17. If 8 castles (that is, rooks) are randomly placed on a chessboard, compute the probability that none of the rooks can capture any of the others. That is, compute the probability that no row or file contains more than one rook.

18. Two cards are randomly selected from an ordinary playing deck. What is the probability that they form a blackjack? That is, what is the probability that one of the cards is an ace and the other one is either a ten, a jack, a queen, or a king?

19. Two symmetric dice have both had two of their sides painted red, two painted black, one painted yellow, and the other painted white. When this pair of dice are flipped, what is the probability that both land on the same color?

20. Suppose that you are playing blackjack against a dealer. In a freshly shuffled deck, what is the probability that neither you nor the dealer is dealt a blackjack?

21. A small community organization consists of 20 families, of which 4 have one child, 8 have two children, 5 have three children, 2 have four children, and 1 has five children.

(a) If one of these families is chosen at random, what is the probability it has i children, $i = 1, 2, 3, 4, 5$?

(b) If one of the children is randomly chosen, what is the probability this child comes from a family having i children, $i = 1, 2, 3, 4, 5$?

22. Consider the following technique for shuffling a deck of n cards. For any initial ordering of the cards, go through the deck one card at a time and at each card flip a fair coin. If the coin comes up heads, then leave the card where it is, and if it comes up tails, then move that card to the end of the deck. After the coin has been flipped n times, say that one round has been completed. For instance, if $n = 4$ and the initial ordering is $1, 2, 3, 4$, then if the successive flips result in the outcome h, t, t, h, then the ordering at the end of the round is $1, 4, 2, 3$. Assuming that all possible outcomes of the sequence of n coin flips are equally likely, what is the probability that the ordering after one round is the same as the initial ordering?

23. A pair of fair dice are rolled. What is the probability that the second die lands on a higher value than does the first?

24. If two dice are rolled, what is the probability that the sum of the upturned faces equals i? Find it for $i = 2, 3, \ldots, 11, 12$.

25. A pair of dice is rolled until a sum of either 5 or 7 appears. Find the probability that a 5 occurs first.

HINT: Let E_n denote the event that a 5 occurs on the nth roll and no 5 or 7 occurs on the first $n - 1$ rolls. Compute $P(E_n)$ and argue that $\sum_{n=1}^{\infty} P(E_n)$ is the desired probability.

26. The game of craps is played as follows: A player rolls two dice. If the sum of the dice is either a 2, 3, or 12, the player loses; if the sum is either a 7 or an 11, he or she wins. If the outcome is anything else, the player continues to roll the dice until he or she rolls either the initial outcome or a 7. If the 7 comes first, the player loses; whereas if the initial outcome reoccurs before the 7, the player wins. Compute the probability of a player winning at craps.

HINT: Let E_i denote the event that the initial outcome is i and the player wins. The desired probability is $\sum_{i=2}^{12} P(E_i)$. To compute $P(E_i)$, define the events $E_{i,n}$ to be the event that the initial sum is i and the player wins on the nth roll. Argue that $P(E_i) = \sum_{n=1}^{\infty} P(E_{i,n})$.

27. An urn contains 3 red and 7 black balls. Players A and B withdraw balls from the urn consecutively until a red ball is selected. Find the probability that A selects the red ball. (A draws the first ball, then B, and so on. There is no replacement of the balls drawn.)

28. An urn contains 5 red, 6 blue, and 8 green balls. If a set of 3 balls is randomly selected, what is the probability that each of the balls will be (a) of the same color; (b) of different colors? Repeat under the assumption that whenever a ball is selected, its color is noted and it is then replaced in the urn before the next selection. This is known as *sampling with replacement*.

29. An urn contains n white and m black balls, where n and m are positive numbers.

(a) If two balls are randomly withdrawn, what is the probability that they are the same color?

(b) If a ball is randomly withdrawn and then replaced before the second one is drawn, what is the probability that the withdrawn balls are the same color?

(c) Show that the probability in part (b) is always larger than the one in part (a).

30. The chess clubs of two schools consist of, respectively, 8 and 9 players. Four members from each club are randomly chosen to participate in a contest between the two schools. The chosen players from one team are then randomly paired with those from the other team, and each pairing plays a game of chess. Suppose that Rebecca and her sister Elise are on the chess clubs at different schools. What is the probability that

(a) Rebecca and Elise will be paired;

(b) Rebecca and Elise will be chosen to represent their schools but will not play each other;

(c) exactly one of Rebecca and Elise will be chosen to represent her school?

31. A 3-person basketball team consists of a guard, a forward, and a center.

(a) If a person is chosen at random from each of three different such teams, what is the probability of selecting a complete team?

(b) What is the probability that all 3 players selected play the same position?

32. A group of individuals containing b boys and g girls is lined up in random order—that is, each of the $(b + g)!$ permutations is assumed to be equally likely. What is the probability that the person in the ith position, $1 \leq i \leq b + g$, is a girl?

33. A forest contains 20 elk, of which 5 are captured, tagged, and then released. A certain time later 4 of the 20 elk are captured. What is the probability that 2 of these 4 have been tagged? What assumptions are you making?

34. The second Earl of Yarborough is reported to have bet at odds of 1000 to 1 that a bridge hand of 13 cards would contain at least one card that is ten or higher. (By *ten or higher* we mean that it is either a ten, a jack, a queen, a king, or an ace.) Nowadays, we call a hand that has no cards higher than 9 *a Yarborough*. What is the probability that a randomly selected bridge hand is a Yarborough?

35. There are 30 psychiatrists and 24 psychologists attending a certain conference. Three of these 54 people are randomly chosen to take part in a panel discussion. What is the probability that at least one psychologist is chosen?

36. Two cards are chosen at random from a deck of 52 playing cards. What is the probability that they

(a) are both aces;

(b) have the same value?

37. An instructor gives her class a set of 10 problems with the information that the final exam will consist of a random selection of 5 of them. If a student has figured out how to do 7 of the problems, what is the probability that he or she will answer correctly.

(a) all 5 problems;

(b) at least 4 of the problems?

38. There are n socks, 3 of which are red, in a drawer. What is the value of n if when 2 of the socks are chosen randomly, the probability that they are both red is $\frac{1}{2}$?

39. There are 5 hotels in a certain town. If 3 people check into hotels in a day, what is the probability they each check into a different hotel? What assumptions are you making?

40. A town contains 4 people who repair televisions. If 4 sets break down, what is the probability that exactly i of the repairers are called? Solve the problem for $i = 1, 2, 3, 4$. What assumptions are you making?

41. If a die is rolled 4 times, what is the probability that 6 comes up at least once?

42. Two dice are thrown n times in succession. Compute the probability that double 6 appears at least once. How large need n be to make this probability at least $\frac{1}{2}$?

43. **(a)** If N people, including A and B, are randomly arranged in a line, what is the probability that A and B are next to each other?

(b) What would the probability be if the people were randomly arranged in a circle?

44. Five people, designated as A, B, C, D, E, are arranged in linear order. Assuming that each possible order is equally likely, what is the probability that
(a) there is exactly one person between A and B;
(b) there are exactly two people between A and B;
(c) there are three people between A and B?

45. A woman has n keys, of which one will open her door.
(a) If she tries the keys at random, discarding those that do not work, what is the probability that she will open the door on her kth try?
(b) What if she does not discard previously tried keys?

46. How many people have to be in a room in order that the probability that at least two of them celebrate their birthday in the same month is at least $\frac{1}{2}$? Assume that all possible monthly outcomes are equally likely.

47. If there are 12 strangers in a room, what is the probability that no two of them celebrate their birthday in the same month?

48. Given 20 people, what is the probability that among the 12 months in the year there are 4 months containing exactly 2 birthdays and 4 containing exactly 3 birthdays?

49. A group of 6 men and 6 women is randomly divided into 2 groups of size 6 each. What is the probability that both groups will have the same number of men?

50. In a hand of bridge, find the probability that you have 5 spades and your partner has the remaining 8.

51. Suppose that n balls are randomly distributed in N compartments. Find the probability that m balls will fall in the first compartment. Assume that all N^n arrangements are equally likely.

52. A closet contains 10 pairs of shoes. If 8 shoes are randomly selected, what is the probability that there will be
(a) no complete pair;
(b) exactly 1 complete pair?

53. If 4 married couples are arranged in a row, find the probability that no husband sits next to his wife.

54. Compute the probability that a bridge hand is void in at least one suit. Note that the answer is not

$$\frac{\binom{4}{1}\binom{39}{13}}{\binom{52}{13}}$$

Why not?

HINT: Use Proposition 4.4.

55. Compute the probability that a hand of 13 cards contains
(a) the ace and king of at least one suit;
(b) all 4 of at least 1 of the 13 denominations.

56. Two players play the following game. Player A chooses one of the three spinners on the next page, and then player B chooses one of the remaining two spinners. Both players then spin their spinner and the one that lands on the higher number is declared the winner. Assuming that each spinner is equally likely to land in any of its 3 regions, would you rather be player A or player B? Explain your answer!

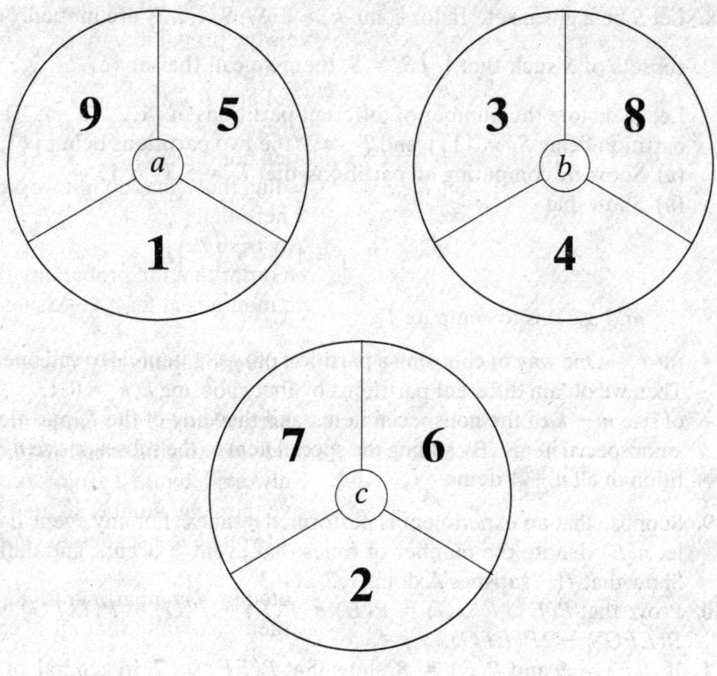

THEORETICAL EXERCISES

Prove the following relations.

1. $EF \subset E \subset E \cup F$.
2. If $E \subset F$, then $F^c \subset E^c$.
3. $F = FE \cup FE^c$, and $E \cup F = E \cup E^cF$.
4. $\left(\bigcup_1^\infty E_i\right)F = \bigcup_1^\infty E_iF$, and $\left(\bigcap_1^\infty E_i\right) \cup F = \bigcap_1^\infty (E_i \cup F)$.
5. For any sequence of events $E_1, E_2, \ldots$, define a new sequence $F_1, F_2, \ldots$ of disjoint events (that is, events such that $F_iF_j = \emptyset$ whenever $i \neq j$) such that for all $n \geq 1$,

$$\bigcup_1^n F_i = \bigcup_1^n E_i$$

6. Let E, F, and G be three events. Find expressions for the events so that of E, F, and G:
   (a) only E occurs;
   (b) both E and G but not F occur;
   (c) at least one of the events occurs;
   (d) at least two of the events occur;
   (e) all three occur;
   (f) none of the events occurs;
   (g) at most one of them occurs;
   (h) at most two of them occur;
   (i) exactly two of them occur;
   (j) at most three of them occur.
7. Find the simplest expression for the following events:
   (a) $(E \cup F)(E \cup F^c)$;
   (b) $(E \cup F)(E^c \cup F)(E \cup F^c)$;
   (c) $(E \cup F)(F \cup G)$.

8. Let S be a given set. If, for some $k > 0$, $S_1, S_2, \ldots, S_k$ are mutually exclusive nonempty subsets of S such that $\bigcup_{i=1}^{k} S_i = S$, then we call the set $\{S_1, S_2, \ldots, S_k\}$ a *partition* of S. Let T_n denote the number of different partitions of $\{1, 2, \ldots, n\}$. Thus, $T_1 = 1$ (the only partition being $S_1 = \{1\}$), and $T_2 = 2$ (the two partitions being $\{\{1, 2,\}\}, \{\{1\}, \{2\}\}$).
(a) Show, by computing all partitions, that $T_3 = 5, T_4 = 15$.
(b) Show that

$$T_{n+1} = 1 + \sum_{k=1}^{n} \binom{n}{k} T_k$$

and use this to compute T_{10}.

HINT: One way of choosing a partition of $n + 1$ items is to call one of the items *special*. Then we obtain different partitions by first choosing $k, k = 0, 1, \ldots, n$, and then a subset of size $n - k$ of the nonspecial items, and then any of the T_k partitions of the remaining k nonspecial items. By adding the special item to the subset of size $n - k$ we obtain a partition of all $n + 1$ items.

9. Suppose that an experiment is performed n times. For any event E of the sample space, let $n(E)$ denote the number of times that event E occurs, and define $f(E) = n(E)/n$. Show that $f(\cdot)$ satisfies Axioms 1, 2, and 3.

10. Prove that $P(E \cup F \cup G) = P(E) + P(F) + P(G) - P(E^cFG) - P(EF^cG) - P(EFG^c) - 2P(EFG)$.

11. If $P(E) = .9$ and $P(F) = .8$, show that $P(EF) \geq .7$. In general, prove Bonferroni's inequality, namely,

$$P(EF) \geq P(E) + P(F) - 1$$

12. Show that the probability that exactly one of the events E or F occurs equals $P(E) + P(F) - 2P(EF)$.

13. Prove that $P(EF^c) = P(E) - P(EF)$.

14. Prove Proposition 4.4 by mathematical induction.

15. An urn contains M white and N black balls. If a random sample of size r is chosen, what is the probability that it contains exactly k white balls?

16. Use induction to generalize Bonferroni's inequality to n events. Namely, show that

$$P(E_1E_2 \cdots E_n) \geq P(E_1) + \cdots + P(E_n) - (n - 1)$$

17. Consider the matching problem, Example 5m, and define A_N to be the number of ways in which the N men can select their hats so that no man selects his own. Argue that

$$A_N = (N - 1)(A_{N-1} + A_{N-2})$$

This formula, along with the boundary conditions $A_1 = 0, A_2 = 1$, can then be solved for A_N, and the desired probability of no matches would be $A_N/N!$.

HINT: After the first man selects a hat that is not his own, there remain $N - 1$ men to select among a set of $N - 1$ hats that does not contain the hat of one of these men. Thus there is one extra man and one extra hat. Argue that we can get no matches either with the extra man selecting the extra hat or with the extra man not selecting the extra hat.

18. Let f_n denote the number of ways of tossing a coin n times such that successive heads never appear. Argue that

$$f_n = f_{n-1} + f_{n-2} \qquad n \geq 2, \text{where } f_0 \equiv 1, f_1 \equiv 2$$

HINT: How many outcomes are there that start with a head, and how many start with a tail?

If P_n denotes the probability that successive heads never appear when a coin is tossed n times, find P_n (in terms of f_n) when all possible outcomes of the n tosses are assumed equally likely. Compute P_{10}.

19. An urn contains n red and m blue balls. They are withdrawn one at a time until a total of r, $r \le n$, red balls have been withdrawn. Find the probability that a total of k balls are withdrawn.

HINT: A total of k balls will be withdrawn if there are $r - 1$ red balls in the first $k - 1$ withdrawals and the kth withdrawal is a red ball.

20. Consider an experiment whose sample space consists of a countably infinite number of points. Show that not all points can be equally likely. Can all points have positive probability of occurring?

***21.** Consider Example 5o, which is concerned with the number of runs of wins obtained when n wins and m losses are randomly permuted. Now consider the total number of runs—that is, win runs plus loss runs—and show that

$$P\{2k \text{ runs}\} = 2\frac{\binom{m-1}{k-1}\binom{n-1}{k-1}}{\binom{m+n}{n}}$$

$$P\{2k+1 \text{ runs}\} = \frac{\binom{m-1}{k-1}\binom{n-1}{k} + \binom{m-1}{k}\binom{n-1}{k-1}}{\binom{m+n}{n}}$$

SELF-TEST PROBLEMS AND EXERCISES

1. A cafeteria offers a 3-course meal. One chooses an entree, a starch, and a dessert. The possible choices are given in the following table.

Course	Choices
Entree	Chicken or roast beef
Starch	Pasta or rice or potatoes
Dessert	Ice cream or Jello or apple pie or a peach

A person is to choose one course from each category.
(a) How many outcomes are in the sample space?
(b) Let A be the event that ice cream is chosen. How many outcomes are in A?
(c) Let B be the event that chicken is chosen. How many outcomes are in B?
(d) List all the outcomes in the event AB.
(e) Let C be the event that rice is chosen. How many outcomes are in C?
(f) List all the outcomes in the event ABC.

2. A customer visiting the suit department of a certain store will purchase a suit with probability .22, a shirt with probability .30, and a tie with probability .28. The customer will purchase both a suit and a shirt with probability .11, both a suit and a tie with probability .14, and both a shirt and a tie with probability .10. A customer will purchase all 3 items with probability .06. What is the probability that a customer purchases
(a) none of these items;
(b) exactly 1 of these items?

3. A deck of cards is dealt out. What is the probability that the fourteenth card dealt is an ace? What is the probability that the first ace occurs on the fourteenth card?

4. Let A denote the event that the midtown temperature in Los Angeles is 70°F, and let B denote the event that the midtown temperature in New York is 70°F. Also, let C denote the event that the maximum of the midtown temperatures in New York and in Los Angeles is 70°F. If $P(A) = .3$, $P(B) = .4$, and $P(C) = .2$, find the probability that the minimum of the two midtown temperatures is 70°F.

5. An ordinary deck of 52 cards is shuffled. What is the probability that the top four cards have
(a) different denominations;
(b) different suits?

6. Urn A contains 3 red and 3 black balls, whereas urn B contains 4 red and 6 black balls. If a ball is randomly selected from each urn, what is the probability that the balls will be the same color?

7. In a state lottery, a player must choose 8 of the numbers from 1 to 40. The lottery commission then performs an experiment that selects 8 of these 40 numbers. Assuming that the choice of the lottery commission is equally likely to be any of the $\binom{40}{8}$ combinations, what is the probability that a player has
(a) all 8 of the numbers selected;
(b) 7 of the numbers selected;
(c) at least 6 of the numbers selected?

8. From a group of 3 freshmen, 4 sophomores, 4 juniors, and 3 seniors a committee of size 4 is randomly selected. Find the probability that the committee will consist of
(a) 1 from each class;
(b) 2 sophomores and 2 juniors;
(c) only sophomores or juniors.

9. For a finite set A, let $N(A)$ denote the number of elements in A.
(a) Show that

$$N(A \cup B) = N(A) + N(B) - N(AB)$$

(b) More generally, show that

$$N\left(\bigcup_{i=1}^{n} A_i\right) = \sum_i N(A_i) - \sum_{i<j}\sum N(A_iA_j) + \cdots + (-1)^{n+1}N(A_1 \cdots A_n)$$

10. Consider an experiment that consists of six horses, numbered 1 through 6, running a race and suppose that the sample space consists of the 6! possible orders in which the horses finish. Let A be the event that the number 1 horse is among the top three finishers, and let B be the event that the number 2 horse comes in second. How many outcomes are in the event $A \cup B$?

11. A 5-card hand is dealt from a well-shuffled deck of 52 playing cards. What is the probability that the hand contains at least one card from each of the four suits?

12. A basketball team consists of 6 frontcourt and 4 backcourt players. If players are divided into roommates at random, what is the probability that there will be exactly two roommate pairs made up of a backcourt and a frontcourt player?

13. Suppose that a person chooses a letter at random from R E S E R V E and then chooses one at random from V E R T I C A L. What is the probability the same letter is chosen?

14. Prove Boole's inequality:

$$P\left(\bigcup_{i=1}^{\infty} A_i\right) \le \sum_{i=1}^{\infty} P(A_i)$$

15. Show that if $P(A_i) = 1$ for all $i \ge 1$, then $P\left(\bigcap_{i=1}^{\infty} A_i\right) = 1$.

16. Let $T_k(n)$ denote the number of partitions of the set $\{1, \ldots, n\}$ into k nonempty subsets, where $1 \le k \le n$. (See Theoretical Exercise 8 for the definition of a partition.) Argue that

$$T_k(n) = kT_k(n-1) + T_{k-1}(n-1)$$

HINT: In how many partitions is $\{1\}$ a subset, and in how many is 1 part of a subset that contains other elements?

17. Five balls are randomly chosen, without replacement, from an urn that contains 5 red, 6 white, and 7 blue balls. Find the probability that at least one ball of each color is chosen.

CHAPTER 3

Conditional Probability and Independence

3.1 INTRODUCTION

In this chapter we introduce one of the most important concepts in probability theory, that of conditional probability. The importance of this concept is twofold. In the first place, we are often interested in calculating probabilities when some partial information concerning the result of the experiment is available; in such a situation the desired probabilities are conditional. Second, even when no partial information is available, conditional probabilities can often be used to compute the desired probabilities more easily.

3.2 CONDITIONAL PROBABILITIES

Suppose that we toss 2 dice and suppose that each of the 36 possible outcomes is equally likely to occur and hence has probability $\frac{1}{36}$. Suppose further that we observe that the first die is a 3. Then, given this information, what is the probability that the sum of the 2 dice equals 8? To calculate this probability, we reason as follows: Given that the initial die is a 3, it follows that there can be at most 6 possible outcomes of our experiment, namely, $(3, 1), (3, 2), (3, 3), (3, 4), (3, 5)$, and $(3, 6)$. Since each of these outcomes originally had the same probability of occurring, the outcomes should still have equal probabilities. That is, given that the first die is a 3, the (conditional) probability of each of the outcomes $(3, 1)$, $(3, 2)$, $(3, 3)$, $(3, 4)$, $(3, 5)$, and $(3, 6)$ is $\frac{1}{6}$, whereas the (conditional) probability of the other 30 points in the sample space is 0. Hence the desired probability will be $\frac{1}{6}$.

If we let E and F denote, respectively, the event that the sum of the dice is 8 and the event that the first die is a 3, then the probability just obtained is called the conditional probability that E occurs given that F has occurred and is denoted by

$$P(E|F)$$

A general formula for $P(E|F)$ that is valid for all events E and F is derived in the same manner: If the event F occurs, then in order for E to occur it is necessary that the actual occurrence be a point in both E and in F; that is, it must be in EF. Now, as we know that F has occurred, it follows that F becomes our new or reduced sample space; hence the probability that the event EF occurs will equal the probability of EF relative to the probability of F. That is, we have the following definition.

Definition

If $P(F) > 0$, then

$$P(E|F) = \frac{P(EF)}{P(F)} \tag{2.1}$$

Example 2a. A coin is flipped twice. If we assume that all four points in the sample space $S = \{(H,H),(H,T),(T,H),(T,T)\}$ are equally likely, what is the conditional probability that both flips result in heads, given that the first flip does?

Solution If $E = \{(H,H)\}$ denotes the event that both flips land heads, and $F = \{(H,H),(H,T)\}$ the event that the first flip lands heads, then the desired probability is given by

$$\begin{aligned} P(E|F) &= \frac{P(EF)}{P(F)} \\ &= \frac{P(\{(H,H)\})}{P(\{(H,H),(H,T)\})} \\ &= \frac{\frac{1}{4}}{\frac{2}{4}} = \frac{1}{2} \end{aligned}$$

■

Example 2b. A bin contains 25 lightbulbs, 5 of which are in good condition and will function for at least thirty days, 10 of which are partially defective and will fail in their second day of use, and 10 of which are totally defective and will not light up. Given that a randomly chosen bulb initially lights, what is the probability it will still be working after one week?

Solution Let G denote the event that the randomly chosen bulb is in good condition, and let T be the event that it is totally defective. From Equation (2.1)

$$P(G|T^c) = \frac{P(GT^c)}{P(T^c)}$$

However, $GT^c = G$, because the bulb will be both in good condition and not totally defective if and only if it is in good condition. Therefore, assuming that each of the 25 bulbs is equally likely to be chosen, we obtain that

$$P(G|T^c) = \frac{5/25}{15/25} = 1/3$$

It should be noted that the preceding probability could have been derived by working with the reduced sample space. That is, because we know that the chosen bulb is not totally defective, the problem reduces to computing the probability that a bulb, chosen at random from a bin containing 10 partially defective and 5 good bulbs, is good. ■

When all outcomes are assumed to be equally likely, it is often easier to compute a conditional probability by a consideration of the reduced sample space, as opposed to a direct application of (2.1).

Example 2c. In the card game bridge the 52 cards are dealt out equally to 4 players—called East, West, North, and South. If North and South have a total of 8 spades among them, what is the probability that East has 3 of the remaining 5 spades?

Solution Probably the easiest way to compute this is to work with the reduced sample space. That is, given that North–South have a total of 8 spades among their 26 cards, there remains a total of 26 cards, exactly 5 of them being spades, to be distributed among the East–West hands. As each distribution is equally likely, it follows that the conditional probability that East will have exactly 3 spades among his or her 13 cards is

$$\frac{\binom{5}{3}\binom{21}{10}}{\binom{26}{13}} \approx .339$$ ■

Example 2d. The organization for which Ms. Jones works is running a dinner for those employees having at least one son. If Jones is known to have two children, what is the conditional probability that they are both boys, given that she is invited to the dinner? Assume that the sample space S is given by $S = \{(b, b), (b, g), (g, b), (g, g)\}$ and all outcomes are equally likely [(b, g) means, for instance, that the older child is a boy and the younger child is a girl].

Solution The knowledge that Jones has been invited to the dinner is equivalent to knowing that she has at least one son. Hence, letting B denote the event that both children are boys and A the event that at least one of them is a boy, we have that the desired probability $P(B|A)$ is given by

$$P(B|A) = \frac{P(BA)}{P(A)}$$

$$= \frac{P(\{(b, b)\})}{P(\{(b, b), (b, g), (g, b)\})} = \frac{\frac{1}{4}}{\frac{3}{4}} = \frac{1}{3}$$

Many readers incorrectly reason that the conditional probability of two boys given at least one is $\frac{1}{2}$, as opposed to the correct $\frac{1}{3}$, since they reason that the Jones child not attending the dinner is equally likely to be a boy or a girl. Their mistake, however, is in assuming that these two possibilities are equally likely. For, initially, there were 4 equally likely outcomes. Now the information that at least one child is a boy is equivalent to knowing that the outcome is not (g, g). Hence we are left with the 3 equally likely outcomes (b, b), (b, g),

(g, b) thus showing that the Jones child not attending the dinner is twice as likely to be a girl as to be a boy. ■

By multiplying both sides of Equation (2.1) by $P(F)$, we obtain

$$P(EF) = P(F)P(E|F) \tag{2.2}$$

In words, Equation (2.2) states that the probability that both E and F occur is equal to the probability that F occurs multiplied by the conditional probability of E given that F occurred. Equation (2.2) is often quite useful in computing the probability of the intersection of events.

Example 2e. Celine is undecided as to whether to take a French course or a chemistry course. She estimates that her probability of receiving an A grade would be $\frac{1}{2}$ in a French course, and $\frac{2}{3}$ in a chemistry course. If Celine decides to base her decision on the flip of a fair coin, what is the probability that she gets an A in chemistry?

Solution If we let C be the event that Celine takes chemistry and A denote the event that she receives an A in whatever course she takes, then the desired probability is $P(CA)$. This is calculated by using Equation (2.2) as follows:

$$\begin{aligned} P(CA) &= P(C)P(A|C) \\ &= \left(\tfrac{1}{2}\right)\left(\tfrac{2}{3}\right) = \tfrac{1}{3} \end{aligned}$$ ■

Example 2f. Suppose that an urn contains 8 red balls and 4 white balls. We draw 2 balls from the urn without replacement. If we assume that at each draw each ball in the urn is equally likely to be chosen, what is the probability that both balls drawn are red?

Solution Let R_1 and R_2 denote, respectively, the events that the first and second balls drawn are red. Now, given that the first ball selected is red, there are 7 remaining red balls and 4 white balls, and so $P(R_2|R_1) = \frac{7}{11}$. As $P(R_1)$ is clearly $\frac{8}{12}$, the desired probability is

$$\begin{aligned} P(R_1R_2) &= P(R_1)P(R_2|R_1) \\ &= \left(\tfrac{2}{3}\right)\left(\tfrac{7}{11}\right) = \tfrac{14}{33} \end{aligned}$$

Of course, this probability could also have been computed by

$$P(R_1R_2) = \frac{\binom{8}{2}}{\binom{12}{2}}$$ ■

A generalization of Equation (2.2), which provides an expression for the probability of the intersection of an arbitrary number of events, is sometimes referred to as the *multiplication rule*.

The multiplication rule

$$P(E_1E_2E_3\cdots E_n) = P(E_1)P(E_2|E_1)P(E_3|E_1E_2)\cdots P(E_n|E_1\cdots E_{n-1})$$

To prove the multiplication rule, just apply the definition of conditional probability to its right-hand side. This gives

$$P(E_1)\frac{P(E_1E_2)}{P(E_1)}\frac{P(E_1E_2E_3)}{P(E_1E_2)}\cdots\frac{P(E_1E_2\cdots E_n)}{P(E_1E_2\cdots E_{n-1})} = P(E_1E_2\cdots E_n)$$

We will now employ the multiplication rule to obtain a second approach to solving Example 5h(b) of Chapter 2.

Example 2g. An ordinary deck of 52 playing cards is randomly divided into 4 piles of 13 cards each. Compute the probability that each pile has exactly 1 ace.

Solution Define events $E_i, i = 1, 2, 3, 4$ as follows:

$$E_1 = \{\text{the ace of spades is in any one of the piles}\}$$

$$E_2 = \{\text{the ace of spades and the ace of hearts are in different piles}\}$$

$$E_3 = \{\text{the aces of spades, hearts, and diamonds are all in different piles}\}$$

$$E_4 = \{\text{all 4 aces are in different piles}\}$$

The probability desired is $P(E_1E_2E_3E_4)$ and by the multiplication rule

$$P(E_1E_2E_3E_4) = P(E_1)P(E_2|E_1)P(E_3|E_1E_2)P(E_4|E_1E_2E_3)$$

Now,

$$P(E_1) = 1$$

since E_1 is the sample space S.

$$P(E_2|E_1) = \tfrac{39}{51}$$

since the pile containing the ace of spades will receive 12 of the remaining 51 cards.

$$P(E_3|E_1E_2) = \tfrac{26}{50}$$

since the piles containing the aces of spades and hearts will receive 24 of the remaining 50 cards; and finally,

$$P(E_4|E_1E_2E_3) = \tfrac{13}{49}$$

Therefore, we obtain that the probability that each pile has exactly 1 ace is

$$P(E_1E_2E_3E_4) = \frac{39 \cdot 26 \cdot 13}{51 \cdot 50 \cdot 49} \approx .105$$

That is, there is approximately a 10.5 percent chance that each pile will contain an ace. (Problem 19 gives another way of using the multiplication rule to solve this problem.) ■

REMARK. Our definition of $P(E|F)$ is consistent with the interpretation of probability as being a long-run relative frequency. To see this, suppose that n repetitions of the experiment are to be performed, where n is large. We claim that if we consider only those experiments in which F occurs, then $P(E|F)$ will equal the long-run proportion of them in which E also occurs. To verify this, note that since $P(F)$ is the long-run proportion of experiments in which F occurs, it follows that in the n repetitions of the experiment F will occur approximately $nP(F)$ times. Similarly, in approximately $nP(EF)$ of these experiments both E and F will occur. Hence, out of the approximately $nP(F)$ experiments in which F occurs, the proportion of them in which E also occurs is approximately equal to

$$\frac{nP(EF)}{nP(F)} = \frac{P(EF)}{P(F)}$$

As this approximation becomes exact as n becomes larger and larger, we see that we have the appropriate definition of $P(E|F)$.

3.3 BAYES' FORMULA

Let E and F be events. We may express E as

$$E = EF \cup EF^c$$

for in order for an outcome to be in E, it must either be in both E and F or be in E but not in F (see Figure 3.1). As EF and EF^c are clearly mutually exclusive, we have by Axiom 3 that

$$\boxed{\begin{aligned} P(E) &= P(EF) + P(EF^c) \\ &= P(E|F)P(F) + P(E|F^c)P(F^c) \\ &= P(E|F)P(F) + P(E|F^c)[1 - P(F)] \end{aligned}} \tag{3.1}$$

Equation (3.1) states that the probability of the event E is a weighted average of the conditional probability of E given that F has occurred and the conditional probability of E given that F has not occurred—each conditional probability being given as

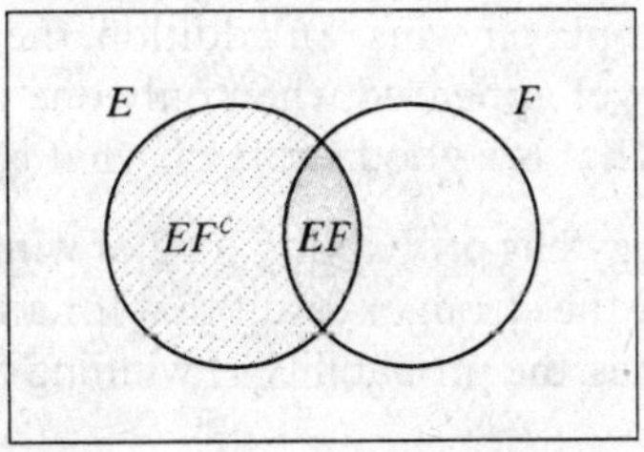

Figure 3.1 $E = EF \cup EF^c$.
EF = Shaded Area;
EF^c = Striped Area.

much weight as the event on which it is conditioned has of occurring. This is an extremely useful formula because its use often enables us to determine the probability of an event by first "conditioning" upon whether or not some second event has occurred. That is, there are many instances where it is difficult to compute the probability of an event directly, but it is straightforward to compute it once we know whether or not some second event has occurred. We illustrate this with some examples.

Example 3a (Part 1). An insurance company believes that people can be divided into two classes: those who are accident prone and those who are not. Their statistics show that an accident-prone person will have an accident at some time within a fixed 1-year period with probability .4, whereas this probability decreases to .2 for a non-accident-prone person. If we assume that 30 percent of the population is accident prone, what is the probability that a new policyholder will have an accident within a year of purchasing a policy?

Solution We shall obtain the desired probability by first conditioning upon whether or not the policyholder is accident prone. Let A_1 denote the event that the policyholder will have an accident within a year of purchase; and let A denote the event that the policyholder is accident prone. Hence the desired probability, $P(A_1)$, is given by

$$\begin{aligned} P(A_1) &= P(A_1|A)P(A) + P(A_1|A^c)P(A^c) \\ &= (.4)(.3) + (.2)(.7) = .26 \end{aligned}$$

■

Example 3a (Part 2). Suppose that a new policyholder has an accident within a year of purchasing a policy. What is the probability that he or she is accident prone?

Solution The desired probability is $P(A|A_1)$, which is given by

$$\begin{aligned} P(A|A_1) &= \frac{P(AA_1)}{P(A_1)} \\ &= \frac{P(A)P(A_1|A)}{P(A_1)} \\ &= \frac{(.3)(.4)}{.26} = \frac{6}{13} \end{aligned}$$

■

Example 3b. Consider the following game played with an ordinary deck of 52 playing cards. The cards are shuffled and then turned over one at a time. At any time the player can guess that the next card to be turned over will be the ace of spades; if it is, then the player wins. In addition, the player is said to win if the ace of spades has not yet appeared when only one card remains and no guess has yet been made. What is a good strategy; what is a bad strategy?

Solution Every stategy has probability $1/52$ of winning! To show this, we will use induction to prove the stronger result that for an n card deck, one of whose cards is the ace of spades, the probability of winning is $1/n$, no matter what strategy is employed. Since this is clearly true for $n = 1$, assume it to be true for an $n - 1$ card deck, and now consider an n card deck. Fix any strategy, and let p denote the probability that this strategy guesses that the first card is the ace of spades. Given that it does, then the player's probability of winning is $1/n$. On the other hand, if the strategy does not guess that the first card is the ace of spades, then the probability that the player wins is the probability that the first card is not the ace of spades, namely $(n - 1)/n$, multiplied by the conditional probability of winning given that the first card is not the ace of spades. But this latter conditional probability is equal to the probability of winning when using an $n - 1$ card deck containing a single ace of spades; it is thus, by the induction hypothesis, $1/(n - 1)$. Hence, given that the strategy does not guess the first card, the probability of winning is

$$\frac{n-1}{n}\frac{1}{n-1} = \frac{1}{n}$$

Thus, letting G be the event that the first card is guessed, we obtain that

$$\begin{aligned}
P\{\text{win}\} &= P\{\text{win}|G\}P(G) + P\{\text{win}|G^c\}(1 - P(G)) \\
&= \frac{1}{n}p + \frac{1}{n}(1 - p) \\
&= \frac{1}{n}
\end{aligned}$$

■

Example 3c. In answering a question on a multiple-choice test, a student either knows the answer or guesses. Let p be the probability that the student knows the answer and $1 - p$ the probability that the student guesses. Assume that a student who guesses at the answer will be correct with probability $1/m$, where m is the number of multiple-choice alternatives. What is the conditional probability that a student knew the answer to a question, given that he or she answered it correctly?

Solution Let C and K denote, respectively, the events that the student answers the question correctly and the event that he or she actually knows the answer. Now

$$\begin{aligned}
P(K|C) &= \frac{P(KC)}{P(C)} \\
&= \frac{P(C|K)P(K)}{P(C|K)P(K) + P(C|K^c)P(K^c)} \\
&= \frac{p}{p + (1/m)(1 - p)} \\
&= \frac{mp}{1 + (m - 1)p}
\end{aligned}$$

For example, if $m = 5$, $p = \frac{1}{2}$, then the probability that a student knew the answer to a question he or she correctly answered is $\frac{5}{6}$. ■

Example 3d. A laboratory blood test is 95 percent effective in detecting a certain disease when it is, in fact, present. However, the test also yields a "false positive" result for 1 percent of the healthy persons tested. (That is, if a healthy person is tested, then, with probability .01, the test result will imply he or she has the disease.) If .5 percent of the population actually has the disease, what is the probability a person has the disease given that the test result is positive?

Solution Let D be the event that the tested person has the disease and E the event that the test result is positive. The desired probability $P(D|E)$ is obtained by

$$\begin{aligned}
P(D|E) &= \frac{P(DE)}{P(E)} \\
&= \frac{P(E|D)P(D)}{P(E|D)P(D) + P(E|D^c)P(D^c)} \\
&= \frac{(.95)(.005)}{(.95)(.005) + (.01)(.995)} \\
&= \frac{95}{294} \approx .323
\end{aligned}$$

Thus only 32 percent of those persons whose test results are positive actually have the disease. As many students are often surprised at this result (as they expected this figure to be much higher, since the blood test seems to be a good one), it is probably worthwhile to present a second argument that, although less rigorous than the preceding one, is probably more revealing. We now do so.

Since .5 percent of the population actually has the disease, it follows that, on the average, 1 person out of every 200 tested will have it. The test will correctly confirm that this person has the disease with probability .95. Thus, on the average out of every 200 persons tested, the test will correctly confirm that .95 persons have the disease. On the other hand, however, out of the (on the average) 199 healthy people, the test will incorrectly state that (199)(.01) of these people have the disease. Hence, for every .95 diseased person that the test correctly states is ill, there are (on the average) (199)(.01) healthy persons that the test incorrectly states are ill. Hence the proportion of time that the test result is correct when it states that a person is ill is

$$\frac{.95}{.95 + (199)(.01)} = \frac{95}{294} \approx .323$$ ■

Equation (3.1) is also useful when one has to reassess one's personal probabilities in the light of additional information. For instance, consider the following examples.

Example 3e. Consider a medical practitioner pondering the following dilemma: "If I'm at least 80 percent certain that my patient has this disease, then I always recommend surgery, whereas if I'm not quite as certain, then I recommend additional tests that are expensive and sometimes painful. Now, initially I was only 60 percent certain that Jones had the disease, so I ordered the series A test, which always gives a positive result when the patient has the disease and almost never does when he is healthy. The test result was positive, and I was all set to recommend surgery when Jones informed me, for the first time, that he is a diabetic. This information complicates matters because, although it doesn't change my original 60 percent estimate of his chances of having the disease, it does affect the interpretation of the results of the A test. This is so because the A test, while never yielding a positive result when the patient is healthy, does unfortunately yield a positive result 30 percent of the time in the case of *diabetic* patients not suffering from the disease. Now what do I do? More tests or immediate surgery?"

Solution In order to decide whether or not to recommend surgery, the doctor should first compute her updated probability that Jones has the disease given that the A test result was positive. Let D denote the event that Jones has the disease, and E the event of a positive A test result. The desired conditional probability $P(D|E)$ is obtained by

$$\begin{aligned} P(D|E) &= \frac{P(DE)}{P(E)} \\ &= \frac{P(D)P(E|D)}{P(E|D)P(D) + P(E|D^c)P(D^c)} \\ &= \frac{(.6)1}{1(.6) + (.3)(.4)} \\ &= .833 \end{aligned}$$

Note that we have computed the probability of a positive test result by conditioning on whether or not Jones has the disease and then using the fact that because Jones is a diabetic his conditional probability of a positive result given he does not have the disease, $P(E|D^c)$, equals .3. Hence, as the doctor should now be over 80 percent certain that Jones has the disease, he should recommend surgery. ■

Example 3f. At a certain stage of a criminal investigation the inspector in charge is 60 percent convinced of the guilt of a certain suspect. Suppose now that a *new* piece of evidence that shows that the criminal has a certain characteristic (such as left-handedness, baldness, or brown hair) is uncovered. If 20 percent of the population possesses this characteristic, how certain of the guilt of the suspect should the inspector now be if it turns out that the suspect has this characteristic?

Solution Letting G denote the event that the suspect is guilty and C the event that he possesses the characteristic of the criminal, we have

$$\begin{aligned}
P(G|C) &= \frac{P(GC)}{P(C)} \\
&= \frac{P(C|G)P(G)}{P(C|G)P(G) + P(C|G^c)P(G^c)} \\
&= \frac{1(.6)}{1(.6) + (.2)(.4)} \\
&\approx .882
\end{aligned}$$

where we have supposed that the probability of the suspect having the characteristic if he is, in fact, innocent is equal to .2, the proportion of the population possessing the characteristic. ■

Example 3g. In the world bridge championships held in Buenos Aires in May 1965 the famous British bridge partnership of Terrence Reese and Boris Schapiro was accused of cheating by using a system of finger signals that could indicate the number of hearts held by the players. Reese and Schapiro denied the accusation, and eventually a hearing was held by the British bridge league. The hearing was in the form of a legal proceeding with a prosecuting and defense team, both having the power to call and cross-examine witnesses. During the course of these proceedings the prosecutor examined specific hands played by Reese and Schapiro and claimed that their playing in these hands was consistent with the hypothesis that they were guilty of having illicit knowledge of the heart suit. At this point, the defense attorney pointed out that their play of these hands was also perfectly consistent with their standard line of play. However, the prosecution then argued that as long as their play was consistent with the hypothesis of guilt, then it must be counted as evidence toward this hypothesis. What do you think of the reasoning of the prosecution?

Solution The problem is basically one of determining how the introduction of new evidence (in the preceding example, the playing of the hands) affects the probability of a particular hypothesis. Now, if we let H denote a particular hypothesis (such as the guilt of Reese and Schapiro), and E the new evidence, then

$$\begin{aligned}
P(H|E) &= \frac{P(HE)}{P(E)} \\
&= \frac{P(E|H)P(H)}{P(E|H)P(H) + P(E|H^c)[1 - P(H)]}
\end{aligned} \tag{3.2}$$

where $P(H)$ is our evaluation of the likelihood of the hypothesis before the introduction of the new evidence. The new evidence will be in support of the hypothesis whenever it makes the hypothesis more likely, that is, whenever $P(H|E) \geq P(H)$. From Equation (3.2), this will be the case whenever

$$P(E|H) \geq P(E|H)P(H) + P(E|H^c)[1 - P(H)]$$

or, equivalently, whenever

$$P(E|H) \geq P(E|H^c)$$

In other words, any new evidence can be considered to be in support of a particular hypothesis only if its occurrence is more likely when the hypothesis is true than when it is false. In fact, the new probability of the hypothesis depends on its initial probability and the ratio of these conditional probabilities, since from Equation (3.2),

$$P(H|E) = \frac{P(H)}{P(H) + [1 - P(H)]\dfrac{P(E|H^c)}{P(E|H)}}$$

Hence, in the problem under consideration, the play of the cards can be considered to support the hypothesis of guilt only if such playing would have been more likely if the partnership were cheating than if they were not. As the prosecutor never made this claim, his assertion that the evidence is in support of the guilt hypothesis is invalid. ■

The change in the probability of a hypothesis when new evidence is introduced can be expressed compactly in terms of the change in the *odds ratio* of this hypothesis, where the concept of odds ratio is defined as follows.

Definition

The odds ratio of an event A is defined by

$$\frac{P(A)}{P(A^c)} = \frac{P(A)}{1 - P(A)}$$

That is, the odds ratio of an event A tells how much more likely it is that the event A occurs than it is that it does not occur. For instance, if $P(A) = \frac{2}{3}$, then $P(A) = 2P(A^c)$, so the odds ratio is 2. If the odds ratio is equal to α, then it is common to say that the odds are "α to 1" in favor of the hypothesis.

Consider now a hypothesis H that is true with probability $P(H)$ and suppose that new evidence E is introduced. Then the conditional probabilities, given the evidence E, that H is true and that H is not true are given by

$$P(H|E) = \frac{P(E|H)P(H)}{P(E)} \qquad P(H^c|E) = \frac{P(E|H^c)P(H^c)}{P(E)}$$

Therefore, the new odds ratio after the evidence E has been introduced is

$$\boxed{\frac{P(H|E)}{P(H^c|E)} = \frac{P(H)}{P(H^c)}\frac{P(E|H)}{P(E|H^c)}} \tag{3.3}$$

That is, the new value of the odds ratio of H is its old value multiplied by the ratio of the conditional probability of the new evidence given that H is true to the conditional probability given that H is not true. This verifies the result of Example 3f, since the odds ratio, and thus the probability of H, increases whenever the new evidence is more likely when H is true than when it is false. Similarly, the odds ratio decreases whenever the new evidence is more likely when H is false than when it is true.

Example 3h. When coin A is flipped it comes up heads with probability $\frac{1}{4}$, whereas when coin B is flipped it comes up heads with probability $\frac{3}{4}$. Suppose that one of these coins is randomly chosen and is flipped twice. If both flips land heads, what is the probability that coin B was the one flipped?

Solution Let B be the event that coin B was the one flipped. Since $P(B) = P(B^c)$, we obtain from Equation (3.3) that

$$\frac{P(B|\text{two heads})}{P(B^c|\text{two heads})} = \frac{\frac{9}{16}}{\frac{1}{16}} = 9$$

Hence the odds are 9:1, or, equivalently the probability is $\frac{9}{10}$, that coin B was the one flipped. ■

Equation (3.1) may be generalized in the following manner: Suppose that $F_1, F_2, \ldots, F_n$ are mutually exclusive events such that

$$\bigcup_{i=1}^{n} F_i = S$$

In other words, exactly one of the events $F_1, F_2, \ldots, F_n$ must occur. By writing

$$E = \bigcup_{i=1}^{n} EF_i$$

and using the fact that the events $EF_i, i = 1, \ldots, n$ are mutually exclusive, we obtain that

$$\begin{aligned} P(E) &= \sum_{i=1}^{n} P(EF_i) \\ &= \sum_{i=1}^{n} P(E|F_i)P(F_i) \end{aligned} \tag{3.4}$$

Thus Equation (3.4) shows how, for given events $F_1, F_2, \ldots, F_n$ of which one and only one must occur, we can compute $P(E)$ by first conditioning on which one of the F_i occurs. That is, Equation (3.4) states that $P(E)$ is equal to a weighted average of $P(E|F_i)$, each term being weighted by the probability of the event on which it is conditioned.

Suppose now that E has occurred and we are interested in determining which one of the F_j also occurred. By Equation (3,4), we have the following proposition.

Proposition 3.1

$$P(F_j|E) = \frac{P(EF_j)}{P(E)} = \frac{P(E|F_j)P(F_j)}{\sum_{i=1}^{n} P(E|F_i)P(F_i)} \tag{3.5}$$

Equation (3.5) is known as Bayes' formula, after the English philosopher Thomas Bayes. If we think of the events F_j as being possible "hypotheses" about some subject matter, then Bayes' formula may be interpreted as showing us how opinions about these hypotheses held before the experiment [that is, the $P(F_j)$] should be modified by the evidence of the experiment.

Example 3i. A plane is missing, and it is presumed that it was equally likely to have gone down in any of 3 possible regions. Let $1 - \beta_i$ denote the probability that the plane will be found upon a search of the ith region when the plane is, in fact, in that region, $i = 1, 2, 3$. (The constants β_i are called overlook probabilities because they represent the probability of overlooking the plane; they are generally attributable to the geographical and environmental conditions of the regions.) What is the conditional probability that the plane is in the ith region, given that a search of region 1 is unsuccessful, $i = 1, 2, 3$?

Solution Let $R_i, i = 1, 2, 3$, be the event that the plane is in region i; and let E be the event that a search of region 1 is unsuccessful. From Bayes' formula we obtain

$$\begin{aligned} P(R_1|E) &= \frac{P(ER_1)}{P(E)} \\ &= \frac{P(E|R_1)P(R_1)}{\sum_{i=1}^{3} P(E|R_i)P(R_i)} \\ &= \frac{(\beta_1)\frac{1}{3}}{(\beta_1)\frac{1}{3} + (1)\frac{1}{3} + (1)\frac{1}{3}} \\ &= \frac{\beta_1}{\beta_1 + 2} \end{aligned}$$

For $j = 2, 3$,

$$P(R_j|E) = \frac{P(E|R_j)P(R_j)}{P(E)}$$

$$= \frac{(1)\frac{1}{3}}{(\beta_1)\frac{1}{3} + \frac{1}{3} + \frac{1}{3}}$$

$$= \frac{1}{\beta_1 + 2} \qquad j = 2, 3$$

It should be noted that the updated (that is, the conditional) probability that the plane is in region j, given the information that a search of region 1 did not find it, is greater than the initial probability that it was in region j when $j \neq 1$ and is less than the initial probability when $j = 1$; which is certainly intuitive, since not finding it when searching region 1 would seem to decrease its chance of being in that region and increase its chance of being elsewhere. Also the conditional probability that the plane is in region 1, given an unsuccessful search of that region, is an increasing function of the overlook probability β_1, which is also intuitive since the larger β_1 is, the more it is reasonable to attribute the unsuccessful search to "bad luck" as opposed to the plane not being there. Similarly, $P(R_j|E)$, $j \neq 1$ is a decreasing function of β_1. ■

The next example has often been used by unscrupulous probability students to win money from their less enlightened friends.

Example 3j. Suppose that we have 3 cards identical in form except that both sides of the first card are colored red, both sides of the second card are colored black, and one side of the third card is colored red and the other side black. The 3 cards are mixed up in a hat, and 1 card is randomly selected and put down on the ground. If the upper side of the chosen card is colored red, what is the probability that the other side is colored black?

Solution Let RR, BB, and RB denote, respectively, the events that the chosen card is all red, all black, or the red–black card. Letting R be the event that the upturned side of the chosen card is red, we have that the desired probability is obtained by

$$P(RB|R) = \frac{P(RB \cap R)}{P(R)}$$

$$= \frac{P(R|RB)P(RB)}{P(R|RR)P(RR) + P(R|RB)P(RB) + P(R|BB)P(BB)}$$

$$= \frac{\left(\frac{1}{2}\right)\left(\frac{1}{3}\right)}{(1)\left(\frac{1}{3}\right) + \left(\frac{1}{2}\right)\left(\frac{1}{3}\right) + 0\left(\frac{1}{3}\right)} = \frac{1}{3}$$

Hence the answer is $\frac{1}{3}$. Some students guess $\frac{1}{2}$ as the answer by incorrectly reasoning that given that a red side appears, there are two equally likely possibilities: that the card is the all red card or the red–black card. Their mistake, however, is in assuming that these two possibilities are equally likely. For, if we think of each card as consisting of two distinct sides, then there are 6 equally likely outcomes of the experiment—namely, $R_1, R_2, B_1, B_2, R_3, B_3$—where the outcome is R_1 if the first side of the all red card is turned face up, R_2 if the second side of the all red card is turned face up, R_3 if the red side of the red–black card is turned face up, and so on. Since the other side of the upturned red side will be black only if the outcome is R_3, we see that the desired probability is the conditional probability of R_3 given that either R_1 or R_2 or R_3 occurred, which obviously equals $\frac{1}{3}$. ■

Example 3k. A new couple, known to have two children, has just moved into town. Suppose that the mother is encountered walking with one of her children. If this child is a girl, what is the probability that both children are girls?

Solution Let us start by defining the following events:

G_1: the first (that is, oldest) child is a girl.
G_2: the second child is a girl.
G: the child seen with the mother is a girl.

Also, let B_1, B_2, B denote similar events except that "girl" is replaced by "boy". Now, the desired probability is $P(G_1G_2|G)$, which can be expressed as follows:

$$\begin{aligned} P(G_1G_2|G) &= \frac{P(G_1G_2G)}{P(G)} \\ &= \frac{P(G_1G_2)}{P(G)} \end{aligned}$$

Also,

$$\begin{aligned} P(G) &= P(G|G_1G_2)P(G_1G_2) + P(G|G_1B_2)P(G_1B_2) \\ &\quad + P(G|B_1G_2)P(B_1G_2) + P(G|B_1B_2)P(B_1B_2) \\ &= P(G_1G_2) + P(G|G_1B_2)P(G_1B_2) + P(G|B_1G_2)P(B_1G_2) \end{aligned}$$

where the final equation used the results $P(G|G_1G_2) = 1$ and $P(G|B_1B_2) = 0$. If we now make the usual assumption that all 4 gender possibilities are equally likely, then we see that

$$\begin{aligned} P(G_1G_2|G) &= \frac{\frac{1}{4}}{\frac{1}{4} + P(G|G_1B_2)/4 + P(G|B_1G_2)/4} \\ &= \frac{1}{1 + P(G|G_1B_2) + P(G|B_1G_2)} \end{aligned}$$

Thus the answer depends on whatever assumptions we want to make about the conditional probabilities that the child seen with the mother is a girl given the event G_1B_2, and that the child seen with the mother is a girl given the event G_2B_1. For instance, if we want to assume that, independent of the genders of the children, the child walking with the mother is the elder child with some probability p, then it would follow that

$$P(G|G_1B_2) = p = 1 - P(G|B_1G_2)$$

implying under this scenario that

$$P(G_1G_2|G) = \frac{1}{2}$$

On the other hand, if we were to assume that if the children are of different genders, then the mother would choose to walk with the girl with probability q, independent of the birth order of the children, then we would have that

$$P(G|G_1B_2) = P(G|B_1G_2) = q$$

implying that

$$P(G_1G_2|G) = \frac{1}{1 + 2q}$$

For instance, if we took $q = 1$, meaning that the mother would always choose to walk with a daughter, then the conditional probability of two daughters would be $\frac{1}{3}$, which is in accord with Example 2d because seeing the mother with a daughter is now equivalent to the event that there is at least one daughter.

Thus, as stated, the problem is incapable of solution. Indeed, even when the usual assumption about equally likely gender probabilities is made, we still need to make additional assumptions before a solution can be given. This is because the sample space of the experiment consists of vectors of the form s_1, s_2, i, where s_1 is the gender of the older child, s_2 is the gender of the younger child, and i identifies the birth order of the child seen with the mother. As a result, to specify the probabilities of the events of the sample space, it is not enough to make assumptions only about the genders of the children, it is also necessary to assume something about the conditional probabilities as to which child is with the mother given the genders of the children. ■

Example 3l. A bin contains 3 different types of disposable flashlights. The probability that a type 1 flashlight will give over 100 hours of use is .7, with the corresponding probabilities for type 2 and type 3 flashlights being .4 and .3, respectively. Suppose that 20 percent of the flashlights in the bin are type 1, 30 percent are type 2, and 50 percent are type 3.

(a) What is the probability that a randomly chosen flashlight will give more than 100 hours of use?

(b) Given the flashlight lasted over 100 hours, what is the conditional probability that it was a type j flashlight, $j = 1, 2, 3$?

Solution (a) Let A denote the event that the flashlight chosen will give over 100 hours of use, and let F_j be the event that a type j flashlight is chosen, $j = 1, 2, 3$. To compute $P(A)$, condition on the type of the flashlight to obtain

$$\begin{aligned} P(A) &= P(A|F_1)P(F_1) + P(A|F_2)P(F_2) + P(A|F_3)P(F_3) \\ &= (.7)(.2) + (.4)(.3) + (.3)(.5) = .41 \end{aligned}$$

There is a 41 percent chance that the flashlight will last for over 100 hours.

(b) The probability is obtained by using Bayes' formula:

$$P(F_j|A) = \frac{P(AF_j)}{P(A)}$$

$$= \frac{P(A|F_j)P(F_j)}{.41}$$

Thus,

$$P(F_1|A) = (.7)(.2)/.41 = 14/41$$
$$P(F_2|A) = (.4)(.3)/.41 = 12/41$$
$$P(F_3|A) = (.3)(.5)/.41 = 15/41$$

For instance, whereas the initial probability that a type 1 flashlight is chosen is only .2, the information that the flashlight has lasted over 100 hours raises the probability of this event to $14/41 \approx .341$. ■

Example 3m. A crime has been committed by a solitary individual, who left a partial amount of his or her DNA at the scene of the crime. Forensic scientists who studied the recovered DNA noted that only five strands could be identified and that each innocent person, independently, would have probability 10^{-5} of having their DNA match on all five strands. The district attorney supposes that the perpetrator of the crime could be any of the one million residents of the town. Ten thousand of these residents have been released from prison within the past ten years; consequently, a sample of their DNA is on file. Before any checking of the DNA file, the district attorney feels that each of the ten thousand ex-criminals has probability α of being guilty of the new crime, while each of the remaining 990,000 residents has probability β, where $\alpha = c\beta$. (That is, she supposes that each recently released convict is c times as likely to be the crime's perpetrator as is each town member who is not a recently released convict.) When the analyzed DNA is compared with the data base of the ten thousand ex-convicts, it turns out that AJ Jones is the only one whose DNA matches the profile. Assuming that the district attorney's estimate of the relationship between α and β is accurate, what is the probability that AJ is guilty?

Solution To begin, note that because probabilities must sum to 1, we have that

$$1 = 10{,}000\alpha + 990{,}000\beta = (10{,}000c + 990{,}000)\beta$$

Thus,

$$\beta = \frac{1}{10{,}000c + 990{,}000}, \qquad \alpha = \frac{c}{10{,}000c + 990{,}000}$$

Now, let G be the event that AJ is guilty, and let M denote the event that AJ is the only one of the ten thousand on file to have a match. Then,

$$P(G|M) = \frac{P(GM)}{P(M)}$$

$$= \frac{P(G)P(M|G)}{P(M|G)P(G) + P(M|G^c)P(G^c)}$$

If AJ is guilty, then he will be the only one to have a DNA match provided that none of the others on file have a match. Therefore,

$$P(M|G) = (1 - 10^{-5})^{9999}$$

On the other hand, if AJ is innocent then in order for him to be the only match, his DNA must match (which will occur with probability 10^{-5}), all others in the database must be innocent, and none of these others can have a match. Now, given that AJ is innocent, the conditional probability that all the others in the database are also innocent is

$$P(\text{all others innocent}|AJ \text{ innocent}) = \frac{P(\text{all in database innocent})}{P(AJ \text{ innocent})}$$

$$= \frac{1 - 10{,}000\alpha}{1 - \alpha}$$

Also, the conditional probability, given their innocence, that none of the others in the database will have a match is $(1 - 10^{-5})^{9999}$. Therefore,

$$P(M|G^c) = 10^{-5}\left(\frac{1 - 10{,}000\alpha}{1 - \alpha}\right)(1 - 10^{-5})^{9999}$$

Because $P(G) = \alpha$, we obtain from the preceding that

$$P(G|M) = \frac{\alpha}{\alpha + 10^{-5}(1 - 10{,}000\alpha)} = \frac{1}{.9 + \frac{10^{-5}}{\alpha}}$$

Thus, if the district attorney, initial feelings were that an arbitrary ex-convict was 100 times more likely to have committed the crime than was a nonconvict, (that is, $c = 100$), then $\alpha = \frac{1}{19{,}900}$ and

$$P(G|M) = \frac{1}{1.099} \approx 0.9099$$

If the district attorney initially felt that the appropriate ratio was $c = 10$, then $\alpha = \frac{1}{109{,}000}$ and

$$P(G|M) = \frac{1}{1.99} \approx 0.5025$$

If the district attorney initially felt that the criminal was equally likely to be any of the members of the town ($c = 1$) then $\alpha = 10^{-6}$ and

$$P(G|M) = \frac{1}{10.9} \approx 0.0917$$

Thus, the probability ranges from approximately nine percent when the district attorney's initial assumption is that all the members of the population have the same chance of being the perpetrator, to approximately 91 percent when she assumes that each ex-convict is 100 times more likely to be the criminal than is a specified townsperson who is not an ex-convict. ■

3.4 INDEPENDENT EVENTS

The previous examples of this chapter show that $P(E|F)$, the conditional probability of E given F, is not generally equal to $P(E)$, the unconditional probability of E. In other words, knowing that F has occurred generally changes the chances of E's occurrence. In the special cases where $P(E|F)$ does in fact equal $P(E)$, we say that E is independent of F. That is, E is independent of F if knowledge that F has occurred does not change the probability that E occurs.

Since $P(E|F) = P(EF)/P(F)$, we see that E is independent of F if

$$P(EF) = P(E)P(F) \tag{4.1}$$

As Equation (4.1) is symmetric in E and F, it shows that whenever E is independent of F, F is also independent of E. We thus have the following definition.

Definition

Two events E and F are said to be *independent* if Equation (4.1) holds. Two events E and F that are not independent are said to be *dependent*.

Example 4a. A card is selected at random from an ordinary deck of 52 playing cards. If E is the event that the selected card is an ace and F is the event that it is a spade, then E and F are independent. This follows because $P(EF) = \frac{1}{52}$, whereas $P(E) = \frac{4}{52}$ and $P(F) = \frac{13}{52}$. ■

Example 4b. Two coins are flipped, and all 4 outcomes are assumed to be equally likely. If E is the event that the first coin lands heads and F the event that the second lands tails, then E and F are independent, since $P(EF) = P(\{(H,T)\}) = \frac{1}{4}$; whereas $P(E) = P(\{(H,H),(H,T)\}) = \frac{1}{2}$ and $P(F) = P(\{(H,T),(T,T)\}) = \frac{1}{2}$. ■

Example 4c. Suppose that we toss 2 fair dice. Let E_1 denote the event that the sum of the dice is 6 and F denote the event that the first die equals 4. Then

$$P(E_1F) = P(\{(4,2)\}) = \tfrac{1}{36}$$

whereas

$$P(E_1)P(F) = \left(\tfrac{5}{36}\right)\left(\tfrac{1}{6}\right) = \tfrac{5}{216}$$

Hence E_1 and F are not independent. Intuitively, the reason for this is clear because if we are interested in the possibility of throwing a 6 (with 2 dice) we shall be quite happy if the first die lands 4 (or any of the numbers 1, 2, 3, 4, 5), for then we shall still have a possibility of getting a total of 6. On the other hand, if the first die landed 6, we would be unhappy because we would no longer have a chance of getting a total of 6. In other words, our chance of getting a total of six depends on the outcome of the first die; hence E_1 and F cannot be independent.

Now, suppose that we let E_2 be the event that the sum of the dice equals 7. Is E_2 independent of F? The answer is yes, since

$$P(E_2F) = P(\{(4,3)\}) = \tfrac{1}{36}$$

whereas

$$P(E_2)P(F) = \left(\tfrac{1}{6}\right)\left(\tfrac{1}{6}\right) = \left(\tfrac{1}{36}\right)$$

We leave it for the reader to present the intuitive argument why the event that the sum of the dice equals seven is independent of the outcome on the first die. ■

Example 4d. If we let E denote the event that the next president is a Republican and F the event that there will be a major earthquake within the next year, then most people would probably be willing to assume that E and F are independent. However, there would probably be some controversy over whether it is reasonable to assume that E is independent of G, where G is the event that there will be a recession within two years after the election. ■

We now show that if E is independent of F, then E is also independent of F^c.

Proposition 4.1

If E and F are independent, then so are E and F^c.

Proof: Assume that E and F are independent. Since $E = EF \cup EF^c$, and EF and EF^c are obviously mutually exclusive, we have that

$$\begin{aligned} P(E) &= P(EF) + P(EF^c) \\ &= P(E)P(F) + P(EF^c) \end{aligned}$$

or, equivalently,

$$\begin{aligned} P(EF^c) &= P(E)[1 - P(F)] \\ &= P(E)P(F^c) \end{aligned}$$

and the result is proved. ✦

Thus, if E is independent of F, then the probability of E's occurrence is unchanged by information as to whether or not F has occurred.

Suppose now that E is independent of F and is also independent of G. Is E then necessarily independent of FG? The answer, somewhat surprisingly, is no. Consider the following example.

Example 4e. Two fair dice are thrown. Let E denote the event that the sum of the dice is 7. Let F denote the event that the first die equals 4 and let G be the event that the second die equals 3. From Example 4c we know that E is independent of F, and the same reasoning as applied there shows that E is also independent of G; but clearly E is not independent of FG [since $P(E|FG) = 1$]. ■

It would appear to follow from Example 4e that an appropriate definition of the independence of three events E, F, and G would have to go further than merely assuming that all of the $\binom{3}{2}$ pairs of events are independent. We are thus led to the following definition.

Definition

The three events E, F, and G are said to be independent if

$$\begin{aligned} P(EFG) &= P(E)P(F)P(G) \\ P(EF) &= P(E)P(F) \\ P(EG) &= P(E)P(G) \\ P(FG) &= P(F)P(G) \end{aligned}$$

It should be noted that if E, F, and G are independent, then E will be independent of any event formed from F and G. For instance, E is independent of $F \cup G$, since

$$\begin{aligned} P[E(F \cup G)] &= P(EF \cup EG) \\ &= P(EF) + P(EG) - P(EFG) \\ &= P(E)P(F) + P(E)P(G) - P(E)P(FG) \\ &= P(E)[P(F) + P(G) - P(FG)] \\ &= P(E)P(F \cup G) \end{aligned}$$

Of course, we may also extend the definition of independence to more than three events. The events $E_1, E_2, \ldots, E_n$ are said to be independent if, for every subset $E_{1'}, E_{2'}, \ldots, E_{r'}, r \leq n$, of these events

$$P(E_{1'}E_{2'} \cdots E_{r'}) = P(E_{1'})P(E_{2'}) \cdots P(E_{r'})$$

Finally, we define an infinite set of events to be independent if every finite subset of these events is independent.

It is sometimes the case that the probability experiment under consideration consists of performing a sequence of subexperiments. For instance, if the experiment consists of continually tossing a coin, we may think of each toss as being a subexperiment. In many cases it is reasonable to assume that the outcomes of any group of the subexperiments have no effect on the probabilities of the outcomes of the other subexperiments. If such is the case, we say that the subexperiments are independent. More formally, we say that the subexperiments are independent if $E_1, E_2, \ldots, E_n, \ldots$ is necessarily an independent sequence of events whenever E_i is an event whose occurrence is completely determined by the outcome of the ith subexperiment.

If each subexperiment is identical—that is, if each subexperiment has the same (sub) sample space and the same probability function on its events—then the subexperiments are called *trials*.

Example 4f. An infinite sequence of independent trials is to be performed. Each trial results in a success with probability p and a failure with probability $1 - p$. What is the probability that

(a) at least 1 success occurs in the first n trials;
(b) exactly k successes occur in the first n trials;
*(c) all trials result in successes?

Solution In order to determine the probability of at least 1 success in the first n trials, it is easiest to compute first the probability of the complementary event, that of no successes in the first n trials. If we let E_i denote the event of a failure on the ith trial, then the probability of no successes is, by independence,

$$P(E_1E_2 \cdots E_n) = P(E_1)P(E_2) \cdots P(E_n) = (1 - p)^n$$

Hence the answer to part (a) is $1 - (1 - p)^n$.

To compute part (b), consider any particular sequence of the first n outcomes containing k successes and $n - k$ failures. Each one of these sequences will, by the assumed independence of trials, occur with probability $p^k(1 - p)^{n-k}$. As there are $\binom{n}{k}$ such sequences (there are $n!/k!\,(n - k)!$ permutations of k successes and $n - k$ failures), the desired probability in part (b) is

$$P\{\text{exactly } k \text{ successes}\} = \binom{n}{k} p^k(1 - p)^{n-k}$$

To answer part (c), we note by part (a) that the probability of the first n trials all resulting in successes is given by

$$P(E_1^c E_2^c \cdots E_n^c) = p^n$$

Hence, using the continuity property of probabilities (Section 2.6), we have that the desired probability $P\left(\bigcap_1^\infty E_i^c\right)$ is given by

$$\begin{aligned}
P\left(\bigcap_{i=1}^{\infty} E_i^c\right) &= P\left(\lim_{n\to\infty} \bigcap_{i=1}^{n} E_i^c\right) \\
&= \lim_{n\to\infty} P\left(\bigcap_{i=1}^{n} E_i^c\right) \\
&= \lim_{n} p^n = \begin{cases} 0 & \text{if } p < 1 \\ 1 & \text{if } p = 1 \end{cases}
\end{aligned}$$

■

Example 4g. A system composed of n separate components is said to be a parallel system if it functions when at least one of the components functions (see Figure 3.2). For such a system, if component i, independent of other components, functions with probability p_i, $i = 1, \ldots, n$, what is the probability that the system functions?

Solution Let A_i denote the event that component i functions. Then

$$\begin{aligned}
P\{\text{system functions}\} &= 1 - P\{\text{system does not function}\} \\
&= 1 - P\{\text{all components do not function}\} \\
&= 1 - P\left(\bigcap_i A_i^c\right) \\
&= 1 - \prod_{i=1}^{n} (1 - p_i) \qquad \text{by independence}
\end{aligned}$$

■

Example 4h. Independent trials, consisting of rolling a pair of fair dice, are performed. What is the probability that an outcome of 5 appears before an outcome of 7 when the outcome of a roll is the sum of the dice?

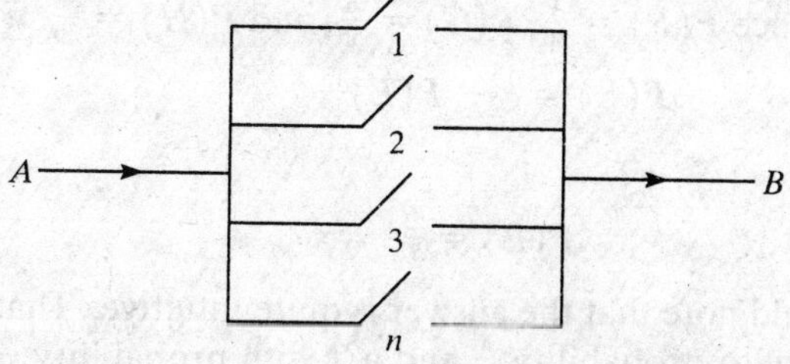

Figure 3.2 Parallel system: functions if current flows from A to B.

Solution If we let E_n denote the event that no 5 or 7 appears on the first $n - 1$ trials and a 5 appears on the nth trial, then the desired probability is

$$P\left(\bigcup_{n=1}^{\infty} E_n\right) = \sum_{n=1}^{\infty} P(E_n)$$

Now, since $P\{5 \text{ on any trial}\} = \frac{4}{36}$ and $P\{7 \text{ on any trial}\} = \frac{6}{36}$, we obtain, by the independence of trials

$$P(E_n) = \left(1 - \tfrac{10}{36}\right)^{n-1} \tfrac{4}{36}$$

and thus

$$\begin{aligned} P\left(\bigcup_{n=1}^{\infty} E_n\right) &= \frac{1}{9} \sum_{n=1}^{\infty} \left(\frac{13}{18}\right)^{n-1} \\ &= \frac{1}{9} \frac{1}{1 - \frac{13}{18}} \\ &= \frac{2}{5} \end{aligned}$$

This result may also have been obtained by using conditional probabilities. If we let E be the event that a 5 occurs before a 7, then we can obtain the desired probability, $P(E)$, by conditioning on the outcome of the first trial, as follows: Let F be the event that the first trial results in a 5; let G be the event that it results in a 7; and let H be the event that the first trial results in neither a 5 nor a 7. Conditioning on which one of these events occurs gives

$$P(E) = P(E|F)P(F) + P(E|G)P(G) + P(E|H)P(H)$$

However,

$$\begin{aligned} P(E|F) &= 1 \\ P(E|G) &= 0 \\ P(E|H) &= P(E) \end{aligned}$$

The first two equalities are obvious. The third follows because if the first outcome results in neither a 5 nor a 7, then at that point the situation is exactly as when the problem first started; namely, the experimenter will continually roll a pair of fair dice until either a 5 or 7 appears. Furthermore, the trials are independent; therefore, the outcome of the first trial will have no effect on subsequent rolls of dice. Since $P(F) = \frac{4}{36}$, $P(G) = \frac{6}{36}$, and $P(H) = \frac{26}{36}$, we see that

$$P(E) = \tfrac{1}{9} + P(E)\tfrac{13}{18}$$

or

$$P(E) = \tfrac{2}{5}$$

The reader should note that the answer is quite intuitive. That is, since a 5 occurs on any roll with probability $\frac{4}{36}$ and a 7 with probability $\frac{6}{36}$, it seems intuitive that the odds that a 5 appears before a 7 should be 6 to 4 against. The probability should be $\frac{4}{10}$, as indeed it is.

The same argument shows that if E and F are mutually exclusive events of an experiment, then, when independent trials of this experiment are performed, the event E will occur before the event F with probability

$$\frac{P(E)}{P(E) + P(F)}$$ ■

The next example presents a problem that occupies an honored place in the history of probability theory. This is the famous *problem of the points*. In general terms, the problem is this: Two players put up stakes and play some game, with the stakes to go to the winner of the game. An interruption requires them to stop before either has won, and when each has some sort of a "partial score." How should the stakes be divided?

This problem was posed to the French mathematician Pascal in 1654 by the Chevalier de Méré, who was a professional gambler at that time. In attacking the problem, Pascal introduced the important idea that the proportion of the prize deserved by the competitors should depend on their respective probabilities of winning if the game were to be continued at that point. Pascal worked out some special cases, and, more important, initiated a correspondence with the famous Frenchman Fermat, who had a great reputation as a mathematician. The resulting exchange of letters led not only to a complete solution to the problem of the points, but also laid the framework for the solution to many other problems connected with games of chance. This celebrated correspondence, dated by some as the birth date of probability theory, was also important in stimulating interest in probability among the mathematicians in Europe, for Pascal and Fermat were both recognized as being among the foremost mathematicians of the time. For instance, within a short time of their correspondence, the young Dutch mathematician Huygens came to Paris to discuss these problems and solutions; and interest and activity in this new field grew rapidly.

Example 4i. ***The problem of the points.*** Independent trials, resulting in a sucess with probability p and a failure with probability $1 - p$, are performed. What is the probability that n successes occur before m failures? If we think of A and B as playing a game such that A gains 1 point when a success occurs and B gains 1 point when a failure occurs, then the desired probability is the probability that A would win if the game were to be continued in a position where A needed n and B needed m more points to win.

Solution We shall present two solutions. The first is due to Pascal and the second to Fermat.

Let us denote by $P_{n,m}$ the probability that n successes occur before m failures. By conditioning on the outcome of the first trial we obtain

$$P_{n,m} = pP_{n-1,m} + (1 - p)P_{n,m-1,m} \qquad n \geq 1, m \geq 1$$

(Why? Reason it out.) By using the obvious boundary conditions $P_{n,0} = 0$, $P_{0,m} = 1$, these equations can be solved for $P_{n,m}$. Rather than go through the tedious details, let us instead consider Fermat's solution.

Fermat argued that in order for n successes to occur before m failures, it is necessary and sufficient that there be at least n successes in the first $m + n - 1$ trials. (Even if the game were to end before a total of $m + n - 1$ trials were completed, we could still imagine that the necessary additional trials were performed.) This is true, for if there are at least n successes in the first $m + n - 1$ trials, there could be at most $m - 1$ failures in those $m + n - 1$ trials; thus n successes would occur before m failures. On the other hand, if there were fewer than n successes in the first $m + n - 1$ trials, there would have to be at least m failures in that same number of trials; thus n successes would not occur before m failures.

Hence, as the probability of exactly k successes in $m + n - 1$ trials is, as shown in Example 4f, $\binom{m+n-1}{k} p^k(1-p)^{m+n-1-k}$, we see that the desired probability of n successes before m failures is

$$P_{n,m} = \sum_{k=n}^{m+n-1} \binom{m+n-1}{k} p^k(1-p)^{m+n-1-k} \qquad \blacksquare$$

The next example deals with a famous problem known as the gambler's ruin problem.*

Example 4j. ***The gambler's ruin problem.*** Two gamblers, A and B, bet on the outcomes of successive flips of a coin. On each flip, if the coin comes up heads, A collects 1 unit from B, whereas if it comes up tails, A pays 1 unit to B. They continue to do this until one of them runs out of money. If it is assumed that the successive flips of the coin are independent and each flip results in a head with probability p, what is the probability that A ends up with all the money if he starts with i units and B starts with $N - i$ units?

Solution Let E denote the event that A ends up with all the money when he starts with i and B with $N - i$, and to make clear the dependence on the initial fortune of A, let $P_i = P(E)$. We shall obtain an expression for $P(E)$ by conditioning on the outcome of the first flip as follows: Let H denote the event that the first flip lands heads; then

$$\begin{aligned} P_i = P(E) &= P(E|H)P(H) + P(E|H^c)P(H^c) \\ &= pP(E|H) + (1-p)P(E|H^c) \end{aligned}$$

Now, given that the first flip lands heads, the situation after the first bet is that A has $i + 1$ units and B has $N - (i + 1)$. Since the successive flips are assumed to be independent with a common probability p of heads, it follows

*The remainder of this section should be considered optional.

that, from that point on, A's probability of winning all the money is exactly the same as if the game were just starting with A having an initial fortune of $i + 1$ and B having an initial fortune of $N - (i + 1)$. Therefore,

$$P(E|H) = P_{i+1}$$

and similarly,

$$P(E|H^c) = P_{i-1}$$

Hence, letting $q = 1 - p$, we obtain

$$P_i = pP_{i+1} + qP_{i-1} \qquad i = 1, 2, \ldots, N - 1 \tag{4.2}$$

By making use of the obvious boundary conditions $P_0 = 0$ and $P_N = 1$, we shall now solve Equation (4.2). Since $p + q = 1$, these equations are equivalent to

$$pP_i + qP_i = pP_{i+1} + qP_{i-1}$$

or

$$P_{i+1} - P_i = \frac{q}{p}(P_i - P_{i-1}) \qquad i = 1, 2, \ldots, N - 1 \tag{4.3}$$

Hence, since $P_0 = 0$, we obtain from Equation (4.3):

$$\begin{aligned}
P_2 - P_1 &= \frac{q}{p}(P_1 - P_0) = \frac{q}{p}P_1 \\
P_3 - P_2 &= \frac{q}{p}(P_2 - P_1) = \left(\frac{q}{p}\right)^2 P_1 \\
&\vdots \\
P_i - P_{i-1} &= \frac{q}{p}(P_{i-1} - P_{i-2}) = \left(\frac{q}{p}\right)^{i-1} P_1 \\
&\vdots \\
P_N - P_{N-1} &= \frac{q}{p}(P_{N-1} - P_{N-2}) = \left(\frac{q}{p}\right)^{N-1} P_1
\end{aligned} \tag{4.4}$$

Adding the first $i - 1$ of Equation (4.4) yields

$$P_i - P_1 = P_1\left[\left(\frac{q}{p}\right) + \left(\frac{q}{p}\right)^2 + \cdots + \left(\frac{q}{p}\right)^{i-1}\right]$$

or

$$P_i = \begin{cases} \dfrac{1 - (q/p)^i}{1 - (q/p)} P_1 & \text{if } \dfrac{q}{p} \neq 1 \\[2ex] iP_1 & \text{if } \dfrac{q}{p} = 1 \end{cases}$$

Upon using the fact that $P_N = 1$, we obtain

$$P_1 = \begin{cases} \dfrac{1 - (q/p)}{1 - (q/p)^N} & \text{if } p \neq \frac{1}{2} \\[2ex] \dfrac{1}{N} & \text{if } p = \frac{1}{2} \end{cases}$$

and hence

$$P_i = \begin{cases} \dfrac{1 - (q/p)^i}{1 - (q/p)^N} & \text{if } p \neq \frac{1}{2} \\[2ex] \dfrac{i}{N} & \text{if } p = \frac{1}{2} \end{cases} \tag{4.5}$$

Let Q_i denote the probability that B winds up with all the money when A starts with i and B with $N - i$. Then, by symmetry with the situation described and on replacing p by q and i by $N - i$, we see that

$$Q_i = \begin{cases} \dfrac{1 - (p/q)^{N-i}}{1 - (p/q)^N} & \text{if } q \neq \frac{1}{2} \\[2ex] \dfrac{N - i}{N} & \text{if } q = \frac{1}{2} \end{cases}$$

Moreover, since $q = \frac{1}{2}$ is equivalent to $p = \frac{1}{2}$, we have when $q \neq \frac{1}{2}$,

$$\begin{aligned} P_i + Q_i &= \frac{1 - (q/p)^i}{1 - (q/p)^N} + \frac{1 - (p/q)^{N-i}}{1 - (p/q)^N} \\ &= \frac{p^N - p^N(q/p)^i}{p^N - q^N} + \frac{q^N - q^N(p/q)^{N-i}}{q^N - p^N} \\ &= \frac{p^N - p^{N-i}q^i - q^N + q^i p^{N-i}}{p^N - q^N} \\ &= 1 \end{aligned}$$

As this result also holds when $p = q = \frac{1}{2}$, we see that

$$P_i + Q_i = 1$$

In words, this equation states that with probability 1 either A or B will wind up with all of the money; or, in other words, the probability that the game continues indefinitely with A's fortune always being between 1 and $N - 1$ is zero. (The reader must be careful because, a priori, there are three possible outcomes of this gambling game, not two. Either A wins or B wins or it goes on forever with nobody winning. We have just shown that this last event has probability 0.)

As a numerical illustration of the preceding result, if A were to start with 5 units and B with 10, then the probability of A's winning would be $\frac{1}{3}$ if p were $\frac{1}{2}$, whereas it would jump to

$$\frac{1 - \left(\frac{2}{3}\right)^5}{1 - \left(\frac{2}{3}\right)^{15}} \approx .87$$

if p were .6.

A special case of the gambler's ruin problem, which is also known as the problem of *duration of play*, was proposed to the Dutch mathematician Christian Huygens by the Frenchman Fermat in 1657. The version he proposed, which was solved by Huygens, was that A and B each have 12 coins. They play for these coins in a game with 3 dice as follows. Whenever 11 is thrown (by either—it makes no difference who rolls the dice), then A gives a coin to B. Whenever 14 is thrown, B gives a coin to A. The person who first wins all the coins wins the game. Since $P\{\text{roll } 11\} = \frac{27}{216}$ and $P\{\text{roll } 14\} = \frac{15}{216}$, we see from Example 4h that for A this is just the gambler's ruin problem (Example 4j) with $p = \frac{15}{42}$, $i = 12$, $N = 24$. The general form of the gambler's ruin problem was solved by the mathematician James Bernoulli and published 8 years after his death in 1713.

For an application of the gambler's ruin problem to drug testing, suppose that two new drugs have been developed for treating a certain disease. Drug i has a cure rate P_i, $i = 1, 2$, in the sense that each patient treated with drug i will be cured with probability P_i. These cure rates are, however, not known, and we are interested in a method for deciding whether $P_1 > P_2$ or $P_2 > P_1$. To decide on one of these alternatives, consider the following test: Pairs of patients are to be treated sequentially, with one member of the pair receiving drug 1 and the other drug 2. The results for each pair are determined, and the testing stops when the cumulative number of cures from one of the drugs exceeds the cumulative number of cures from the other by some fixed, predetermined number. More formally, let

$$X_j = \begin{cases} 1 & \text{if the patient in the } j\text{th pair that receives drug 1 is cured} \\ 0 & \text{otherwise} \end{cases}$$

$$Y_j = \begin{cases} 1 & \text{if the patient in the } j\text{th pair that receives drug 2 is cured} \\ 0 & \text{otherwise} \end{cases}$$

For a predetermined positive integer M, the test stops after pair N, where N is the first value of n such that either

$$X_1 + \cdots + X_n - (Y_1 + \cdots + Y_n) = M$$

or

$$X_1 + \cdots + X_n - (Y_1 + \cdots + Y_n) = -M$$

In the former case we then assert that $P_1 > P_2$ and in the latter that $P_2 > P_1$.

In order to help ascertain whether the foregoing is a good test, one thing we would like to know is the probability that it leads to an incorrect decision. That is, for given P_1 and P_2, where $P_1 > P_2$, what is the probability that the test will incorrectly assert that $P_2 > P_1$? To determine this probability, note that after each pair is checked, the cumulative difference of cures using drug 1 versus drug 2 will go up by 1 with probability $P_1(1 - P_2)$

—since this is the probability that drug 1 leads to a cure and drug 2 does not—or go down by 1 with probability $(1 - P_1)P_2$ or remain the same with probability $P_1P_2 + (1 - P_1)(1 - P_2)$. Hence, if we consider only those pairs in which the cumulative difference changes, then the difference will go up by 1 with probability

$$P = P\{\text{up } 1 | \text{up } 1 \text{ or down } 1\}$$
$$= \frac{P_1(1 - P_2)}{P_1(1 - P_2) + (1 - P_1)P_2}$$

and down by 1 with probability

$$1 - P = \frac{P_2(1 - P_1)}{P_1(1 - P_2) + (1 - P_1)P_2}$$

Hence the probability that the test will assert that $P_2 > P_1$ is equal to the probability that a gambler who wins each (one unit) bet with probability P will go down M before going up M. But Equation (4.5), with $i = M$, $N = 2M$, shows that this probability is given by

$$P\{\text{test asserts that } P_2 > P_1\}$$
$$= 1 - \frac{1 - \left(\dfrac{1 - P}{P}\right)^M}{1 - \left(\dfrac{1 - P}{P}\right)^{2M}}$$
$$= 1 - \frac{1}{1 + \left(\dfrac{1 - P}{P}\right)^M}$$
$$= \frac{1}{1 + \gamma^M}$$

where

$$\gamma = \frac{P}{1 - P} = \frac{P_1(1 - P_2)}{P_2(1 - P_1)}$$

For instance, if $P_1 = .6$ and $P_2 = .4$, then the probability of an incorrect decision is .017 when $M = 5$ and reduces to .0003 when $M = 10$. ■

Suppose that we are presented with a set of elements and we want to determine whether at least one member of this set has a certain property. We can attack this question probabilistically by randomly choosing an element of the set in such a way that each element has a positive probability of being selected. Then the original question can be answered by considering the probability that the randomly selected element has the property of interest. If that probability is positive, then we have established that at least one element of the set has the property, and if it is zero, then none of them do.

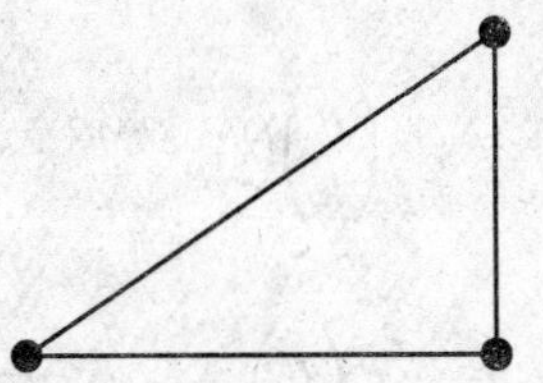

Figure 3.3

The final example of this section illustrates this technique.

Example 4k. The complete graph having n vertices is defined to be a set of n points (called vertices) in the plane and the $\binom{n}{2}$ lines (called edges) connecting each pair of vertices. The complete graph having 3 vertices is shown in Figure 3.3. Suppose now that each edge in a complete graph on n vertices is to be colored either red or blue. For a fixed integer k, a question of interest is whether there is a way of coloring the edges so that no set of k vertices has all of its $\binom{k}{2}$ connecting edges the same color. It can be shown, by a probabilistic argument, that if n is not too large, then the answer is yes.

The argument runs as follows. Suppose that each edge is, independently, equally likely to be colored either red or blue. That is, each edge is red with probability $\frac{1}{2}$. Number the $\binom{n}{k}$ sets of k vertices and define the events E_i, $i = 1, \ldots, \binom{n}{k}$ as follows:

$$E_i = \{\text{all of the connecting edges of the } i\text{th set of } k \text{ vertices are the same color}\}$$

Now, since each of the $\binom{k}{2}$ connecting edges of a set of k vertices is equally likely to be either red or blue, it follows that the probability that they are all the same color is

$$P(E_i) = 2\left(\tfrac{1}{2}\right)^{k(k-1)/2}$$

Therefore, since

$$P\left(\bigcup_i E_i\right) \le \sum P(E_i) \qquad \text{(Boole's inequality)}$$

we obtain that $P\left(\bigcup_i E_i\right)$, the probability that there is a set of k vertices all of whose connecting edges are similarly colored, satisfies

$$P\left(\bigcup_i E_i\right) \le \binom{n}{k}\left(\tfrac{1}{2}\right)^{k(k-1)/2-1}$$

Hence if

$$\binom{n}{k}\left(\frac{1}{2}\right)^{k(k-1)/2-1} < 1$$

or, equivalently, if

$$\binom{n}{k} < 2^{k(k-1)/2-1}$$

then the probability that at least one of the $\binom{n}{k}$ sets of k vertices has all of its connecting edges the same color is less than 1. Therefore, under the preceding condition on n and k, it follows that there is a positive probability that no set of k vertices has all of its connecting edges the same color. But this implies that there is at least one way of coloring the edges for which no set of k vertices has all of its connecting edges the same color. ■

REMARKS. (a) It should be noted that whereas the preceding argument established a condition on n and k that guarantees the existence of a coloring scheme satisfying the desired property, it gives no information about how to obtain such a scheme. (Although one possibility would simply be to choose the colors at random, check to see if the resulting coloring satisfies the property, and repeat this until it does.)

(b) The method of introducing probability to a problem whose statement is purely deterministic has been called the *probabilistic method*.† Other examples of this method are given in Theoretical Exercise 22 and Example 2q of Chapter 7.

3.5 $P(\cdot\,|F)$ IS A PROBABILITY

Conditional probabilities satisfy all of the properties of ordinary probabilities. This is proved by Proposition 5.1, which shows that $P(E|F)$ satisfies the three axioms of a probability.

Proposition 5.1

(a) $0 \leq P(E|F) \leq 1$.

(b) $P(S|F) = 1$.

(c) If $E_i, i = 1, 2, \ldots$ *are mutually exclusive events, then*

$$P\left(\bigcup_1^\infty E_i|F\right) = \sum_1^\infty P(E_i|F)$$

†See N. Alon, J. Spencer, and P. Erdos, *The Probabilistic Method* (New York: John Wiley & Sons, Inc., 1992).

Proof: To prove part (a), we must show that $0 \le P(EF)/P(F) \le 1$. The left-side inequality is obvious, whereas the right side follows because $EF \subset F$, which implies that $P(EF) \le P(F)$. Part (b) follows because

$$P(S|F) = \frac{P(SF)}{P(F)} = \frac{P(F)}{P(F)} = 1$$

Part (c) follows since

$$\begin{aligned}
P\left(\bigcup_{i=1}^{\infty} E_i|F\right) &= \frac{P\left(\left(\bigcup_{i=1}^{\infty} E_i\right)F\right)}{P(F)} \\
&= \frac{P\left(\bigcup_{1}^{\infty} E_iF\right)}{P(F)} \qquad \text{since} \qquad \left(\bigcup_{1}^{\infty} E_i\right)F = \bigcup_{1}^{\infty} E_iF \\
&= \frac{\sum_{1}^{\infty} P(E_iF)}{P(F)} \\
&= \sum_{1}^{\infty} P(E_i|F)
\end{aligned}$$

where the next-to-last equality follows because $E_iE_j = \emptyset$ implies that $E_iFE_jF = \emptyset$. ✦

If we define $Q(E) = P(E|F)$, then it follows from Proposition 5.1 that $Q(E)$ may be regarded as a probability function on the events of S. Hence all of the propositions previously proved for probabilities apply to it. For instance, we have

$$Q(E_1 \cup E_2) = Q(E_1) + Q(E_2) - Q(E_1E_2)$$

or, equivalently,

$$P(E_1 \cup E_2|F) = P(E_1|F) + P(E_2|F) - P(E_1E_2|F)$$

Also, if we define the conditional probability $Q(E_1|E_2)$ by $Q(E_1|E_2) = Q(E_1E_2)/Q(E_2)$, then from Equation (3.1) we see that

$$Q(E_1) = Q(E_1|E_2)Q(E_2) + Q(E_1|E_2^c)Q(E_2^c) \tag{5.1}$$

Since

$$\begin{aligned}
Q(E_1|E_2) &= \frac{Q(E_1E_2)}{Q(E_2)} \\
&= \frac{P(E_1E_2|F)}{P(E_2|F)} \\
&= \frac{\dfrac{P(E_1E_2F)}{P(F)}}{\dfrac{P(E_2F)}{P(F)}} \\
&= P(E_1|E_2F)
\end{aligned}$$

we see that Equation (5.1) is equivalent to

$$P(E_1|F) = P(E_1|E_2F)P(E_2|F) + P(E_1|E_2^cF)P(E_2^c|F)$$

Example 5a. Consider Example 3a, which is concerned with an insurance company that believes that people can be divided into two distinct classes—those who are accident prone and those who are not. During any given year an accident-prone person will have an accident with probability .4, whereas the corresponding figure for a non-accident-prone person is .2. What is the conditional probability that a new policyholder will have an accident in his or her second year of policy ownership, given that the policyholder has had an accident in the first year?

Solution If we let A be the event that the policyholder is accident prone and we let $A_i, i = 1, 2$, be the event that he or she has had an accident in the ith year, then the desired probability $P(A_2|A_1)$ may be obtained by conditioning on whether or not the policyholder is accident prone, as follows:

$$P(A_2|A_1) = P(A_2|AA_1)P(A|A_1) + P(A_2|A^cA_1)P(A^c|A_1)$$

Now,

$$P(A|A_1) = \frac{P(A_1A)}{P(A_1)} = \frac{P(A_1|A)P(A)}{P(A_1)}$$

However, $P(A)$ is assumed to equal $\frac{3}{10}$, and it was shown in Example 3a that $P(A_1) = .26$. Hence

$$P(A|A_1) = \frac{(.4)(.3)}{.26} = \frac{6}{13}$$

and thus

$$P(A^c|A_1) = 1 - P(A|A_1) = \frac{7}{13}$$

Since $P(A_2|AA_1) = .4$ and $P(A_2|A^cA_1) = .2$, we see that

$$P(A_2|A_1) = (.4)\tfrac{6}{13} + (.2)\tfrac{7}{13} \approx .29$$ ■

The next example deals with a problem in the theory of runs.

Example 5b. Independent trials, each resulting in a success with probability p or a failure with probability $q = 1 - p$, are performed. We are interested in computing the probability that a run of n consecutive successes occurs before a run of m consecutive failures.

Solution Let E be the event that a run of n consecutive successes occurs before a run of m consecutive failures. To obtain $P(E)$, we start by conditioning on the outcome of the first trial. That is, letting H denote the event that the first trial results in a success, we obtain

$$P(E) = pP(E|H) + qP(E|H^c) \tag{5.2}$$

Now, given that the first trial was successful, one way we can get a run of n successes before a run of m failures would be to have the next $n - 1$ trials all result in successes. So, let us condition on whether or not that occurs. That is, letting F be the event that trials 2 through n all are successes, we obtain

$$P(E|H) = P(E|FH)P(F|H) + P(E|F^cH)P(F^c|H) \tag{5.3}$$

Clearly, $P(E|FH) = 1$; on the other hand, if the event F^cH occurs, then the first trial would result in a success, but there would be a failure some time during the next $n - 1$ trials. However, when this failure occurs, it would wipe out all of the previous successes, and the situation would be exactly as if we started out with a failure. Hence

$$P(E|F^cH) = P(E|H^c)$$

As the independence of trials implies that F and H are independent and as $P(F) = p^{n-1}$, we obtain from (5.3)

$$P(E|H) = p^{n-1} + (1 - p^{n-1})P(E|H^c) \tag{5.4}$$

We now obtain an expression for $P(E|H^c)$ in a similar manner. That is, we let G denote the event that trials 2 through m are all *failures*. Then

$$P(E|H^c) = P(E|GH^c)P(G|H^c) + P(E|G^cH^c)P(G^c|H^c) \tag{5.5}$$

Now, GH^c is the event that the first m trials all result in failures, so $P(E|GH^c) = 0$. Also, if G^cH^c occurs, the first trial is a failure, but there is at least one success in the next $m - 1$ trials. Hence, as this success wipes out all previous failures, we see that

$$P(E|G^cH^c) = P(E|H)$$

Thus, because $P(G^c|H^c) = P(G^c) = 1 - q^{m-1}$, we obtain from (5.5)

$$P(E|H^c) = (1 - q^{m-1})P(E|H) \tag{5.6}$$

Solving Equations (5.4) and (5.6) yields

$$P(E|H) = \frac{p^{n-1}}{p^{n-1} + q^{m-1} - p^{n-1}q^{m-1}}$$

and

$$P(E|H^c) = \frac{(1 - q^{m-1})p^{n-1}}{p^{n-1} + q^{m-1} - p^{n-1}q^{m-1}}$$

and thus

$$\begin{aligned} P(E) &= pP(E|H) + qP(E|H^c) \\ &= \frac{p^n + qp^{n-1}(1 - q^{m-1})}{p^{n-1} + q^{m-1} - p^{n-1}q^{m-1}} \\ &= \frac{p^{n-1}(1 - q^m)}{p^{n-1} + q^{m-1} - p^{n-1}q^{m-1}} \end{aligned} \tag{5.7}$$

It is interesting to note that by the symmetry of the problem, the probability of obtaining a run of m failures before a run of n successes would be given by Equation (5.7) with p and q interchanged and n and m interchanged. Hence this probability would equal

$$P\{\text{run of } m \text{ failures before a run of } n \text{ successes}\}$$
$$= \frac{q^{m-1}(1 - p^n)}{q^{m-1} + p^{n-1} - q^{m-1}p^{n-1}} \tag{5.8}$$

Since Equations (5.7) and (5.8) sum to 1, it follows that, with probability 1, either a run of n successes or a run of m failures will eventually occur.

As an example of Equation (5.7) we note that in tossing a fair coin the probability that a run of 2 heads will precede a run of 3 tails is $\frac{7}{10}$, for 2 consecutive heads before 4 consecutive tails the probability rises to $\frac{5}{6}$. ■

In our next example we return to the matching problem (Example 5m, Chapter 2) and this time obtain a solution by using conditional probabilities.

Example 5c. At a party n men take off their hats. The hats are then mixed up, and each man randomly selects one. We say that a match occurs if a man selects his own hat. What is the probability of
(a) no matches;
(b) exactly k matches?

Solution (a) Let E denote the event that no matches occur, and to make explicit the dependence on n write $P_n = P(E)$. We start by conditioning on whether or not the first man selects his own hat—call these events M and M^c. Then

$$P_n = P(E) = P(E|M)P(M) + P(E|M^c)P(M^c)$$

Clearly, $P(E|M) = 0$, so

$$P_n = P(E|M^c)\frac{n-1}{n} \tag{5.9}$$

Now, $P(E|M^c)$ is the probability of no matches when $n - 1$ men select from a set of $n - 1$ hats that does not contain the hat of one of these men. This can happen in either of two mutually exclusive ways. Either there are no matches and the extra man does not select the extra hat (this being the hat of the man who chose first), or there are no matches and the extra man does select the extra hat. The probability of the first of these events is just P_{n-1}, which is seen by regarding the extra hat as "belonging" to the extra man. As the second event has probability $[1/(n-1)]P_{n-2}$, we have

$$P(E|M^c) = P_{n-1} + \frac{1}{n-1}P_{n-2}$$

and thus, from Equation (5.9),

$$P_n = \frac{n-1}{n}P_{n-1} + \frac{1}{n}P_{n-2}$$

or, equivalently,

$$P_n - P_{n-1} = -\frac{1}{n}(P_{n-1} - P_{n-2}) \tag{5.10}$$

However, as P_n is the probability of no matches when n men select among their own hats, we have

$$P_1 = 0 \qquad P_2 = \tfrac{1}{2}$$

So, from Equation (5.10)

$$P_3 - P_2 = -\frac{(P_2 - P_1)}{3} = -\frac{1}{3!} \qquad \text{or} \qquad P_3 = \frac{1}{2!} - \frac{1}{3!}$$

$$P_4 - P_3 = -\frac{(P_3 - P_2)}{4} = \frac{1}{4!} \qquad \text{or} \qquad P_4 = \frac{1}{2!} - \frac{1}{3!} + \frac{1}{4!}$$

and, in general, we see that

$$P_n = \frac{1}{2!} - \frac{1}{3!} + \frac{1}{4!} - \cdots + \frac{(-1)^n}{n!}$$

(b) To obtain the probability of exactly k matches, we consider any fixed group of k men. The probability that they, and only they, select their own hats is

$$\frac{1}{n}\frac{1}{n-1}\cdots\frac{1}{n-(k-1)}P_{n-k} = \frac{(n-k)!}{n!}P_{n-k}$$

where P_{n-k} is the conditional probability that the other $n-k$ men, selecting among their own hats, have no matches. As there are $\binom{n}{k}$ choices of a set of k men, the desired probability of exactly k matches is

$$\frac{P_{n-k}}{k!} = \frac{\frac{1}{2!} - \frac{1}{3!} + \cdots + \frac{(-1)^{n-k}}{(n-k)!}}{k!}$$ ∎

An important concept in probability theory is that of the conditional independence of events. We say that the events E_1 and E_2 are *conditionally independent* given F if, given that F occurs, the conditional probability that E_1 occurs is unchanged by information as to whether or not E_2 occurs. More formally, E_1 and E_2 are said to be conditionally independent given F if

$$P(E_1|E_2F) = P(E_1|F) \tag{5.11}$$

or, equivalently,

$$P(E_1E_2|F) = P(E_1|F)P(E_2|F) \tag{5.12}$$

The notion of conditional independence can easily be extended to more than two events and this extension is left as an exercise.

The reader should note that the concept of conditional independence was implicitly employed in Example 5a, where it was implicitly assumed that the events that a policyholder had an accident in his or her ith year, $i = 1, 2, \ldots$, were conditionally independent given whether or not the person was accident prone. [This was used to evaluate $P(A_2|AA_1)$ and $P(A_2|A^c A_1)$ as, respectively, .4 and .2.] The following example, sometimes referred to as Laplace's rule of succession, further illustrates the concept of conditional independence

Example 5d. ***Laplace's rule of succession.*** There are $k + 1$ coins in a box. The ith coin will, when flipped, turn up heads with probability i/k, $i = 0, 1, \ldots, k$. A coin is randomly selected from the box and is then repeatedly flipped. If the first n flips all result in heads, what is the conditional probability that the $(n + 1)$st flip will do likewise?

Solution Let C_i denote the event that the ith coin is initially selected, $i = 0, 1, \ldots, k$; let F_n denote the event that the first n flips all result in heads; and let H be the event that the $(n + 1)$st flip is a head. The desired probability, $P(H|F_n)$, is now obtained as follows:

$$P(H|F_n) = \sum_{i=0}^{k} P(H|F_n C_i)\, P(C_i|F_n)$$

Now, given that the ith coin is selected, it is reasonable to assume that the outcomes will be conditionally independent, with each one resulting in a head with probability i/k. Hence

$$P(H|F_n C_i) = P(H|C_i) = \frac{i}{k}$$

Also,

$$\begin{aligned} P(C_i|F_n) &= \frac{P(C_i F_n)}{P(F_n)} \\ &= \frac{P(F_n|C_i)P(C_i)}{\sum_{j=0}^{k} P(F_n|C_j)P(C_j)} \\ &= \frac{(i/k)^n[1/(k+1)]}{\sum_{j=0}^{k} (j/k)^n[1/(k+1)]} \end{aligned}$$

Hence

$$P(H|F_n) = \frac{\sum_{i=0}^{k} (i/k)^{n+1}}{\sum_{j=0}^{k} (j/k)^n}$$

But if k is large, we can use the integral approximations

$$\frac{1}{k}\sum_{i=0}^{k}\left(\frac{i}{k}\right)^{n+1} \approx \int_0^1 x^{n+1}\,dx = \frac{1}{n+2}$$

$$\frac{1}{k}\sum_{j=0}^{k}\left(\frac{j}{k}\right)^{n} \approx \int_0^1 x^{n}\,dx = \frac{1}{n+1}$$

so, for k large,

$$P(H|F_n) \approx \frac{n+1}{n+2}$$ ■

SUMMARY

For events E and F, the conditional probability of E given that F has occurred is denoted by $P(E|F)$ and is defined by

$$P(E|F) = \frac{P(EF)}{P(F)}$$

The identity

$$P(E_1E_2\cdots E_n) = P(E_1)P(E_2|E_1)\cdots P(E_n|E_1\cdots E_{n-1})$$

is known as the *multiplication rule* of probability.

A useful identity is that

$$P(E) = P(E|F)P(F) + P(E|F^c)P(F^c)$$

It can be used to compute $P(E)$ by "conditioning" on whether F occurs.

$P(H)/P(H^c)$ is called the *odds ratio* of the event H. The identity

$$\frac{P(H|E)}{P(H^c|E)} = \frac{P(H)\;P(E|H)}{P(H^c)P(E|H^c)}$$

shows that when new evidence E is obtained, the value of the odds ratio of H becomes its old value multiplied by the ratio of the conditional probability of the new evidence when H is true to its conditional probability when H is not true.

Let $F_i, i = 1, \ldots, n$, be mutually exclusive events whose union is the entire sample space. The identity

$$P(F_j|E) = \frac{P(E|F_j)P(F_j)}{\sum_{i=1}^{n} P(E|F_i)P(F_i)}$$

is known as *Bayes' formula*. If the events $F_i, i = 1, \ldots, n$, are competing hypotheses, then Bayes' formula shows how to compute the conditional probabilities of these hypotheses when additional evidence E becomes available.

If $P(EF) = P(E)P(F)$, then we say that the events E and F are *independent*. This condition is equivalent to $P(E|F) = P(E)$ and to $P(F|E) = P(F)$. Thus E and F are independent if knowledge of the occurrence of one of them does not affect the probability of the other.

The events $E_1, \ldots, E_n$ are said to be independent if for any subset $E_{i_1}, \ldots, E_{i_r}$ of them,

$$P(E_{i_1} \cdots E_{i_r}) = P(E_{i_1}) \cdots P(E_{i_r})$$

For a fixed event F, $P(E|F)$ can be considered to be a probability function on the events E of the sample space.

PROBLEMS

1. Two fair dice are rolled. What is the conditional probability that at least one lands on 6 given that the dice land on different numbers?

2. If two fair dice are rolled, what is the conditional probability that the first one lands on 6 given that the sum of the dice is i? Compute for all values of i between 2 and 12.

3. Use Equation (2.1) to compute, in a hand of bridge, the conditional probability that East has 3 spades given that North–South have a combined total of 8 spades.

4. What is the probability that at least one of a pair of fair dice lands on 6, given that the sum of the dice is $i, i = 2, 3, \ldots, 12$?

5. An urn contains 6 white and 9 black balls. If 4 balls are to be randomly selected without replacement, what is the probability that the first 2 selected are white and the last 2 black?

6. Consider an urn containing 12 balls of which 8 are white. A sample of size 4 is to be drawn with replacement (without replacement). What is the conditional probability (in each case) that the first and third balls drawn will be white, given that the sample drawn contains exactly 3 white balls?

7. The king comes from a family of 2 children. What is the probability that the other child is his sister?

8. A couple has 2 children. What is the probability that both are girls if the eldest is a girl?

9. Consider 3 urns. Urn A contains 2 white and 4 red balls; urn B contains 8 white and 4 red balls; and urn C contains 1 white and 3 red balls. If 1 ball is selected from each urn, what is the probability that the ball chosen from urn A was white, given that exactly 2 white balls were selected?

10. Three cards are randomly selected, without replacement, from an ordinary deck of 52 playing cards. Compute the conditional probability that the first card selected is a spade, given that the second and third cards are spades.

11. An ectopic pregnancy is twice as likely to develop when the pregnant woman is a smoker as it is when she is a nonsmoker. If 32 percent of women of childbearing age are smokers, what percentage of women having ectopic pregnancies are smokers?

12. Ninety-eight percent of all babies survive delivery. However, 15 percent of all births involve Cesarean (C) sections, and when a C section is performed the baby survives 96 percent of the time. If a randomly chosen pregnant woman does not have a C section, what is the probability that her baby survives?

13. In a certain community, 36 percent of the families own a dog, and 22 percent of the families that own a dog also own a cat. In addition, 30 percent of the families own a cat. What is

(a) the probability that a randomly selected family owns both a dog and a cat;

(b) the conditional probability that a randomly selected family owns a dog given that it owns a cat?

14. A total of 46 percent of the voters in a certain city classify themselves as Independents, whereas 30 percent classify themselves as Liberals and 24 percent as Conservatives. In a recent local election, 35 percent of the Independents, 62 percent of the Liberals, and 58 percent of the Conservatives voted. A voter is chosen at random. Given that this person voted in the local election, what is the probability that he or she is
(a) an Independent;
(b) a Liberal;
(c) a Conservative?
(d) What fraction of voters participated in the local election?

15. A total of 48 percent of the women and 37 percent of the men that took a certain "quit smoking" class remained nonsmokers for at least one year after completing the class. These people then attended a success party at the end of a year. If 62 percent of the original class were male,
(a) what percentage of those attending the party were women?
(b) what percentage of the original class attended the party?

16. Fifty-two percent of the students at a certain college are females. Five percent of the students in this college are majoring in computer science. Two percent of the students are women majoring in computer science. If a student is selected at random, find the conditional probability that
(a) this student is female, given that the student is majoring in computer science;
(b) this student is majoring in computer science, given that the student is female.

17. A total of 500 married working couples were polled about their annual salaries, with the following information resulting.

	Husband	
Wife	Less than $25,000	More than $25,000
Less than $25,000	212	198
More than $25,000	36	54

Thus, for instance, in 36 of the couples the wife earned more and the husband earned less than $25,000. If one of the couples is randomly chosen, what is
(a) the probability that the husband earns less than $25,000;
(b) the conditional probability that the wife earns more than $25,000 given that the husband earns more than this amount;
(c) the conditional probability that the wife earns more than $25,000 given that the husband earns less than this amount?

18. A recent college graduate is planning to take the first three actuarial examinations in the coming summer. She will take the first actuarial exam in June. If she passes that exam, then she will take the second exam in July, and if she also passes that one, then she will take the third exam in September. If she fails an exam, then she is not allowed to take any others. The probability that she passes the first exam is .9. If she passes the first exam, then the conditional probability that she passes the second one is .8, and if she passes both the first and the second exams, then the conditional probability that she passes the third exam is .7.
(a) What is the probability that she passes all three exams?
(b) Given that she did not pass all three exams, what is the conditional probability that she failed the second exam?

19. Suppose that an ordinary deck of 52 cards (which contains 4 aces) is randomly divided into 4 hands of 13 cards each. We are interested in determining p, the probability that each hand has an ace. Let E_i be the event that the ith hand has exactly one ace. Determine $p = P(E_1E_2E_3E_4)$ by using the multiplication rule.

20. An urn initially contains 5 white and 7 black balls. Each time a ball is selected, its color is noted and it is replaced in the urn along with 2 other balls of the same color. Compute the probability that
(a) the first 2 balls selected are black and the next 2 white;
(b) of the first 4 balls selected, exactly 2 are black.

21. Urn I contains 2 white and 4 red balls, whereas urn II contains 1 white and 1 red ball. A ball is randomly chosen from urn I and put into urn II, and a ball is then randomly selected from urn II. What is
(a) the probability that the ball selected from urn II is white;
(b) the conditional probability that the transferred ball was white, given that a white ball is selected from urn II?

22. Each of 2 balls is painted either black or gold and then placed in an urn. Suppose that each ball is colored black with probability $\frac{1}{2}$, and that these events are independent.
(a) Suppose that you obtain information that the gold paint has been used (and thus at least one of the balls is painted gold). Compute the conditional probability that both balls are painted gold.
(b) Suppose, now, that the urn tips over and 1 ball falls out. It is painted gold. What is the probability that both balls are gold in this case? Explain.

23. The following method was proposed to estimate the number of people over the age of 50 that reside in a town of known population 100,000. "As you walk along the streets, keep a running count of the percentage of people that you encounter who are over 50. Do this for a few days; then multiply the obtained percentage by 100,000 to obtain the estimate." Comment on this method.

HINT: Let p denote the proportion of people in this town who are over 50. Furthermore, let α_1 denote the proportion of time that a person under the age of 50 spends in the streets, and let α_2 be the corresponding value for those over 50. What quantity does the method suggested estimate? When is it approximately equal to p?

24. Suppose that 5 percent of men and .25 percent of women are colorblind. A colorblind person is chosen at random. What is the probability of this person being male? Assume that there are an equal number of males and females. What if the population consisted of twice as many males as females?

25. All the workers at a certain company drive to work and park in the company's lot. The company is interested in estimating the average number of workers in a car. Which of the following methods will enable the company to estimate this quantity? Explain your answer.
1. Randomly choose n workers, find out how many were in the cars in which they were driven, and take the average of the n values.
2. Randomly choose n cars in the lot, find out how many were driven in those cars, and take the average of the n values.

26. Suppose that an ordinary deck of 52 cards is shuffled and the cards are then turned over one at a time until the first ace appears. Given that the first ace is the 20th card to appear, what is the conditional probability that the card following it is the
(a) ace of spades; **(b)** two of clubs?

27. There are 15 tennis balls in a box, of which 9 have not previously been used. Three of the balls are randomly chosen, played with, and then returned to the box. Later, another 3 balls are randomly chosen from the box. Find the probability that none of these balls has ever been used.

28. Consider two boxes, one containing 1 black and 1 white marble, the other 2 black and 1 white marble. A box is selected at random, and a marble is drawn at random from the selected box. What is the probability that the marble is black? What is the probability that the first box was the one selected, given that the marble is white?

29. English and American spellings are *rigour* and *rigor*, respectively. A man staying at a Parisian hotel writes this word, and a letter taken at random from his spelling is found to be a vowel. If 40 percent of the English-speaking men at the hotel are English and 60 percent are Americans, what is the probability that the writer is an Englishman?

30. In Example 3f, suppose that the new evidence is subject to different possible interpretations and in fact shows only that it is 90 percent likely that the criminal possesses this certain characteristic. In this case how likely would it be that the suspect is guilty (assuming, as before, that he has this characteristic)?

31. One probability class of 30 students contains 15 that are good, 10 that are fair, and 5 that are of poor quality. A second probability class, also of 30 students, contains 5 that are good, 10 that are fair, and 15 that are poor. You (the expert) are aware of these numbers, but you have no idea which class is which. If you examine one student selected at random from each class and find that the student from class A is a fair student whereas the student from class B is a poor student, what is the probability that class A is the superior class?

32. Stores A, B, and C have 50, 75, and 100 employees and, respectively, 50, 60, and 70 percent of these are women. Resignations are equally likely among all employees, regardless of sex. One employee resigns, and this is a woman. What is the probability that she works in store C?

33. **(a)** A gambler has in his pocket a fair coin and a two-headed coin. He selects one of the coins at random; when he flips it, it shows heads. What is the probability that it is the fair coin?

(b) Suppose that he flips the same coin a second time and again it shows heads. What is now the probability that it is the fair coin?

(c) Suppose that he flips the same coin a third time and it shows tails. What is now the probability that it is the fair coin?

34. Urn A has 5 white and 7 black balls. Urn B has 3 white and 12 black balls. We flip a fair coin. If the outcome is heads, then a ball from urn A is selected, whereas if the outcome is tails, then a ball from urn B is selected. Suppose that a white ball is selected. What is the probability that the coin landed tails?

35. In Example 3a, what is the probability that someone has an accident in the second year, given that he or she has had no accidents in the first year?

36. Consider a sample of size 3 drawn in the following manner: We start with an urn containing 5 white and 7 red balls. At each stage a ball is drawn and its color is noted. The ball is then returned to the urn along with an additional ball of the same color. Find the probability that the sample will contain exactly

(a) 0 white balls; **(b)** 1 white ball;

(c) 3 white balls; **(d)** 2 white balls.

37. A deck of cards is shuffled and then divided into two halves of 26 cards each. A card is drawn from one of the halves; it turns out to be an ace. The ace is then placed in the second half-deck. The half is then shuffled, and a card is drawn from it. Compute the probability that this drawn card is an ace.

HINT: Condition on whether or not the interchanged card is selected.

38. Three cooks, A, B, and C, bake a special kind of cake, and with respective probabilities .02, .03, and .05 it fails to rise. In the restaurant where they work, A bakes 50 percent of these cakes, B 30 percent, and C 20 percent. What proportion of "failures" is caused by A?

39. There are 3 coins in a box. One is a two-headed coin; another is a fair coin; and the third is a biased coin that comes up heads 75 percent of the time. When one of the 3 coins is selected at random and flipped, it shows heads. What is the probability that it was the two-headed coin?

40. Three prisoners are informed by their jailer that one of them has been chosen at random to be executed, and the other two are to be freed. Prisoner A asks the jailer to tell him privately which of his fellow prisoners will be set free, claiming that there would be no harm in divulging this information because he already knows that at least one of the two will go free. The jailer refuses to answer this question, pointing out that if A knew which of his fellow prisoners were to be set free, then his own probability of being executed would rise from $\frac{1}{3}$ to $\frac{1}{2}$ because he would then be one of two prisoners. What do you think of the jailer's reasoning?

41. Suppose we have 10 coins such that if the ith coin is flipped, heads will appear with probability $i/10$, $i = 1, 2, \ldots, 10$. When one of the coins is randomly selected and flipped, it shows heads. What is the conditional probability that it was the fifth coin?

42. In any given year a male automobile policyholder will make a claim with probability p_m, and a female policyholder will make a claim with probability p_f, where $p_f \neq p_m$. The fraction of the policyholders that are male is α, $0 < \alpha < 1$. A policyholder is randomly chosen. If A_i denotes the event that this policyholder will make a claim in year i, show that

$$P(A_2|A_1) > P(A_1)$$

Give an intuitive explanation why the preceding inequality is true.

43. An urn contains 5 white and 10 black balls. A fair die is rolled and that number of balls is randomly chosen from the urn. What is the probability that all of the balls selected are white? What is the conditional probability that the die landed on 3 if all the balls selected are white?

44. Each of 2 cabinets identical in appearance has 2 drawers. Cabinet A contains a silver coin in each drawer, and cabinet B contains a silver coin in one of its drawers and a gold coin in the other. A cabinet is randomly selected, one of its drawers is opened, and a silver coin is found. What is the probability that there is a silver coin in the other drawer?

45. Suppose that there was a cancer diagnostic test that was 95 percent accurate both on those that do and those that do not have the disease. If .4 percent of the population have cancer, compute the probability that a tested person has cancer, given that his or her test result indicates so.

46. Suppose that an insurance company classifies people into one of three classes: good risks, average risks, and bad risks. Their records indicate that the probabilities that good, average, and bad risk persons will be involved in an accident over a 1-year span are, respectively, .05, .15, and .30. If 20 percent of the population are good risks, 50 percent are average risks, and 30 percent are bad risks, what proportion of people have accidents in a fixed year? If policyholder A had no accidents in 1997, what is the probability that he or she is a good (average) risk?

47. A worker has asked her supervisor for a letter of recommendation for a new job. She estimates that there is an 80 percent chance that she will get the job if she receives a strong recommendation, a 40 percent chance if she receives a moderately good recommendation, and a 10 percent chance if she receives a weak recommendation. She further estimates that the probabilities that the recommendation will be strong, moderate, or weak are .7, .2, and .1, respectively.

(a) How certain is she that she will receive the new job offer?
(b) Given that she does receive the offer, how likely should she feel that she received a strong recommendation; a moderate recommendation; a weak recommendation?
(c) Given that she does not receive the job offer, how likely should she feel that she received a strong recommendation; a moderate recommendation; a weak recommendation?

48. A high school student is anxiously waiting to receive mail telling her whether she has been accepted to a certain college. She estimates that the conditional probabilities, given that she is accepted and that she is rejected, of receiving notification on each day of next week are as follows:

Day	P(mail | accepted)	P(mail | rejected)
Monday	.15	.05
Tuesday	.20	.10
Wednesday	.25	.10
Thursday	.15	.15
Friday	.10	.20

She estimates that her probability of being accepted is .6.
(a) What is the probability that mail is received on Monday?
(b) What is the conditional probability that mail is received on Tuesday given that it is not received on Monday?
(c) If there is no mail through Wednesday, what is the conditional probability that she will be accepted?
(d) What is the conditional probability that she will be accepted if mail comes on Thursday?
(e) What is the conditional probability that she will be accepted if no mail arrives that week?

49. A parallel system functions whenever at least one of its components works. Consider a parallel system of n components and suppose that each component independently works with probability $\frac{1}{2}$. Find the conditional probability that component 1 works given that the system is functioning.

50. If you had to construct a mathematical model for events E and F, as described in parts (a) through (e), would you assume that they were independent events? Explain your reasoning.
(a) E is the event that a businesswoman has blue eyes, and F is the event that her secretary has blue eyes.
(b) E is the event that a professor owns a car, and F is the event that he is listed in the telephone book.
(c) E is the event that a man is under 6 feet tall, and F is the event that he weighs over 200 pounds.
(d) E is the event that a woman lives in the United States, and F is the event that she lives in the western hemisphere.
(e) E is the event that it will rain tomorrow, and F is the event that it will rain the day after tomorrow.

51. In a class there are 4 freshman boys, 6 freshman girls, and 6 sophomore boys. How many sophomore girls must be present if sex and class are to be independent when a student is selected at random?

52. Suppose that you continually collect coupons and that there are m different types. Suppose also that each time a new coupon is obtained it is a type i coupon with probability $p_i, i = 1, \ldots, m$. Suppose that you have just collected your nth coupon. What is the probability that it is a new type?

HINT: Condition on the type of this coupon.

53. A simplified model for the movement of the price of a stock supposes that on each day the stock's price either moves up 1 unit with probability p or it moves down 1 unit with probability $1 - p$. The changes on different days are assumed to be independent.
(a) What is the probability that after 2 days the stock will be at its original price?
(b) What is the probability that after 3 days the stock's price will have increased by 1 unit?
(c) Given that after 3 days the stock's price has increased by 1 unit, what is the probability that it went up on the first day?

54. Suppose that we want to generate the outcome of the flip of a fair coin but that all we have at our disposal is a biased coin which lands on heads with some unknown probability p that need not be equal to $\frac{1}{2}$. Consider the following procedure for accomplishing our task.
1. Flip the coin.
2. Flip the coin again.
3. If both flips land heads or both land tails, return to step 1.
4. Let the result of the last flip be the result of the experiment.
(a) Show that the result is equally likely to be either heads or tails.
(b) Could we use a simpler procedure that continues to flip the coin until the last two flips are different and then lets the result be the outcome of the final flip?

55. Independent flips of a coin that lands on heads with probability p are made. What is the probability that the first four outcomes are
(a) H, H, H, H;
(b) T, H, H, H?
(c) What is the probability that the pattern T, H, H, H occurs before the pattern H, H, H, H?
HINT FOR PART (C): How can the pattern H, H, H, H occur first?

56. The color of a person's eyes is determined by a single pair of genes. If they are both blue-eyed genes, then the person will have blue eyes; if they are both brown-eyed genes, then the person will have brown eyes; and if one of them is a blue-eyed gene and the other a brown-eyed gene, then the person will have brown eyes. (Because of the latter fact we say that the brown-eyed gene is *dominant* over the blue-eyed one.) A newborn child independently receives one eye gene from each of its parents and the gene it receives from a parent is equally likely to be either of the two eye genes of that parent. Suppose that Smith and both of his parents have brown eyes, but Smith's sister has blue eyes.
(a) What is the probability that Smith possesses a blue-eyed gene?
Suppose that Smith's wife has blue eyes.
(b) What is the probability that their first child will have blue eyes?
(c) If their first child has brown eyes, what is the probability that their next child will also have brown eyes?

57. Genes relating to albinism are denoted by A and a. Only those people who receive the a gene from both parents will be albino. Persons having the gene pair A, a are normal in appearance and, because they can pass on the trait to their offspring, are called carriers. Suppose that a normal couple has two children, exactly one of whom is an albino. Suppose that the nonalbino child mates with a person who is known to be a carrier for albinism.
(a) What is the probability that their first offspring is an albino?
(b) What is the conditional probability that their second offspring is an albino given that their firstborn is not?

58. Barbara and Dianne go target shooting. Suppose that each of Barbara's shots hits the wooden duck target with probability p_1, while each shot of Dianne's hits it with probability p_2. Suppose that they shoot simultaneously at the same target. If the wooden duck is knocked over (indicating that it was hit), what is the probability that

(a) both shots hit the duck; **(b)** Barbara's shot hit the duck?
What independence assumptions have you made?

59. A and B are involved in a duel. The rules of the duel are that they are to pick up their guns and shoot at each other simultaneously. If one or both are hit, then the duel is over. If both shots miss, then they repeat the process. Suppose that the results of the shots are independent and that each shot of A will hit B with probability p_A, and each shot of B will hit A with probability p_B. What is
(a) the probability that A is not hit;
(b) the probability that both duelists are hit;
(c) the probability that the duel ends after the nth round of shots;
(d) the conditional probability that the duel ends after the nth round of shots given that A is not hit;
(e) the conditional probability that the duel ends after the nth round of shots given that both duelists are hit?

60. A true–false question is to be posed to a husband and wife team on a quiz show. Both the husband and the wife will, independently, give the correct answer with probability p. Which of the following is a better strategy for this couple?
(a) Choose one of them and let that person answer the question; or
(b) have them both consider the question and then either give the common answer if they agree or, if they disagree, flip a coin to determine which answer to give?

61. In Problem 60, if $p = .6$ and the couple uses the strategy in part (b), what is the conditional probability that the couple gives the correct answer given that they (a) agree; (b) disagree?

62. The probability of the closing of the ith relay in the circuits shown is given by $p_i, i = 1, 2, 3, 4, 5$. If all relays function independently, what is the probability that a current flows between A and B for the respective circuits?

(a)

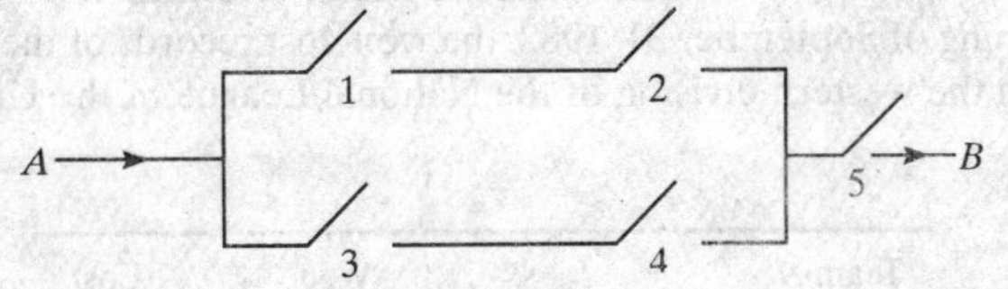

(b)

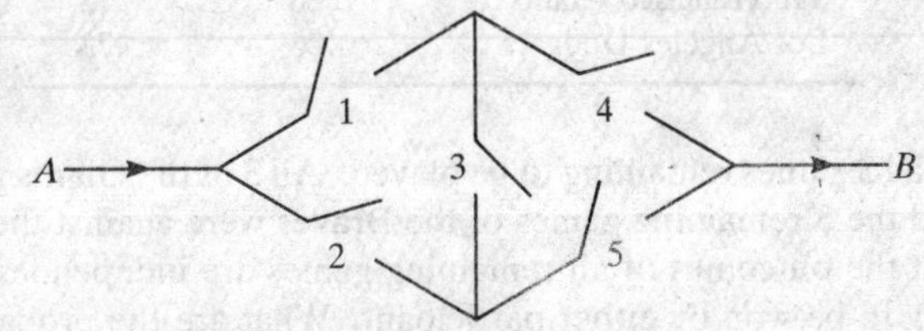

HINT FOR (B): Condition on whether relay 3 closes.

63. An engineering system consisting of n components is said to be a k-out-of-n system ($k \leq n$) if the system functions if and only if at least k of the n components function. Suppose that all components function independently of each other.

(a) If the ith component functions with probability $P_i, i = 1, 2, 3, 4$, compute the probability that a 2-out-of-4 system functions.
(b) Repeat part (a) for a 3-out-of-5 system.
(c) Repeat foı a k-out-of-n system when all the P_i equal p (that is, $P_i = p, i = 1, 2, \ldots, n$).

64. In Problem 62a, find the conditional probability that relays 1 and 2 are both closed given that a current flows from A to B.

65. A certain organism possesses a pair of each of 5 different genes (which we will designate by the first 5 letters of the English alphabet). Each gene appears in 2 forms (which we designate by lowercase and capital letters). The capital letter will be assumed to be the dominant gene in the sense that if an organism possesses the gene pair xX, then it will outwardly have the appearance of the X gene. For instance, if X stands for brown eyes and x for blue eyes, then an individual having either gene pair XX or xX will have brown eyes, whereas one having gene pair xx will have blue eyes. The characteristic appearance of an organism is called its phenotype, whereas its genetic constitution is called its genotype. (Thus 2 organisms with respective genotypes aA, bB, cc, dD, ee and AA, BB, cc, DD, ee would have different genotypes but the same phenotype.) In a mating between 2 organisms each one contributes, at random, one of its gene pairs of each type. The 5 contributions of an organism (one of each of the 5 types) are assumed to be independent and are also independent of the contributions of its mate. In a mating between organisms having genotypes aA, bB, cC, dD, eE and aa, bB, cc, Dd, ee what is the probability that the progeny will (i) phenotypically and (ii) genotypically resemble
(a) the first parent; **(b)** the second parent;
(c) either parent; **(d)** neither parent?

66. There is a 50–50 chance that the queen carries the gene for hemophilia. If she is a carrier, then each prince has a 50–50 chance of having hemophilia. If the queen has had three princes without the disease, what is the probability the queen is a carrier? If there is a fourth prince, what is the probability that he will have hemophilia?

67. On the morning of September 31, 1982, the won–lost records of the three leading baseball teams in the western division of the National League of the United States were as follows:

Team	Won	Lost
Atlanta Braves	87	72
San Francisco Giants	86	73
Los Angeles Dodgers	86	73

Each team had 3 games remaining to be played. All 3 of the Giants games were with the Dodgers, and the 3 remaining games of the Braves were against the San Diego Padres. Suppose that the outcomes of all remaining games are independent and each game is equally likely to be won by either participant. What are the probabilities that each of the teams wins the division? If two teams tie for first place, they have a playoff game, which each team has an equal chance of winning.

68. A town council of 7 members contains a steering committee of size 3. New ideas for legislation go first to the steering committee and then on to the council as a whole if at least 2 of the 3 committee members approve the legislation. Once at the full council, the legislation requires a majority vote (of at least 4) to pass. Consider now a new piece of legislation and suppose that each town council member will approve it, independently, with probability p. What is the probability that a given steering committee member's vote is decisive in the sense that if that person's vote were reversed, then the final fate of the legislation would be reversed? What is the corresponding probability for a given council member not on the steering committee?

69. Suppose that each child born to a couple is equally likely to be a boy or a girl independent of the sex distribution of the other children in the family. For a couple having 5 children, compute the probabilities of the following events:
(a) All children are of the same sex.
(b) The 3 eldest are boys and the others girls.
(c) Exactly 3 are boys.
(d) The 2 oldest are girls.
(e) There is at least 1 girl.

70. A and B alternate rolling a pair of dice, stopping either when A rolls the sum 9 or when B rolls the sum 6. Assuming that A rolls first, find the probability that the final roll is made by A.

71. In a certain village it is traditional for the eldest son and his wife to be responsible for taking care of his parents as they age. In recent years, however, the women of this village, not wanting this responsibility, have not looked favorably upon marrying an eldest son.
(a) If every family in the village has two children, what proportion of all sons are eldest sons?
(b) If every family in the village has three children, what proportion of all sons are eldest sons?
Assume that each child is, independently, equally likely to be either a boy or a girl.

72. Suppose that E and F are mutually exclusive events of an experiment. Show that if independent trials of this experiment are performed, then E will occur before F with probability $P(E)/[P(E) + P(F)]$.

73. A and B play a series of games. Each game is independently won by A with probability p and by B with probability $1 - p$. They stop when the total number of wins of one of the players is two greater than that of the other player. The player with the greater number of total wins is declared the match winner.
(a) Find the probability that a total of 4 games are played.
(b) Find the probability that A is the match winner.

74. In successive rolls of a pair of fair dice, what is the probability of getting 2 sevens before 6 even numbers?

75. Players are of equal skill, and in a contest the probability is $\frac{1}{2}$ that a specified one of the two contestants will be the victor. A group of 2^n players are paired off against each other at random. The 2^{n-1} winners are again paired off randomly, and so on, until a single winner remains. Consider two specified contestants, A and B, and define the events $A_i, i \leq n, E$ by

A_i: A plays in exactly i contests:
E: A and B ever play each other.

(a) Find $P(A_i), i = 1, \ldots, n$. **(b)** Find $P(E)$.
(c) Let $P_n = P(E)$. Show that

$$P_n = \frac{1}{2^n - 1} + \frac{2^n - 2}{2^n - 1}\left(\frac{1}{2}\right)^2 P_{n-1}$$

and use this to check your answer obtained in part (b).

HINT: Find $P(E)$ by conditioning on which of the events $A_i, i = 1, \ldots, n$ occur. In simplifying your answer use the algebraic identity

$$\sum_{i=1}^{n-1} i x^{i-1} = \frac{1 - nx^{n-1} + (n-1)x^n}{(1-x)^2}$$

For another approach for solving this problem, note that there are a total of $2^n - 1$ games played.

(d) Explain why a total of $2^n - 1$ games are played.
Number these games and let B_i denote the event that A and B play each other in game $i, i = 1, \ldots, 2^n - 1$.
(e) What is $P(B_i)$?
(f) Use part (e) to find $P(E)$.

76. A stock market investor owns shares in a stock whose present value is 25. She has decided that she must sell her stock if it either goes down to 10 or up to 40. If each change of price is either up 1 point with probability .55 or down 1 point with probability .45, and the successive changes are independent, what is the probability that the investor retires a winner?

77. A and B flip coins. A starts and continues flipping until a tail occurs. At this point B starts flipping and continues until there is a tail, then A takes over, and so on. Let P_1 be the probability of the coin's landing heads when A flips, and P_2 when B flips. The winner of the game is the first one to get
(a) 2 heads in a row; **(b)** a total of 2 heads;
(c) 3 heads in a row; **(d)** a total of 3 heads.
In each case, find the probability that A wins.

78. Die A has 4 red and 2 white faces, whereas die B has 2 red and 4 white faces. A fair coin is flipped once. If it lands on heads, the game continues with die A; if it lands tails, then die B is to be used.
(a) Show that the probability of red at any throw is $\frac{1}{2}$.
(b) If the first two throws result in red, what is the probability of red at the third throw?
(c) If red turns up at the first two throws, what is the probability that it is die A that is being used?

79. An urn contains 12 balls, of which 4 are white. Three players—A, B, C—successively draw from the urn, A first, then B, then C, then A, and so on. The winner is the first one to draw a white ball. Find the win probabilities for each player if
(a) each ball is replaced after it is drawn;
(b) the withdrawn balls are not replaced.

80. Repeat Problem 79 when each of the 3 players selects from his own urn. That is, suppose that there are 3 different urns of 12 balls with 4 white in each.

81. Let $S = \{1, 2, \ldots, n\}$ and suppose that A and B are, independently, equally likely to be any of the 2^n subsets (including the null set and S itself) of S.
(a) Show that

$$P\{A \subset B\} = \left(\tfrac{3}{4}\right)^n$$

HINT: Let $N(B)$ denote the number of elements in B. Use

$$P\{A \subset B\} = \sum_{i=0}^{n} P\{A \subset B | N(B) = i\} P\{N(B) = i\}$$

(b) Show that $P\{AB = \emptyset\} = \left(\frac{3}{4}\right)^n$.

82. In Example 5d, what is the conditional probability that the ith coin was selected given that the first n trials all result in heads?

83. In Laplace's rule of succession, Example 5d, are the outcomes of the successive flips independent? Explain.

84. A person tried by a 3-judge panel is declared guilty if at least 2 judges cast votes of guilty. Suppose that when the defendant is, in fact, guilty, each judge will independently vote guilty with probability .7, whereas when the defendant is, in fact, innocent, this probabil-

ity drops to .2. If 70 percent of defendants are guilty, compute the conditional probability that judge number 3 votes guilty given that
(a) judges 1 and 2 vote guilty;
(b) judges 1 and 2 cast 1 guilty and 1 not guilty vote;
(c) judges 1 and 2 both cast not guilty votes.
Let $E_i, i = 1, 2, 3$ denote the event that judge i casts a guilty vote. Are these events independent. Are they conditionally independent? Explain.

85. Suppose that n independent trials, each of which results in any of the outcomes 0, 1, or 2 with respective probabilities, p_0, p_1, and p_2, $\sum_{i=0}^{2} p_i = 1$, are performed. Find the probability that outcomes 1 and 2 both occur at least once.

THEORETICAL EXERCISES

1. If $P(A) > 0$ show that

$$P(AB|A) \geq P(AB|A \cup B)$$

2. If $A \subset B$, express the following probabilities as simply as possible:

$$P(A|B), \qquad P(A|B^c), \qquad P(B|A), \qquad P(B|A^c)$$

3. Consider a school community of m families, with n_i of them having i children, $i = 1, \ldots, k$, $\sum_{i=1}^{k} n_i = m$. Consider the following two methods for choosing a child.

1. Choose one of the m families at random and then randomly choose a child from that family.
2. Choose one of the $\sum_{i=1}^{k} in_i$ children at random.

Show that method 1 is more likely than method 2 to result in the choice of a first-born child.

HINT: In solving this problem you will need to show that

$$\sum_{i=1}^{k} in_i \sum_{j=1}^{k} \frac{n_j}{j} \geq \sum_{i=1}^{k} n_i \sum_{j=1}^{k} n_j$$

To show the preceding, multiply the sums and show that for all pairs i,j, the coefficient of the term $n_i n_j$ is greater in the expression on the left than in the one on the right.

4. A ball is in any one of n boxes. It is in the ith box with probability P_i. If the ball is in box i, a search of that box will uncover it with probability α_i. Show that the conditional probability that the ball is in box j, given that a search of box i did not uncover it, is

$$\frac{P_j}{1 - \alpha_i P_i} \quad \text{if } j \neq i$$

$$\frac{(1 - \alpha_i)P_i}{1 - \alpha_i P_i} \quad \text{if } j = i$$

5. An event F is said to carry negative information about an event E, and we write $F \searrow E$ if

$$P(E|F) \leq P(E)$$

Prove or give counterexamples to the following assertions:
(a) If $F \searrow E$, then $E \searrow F$. **(b)** If $F \searrow E$ and $E \searrow G$, then $F \searrow G$.
(c) If $F \searrow E$ and $G \searrow E$, then $FG \searrow E$.
Repeat parts (a), (b), and (c) when $\searrow$ is replaced by $\nearrow$, where we say that F carries positive information about E, written $F \nearrow E$, when $P(E|F) \geq P(E)$.

6. Prove that if $E_1, E_2, \ldots, E_n$ are independent events, then

$$P(E_1 \cup E_2 \cup \cdots \cup E_n) = 1 - \prod_{i=1}^{n} [1 - P(E_i)]$$

7. (a) An urn contains n white and m black balls. The balls are withdrawn one at a time until only those of the same color are left. Show that with probability $n/(n + m)$ they are all white.

HINT: Imagine that the experiment continues until all the balls are removed and consider the last ball withdrawn.

(b) A pond contains 3 distinct species of fish, which we will call the Red, Blue, and Green fish. There are r Red, b Blue, and g Green fish. Suppose that the fish are removed from the pond in a random order (that is, each selection is equally likely to be any of the remaining fish). What is the probability that the Red fish are the first species to become extinct in the pond?

HINT: Write $P\{R\} = P\{RBG\} + P\{RGB\}$, and compute the probabilities on the right by first conditioning on the last species to be removed.

8. Consider two independent tosses of a fair coin. Let A be the event that the first toss lands heads, let B be the event that the second toss lands heads, and let C be the event that both land on the same side. Show that the events A, B, C are pairwise independent—that is, A and B are independent, A and C are independent, and B and C are independent—but not independent.

9. Consider a collection of n individuals. Assume that each person's birthday is equally likely to be any of the 365 days of the year and also that the birthdays are independent. Let $A_{i,j}$ denote the event that persons i and j have the same birthday, $i \neq j$. Show that these events are pairwise independent. That is, $A_{i,j}$ and $A_{r,s}$ are independent but the $\binom{n}{2}$ events $A_{i,j}, i \neq j$ are not independent.

10. A total of n independent tosses of a coin that lands on heads with probability p are made. How large need n be so that the probability of obtaining at least one head is at least $\frac{1}{2}$?

11. If $0 \leq a_i \leq 1, i = 1, 2, \ldots$, show that

$$\sum_{i=1}^{\infty} \left[a_i \prod_{j=1}^{i-1} (1 - a_j) \right] + \prod_{i=1}^{\infty} (1 - a_i) = 1$$

HINT: Suppose that an infinite number of coins are to be flipped. Let a_i be the probability that the ith coin lands heads, and consider when the first head occurs.

12. The probability of getting a head on a single toss of a coin is p. Consider that A starts and continues to flip the coin until a tail shows up, at which point B starts flipping. Then B continues to flip until a tail comes up, at which point A takes over, and so on. Let $P_{n,m}$ denote the probability that A accumulates a total of n heads before B accumulates m. Show that

$$P_{n,m} = pP_{n-1,m} + (1 - p)(1 - P_{m,n})$$

***13.** Suppose that you are gambling against an infinitely rich adversary and at each stage you either win or lose 1 unit with respective probabilities p and $1 - p$. Show that the probability that you eventually go broke is

$$\begin{array}{ll} 1 & \text{if } p \le \frac{1}{2} \\ (q/p)^i & \text{if } p > \frac{1}{2} \end{array}$$

where $q = 1 - p$ and where i is your initial fortune.

14. Independent trials that result in a success with probability p are successively performed until a total of r successes is obtained. Show that the probability that exactly n trials are required is

$$\binom{n-1}{r-1} p^r (1-p)^{n-r}$$

Use this result to solve the problem of the points (Example 4i).

HINT: In order for it to take n trials to obtain r successes, how many successes must occur in the first $n - 1$ trials?

15. Independent trials that result in a success with probability p and a failure with probability $1 - p$ are called Bernoulli trials. Let P_n denote the probability that n Bernoulli trials result in an even number of successes (0 being considered an even number). Show that

$$P_n = p(1 - P_{n-1}) + (1 - p)P_{n-1} \qquad n \ge 1$$

and use this to prove (by induction) that

$$P_n = \frac{1 + (1 - 2p)^n}{2}$$

16. Let Q_n denote the probability that in n tosses of a fair coin no run of 3 consecutive heads appears. Show that

$$Q_n = \tfrac{1}{2}Q_{n-1} + \tfrac{1}{4}Q_{n-2} + \tfrac{1}{8}Q_{n-3}$$
$$Q_0 = Q_1 = Q_2 = 1$$

Find Q_8.

HINT: Condition on the first tail.

17. Consider the gambler's ruin problem with the exception that A and B agree to play no more than n games. Let $P_{n,i}$ denote the probability that A winds up with all the money when A starts with i and B with $N - i$. Derive an equation for $P_{n,i}$ in terms of $P_{n-1,i+1}$ and $P_{n-1,i-1}$ and compute $P_{7,3}$, $N = 5$.

18. Consider two urns, each containing both white and black balls. The probabilities of drawing white balls from the first and second urns are, respectively, p and p'. Balls are sequentially selected with replacement as follows: With probability α a ball is initially chosen from the first urn, and with probability $1 - \alpha$ it is chosen from the second urn. The subsequent selections are then made according to the rule that whenever a white ball is drawn (and replaced), the next ball is drawn from the same urn; but when a black ball is drawn, the next ball is taken from the other urn. Let α_n denote the probability that the nth ball is chosen from the first urn. Show that

$$\alpha_{n+1} = \alpha_n(p + p' - 1) + 1 - p' \qquad n \ge 1$$

and use this to prove that

$$\alpha_n = \frac{1 - p'}{2 - p - p'} + \left(\alpha - \frac{1 - p'}{2 - p - p'}\right)(p + p' - 1)^{n-1}$$

Let P_n denote the probability that the nth ball selected is white. Find P_n. Also compute $\lim_{n\to\infty} \alpha_n$ and $\lim_{n\to\infty} P_n$.

19. *The Ballot Problem.* In an election, candidate A receives n votes and candidate B receives m votes, where $n > m$. Assuming that all of the $(n + m)!/n!\,m!$ orderings of the votes are equally likely, let $P_{n,m}$ denote the probability that A is always ahead in the counting of the votes.

(a) Compute $P_{2,1}, P_{3,1}, P_{3,2}, P_{4,1}, P_{4,2}, P_{4,3}$.

(b) Find $P_{n,1}, P_{n,2}$.

(c) Based on your results in parts (a) and (b), conjecture the value of $P_{n,m}$.

(d) Derive a recursion for $P_{n,m}$ in terms of $P_{n-1,m}$ and $P_{n,m-1}$ by conditioning on who receives the last vote.

(e) Use part (d) to verify your conjecture in part (c) by an induction proof on $n + m$.

20. As a simplified model for weather forecasting, suppose that the weather (either wet or dry) tomorrow will be the same as the weather today with probability p. If the weather is dry on January 1, show that P_n, the probability that it will be dry n days later, satisfies

$$P_n = (2p - 1)P_{n-1} + (1 - p) \qquad n \geq 1$$
$$P_0 = 1$$

Prove that

$$P_n = \tfrac{1}{2} + \tfrac{1}{2}(2p - 1)^n \qquad n \geq 0$$

21. A bag contains a white and b black balls. Balls are chosen from the bag according to the following method:

1. A ball is chosen at random and is discarded.
2. A second ball is then chosen. If its color is different from that of the preceding ball, it is replaced in the bag, and the process is repeated from the beginning. If its color is the same, it is discarded, and we start from step 2.

In other words, balls are sampled and discarded until a change in color occurs, at which point the last ball is returned to the urn and the process starts anew. Let $P_{a,b}$ denote the probability that the last ball in the bag is white. Prove that

$$P_{a,b} = \tfrac{1}{2}$$

HINT: Use induction on $k \equiv a + b$.

***22.** A round-robin tournament of n contestants is one in which each of the $\binom{n}{2}$ pairs of contestants play each other exactly once, with the outcome of any play being that one of the contestants wins and the other loses. For a fixed integer k, $k < n$, a question of interest is whether it is possible that the tournament outcome is such that for every set of k players there is a player who beat each member of this set. Show that if

$$\binom{n}{k}\left[1 - \left(\tfrac{1}{2}\right)^k\right]^{n-k} < 1$$

then such an outcome is possible.

HINT: Suppose that the results of the games are independent and that each game is equally likely to be won by either contestant. Number the $\binom{n}{k}$ sets of k contestants, and let B_i denote the event that no contestant beat all of the k players in the ith set. Then use Boole's inequality to bound $P\left(\bigcup_i B_i\right)$.

23. Prove directly that

$$P(E|F) = P(E|FG)P(G|F) + P(E|FG^c)P(G^c|F)$$

24. Prove the equivalence of Equations (5.11) and (5.12).

25. Extend the definition of conditional independence to more than 2 events.

26. Prove or give a counterexample. If E_1 and E_2 are independent, then they are conditionally independent given F.

27. In Laplace's rule of succession (Example 5d) show that if the first n flips all result in heads, then the conditional probability that the next m flips also result in all heads is $(n + 1)/(n + m + 1)$.

28. In Laplace's rule of succession, suppose that the first n flips resulted in r heads and $n - r$ tails. Show that the probability that the $(n + 1)$st flip turns up heads is $(r + 1)/(n + 2)$. To do so, you will have to prove and use the identity

$$\int_0^1 y^n(1 - y)^m\,dy = \frac{n!\,m!}{(n + m + 1)!}$$

HINT: To prove the identity, let $C(n, m) = \int_0^1 y^n(1 - y)^m\,dy$. Integrating by parts yields that

$$C(n, m) = \frac{m}{n + 1}C(n + 1, m - 1)$$

Starting with $C(n, 0) = 1/(n + 1)$, prove the identity by induction on m.

29. Suppose that a nonmathematical but philosophically minded friend of yours claims that Laplace's rule of succession must be incorrect because it can lead to ridiculous conclusions. "For instance," says he, "if a boy is 10 years old, the rule states that having lived 10 years, the boy has probability $\frac{11}{12}$ of living another year. On the other hand, if the boy has an 80-year-old grandfather, then by Laplace's rule the grandfather has probability $\frac{81}{82}$ of surviving another year. However, this is ridiculous. Clearly, the boy is more likely to survive an additional year than is the grandfather." How would you answer your friend?

SELF-TEST PROBLEMS AND EXERCISES

1. In a game of bridge, West has no aces. What is the probability of his partner's having (a) no aces, (b) 2 or more aces? (c) What would the probabilities be if West had exactly 1 ace?

2. The probability that a new car battery functions for over 10,000 miles is .8, the probability that it functions for over 20,000 miles is .4, and the probability that it functions for over 30,000 miles is .1. If a new car battery is still working after 10,000 miles, what is the probability that

(a) its total life will exceed 20,000 miles;

(b) its additional life will exceed 20,000 miles?

3. How can 20 balls, 10 white and 10 black, be put into two urns so as to maximize the probability of drawing a white ball if an urn is selected at random and a ball is drawn at random from it?

4. Urn A contains 2 white balls and 1 black ball, whereas urn B contains 1 white ball and 5 black balls. A ball is drawn at random from urn A and placed in urn B. A ball is then drawn from urn B. It happens to be white. What is the probability that the ball transferred was white?

5. An urn contains b black balls and r red balls. One of the balls is drawn at random, but when it is put back in the urn, c additional balls of the same color are put in with it. Now, suppose that we draw another ball. Show that the probability that the first ball was black, given that the second ball drawn was red, is $b/(b + r + c)$.

6. A friend randomly chooses two cards, without replacement, from an ordinary deck of 52 playing cards. In each of the following situations, determine the conditional probability that both cards are aces.

(a) You ask your friend if one of the cards is the ace of spades and your friend answers in the affirmative.

(b) You ask your friend if the first card selected is an ace and your friend answers in the affirmative.

(c) You ask your friend if the second card selected is an ace and your friend answers in the affirmative.

(d) You ask your friend if either of the cards selected is an ace and your friend answers in the affirmative.

7. Show that

$$\frac{P(H|E)}{P(G|E)} = \frac{P(H)\,P(E|H)}{P(G)\,P(E|G)}$$

Suppose that before observing new evidence the hypothesis H is three times as likely to be true as is the hypothesis G. If the new evidence is twice as likely when G is true than it is when H is true, which hypothesis is more likely after the evidence has been observed?

8. You ask your neighbor to water a sickly plant while you are on vacation. Without water it will die with probability .8; with water it will die with probability .15. You are 90 percent certain that your neighbor will remember to water the plant.

(a) What is the probability that the plant will be alive when you return?

(b) If it is dead, what is the probability your neighbor forgot to water it?

9. In a certain species of rats, black dominates over brown. Suppose that a black rat with two black parents has a brown sibling.

(a) What is the probability that this rat is a pure black rat (as opposed to being a hybrid with one black and one brown gene)?

(b) Suppose that when the black rat is mated with a brown rat, all 5 of their offspring are black. Now, what is the probability that the rat is a pure black rat?

10. (a) In Problem 62b, find the probability that a current flows from A to B by conditioning on whether relay 1 closes.

(b) Find the conditional probability that relay 3 is closed, given that a current flows from A to B.

11. For the k-out-of-n system described in Problem 63, assume that each component independently works with probability $\frac{1}{2}$. Find the conditional probability that component 1 is working given that the system works, when

(a) $k = 1, n = 2$; **(b)** $k = 2, n = 3$.

12. Mr. Jones has devised a gambling system for winning at roulette. When he bets, he bets on red, and places a bet only when the 10 previous spins of the roulette have landed on a black number. He reasons that his chance of winning is quite large because the probability of 11 consecutive spins resulting in black is quite small. What do you think of this system?

13. Three players simultaneously toss coins. The coin tossed by A (B) $[C]$ turns up heads with probability $P_1(P_2)[P_3]$. If one person gets an outcome different from those of the other two, then he is the odd man out. If there is no odd man out, the players flip again and continue to do so until they get an odd man out. What is the probability that A will be the odd man?

14. Suppose that there are n possible outcomes of a trial, with outcome i resulting with probability p_i, $i = 1, \ldots, n$, $\sum_{i=1}^{n} p_i = 1$. If two independent trials are observed, what is the probability that the result of the second trial is larger than that of the first?

15. If A flips $n + 1$ and B flips n fair coins, show that the probability that A gets more heads than B is $\frac{1}{2}$.

HINT: Condition on which player has more heads after each has flipped n coins. (There are three possibilities.)

16. Prove or give counterexamples to the following statements:
(a) If E is independent of F and E is independent of G, then E is independent of $F \cup G$.
(b) If E is independent of F, and E is independent of G, and $FG = \emptyset$, then E is independent of $F \cup G$.
(c) If E is independent of F, and F is independent of G, and E is independent of FG, then G is independent of EF.

17. Let A and B be events having positive probability. State whether each of the following statements is (i) necessarily true, (ii) necessarily false, or (iii) possibly true.
(a) If A and B are mutually exclusive, then they are independent.
(b) If A and B are independent, then they are mutually exclusive.
(c) $P(A) = P(B) = .6$, and A and B are mutually exclusive.
(d) $P(A) = P(B) = .6$, and A and B are independent.

18. Rank the following from most likely to least likely to occur.
1. A fair coin lands on heads.
2. Three independent trials, each of which is a success with probability .8, all result in successes.
3. Seven independent trials, each of which is a success with probability .9, all result in successes.

19. There are two local factories that produce radios. Each radio produced at factory A is defective with probability .05, whereas each one produced at factory B is defective with probability .01. Suppose you purchase two radios that were produced at the same factory, which is equally likely to have been either factory A or factory B. If the first radio that you check is defective, what is the conditional probability that the other one is also defective?

20. Show that if $P(A|B) = 1$, then $P(B^c|A^c) = 1$.

21. An urn initially contains 1 red and 1 blue ball. At each stage a ball is randomly withdrawn and replaced by two other balls of the same color. (For instance, if the red ball is initially chosen, then there would be 2 red and 1 blue ball in the urn when the next selection occurs.) Show by mathematical induction that the probability that there are exactly i red balls in the urn after n stages have been completed is $\dfrac{1}{n+1}$, $1 \le i \le n + 1$.

22. A total of $2n$ cards, of which 2 are aces, are to be randomly divided among two players, with each player receiving n cards. Each player is then to declare, in sequence, whether he or she has received any aces. What is the conditional probability that the second player has no aces given that the first player declares in the affirmative, when (a) $n = 2$, (b) $n = 10$, (c) $n = 100$? What does the probability converge to as n goes to infininty? Why?

CHAPTER 4

Random Variables

4.1 RANDOM VARIABLES

It is frequently the case when an experiment is performed that we are mainly interested in some function of the outcome as opposed to the actual outcome itself. For instance, in tossing dice we are often interested in the sum of the two dice and are not really concerned about the separate values of each die. That is, we may be interested in knowing that the sum is 7 and not be concerned over whether the actual outcome was $(1,6)$ or $(2,5)$ or $(3,4)$ or $(4,3)$ or $(5,2)$ or $(6,1)$. Also, in coin flipping, we may be interested in the total number of heads that occur and not care at all about the actual head–tail sequence that results. These quantities of interest, or more formally, these real-valued functions defined on the sample space, are known as *random variables*.

Because the value of a random variable is determined by the outcome of the experiment, we may assign probabilities to the possible values of the random variable.

Example 1a. Suppose that our experiment consists of tossing 3 fair coins. If we let Y denote the number of heads appearing, then Y is a random variable taking on one of the values 0, 1, 2, 3 with respective probabilities

$$P\{Y = 0\} = P\{(T,T,T)\} = \tfrac{1}{8}$$
$$P\{Y = 1\} = P\{(T,T,H),(T,H,T),(H,T,T)\} = \tfrac{3}{8}$$
$$P\{Y = 2\} = P\{(T,H,H),(H,T,H),(H,H,T)\} = \tfrac{3}{8}$$
$$P\{Y = 3\} = P\{(H,H,H)\} = \tfrac{1}{8}$$

Since Y must take on one of the values 0 through 3, we must have

$$1 = P\left(\bigcup_{i=0}^{3}\{Y = i\}\right) = \sum_{i=0}^{3} P\{Y = i\}$$

which, of course, is in accord with the preceding probabilities. ■

Example 1b. Three balls are to be randomly selected without replacement from an urn containing 20 balls numbered 1 through 20. If we bet that at least one of the drawn balls has a number as large as or larger than 17, what is the probability that we win the bet?

Solution Let X denote the largest number selected. Then X is a random variable taking on one of the values $3, 4, \ldots, 20$. Furthermore, if we suppose that each of the $\binom{20}{3}$ possible selections are equally likely to occur, then

$$P\{X = i\} = \frac{\binom{i-1}{2}}{\binom{20}{3}} \qquad i = 3, \ldots, 20 \tag{1.1}$$

Equation (1.1) follows because the number of selections that result in the event $\{X = i\}$ is just the number of selections that result in ball numbered i and two of the balls numbered 1 through $i - 1$ being chosen. As there are clearly $\binom{1}{1}\binom{i-1}{2}$ such selections, we obtain the probabilities expressed in Equation (1.1). From this equation we see that

$$P\{X = 20\} = \frac{\binom{19}{2}}{\binom{20}{3}} = \frac{3}{20} = .150$$

$$P\{X = 19\} = \frac{\binom{18}{2}}{\binom{20}{3}} = \frac{51}{380} \approx .134$$

$$P\{X = 18\} = \frac{\binom{17}{2}}{\binom{20}{3}} = \frac{34}{285} \approx .119$$

$$P\{X = 17\} = \frac{\binom{16}{2}}{\binom{20}{3}} = \frac{2}{19} \approx .105$$

Hence, as the event $\{X \geq 17\}$ is the union of the disjoint events $\{X = i\}$, $i = 17, 18, 19, 20$, it follows that the probability of our winning the bet is given by

$$P\{X \geq 17\} \approx .105 + .119 + .134 + .150 = .508$$

■

Example 1c. Independent trials, consisting of the flipping of a coin having probability p of coming up heads, are continually performed until either a head occurs or a total of n flips is made. If we let X denote the number of times the coin is flipped, then X is a random variable taking on one of the values $1, 2, 3, \ldots, n$ with respective probabilities

$$\begin{aligned}
P\{X = 1\} &= P\{H\} = p \\
P\{X = 2\} &= P\{(T, H)\} = (1 - p)p \\
P\{X = 3\} &= P\{(T, T, H)\} = (1 - p)^2 p \\
&\vdots \\
P\{X = n - 1\} &= P\{(\underbrace{T, T, \ldots, T}_{n-2}, H)\} = (1 - p)^{n-2} p \\
P\{X = n\} &= P\{(\underbrace{T, T, \ldots, T}_{n-1}, T), (\underbrace{T, T, \ldots, T}_{n-1}, H)\} = (1 - p)^{n-1}
\end{aligned}$$

As a check, note that

$$\begin{aligned}
P\left(\bigcup_{i=1}^{n} \{X = i\}\right) &= \sum_{i=1}^{n} P\{X = i\} \\
&= \sum_{i=1}^{n-1} p(1 - p)^{i-1} + (1 - p)^{n-1} \\
&= p\left[\frac{1 - (1 - p)^{n-1}}{1 - (1 - p)}\right] + (1 - p)^{n-1} \\
&= 1 - (1 - p)^{n-1} + (1 - p)^{n-1} \\
&= 1
\end{aligned}$$

■

Example 1d. Three balls are randomly chosen from an urn containing 3 white, 3 red, and 5 black balls. Suppose that we win \$1 for each white ball selected and lose \$1 for each red selected. If we let X denote our total winnings from the experiment, then X is a random variable taking on the possible values 0, ±1, ±2, ±3 with respective probabilities

$$P\{X = 0\} = \frac{\binom{5}{3} + \binom{3}{1}\binom{3}{1}\binom{5}{1}}{\binom{11}{3}} = \frac{55}{165}$$

$$P\{X = 1\} = P\{X = -1\} = \frac{\binom{3}{1}\binom{5}{2} + \binom{3}{2}\binom{3}{1}}{\binom{11}{3}} = \frac{39}{165}$$

$$P\{X = 2\} = P\{X = -2\} = \frac{\binom{3}{2}\binom{5}{1}}{\binom{11}{3}} = \frac{15}{165}$$

$$P\{X = 3\} = P\{X = -3\} = \frac{\binom{3}{3}}{\binom{11}{3}} = \frac{1}{165}$$

These probabilities are obtained, for instance, by noting that in order for X to equal 0, either all 3 balls selected must be black or 1 ball of each color must be selected. Similarly, the event $\{X = 1\}$ occurs either if 1 white and 2 black balls are selected or if 2 white and 1 red is selected. As a check we note that

$$\sum_{i=0}^{3} P\{X = i\} + \sum_{i=1}^{3} P\{X = -i\} = \frac{55 + 39 + 15 + 1 + 39 + 15 + 1}{165} = 1$$

The probability that we win money is given by

$$\sum_{i=1}^{3} P\{X = i\} = \tfrac{55}{165} = \tfrac{1}{3}$$ ■

Example 1e. Suppose that there are N distinct types of coupons and each time one obtains a coupon it is, independent of prior selections, equally likely to be any one of the N types. One random variable of interest is T, the number of coupons that needs to be collected until one obtains a complete set of at least one of each type. Rather than derive $P\{T = n\}$ directly, let us start by considering the probability that T is greater than n. To do so, fix n and define the events A_1, $A_2, \ldots, A_N$ as follows: A_j is the event that no type j coupon is contained among the first n, $j = 1, \ldots, N$. Hence

$$\begin{aligned} P\{T > n\} &= P\left(\bigcup_{j=1}^{N} A_j\right) \\ &= \sum_j P(A_j) - \sum\sum_{j_1<j_2} P(A_{j_1}A_{j_2}) + \cdots \\ &\quad + (-1)^{k+1} \sum\sum_{j_1<j_2<\cdots<j_k}\sum P(A_{j_1}A_{j_2}\cdots A_{j_k}) \cdots \\ &\quad + (-1)^{N+1} P(A_1A_2\cdots A_N) \end{aligned}$$

Now A_j will occur if each of the n coupons is not of type j. As each of the coupons will not be of type j with probability $(N-1)/N$, we have, by the assumed independence of the types of successive coupons, that

$$P(A_j) = \left(\frac{N-1}{N}\right)^n$$

Also the event $A_{j_1}A_{j_2}$ will occur if none of the first n is of either type j_1 or j_2. Thus, again using independence, we see that

$$P(A_{j_1}A_{j_2}) = \left(\frac{N-2}{N}\right)^n$$

The same reasoning gives that

$$P(A_{j_1}A_{j_2}\cdots A_{j_k}) = \left(\frac{N-k}{N}\right)^n$$

and we see that for $n > 0$,

$$\begin{aligned} P\{T > n\} &= N\left(\frac{N-1}{N}\right)^n - \binom{N}{2}\left(\frac{N-2}{N}\right)^n + \binom{N}{3}\left(\frac{N-3}{N}\right)^n - \cdots \\ &\quad + (-1)^N\binom{N}{N-1}\left(\frac{1}{N}\right)^n \\ &= \sum_{i=1}^{N-1}\binom{N}{i}\left(\frac{N-i}{N}\right)^n(-1)^{i+1} \qquad (1.2) \end{aligned}$$

The probability that T equals n can now be obtained from the preceding by using

$$P\{T > n-1\} = P\{T = n\} + P\{T > n\}$$

or, equivalently,

$$P\{T = n\} = P\{T > n-1\} - P\{T > n\}$$

Another random variable of interest is the number of distinct types of coupons that are contained in the first n selections—call this random variable D_n. To compute $P\{D_n = k\}$, let us start by fixing attention on a particular set of k distinct types, and let us then determine the probability that this set constitutes the set of distinct types obtained in the first n selections. Now, in order for this to be the situation, it is necessary and sufficient that of the first n coupons obtained

A: each is one of these k types.
B: each of these k types is represented.

Now each coupon selected will be one of the k types with probability k/N, and so the probability that A will be valid is $(k/N)^n$. Also, given that a coupon is of one of the k types under consideration, it is easy to see that it is equally likely to be of any one of these k types. Hence the conditional probability of B given that A occurs is the same as the probability that a set of n coupons, each equally likely to be any of k possible types, contains a complete set of all k types. But this is just the probability that the number needed to amass a complete set, when choosing among k types, is less than or equal to n and is thus obtainable from Equation (1.2) with k replacing N. Hence we see that

$$P(A) = \left(\frac{k}{N}\right)^n$$

$$P(B|A) = 1 - \sum_{i=1}^{k-1} \binom{k}{i} \left(\frac{k-i}{k}\right)^n (-1)^{i+1}$$

Finally, as there are $\binom{N}{k}$ possible choices for the set of k types, we arrive at

$$P\{D_n = k\} = \binom{N}{k} P(AB)$$
$$= \binom{N}{k} \left(\frac{k}{N}\right)^n \left[1 - \sum_{i=1}^{k-1} \binom{k}{i} \left(\frac{k-i}{k}\right)^n (-1)^{i+1}\right]$$

REMARK. Since one must collect at least N coupons to obtain a compete set, it follows that $P\{T > n\} = 1$ if $n < N$. Thus, from Equation (1.2) we obtain the interesting combinatorial identity that for integers $1 \le n < N$

$$\sum_{i=1}^{N-1} \binom{N}{i} \left(\frac{N-i}{N}\right)^n (-1)^{i+1} = 1$$

which can be written as

$$\sum_{i=0}^{N-1} \binom{N}{i} \left(\frac{N-i}{N}\right)^n (-1)^{i+1} = 0$$

or, upon multiplying by $(-1)^N N^n$ and letting $j = N - i$,

$$\sum_{j=1}^{N} \binom{N}{j} j^n (-1)^{j-1} = 0 \qquad 1 \le n < N$$

■

For a random variable X, the function F defined by

$$F(x) = P\{X \le x\} \qquad -\infty < x < \infty$$

is called the *cumulative distribution function* or, more simply, the *distribution function* of X. Thus the distribution function specifies, for all real values x, the probability that the random variable is less than or equal to x.

Suppose that $a \le b$. Because the event $\{X \le a\}$ is contained in the event $\{X \le b\}$, it follows that $F(a)$, the probability of the former, is less than or equal to $F(b)$, the probability of the latter. In other words, $F(x)$ is a nondecreasing function of x. Other general properties of the distribution function are given in Section 4.9.

4.2 DISCRETE RANDOM VARIABLES

A random variable that can take on at most a countable number of possible values is said to be discrete. For a discrete random variable X, we define the probability mass function $p(a)$ of X by

$$p(a) = P\{X = a\}$$

The probability mass function $p(a)$ is positive for at most a countable number of values of a. That is, if X must assume one of the values $x_1, x_2, \ldots$, then

$$p(x_i) \geq 0 \qquad i = 1, 2, \ldots$$
$$p(x) = 0 \qquad \text{all other values of } x$$

Since X must take on one of the values x_i, we have

$$\sum_{i=1}^{\infty} p(x_i) = 1$$

It is often instructive to present the probability mass function in a graphical format by plotting $p(x_i)$ on the y-axis against x_i on the x-axis. For instance, if the probability mass function of X is

$$p(0) = \tfrac{1}{4} \qquad p(1) = \tfrac{1}{2} \qquad p(2) = \tfrac{1}{4}$$

we can represent this graphically as shown in Figure 4.1. Similarly, a graph of the probability mass function of the random variable representing the sum when two dice are rolled looks like the one shown in Figure 4.2.

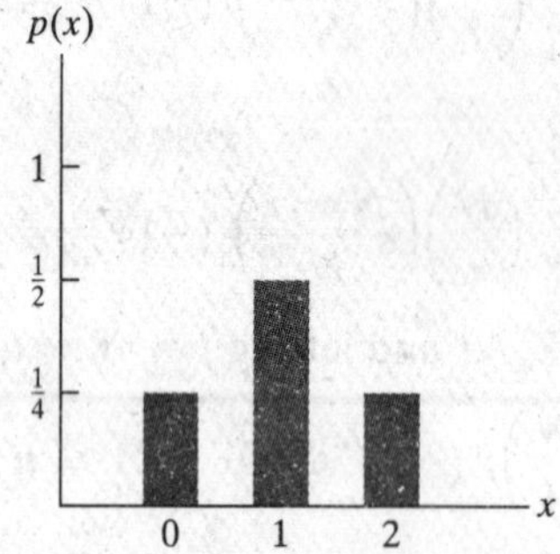

Figure 4.1

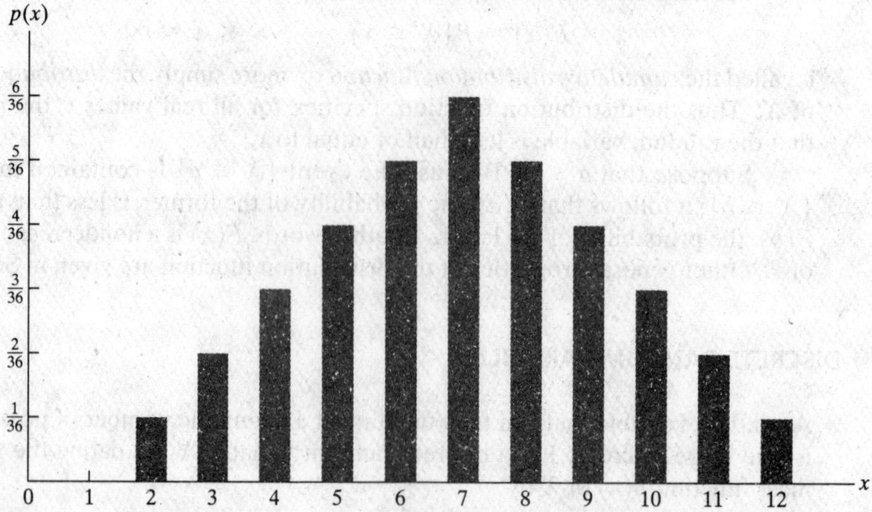

Figure 4.2

Example 2a. The probability mass function of a random variable X is given by $p(i) = c\lambda^i/i!$, $i = 0, 1, 2, \ldots$, where λ is some positive value. Find (a) $P\{X = 0\}$ and (b) $P\{X > 2\}$.

Solution Since $\sum_{i=0}^{\infty} p(i) = 1$, we have that

$$c \sum_{i=0}^{\infty} \frac{\lambda^i}{i!} = 1$$

implying, because $e^x = \sum_{i=0}^{\infty} x^i/i!$, that

$$ce^{\lambda} = 1 \qquad \text{or} \qquad c = e^{-\lambda}$$

Hence

(a) $P\{X = 0\} = e^{-\lambda}\lambda^0/0! = e^{-\lambda}$

(b) $P\{X > 2\} = 1 - P\{X \leq 2\} = 1 - P\{X = 0\} - P\{X = 1\}$
$\qquad - P\{X = 2\}$
$\qquad = 1 - e^{-\lambda} - \lambda e^{-\lambda} - \dfrac{\lambda^2 e^{-\lambda}}{2}$ ■

The cumulative distribution function F can be expressed in terms of $p(a)$ by

$$F(a) = \sum_{\text{all } x \leq a} p(x)$$

If X is a discrete random variable whose possible values are $x_1, x_2, x_3, \ldots$, where $x_1 < x_2 < x_3 < \cdots$, then its distribution function F is a step function. That is, the value of F is constant in the intervals $[x_{i-1}, x_i)$ and then takes a step (or jump) of size $p(x_i)$ at x_i. For instance, if X has a probability mass function given by

$$p(1) = \tfrac{1}{4} \qquad p(2) = \tfrac{1}{2} \qquad p(3) = \tfrac{1}{8} \qquad p(4) = \tfrac{1}{8}$$

then its cumulative distribution function is given by

$$F(a) = \begin{cases} 0 & a < 1 \\ \frac{1}{4} & 1 \leq a < 2 \\ \frac{3}{4} & 2 \leq a < 3 \\ \frac{7}{8} & 3 \leq a < 4 \\ 1 & 4 \leq a \end{cases}$$

This is graphically depicted in Figure 4.3.

The reader should note that the size of the step at any of the values 1, 2, 3, 4 is equal to the probability that X assumes that particular value.

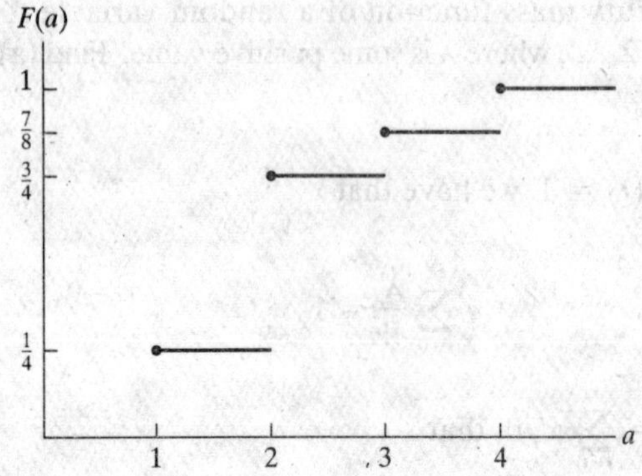

Figure 4.3

4.3 EXPECTED VALUE

One of the most important concepts in probability theory is that of the expectation of a random variable. If X is a discrete random variable having a probability mass function $p(x)$, the *expectation* or the *expected value* of X, denoted by $E[X]$, is defined by

$$E[X] = \sum_{x:p(x)>0} xp(x)$$

In words, the expected value of X is a weighted average of the possible values that X can take on, each value being weighted by the probability that X assumes it. For instance, if the probability mass function of X is given by

$$p(0) = \tfrac{1}{2} = p(1)$$

then

$$E[X] = 0\left(\tfrac{1}{2}\right) + 1\left(\tfrac{1}{2}\right) = \tfrac{1}{2}$$

is just the ordinary average of the two possible values 0 and 1 that X can assume. On the other hand, if

$$p(0) = \tfrac{1}{3} \qquad p(1) = \tfrac{2}{3}$$

then

$$E[X] = 0\left(\tfrac{1}{3}\right) + 1\left(\tfrac{2}{3}\right) = \tfrac{2}{3}$$

is a weighted average of the two possible values 0 and 1, where the value 1 is given twice as much weight as the value 0, since $p(1) = 2p(0)$.

Another motivation of the definition of expectation is provided by the frequency interpretation of probabilities. This interpretation (partially justified by the strong law of large numbers, to be presented in Chapter 8) assumes that if an infinite sequence of independent replications of an experiment is performed, then for any event E, the proportion of time that E occurs will be $P(E)$. Now, consider a random variable X that must take on one of the values $x_1, x_2, \ldots x_n$ with respective probabilities $p(x_1), p(x_2), \ldots, p(x_n)$; and think of X as representing our winnings in a single game of chance. That is, with probability $p(x_i)$ we shall win x_i units $i = 1, 2, \ldots, n$.

Now by the frequency interpretation, if follows that if we continually play this game, then the proportion of time that we win x_i will be $p(x_i)$. As this is true for all i, $i = 1, 2, \ldots, n$, it follows that our average winnings per game will be

$$\sum_{i=1}^{n} x_i p(x_i) = E[X]$$

Example 3a. Find $E[X]$ where X is the outcome when we roll a fair die.

Solution Since $p(1) = p(2) = p(3) = p(4) = p(5) = p(6) = \frac{1}{6}$, we obtain that

$$E[X] = 1\left(\tfrac{1}{6}\right) + 2\left(\tfrac{1}{6}\right) + 3\left(\tfrac{1}{6}\right) + 4\left(\tfrac{1}{6}\right) + 5\left(\tfrac{1}{6}\right) + 6\left(\tfrac{1}{6}\right) = \tfrac{7}{2}$$ ■

Example 3b. We say that I is an indicator variable for the event A if

$$I = \begin{cases} 1 & \text{if } A \text{ occurs} \\ 0 & \text{if } A^c \text{ occurs} \end{cases}$$

Find $E[I]$.

Solution Since $p(1) = P(A)$, $p(0) = 1 - P(A)$, we have that

$$E[I] = P(A)$$

That is, the expected value of the indicator variable for the event A is equal to the probability that A occurs. ■

Example 3c. A contestant on a quiz show is presented with two questions, questions 1 and 2, which he is to attempt to answer in some order chosen by him. If he decides to try question i first, then he will be allowed to go on to question j, $j \neq i$, only if his answer to i is correct. If his initial answer is incorrect, he is not allowed to answer the other question. The contestant is to receive V_i dollars if he answers question i correctly, $i = 1, 2$. Thus, for instance, he will receive $V_1 + V_2$ dollars if both questions are correctly answered. If the probability that he knows the answer to question i is P_i, $i = 1, 2$, which question should he attempt first so as to maximize his expected winnings? Assume that the events E_i, $i = 1, 2$, that he knows the answer to question i, are independent events.

Solution If he attempts question 1 first, then he will win

$$\begin{array}{ll} 0 & \text{with probability } 1 - P_1 \\ V_1 & \text{with probability } P_1(1 - P_2) \\ V_1 + V_2 & \text{with probability } P_1 P_2 \end{array}$$

Hence his expected winnings in this case will be

$$V_1 P_1(1 - P_2) + (V_1 + V_2)P_1 P_2$$

On the other hand, if he attempts question 2 first, his expected winnings will be

$$V_2 P_2(1 - P_1) + (V_1 + V_2)P_1 P_2$$

Therefore, it is better to try question 1 first if

$$V_1P_1(1 - P_2) \geq V_2P_2(1 - P_1)$$

or, equivalently, if

$$\frac{V_1P_1}{1 - P_1} \geq \frac{V_2P_2}{1 - P_2}$$

Thus, for instance, if he is 60 percent certain of answering question 1, worth \$200, correctly and he is 80 percent certain of answering question 2, worth \$100, correctly, then he should attempt question 2 first because

$$400 = \frac{(100)(.8)}{.2} > \frac{(200)(.6)}{.4} = 300$$ ■

Example 3d. A school class of 120 students are driven in 3 buses to a symphonic performance. There are 36 students in one of the buses, 40 in another, and 44 in the third bus. When the buses arrive, one of the 120 students is randomly chosen. Let X denote the number of students on the bus of that randomly chosen student, and find $E[X]$.

Solution Since the randomly chosen student is equally likely to be any of the 120 students, it follows that

$$P\{X = 36\} = \frac{36}{120} \qquad P\{X = 40\} = \frac{40}{120} \qquad P\{X = 44\} = \frac{44}{120}$$

Hence

$$E[X] = 36\left(\frac{3}{10}\right) + 40\left(\frac{1}{3}\right) + 44\left(\frac{11}{30}\right) = \frac{1208}{30} = 40.2667$$

On the other hand, the average number of students on a bus is $120/3 = 40$, showing that the expected number of students on the bus of a randomly chosen student is larger than the average number of students on a bus. This is a general phenomenon and occurs because the more students there are on a bus, then the more likely a randomly chosen student would have been on that bus. As a result, buses with many students are given more weight than those with fewer students (see Self-Test Problem 4). ■

REMARK. The concept of expectation is analogous to the physical concept of the *center of gravity* of a distribution of mass. Consider a discrete random variable X having probability mass function $p(x_i), i \geq 1$. If we now imagine a weightless rod in which weights with mass $p(x_i), i \geq 1$, are located at the points $x_i, i \geq 1$ (see Figure 4.4), then the point at which the rod would be in balance is known as the center of gravity. For those readers acquainted with elementary statics it is now a simple matter to show that this point is at $E[X]$.† ■

†To prove this, we must show that the sum of the torques tending to turn the point around $E[X]$ is equal to 0. That is, we must show that $0 = \sum_i (x_i - E[X])p(x_i)$, which is immediate.

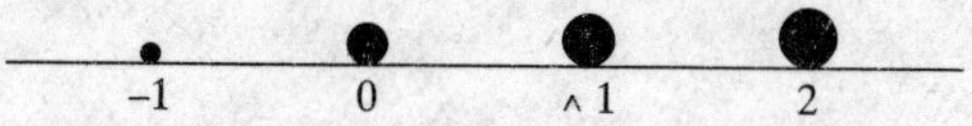

$p(-1) = .10,\quad p(0) = .25,\quad p(1) = .30,\quad p(2) = .35$

∧ = center of gravity = .9

Figure 4.4

4.4 EXPECTATION OF A FUNCTION OF A RANDOM VARIABLE

Suppose that we are given a discrete random variable along with its probability mass function, and that we want to compute the expected value of some function of X, say $g(X)$. How can we accomplish this? One way is as follows: Since $g(X)$ is itself a discrete random variable, it has a probability mass function, which can be determined from the probability mass function of X. Once we have determined the probability mass function of $g(X)$ we can then compute $E[g(X)]$ by using the definition of expected value.

Example 4a. Let X denote a random variable that takes on any of the values $-1, 0, 1$ with respective probabilities

$$P\{X = -1\} = .2 \qquad P\{X = 0\} = .5 \qquad P\{X = 1\} = .3$$

Compute $E[X^2]$.

Solution Letting $Y = X^2$, it follows that the probability mass function of Y is given by

$$P\{Y = 1\} = P\{X = -1\} + P\{X = 1\} = .5$$
$$P\{Y = 0\} = P\{X = 0\} = .5$$

Hence

$$E[X^2] = E[Y] = 1(.5) + 0(.5) = .5$$

The reader should note that

$$5 = E[X^2] \neq (E[X])^2 = .01$$ ■

Although the preceding procedure will always enable us to compute the expected value of any function of X from a knowledge of the probability mass function of X, there is another way of thinking about $E[g(X)]$. That is, noting that $g(X)$ will equal $g(x)$ whenever X is equal to x, it seems reasonable that $E[g(X)]$ should just be a weighted average of the values $g(x)$ with $g(x)$ being weighted by the probability that X is equal to x. That is, the following result is quite intuitive.

Proposition 4.1

If X is a discrete random variable that takes on one of the values $x_i, i \geq 1$, with respective probabilities $p(x_i)$, then for any real-valued function g

$$E[g(X)] = \sum_i g(x_i)p(x_i)$$

Before proving this proposition let us check that it is in accord with the results of Example 5a. Applying it to this example yields

$$\begin{aligned} E\{X^2\} &= (-1)^2(.2) + 0^2(.5) + 1^2(.3) \\ &= 1(.2 + .3) + 0(.5) \\ &= .5 \end{aligned}$$

which is in agreement with the result given in Example 4a.

Proof of Proposition 4.1: The proof of Proposition 4.1 proceeds, as in the preceding verification, by grouping together all the terms in $\sum g(x_i)p(x_i)$ having the same value of $g(x_i)$. Specifically, suppose that $y_j, j \geq 1$ represent the different values of $g(x_i), i \geq 1$. Then, grouping all the $g(x_i)$ having the same value gives that

$$\begin{aligned} \sum_i g(x_i)p(x_i) &= \sum_j \sum_{i:g(x_i)=y_j} g(x_i)p(x_i) \\ &= \sum_j y_j \sum_{i:g(x_i)=y_j} p(x_i) \\ &= \sum_j y_j P\{g(X) = y_j\} \\ &= E[g(X)] \end{aligned}$$

✦

Example 4b. A product, sold seasonally, yields a net profit of b dollars for each unit sold and a net loss of ℓ dollars for each unit left unsold when the season ends. The number of units of the product that are ordered at a specific department store during any season is a random variable having probability mass function $p(i), i \geq 0$. If the store must stock this product in advance, determine the number of units the store should stock so as to maximize its expected profit.

Solution Let X denote the number of units ordered. If s units are stocked, then the profit, call it $P(s)$, can be expressed as

$$\begin{aligned} P(s) &= bX - (s - X)\ell \qquad &\text{if } X \leq s \\ &= sb &\text{if } X > s \end{aligned}$$

Hence the expected profit equals

$$
\begin{aligned}
E[P(s)] &= \sum_{i=0}^{s} [bi - (s - i)\ell]p(i) + \sum_{i=s+1}^{\infty} sbp(i) \\
&= (b + \ell)\sum_{i=0}^{s} ip(i) - s\ell \sum_{i=0}^{s} p(i) + sb\left[1 - \sum_{i=0}^{s} p(i)\right] \\
&= (b + \ell)\sum_{i=0}^{s} ip(i) - (b + \ell)s \sum_{i=0}^{s} p(i) + sb \\
&= sb + (b + \ell)\sum_{i=0}^{s} (i - s)p(i)
\end{aligned}
$$

To determine the optimum value of s, let us investigate what happens to our profit when we increase s by 1 unit. By substitution we see that the expected profit in this case is given by

$$
\begin{aligned}
E[P(s + 1)] &= b(s + 1) + (b + \ell)\sum_{i=0}^{s+1} (i - s - 1)p(i) \\
&= b(s + 1) + (b + \ell)\sum_{i=0}^{s} (i - s - 1)p(i)
\end{aligned}
$$

Therefore,

$$
E[P(s + 1)] - E[P(s)] = b - (b + \ell)\sum_{i=0}^{s} p(i)
$$

Hence stocking $s + 1$ units will be better than stocking s units whenever

$$
\sum_{i=0}^{s} p(i) < \frac{b}{b + \ell} \tag{4.1}
$$

As the left-hand side of Equation (4.1) is increasing in s while the right-hand side is constant, it follows that the inequality will be satisfied for all values of $s \le s^*$ where s^* is the largest value of s satisfying Equation (5.1), see page 212. Since

$$
E[P(0)] < \cdots < E[P(s^*)] < E[P(s^* + 1)] > E[P(s^* + 2)] > \cdots
$$

it follows that stocking $s^* + 1$ items will lead to a maximum expected profit. ■

Example 4c. ***Utility*** Suppose that you must choose one of two possible actions, each of which can result in any of n consequences, denoted as $C_1, \dots, C_n$. Suppose that if the first action is chosen, then consequence C_i will result with probability $p_i, i = 1, \dots, n$; whereas if the second action is chosen, then consequence C_i will result with probability $q_i, i = 1, \dots, n$, where $\sum_{i=1}^{n} p_i = \sum_{i=1}^{n} q_i = 1$. The following approach can be used to determine which action to choose. It starts by assigning numerical values to the different consequences in the following manner. First, identify the least and the most desirable consequence, call them c and C, respectively; give the consequence c the value 0 and give C the value 1. Now consider any of the other $n - 2$ consequences, say C_i. To value this consequence, imagine that you are given the choice between either receiving C_i or taking part in a random experiment that either earns you consequence C with probability u or consequence c with probability $1 - u$. Clearly your choice will depend on the value of u. If $u = 1$, then the experiment is certain to result in consequence C; since C is the most desirable consequence, you will clearly prefer the experiment to receiving C_i. On the other hand, if $u = 0$, then the experiment will result in the least desirable consequence, namely c, and so in this case you will clearly prefer the consequence C_i to the experiment. Now, as u decreases from 1 down to 0, it seems reasonable that your choice will at some point switch from the experiment to the certain return of C_i, and at that critical switch point you will be indifferent between the two alternatives. Take that indifference probability u as the value of the consequence C_i. In other words, the value of C_i is that probability u such that you are indifferent between either receiving the consequence C_i or taking part in an experiment that returns consequence C with probability u or consequence c with probability $1 - u$. We call this indifference probability the *utility* of the consequence C_i, and we designate it as $u(C_i)$.

To determine which action is superior, we need to evaluate each one. Consider the first action, which results in consequence C_i with probability p_i, $i = 1, \ldots, n$. We can think of the result of this action as being determined by a two-stage experiment. In the first stage, one of the values $1, \ldots, n$ is chosen according to the probabilities $p_1, \ldots, p_n$; if value i is chosen, you receive consequence C_i. However, since C_i is equivalent to obtaining consequence C with probability $u(C_i)$ or consequence c with probability $1 - u(C_i)$, it follows that the result of the two-stage experiment is equivalent to an experiment in which either consequence C or c is obtained, with C being obtained with probability

$$\sum_{i=1}^{n} p_i u(C_i)$$

Similarly, the result of choosing the second action is equivalent to taking part in an experiment in which either consequence C or c is obtained, with C being obtained with probability

$$\sum_{i=1}^{n} q_i u(C_i)$$

Since C is preferable to c, it thus follows that the first action is preferable to the second action if

$$\sum_{i=1}^{n} p_i u(C_i) > \sum_{i=1}^{n} q_i u(C_i)$$

In other words, the worth of an action can be measured by the expected value of the utility of its consequence, and the action with the largest expected utility is most preferable. ■

A simple corollary of Proposition 4.1 is Corollary 4.1.

Corollary 4.1

If a and b are constants, then

$$E[aX + b] = aE[X] + b$$

Proof

$$\begin{aligned} E[aX + b] &= \sum_{x:p(x)>0} (ax + b)p(x) \\ &= a \sum_{x:p(x)>0} xp(x) + b \sum_{x:p(x)>0} p(x) \\ &= aE[X] + b \end{aligned}$$

✦

The expected value of a random variable X, $E[X]$ is also referred to as the mean or the first moment of X. The quantity $E[X^n]$, $n \geq 1$, is called the nth moment of X. By Proposition 4.1 we note that

$$E[X^n] = \sum_{x:p(x)>0} x^n p(x)$$

4.5 VARIANCE

Given a random variable X along with its distribution function F, it would be extremely useful if we were able to summarize the essential properties of F by certain suitably defined measures. One such measure would be $E[X]$, the expected value of X. However, although $E[X]$ yields the weighted average of the possible values of X, it does not tell us anything about the variation, or spread, of these values. For instance, although random variables W, Y, and Z, having probability mass functions determined by

$$W = 0 \quad \text{with probability } 1$$

$$Y = \begin{cases} -1 & \text{with probability } \frac{1}{2} \\ +1 & \text{with probability } \frac{1}{2} \end{cases}$$

$$Z = \begin{cases} -100 & \text{with probability } \frac{1}{2} \\ +100 & \text{with probability } \frac{1}{2} \end{cases}$$

all have the same expectation—namely, 0—there is much greater spread in the possible value of Y than in those of W (which is a constant) and in the possible values of Z than in those of Y.

As we expect X to take on values around its mean $E[X]$, it would appear that a reasonable way of measuring the possible variation of X would be to look at how far apart X would be from its mean on the average. One possible way to measure this would be to consider the quantity $E[|X - \mu|]$, where $\mu = E[X]$. However, it turns out to be mathematically inconvenient to deal with this quantity, and so a more tractable quantity is usually considered—namely, the expectation of the square of the difference between X and its mean. We thus have the following definition.

Definition

If X is a random variable with mean μ, then the variance of X, denoted by $\text{Var}(X)$, is defined by

$$\text{Var}(X) = E[(X - \mu)^2]$$

An alternative formula for $\text{Var}(X)$ is derived as follows:

$$\begin{aligned}
\text{Var}(X) &= E[X - \mu)^2] \\
&= \sum_x (x - \mu)^2 p(x) \\
&= \sum_x (x^2 - 2\mu x + \mu^2) p(x) \\
&= \sum_x x^2 p(x) - 2\mu \sum_x x p(x) + \mu^2 \sum_x p(x) \\
&= E[X^2] - 2\mu^2 + \mu^2 \\
&= E[X^2] - \mu^2
\end{aligned}$$

That is,

$$\boxed{\text{Var}(X) = E[X^2] - (E[X])^2}$$

In words, the variance of X is equal to the expected value of X^2 minus the square of its expected value. This is, in practice, often the easiest way to compute $\text{Var}(X)$.

Example 5a. Calculate $\text{Var}(X)$ if X represents the outcome when a fair die is rolled.

Solution It was shown in Example 3a that $E[X] = \frac{7}{2}$. Also,

$$\begin{aligned}
E[X^2] &= 1^2\left(\tfrac{1}{6}\right) + 2^2\left(\tfrac{1}{6}\right) + 3^2\left(\tfrac{1}{6}\right) + 4^2\left(\tfrac{1}{6}\right) + 5^2\left(\tfrac{1}{6}\right) + 6^2\left(\tfrac{1}{6}\right) \\
&= \left(\tfrac{1}{6}\right)(91)
\end{aligned}$$

Hence

$$\text{Var}(X) = \tfrac{91}{6} - \left(\tfrac{7}{2}\right)^2 = \tfrac{35}{12} \qquad \blacksquare$$

A useful identity is that for any constants a and b.

$$\text{Var}(aX + b) = a^2\,\text{Var}(X)$$

To prove the preceding, let $\mu = E[X]$ and note that from Corollary 4.1, $E[aX + b] = a\mu + b$. Hence

$$\begin{aligned}\text{Var}(aX + b) &= E[(aX + b - a\mu - b)^2] \\ &= E[a^2(X - \mu)^2] \\ &= a^2E[(X - \mu)^2] \\ &= a^2\,\text{Var}(X)\end{aligned}$$

REMARKS. (a) Analogous to the mean being the center of gravity of a distribution of mass, the variance represents, in the terminology of mechanics, the moment of inertia.

(b) The square root of the $\text{Var}(X)$ is called the *standard deviation* of X, and we denote it by $\text{SD}(X)$. That is,

$$\text{SD}(X) = \sqrt{\text{Var}(X)}$$

Discrete random variables are often classified according to their probability mass function. In the next few sections we consider some of the more common types.

4.6 THE BERNOULLI AND BINOMIAL RANDOM VARIABLES

Suppose that a trial, or an experiment, whose outcome can be classified as either a *success* or a *failure* is performed. If we let $X = 1$ when the outcome is a success and $X = 0$ when it is a failure, then the probability mass function of X is given by

$$\begin{aligned}p(0) &= P\{X = 0\} = 1 - p \\ p(1) &= P\{X = 1\} = p\end{aligned} \qquad (6.1)$$

where $p, 0 \le p \le 1$, is the probability that the trial is a success.

A random variable X is said to be a Bernoulli random variable (after the Swiss mathematician James Bernoulli) if its probability mass function is given by Equations (6.1) for some $p \in (0, 1)$.

Suppose now that n independent trials, each of which results in a success with probability p and in a failure with probability $1 - p$, are to be performed. If X represents the number of successes that occur in the n trials, then X is said to be a *binomial* random variable with parameters (n, p). Thus a Bernoulli random variable is just a binomial random variable with parameters $(1, p)$.

The probability mass function of a binomial random variable having parameters (n, p) is given by

$$p(i) = \binom{n}{i} p^i (1 - p)^{n-i} \qquad i = 0, 1, \ldots, n \tag{6.2}$$

The validity of Equation (6.2) may be verified by first noting that the probability of any particular sequence of n outcomes containing i successes and $n - i$ failures is, by the assumed independence of trials, $p^i(1 - p)^{n-i}$. Equation (6.2) then follows, since there are $\binom{n}{i}$ different sequences of the n outcomes leading to i successes and $n - i$ failures. This perhaps can most easily be seen by noting that there are $\binom{n}{i}$ different choices of the i trials that result in successes. For instance, if $n = 4, i = 2$, then there are $\binom{4}{2} = 6$ ways in which the four trials can result in two successes, namely, any of the outcomes (s, s, f, f), (s, f, s, f), (s, f, f, s), (f, s, s, f), (f, s, f, s), or (f, f, s, s), where the outcome (s, s, f, f) means, for instance, that the first two trials are successes and the last two failures. Since each of these outcomes has probability $p^2(1 - p)^2$ of occurring, the desired probability of two successes in the four trials is thus $\binom{4}{2} p^2(1 - p)^2$.

Note that by the binomial theorem, the probabilities sum to 1; that is,

$$\sum_{i=0}^{\infty} p(i) = \sum_{i=0}^{n} \binom{n}{i} p^i (1 - p)^{n-i} = [p + (1 - p)]^n = 1$$

Example 6a. Five fair coins are flipped. If the outcomes are assumed independent, find the probability mass function of the number of heads obtained.

Solution If we let X equal the number of heads (successes) that appear, then X is a binomial random variable with parameters $\left(n = 5, p = \frac{1}{2}\right)$. Hence, by Equation (6.2),

$$P\{X = 0\} = \binom{5}{0}\left(\frac{1}{2}\right)^0\left(\frac{1}{2}\right)^5 = \frac{1}{32}$$

$$P\{X = 1\} = \binom{5}{1}\left(\frac{1}{2}\right)^1\left(\frac{1}{2}\right)^4 = \frac{5}{32}$$

$$P\{X = 2\} = \binom{5}{2}\left(\frac{1}{2}\right)^2\left(\frac{1}{2}\right)^3 = \frac{10}{32}$$

$$P\{X = 3\} = \binom{5}{3}\left(\frac{1}{2}\right)^3\left(\frac{1}{2}\right)^2 = \frac{10}{32}$$

$$P\{X = 4\} = \binom{5}{4}\left(\frac{1}{2}\right)^4\left(\frac{1}{2}\right)^1 = \frac{5}{32}$$

$$P\{X = 5\} = \binom{5}{5}\left(\frac{1}{2}\right)^5\left(\frac{1}{2}\right)^0 = \frac{1}{32}$$

■

Example 6b. It is known that screws produced by a certain company will be defective with probability .01 independently of each other. The company sells the screws in packages of 10 and offers a money-back guarantee that at most 1 of the 10 screws is defective. What proportion of packages sold must the company replace?

Solution If X is the number of defective screws in a package, then X is a binomial random variable with parameters (10, .01). Hence the probability that a package will have to be replaced is

$$1 - P\{X = 0\} - P\{X = 1\} = 1 - \binom{10}{0}(.01)^0(.99)^{10} - \binom{10}{1}(.01)^1(.99)^9$$
$$\approx .004$$

Hence only .4 percent of the packages will have to be replaced. ■

Example 6c. The following gambling game, known as the wheel of fortune (or chuck-a-luck), is quite popular at many carnivals and gambling casinos: A player bets on one of the numbers 1 through 6. Three dice are then rolled, and if the number bet by the player appears i times, $i = 1, 2, 3$, then the player wins i units; on the other hand, if the number bet by the player does not appear on any of the dice, then the player loses 1 unit. Is this game fair to the player? (Actually, the game is played by spinning a wheel that comes to rest on a slot labeled by three of the numbers 1 through 6, but it is mathematically equivalent to the dice version.)

Solution If we assume that the dice are fair and act independently of each other, then the number of times that the number bet appears is a binomial random variable with parameters $\left(3, \frac{1}{6}\right)$. Hence, letting X denote the player's winnings in the game, we have

$$P\{X = -1\} = \binom{3}{0}\left(\frac{1}{6}\right)^0\left(\frac{5}{6}\right)^3 = \frac{125}{216}$$

$$P\{X = 1\} = \binom{3}{1}\left(\frac{1}{6}\right)^1\left(\frac{5}{6}\right)^2 = \frac{75}{216}$$

$$P\{X = 2\} = \binom{3}{2}\left(\frac{1}{6}\right)^2\left(\frac{5}{6}\right)^1 = \frac{15}{216}$$

$$P\{X = 3\} = \binom{3}{3}\left(\frac{1}{6}\right)^3\left(\frac{5}{6}\right)^0 = \frac{1}{216}$$

In order to determine whether or not this is a fair game for the player, let us determine $E[X]$. From the preceding probabilities we obtain

$$E[X] = \frac{-125 + 75 + 30 + 3}{216}$$
$$= \frac{-17}{216}$$

Hence, in the long run, the player will lose 17 units per every 216 games he plays. ∎

In the next example we consider the simplest form of the theory of inheritance as developed by G. Mendel (1822–1884).

Example 6d. Suppose that a particular trait (such as eye color or left handedness) of a person is classified on the basis of one pair of genes and suppose that d represents a dominant gene and r a recessive gene. Thus a person with dd genes is pure dominance, one with rr is pure recessive, and one with rd is hybrid. The pure dominance and the hybrid are alike in appearance. Children receive 1 gene from each parent. If, with respect to a particular trait, 2 hybrid parents have a total of 4 children, what is the probability that 3 of the 4 children have the outward appearance of the dominant gene?

Solution If we assume that each child is equally likely to inherit either of 2 genes from each parent, the probabilities that the child of 2 hybrid parents will have dd, rr, or rd pairs of genes are, respectively, $\frac{1}{4}, \frac{1}{4}, \frac{1}{2}$. Hence, as an offspring will have the outward appearance of the dominant gene if its gene pair is either dd or rd, it follows that the number of such children is binomially distributed with parameters $\left(4, \frac{3}{4}\right)$. Thus the desired probability is

$$\binom{4}{3}\left(\frac{3}{4}\right)^3\left(\frac{1}{4}\right)^1 = \frac{27}{64}$$ ∎

Example 6e. Consider a jury trial in which it takes 8 of the 12 jurors to convict; that is, in order for the defendant to be convicted, at least 8 of the jurors must vote him guilty. If we assume that jurors act independently and each makes the right decision with probability θ, what is the probability that the jury renders a correct decision?

Solution The problem, as stated, is incapable of solution, for there is not yet enough information. For instance, if the defendant is innocent, the probability of the jury's rendering a correct decision is

$$\sum_{i=5}^{12} \binom{12}{i} \theta^i (1-\theta)^{12-i}$$

whereas, if he is guilty, the probability of a correct decision is

$$\sum_{i=8}^{12} \binom{12}{i} \theta^i (1-\theta)^{12-i}$$

Therefore, if α represents the probability that the defendant is guilty, then, by conditioning on whether or not he is guilty, we obtain that the probability that the jury renders a correct decision is

$$\alpha \sum_{i=8}^{12} \binom{12}{i} \theta^i (1-\theta)^{12-i} + (1-\alpha) \sum_{i=5}^{12} \binom{12}{i} \theta^i (1-\theta)^{12-i}$$ ∎

Example 6f. A communication system consists of n components, each of which will, independently, function with probability p. The total system will be able to operate effectively if at least one-half of its components function.

(a) For what values of p is a 5-component system more likely to operate effectively than a 3-component system?
(b) In general, when is a $(2k + 1)$-component system better than a $(2k - 1)$-component system?

Solution (a) As the number of functioning components is a binomial random variable with parameters (n, p), it follows that the probability that a 5-component system will be effective is

$$\binom{5}{3}p^3(1 - p)^2 + \binom{5}{4}p^4(1 - p) + p^5$$

whereas the corresponding probability for a 3-component system is

$$\binom{3}{2}p^2(1 - p) + p^3$$

Hence, the 5-component system is better if

$$10p^3(1 - p)^2 + 5p^4(1 - p) + p^5 > 3p^2(1 - p) + p^3$$

which reduces to

$$3(p - 1)^2(2p - 1) > 0$$

or

$$p > \tfrac{1}{2}$$

(b) In general, a system with $2k + 1$ components will be better than one with $2k - 1$ components if (and only if) $p > \frac{1}{2}$. To prove this, consider a system of $2k + 1$ components and let X denote the number of the first $2k - 1$ that function. Then

$$\begin{aligned} P_{2k+1}(\text{effective}) &= P\{X \geq k + 1\} + P\{X = k\}(1 - (1 - p)^2) \\ &\quad + P\{X = k - 1\}p^2 \end{aligned}$$

which follows since the $(2k + 1)$-component system will be effective if either

(i) $X \geq k + 1$;
(ii) $X = k$ and at least one of the remaining 2 components function; or
(iii) $X = k - 1$ and both of the next 2 function.

As

$$\begin{aligned} P_{2k-1}(\text{effective}) &= P\{X \geq k\} \\ &= P\{X = k\} + P\{X \geq k + 1\} \end{aligned}$$

we obtain

$$
\begin{aligned}
&P_{2k+1}(\text{effective}) - P_{2k-1}(\text{effective}) \\
&\quad = P\{X = k-1\}p^2 - (1-p)^2 P\{X = k\} \\
&\quad = \binom{2k-1}{k-1} p^{k-1}(1-p)^k p^2 - (1-p)^2 \binom{2k-1}{k} p^k (1-p)^{k-1} \\
&\quad = \binom{2k-1}{k} p^k (1-p)^k [p - (1-p)] \quad \text{since} \binom{2k-1}{k-1} = \binom{2k-1}{k} \\
&\quad > 0 \Leftrightarrow p > \tfrac{1}{2}
\end{aligned}
$$

■

4.6.1 Properties of Binomial Random Variables

We will now examine the properties of a binomial random variable with parameters n and p. To begin, let us compute its expected value and variance. Now,

$$
\begin{aligned}
E[X^k] &= \sum_{i=0}^{n} i^k \binom{n}{i} p^i (1-p)^{n-i} \\
&= \sum_{i=1}^{n} i^k \binom{n}{i} p^i (1-p)^{n-i}
\end{aligned}
$$

Using the identity

$$
i\binom{n}{i} = n\binom{n-1}{i-1}
$$

gives that

$$
\begin{aligned}
E[X^k] &= np \sum_{i=1}^{n} i^{k-1} \binom{n-1}{i-1} p^{i-1}(1-p)^{n-i} \\
&= np \sum_{j=0}^{n-1} (j+1)^{k-1} \binom{n-1}{j} p^j (1-p)^{n-1-j} \quad \text{by letting } j = i-1 \\
&= npE[(Y+1)^{k-1}]
\end{aligned}
$$

where Y is a binomial random variable with parameters $n-1, p$. Setting $k = 1$ in the preceding equation yields

$$
E[X] = np
$$

That is, the expected number of successes that occur in n independent trials when each is a success with probability p is equal to np. Setting $k = 2$ in the preceding equation, and using the preceding formula for the expected value of a binomial random variable, gives that

$$
\begin{aligned}
E[X^2] &= npE[Y+1] \\
&= np[(n-1)p + 1]
\end{aligned}
$$

Since $E[X] = np$ we obtain

$$\begin{aligned}\text{Var}(X) &= E[X^2] - (E[X])^2 \\ &= np[(n-1)p + 1] - (np)^2 \\ &= np(1-p)\end{aligned}$$

Summing up, we have shown the following:

If X is a binomial random variable with parameters n and p, then

$$E[X] = np$$
$$\text{Var}(X) = np(1-p)$$

The following proposition details how the binomial probability mass function first increases and then decreases.

Proposition 6.1

If X is a binomial random variable with parameters (n, p), where $0 < p < 1$, then as k goes from 0 to n, $P\{X = k\}$ first increases monotonically and then decreases monotonically, reaching its largest value when k is the largest integer less than or equal to $(n + 1)p$.

Proof: We prove the proposition by considering $P\{X = k\}/P\{X = k - 1\}$ and determining for what values of k it is greater or less than 1. Now,

$$\frac{P\{X = k\}}{P\{X = k-1\}} = \frac{\dfrac{n!}{(n-k)!k!}p^k(1-p)^{n-k}}{\dfrac{n!}{(n-k+1)!(k-1)!}p^{k-1}(1-p)^{n-k+1}}$$
$$= \frac{(n-k+1)p}{k(1-p)}$$

Hence $P\{X = k\} \geq P\{X = k - 1\}$ if and only if

$$(n - k + 1)p \geq k(1 - p)$$

or, equivalently, if and only if

$$k \leq (n + 1)p$$

and the proposition is proved. ✦

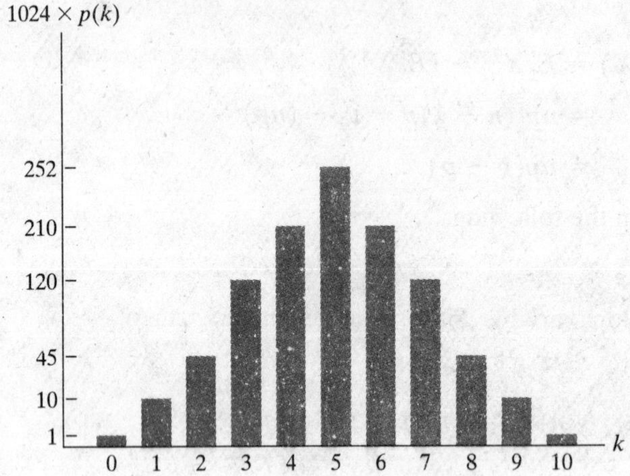

Figure 4.5 Graph of $p(k) = \binom{10}{k}\left(\frac{1}{2}\right)^{10}$

As an illustration of Proposition 6.1 consider Figure 4.5, the graph of the probability mass function of a binomial random variable with parameters $\left(10, \frac{1}{2}\right)$.

Example 6g. In a U.S. presidential election the candidate who gains the maximum number of votes in a state is awarded the total number of electoral college votes allocated to that state. The number of electoral college votes of a given state is roughly proportional to the population of that state—that is, a state of population size n has roughly nc electoral votes. (Actually, it is closer to $nc + 2$ as a state is given an electoral vote for each member of the House of Representatives, the number of such representatives being roughly proportional to its population, and one electoral college vote for each of its two senators.) Let us determine the average power in a close presidential election of a citizen in a state of size n, where by *average power* in a close election we mean the following: A voter in a state of size $n = 2k + 1$ will be decisive if the other $n - 1$ voters split their votes evenly between the two candidates. (We are assuming here that n is odd, but the case where n is even is quite similar.) As the election is close, we shall suppose that each of the other $n - 1 = 2k$ voters acts independently and is equally likely to vote for either candidate. Hence the probability that a voter in a state of size $n = 2k + 1$ will make a difference to the outcome is the same as the probability that $2k$ tosses of a fair coin lands heads and tails an equal number of times. That is,

$$
\begin{aligned}
P\{\text{voter in state of size } 2k + 1 \text{ makes a difference}\} \\
&= \binom{2k}{k}\left(\frac{1}{2}\right)^k\left(\frac{1}{2}\right)^k \\
&= \frac{(2k)!}{k!\,k!\,2^{2k}}
\end{aligned}
$$

To approximate the preceding, we make use of Stirling's approximation, which says that for k large,

$$k! \sim k^{k+1/2}e^{-k}\sqrt{2\pi}$$

where we say that $a_k \sim b_k$ when the ratio a_k/b_k approaches 1 as k approaches ∞. Hence we see that

$$P\{\text{voter in state of size } 2k + 1 \text{ makes a difference}\}$$
$$\sim \frac{(2k)^{2k+1/2}e^{-2k}\sqrt{2\pi}}{k^{2k+1}e^{-2k}(2\pi)2^{2k}} = \frac{1}{\sqrt{k\pi}}$$

As such a voter will, if he or she makes a difference, affect nc electoral votes, we see that the expected number of electoral votes a voter in a state of size n will affect—or the voter's average power—is given by

$$\begin{aligned}\text{average power} &= ncP\{\text{makes a difference}\}\\ &\sim \frac{nc}{\sqrt{n\pi/2}}\\ &= c\sqrt{2n/\pi}\end{aligned}$$

Hence the average power of a voter in a state of size n is proportional to the square root of n, thus showing that in presidential elections, voters in large states have more power than do those in smaller states. ∎

4.6.2 Computing the Binomial Distribution Function

Suppose that X is binomial with parameters (n, p). The key to computing its distribution function

$$P\{X \leq i\} = \sum_{k=0}^{i} \binom{n}{k} p^k(1 - p)^{n-k} \qquad i = 0, 1, \ldots, n$$

is to utilize the following relationship between $P\{X = k + 1\}$ and $P\{X = k\}$, which was established in the proof of Proposition 6.1:

$$P\{X = k + 1\} = \frac{p}{1 - p}\frac{n - k}{k + 1}P\{X = k\} \tag{6.3}$$

Example 6h. Let X be a binomial random variable with parameters $n = 6$, $p = .4$. Then, starting with $P\{X = 0\} = (.6)^6$ and recursively employing Equation (6.3), we obtain

$$P\{X = 0\} = (.6)^6 \approx .0467$$

$$P\{X = 1\} = \tfrac{4}{6}\tfrac{6}{1}P\{X = 0\} \approx .1866$$
$$P\{X = 2\} = \tfrac{4}{6}\tfrac{5}{2}P\{X = 1\} \approx .3110$$
$$P\{X = 3\} = \tfrac{4}{6}\tfrac{4}{3}P\{X = 2\} \approx .2765$$
$$P\{X = 4\} = \tfrac{4}{6}\tfrac{3}{4}P\{X = 3\} \approx .1382$$
$$P\{X = 5\} = \tfrac{4}{6}\tfrac{2}{5}P\{X = 4\} \approx .0369$$
$$P\{X = 6\} = \tfrac{4}{6}\tfrac{1}{6}P\{X = 5\} \approx .0041$$

■

A computer program that utilizes the recursion (6.3) to compute the binomial distribution function is easily written. To compute $P\{X \le i\}$ the program should compute first $P\{X = i\}$ and then use the recursion to compute successively $P\{X = i - 1\}, P\{X = i - 2\}$, and so on. Such a program is on the website. In utilizing it, one enters the binomial parameters n and p and a value i and the program computes the probabilities that a binomial (n, p) random variable is equal to and is less than or equal to i.

HISTORICAL NOTE

Independent trials having a common success probability p were first studied by the Swiss mathematician Jacques Bernoulli (1654–1705). In his book *Ars Conjectandi* (the Art of Conjecturing), published by his nephew Nicholas eight years after his death in 1713, Bernoulli showed that if the number of such trials were large, then the proportion of them that were successes would be close to p with a probability near 1.

Jacques Bernoulli was from the first generation of the most famous mathematical family of all time. Altogether there were between eight and twelve Bernoullis, spread over three generations, who made fundamental contributions to probability, statistics, and mathematics. One difficulty in knowing their exact number is the fact that several had the same name. (For example, two of the sons of Jacques's brother Jean were named Jacques and Jean.) Another difficulty is that several of the Bernoullis were known by different names in different places. Our Jacques (sometimes written Jaques) was, for instance, also known as Jakob (sometimes written Jacob) and as James Bernoulli. But whatever their number, their influence and output were prodigious. Like the Bachs of music, the Bernoullis of mathematics were a family for the ages!

Example 6i. If X is a binomial random variable with parameters $n = 100$, $p = .75$, find $P\{X = 70\}$ and $P\{X \leq 70\}$.

Solution The answer shown below can be found on the website.

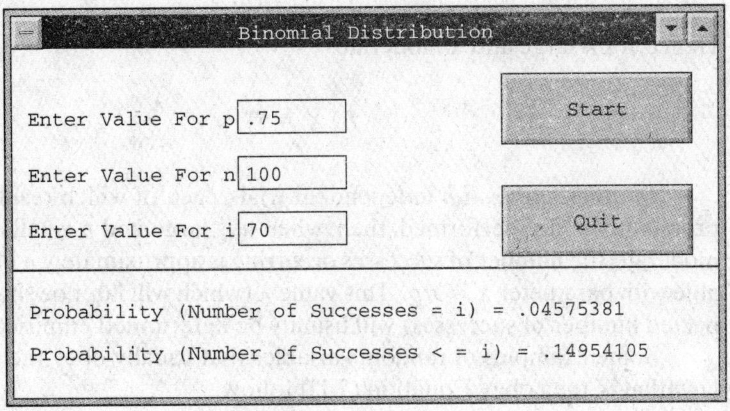

4.7 THE POISSON RANDOM VARIABLE

A random variable X, taking on one of the values $0, 1, 2, \ldots$, is said to be a *Poisson* random variable with parameter λ if for some $\lambda > 0$,

$$p(i) = P\{X = i\} = e^{-\lambda}\frac{\lambda^i}{i!} \qquad i = 0, 1, 2, \ldots \tag{7.1}$$

Equation (7.1) defines a probability mass function, since

$$\sum_{i=0}^{\infty} p(i) = e^{-\lambda}\sum_{i=0}^{\infty}\frac{\lambda^i}{i!} = e^{-\lambda}e^{\lambda} = 1$$

The Poisson probability distribution was introduced by S. D. Poisson in a book he wrote regarding the application of probability theory to lawsuits, criminal trials, and the like. This book, published in 1837, was entitled *Recherchés sur la probabilité des jugements en matière criminelle et en matière civile*.

The Poisson random variable has a tremendous range of applications in diverse areas because it may be used as an approximation for a binomial random variable with parameters (n, p) when n is large and p is small enough so that np is a moderate size. To see this, suppose that X is a binomial random variable with parameters (n, p) and let $\lambda = np$. Then

$$\begin{aligned}
P\{X = i\} &= \frac{n!}{(n-i)!\,i!}p^i(1-p)^{n-i} \\
&= \frac{n!}{(n-i)!\,i!}\left(\frac{\lambda}{n}\right)^i\left(1-\frac{\lambda}{n}\right)^{n-i} \\
&= \frac{n(n-1)\cdots(n-i+1)}{n^i}\,\frac{\lambda^i}{i!}\,\frac{(1-\lambda/n)^n}{(1-\lambda/n)^i}
\end{aligned}$$

Now, for n large and λ moderate,

$$\left(1 - \frac{\lambda}{n}\right)^n \approx e^{-\lambda} \qquad \frac{n(n-1)\cdots(n-i+1)}{n^i} \approx 1 \qquad \left(1 - \frac{\lambda}{n}\right)^i \approx 1$$

Hence, for n large and λ moderate,

$$P\{X = i\} \approx e^{-\lambda}\frac{\lambda^i}{i!}$$

In other words, if n independent trials, each of which results in a success with probability p, are performed, then, when n is large and p small enough to make np moderate, the number of successes occurring is approximately a Poisson random variable with parameter $\lambda = np$. This value λ (which will later be shown to equal the expected number of successes) will usually be determined empirically.

Some examples of random variables that usually obey the Poisson probability law [that is, they obey Equation (7.1)] follow:

1. The number of misprints on a page (or a group of pages) of a book
2. The number of people in a community living to 100 years of age
3. The number of wrong telephone numbers that are dialed in a day
4. The number of packages of dog biscuits sold in a particular store each day
5. The number of customers entering a post office on a given day
6. The number of vacancies occurring during a year in the federal judicial system
7. The number of α-particles discharged in a fixed period of time from some radioactive material

Each of the preceding, and numerous other random variables, are approximately Poisson for the same reason—namely, because of the Poisson approximation to the binomial. For instance, we can suppose that there is a small probability p that each letter typed on a page will be misprinted. Hence the number of misprints on a page will be approximately Poisson with $\lambda = np$, where n is the number of letters on a page. Similarly, we can suppose that each person in a community has some small probability of reaching age 100. Also, each person entering a store may be thought of as having some small probability of buying a package of dog biscuits, and so on.

Example 7a. Suppose that the number of typographical errors on a single page of this book has a Poisson distribution with parameter $\lambda = \frac{1}{2}$. Calculate the probability that there is at least one error on this page.

Solution Letting X denote the number of errors on this page, we have

$$P\{X \geq 1\} = 1 - P\{X = 0\} = 1 - e^{-1/2} \approx .393 \qquad \blacksquare$$

Example 7b. Suppose that the probability that an item produced by a certain machine will be defective is .1. Find the probability that a sample of 10 items will contain at most 1 defective item.

Solution The desired probability is $\binom{10}{0}(.1)^0(.9)^{10} + \binom{10}{1}(.1)^1(.9)^9 = .7361$, whereas the Poisson approximation yields the value $e^{-1} + e^{-1} \approx .7358$. ∎

Example 7c. Consider an experiment that consists of counting the number of α-particles given off in a 1-second interval by 1 gram of radioactive material. If we know from past experience that, on the average, 3.2 such α-particles are given off, what is a good approximation to the probability that no more than 2 α-particles will appear?

Solution If we think of the gram of radioactive material as consisting of a large number n of atoms, each of which has probability $3.2/n$ of disintegrating and sending off an α-particle during the second considered, then we see that, to a very close approximation, the number of α-particles given off will be a Poisson random variable with parameter $\lambda = 3.2$. Hence the desired probability is

$$
\begin{aligned}
P\{X \leq 2\} &= e^{-3.2} + 3.2e^{-3.2} + \frac{(3.2)^2}{2}e^{-3.2} \\
&\approx .3799
\end{aligned}
$$
∎

Before computing the expected value and variance of the Poisson random variable with parameter λ, recall that this random variable approximates a binomial random variable with parameters n and p when n is large, p is small, and $\lambda = np$. Since such a binomial random variable has expected value $np = \lambda$ and variance $np(1-p) = \lambda(1-p) \approx \lambda$ (since p is small), it would seem that both the expected value and the variance of a Poisson random variable would equal its parameter λ. We now verify this result.

$$
\begin{aligned}
E[X] &= \sum_{i=0}^{\infty} \frac{ie^{-\lambda}\lambda^i}{i!} \\
&= \lambda \sum_{i=1}^{\infty} \frac{e^{-\lambda}\lambda^{i-1}}{(i-1)!} \\
&= \lambda e^{-\lambda} \sum_{j=0}^{\infty} \frac{\lambda^j}{j!} \qquad \text{by letting } j = i - 1 \\
&= \lambda \qquad \text{since } \sum_{j=0}^{\infty} \frac{\lambda^j}{j!} = e^{\lambda}
\end{aligned}
$$

Thus the expected value of a Poisson random variable X is indeed equal to its parameter λ. To determine its variance we first compute $E[X^2]$.

$$
\begin{aligned}
E[X^2] &= \sum_{i=0}^{\infty} \frac{i^2 e^{-\lambda}\lambda^i}{i!} \\
&= \lambda \sum_{i=1}^{\infty} \frac{i e^{-\lambda}\lambda^{i-1}}{(i-1)!} \\
&= \lambda \sum_{j=0}^{\infty} \frac{(j+1)e^{-\lambda}\lambda^j}{j!} \qquad \text{by letting } j = i - 1 \\
&= \lambda \left[\sum_{j=0}^{\infty} \frac{j e^{-\lambda}\lambda^j}{j!} + \sum_{j=0}^{\infty} \frac{e^{-\lambda}\lambda^j}{j!} \right] \\
&= \lambda(\lambda + 1)
\end{aligned}
$$

where the final equality follows since the first sum is the expected value of a Poisson random variable with parameter λ and the second is the sum of the probabilities of this random variable. Therefore, since we have shown that $E[X] = \lambda$, we obtain that

$$
\begin{aligned}
\mathrm{Var}(X) &= E[X^2] - (E[X])^2 \\
&= \lambda
\end{aligned}
$$

> The expected value and variance of a Poisson random variable are both equal to its parameter λ.

We have shown that the Poisson with parameter np is a very good approximation to the distribution of the number of successes in n independent trials when each trial has probability p of being a success, provided that n is large and p small. In fact, it remains a good approximation even when the trials are not independent, provided that their dependence is weak. For instance, recall the matching problem (Example 5m of Chapter 2) where n men randomly select hats from a set consisting of one hat from each person. From the point of view of the number of men that select their own hat, we may regard the random selection as the result of n trials where we say that trial i is a success if person i selects his own hat, $i = 1, \ldots, n$. Defining the events $E_i, i = 1, \ldots, n$, by

$$E_i = \{\text{trial } i \text{ is a success}\}$$

then it is easy to see that

$$P\{E_i\} = \frac{1}{n} \quad \text{and} \quad P\{E_i|E_j\} = \frac{1}{n-1}, j \neq i$$

Thus, we see that while the events $E_i, i = 1, \ldots, n$ are not independent, their dependence, for large n, appears to be weak. Based on this, it seems reasonable to expect that the number of successes will approximately have a Poisson distribution with parameter $n \times 1/n = 1$, and indeed this is verified in Example 5m of Chapter 2.

For a second illustration of the strength of the Poisson approximation when the trials are weakly dependent, let us reconsider the birthday problem presented in Example 5i of Chapter 2. In this example we suppose that each of n people is equally likely to have any of the 365 days of the year as their birthday, and the problem is to determine the probability that a set of n independent people all have different birthdays. A combinatorial argument was used to determine this probability and it was then computed that when $n = 23$ this probability was less than $\frac{1}{2}$.

We can approximate the above probability by using the Poisson approximation as follows. Imagine that we have a trial for each of the $\binom{n}{2}$ pairs of individuals i and $j, i \neq j$, and say that trial i, j is a success if persons i and j have the same birthday. If we let E_{ij} denote the event that trial i, j is a success, then whereas the $\binom{n}{2}$ events E_{ij}, $1 \leq i < j \leq n$ are not independent (see Theoretical Exercise 21) their dependence appears to be rather weak. (Indeed, these events are even *pairwise independent* in that any 2 of the events E_{ij} and E_{kl} are independent—again see Theoretical Exercise 21). As $P(E_{ij}) = 1/365$ it is thus reasonable to suppose that the number of successes should approximately have a Poisson distribution with mean $\binom{n}{2}\Big/365 = n(n-1)/730$. Therefore,

$$P\{\text{no 2 people have the same birthday}\} = P\{0 \text{ successes}\}$$
$$\approx \exp\left\{\frac{-n(n-1)}{730}\right\}$$

To determine the smallest integer n for which this probability is less than $\frac{1}{2}$ note that

$$\exp\left\{\frac{-n(n-1)}{730}\right\} \leq \frac{1}{2}$$

is equivalent to

$$\exp\left\{\frac{n(n-1)}{730}\right\} \geq 2$$

or, taking logarithms of both sides, we obtain

$$n(n-1) \geq 730 \log 2$$
$$\approx 505.997$$

which yields the solution $n = 23$, in agreement with the result of Example 5i of Chapter 2.

Suppose now that we wanted the probability that among the n people no 3 of them have their birthday on the same day. Whereas this now becomes a difficult combinatorial problem, it is a simple matter to obtain a good approximation. To begin, imagine that we have a trial for each of the $\binom{n}{3}$ triplets i, j, k where

$1 \le i < j < k \le n$, and call the i, j, k trial a success if persons i, j, and k all have their birthday on the same day. As above, we can then conclude that the number of successes is approximately a Poisson random variable with parameter

$$\binom{n}{3} P\{i, j, k \text{ have the same birthday}\} = \binom{n}{3}\left(\frac{1}{365}\right)^2 = \frac{n(n-1)(n-2)}{6 \times (365)^2}$$

Hence

$$P\{\text{no 3 have the same birthday}\} \approx \exp\left\{\frac{-n(n-1)(n-2)}{799350}\right\}$$

This probability will be less than $\frac{1}{2}$ when n is such that

$$n(n-1)(n-2) \ge 799350 \log 2 \approx 554067.1$$

which is equivalent to $n \ge 84$. Thus, the approximate probability that at least 3 people in a group of size 84 or larger will have the same birthday exceeds $\frac{1}{2}$.

Another use of the Poisson probability distribution arises in situations where "events" occur at certain points in time. One example of this is that an event is the occurrence of an earthquake; another possibility would be for events to correspond to people entering a particular establishment (bank, post office, gas station, and so on); and a third possibility is for an event to occur whenever a war starts. Let us suppose that events are indeed occurring at certain (random) points of time, and let us assume that for some positive constant λ the following assumptions hold true:

1. The probability that exactly 1 event occurs in a given interval of length h is equal to $\lambda h + o(h)$, where $o(h)$ stands for any function $f(h)$ that is such that $\lim_{h \to 0} f(h)/h = 0$. [For instance, $f(h) = h^2$ is $o(h)$, whereas $f(h) = h$ is not.]
2. The probability that 2 or more events occur in an interval of length h is equal to $o(h)$.
3. For any integers $n, j_1, j_2, \ldots, j_n$, and any set of n nonoverlapping intervals, if we define E_i to be the event that exactly j_i of the events under consideration occur in the ith of these intervals, then events $E_1, E_2, \ldots, E_n$ are independent.

Loosely put, assumptions 1 and 2 state that for small values of h, the probability that exactly 1 event occurs in an interval of size h equals λh plus something that is small compared to h, whereas the probability that 2 or more events occur is small compared to h. Assumption 3 states that whatever occurs in one interval has no (probability) effect on what will occur in other nonoverlapping intervals.

Under assumptions 1, 2, and 3, we shall now show that the number of events occurring in any interval of length t is a Poisson random variable with parameter λt. To be precise, let us call the interval $[0, t]$ and denote by $N(t)$ the number of events occurring in that interval. To obtain an expression for $P\{N(t) = k\}$, we start by breaking the interval $[0, t]$ into n nonoverlapping subintervals each of length t/n (Figure 4.6)

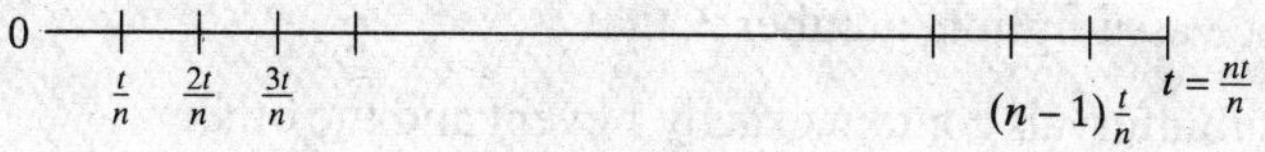

Figure 4.6

Now,

$$P\{N(t) = k\} = P\{k \text{ of the } n \text{ subintervals contain exactly 1 event and the other } n - k \text{ contain 0 events}\} + P\{N(t) = k \text{ and at least 1 subinterval contains 2 or more events}\} \tag{7.2}$$

This follows because the event on the left side of Equation (7.2), that is, $\{N(t) = k\}$, is clearly equal to the union of the two mutually exclusive events on the right side of the equation. Letting A and B denote the two mutually exclusive events on the right side of Equation (7.2), we have

$$\begin{aligned}
P(B) &\leq P\{\text{at least one subinterval contains 2 or more events}\} \\
&= P\left(\bigcup_{i=1}^{n} \{i\text{th subinterval contains 2 or more events}\}\right) \\
&\leq \sum_{i=1}^{n} P\{i\text{th subinterval contains 2 or more events}\} && \text{by Boole's inequality} \\
&= \sum_{i=1}^{n} o\left(\frac{t}{n}\right) && \text{by assumption 2} \\
&= no\left(\frac{t}{n}\right) \\
&= t\left[\frac{o(t/n)}{t/n}\right]
\end{aligned}$$

Now, for any t, $t/n \to 0$ as $n \to \infty$ and so $o(t/n)/(t/n) \to 0$ as $n \to \infty$ by the definition of $o(h)$. Hence

$$P(B) \to 0 \quad \text{as} \quad n \to \infty \tag{7.3}$$

On the other hand, since assumptions 1 and 2 imply that†

$$P\{0 \text{ events occur in an interval of length } h\} = 1 - [\lambda h + o(h) + o(h)] = 1 - \lambda h - o(h)$$

†The sum of two functions both of which are $o(h)$ is also $o(h)$. This is so because if $\lim_{h\to 0} f(h)/h = \lim_{h\to 0} g(h)/h = 0$, then $\lim_{h\to 0} [f(h) + g(h)]/h = 0$.

we see from the independence assumption, number 3, that

$$P(A) = P\{k \text{ of the subintervals contain exactly 1 event and the other } n-k \text{ contain 0 events}\}$$
$$= \binom{n}{k}\left[\frac{\lambda t}{n} + o\left(\frac{t}{n}\right)\right]^k \left[1 - \left(\frac{\lambda t}{n}\right) - o\left(\frac{t}{n}\right)\right]^{n-k}$$

However, since

$$n\left[\frac{\lambda t}{n} + o\left(\frac{t}{n}\right)\right] = \lambda t + t\left[\frac{o(t/n)}{t/n}\right] \rightarrow \lambda t \quad \text{as} \quad n \rightarrow \infty$$

it follows, by the same argument that verified the Poisson approximation to the binomial, that

$$P(A) \rightarrow e^{-\lambda t}\frac{(\lambda t)^k}{k!} \quad \text{as} \quad n \rightarrow \infty \tag{7.4}$$

Thus, from Equations (7.2), (7.3), and (7.4), we obtain, by letting $n \rightarrow \infty$,

$$P\{N(t) = k\} = e^{-\lambda t}\frac{(\lambda t)^k}{k!} \qquad k = 0, 1, \ldots \tag{7.5}$$

Hence, if assumptions 1, 2, and 3 are satisfied, the number of events occurring in any fixed interval of length t is a Poisson random variable with mean λt, and we say that the events occur in accordance with a Poisson process having rate λ. The value λ, which can be shown to equal the rate per unit time at which events occur, is a constant that must be empirically determined.

The preceding discussion explains why a Poisson random variable is usually a good approximation for such diverse phenomena as the following:

1. The number of earthquakes occurring during some fixed time span
2. The number of wars per year
3. The number of electrons emitted from a heated cathode during a fixed time period
4. The number of deaths in a given period of time of the policyholders of a life insurance company

Example 7d. Suppose that earthquakes occur in the western portion of the United States in accordance with assumptions 1, 2, and 3 with $\lambda = 2$ and with 1 week as the unit of time. (That is, earthquakes occur in accordance with the three assumptions at a rate of 2 per week.)

(a) Find the probability that at least 3 earthquakes occur during the next 2 weeks.
(b) Find the probability distribution of the time, starting from now, until the next earthquake.

Solution (a) From Equation (7.5) we have

$$P\{N(2) \geq 3\} = 1 - P\{N(2) = 0\} - P\{N(2) = 1\} - P\{N(2) = 2\}$$

$$= 1 - e^{-4} - 4e^{-4} - \frac{4^2}{2}e^{-4}$$

$$= 1 - 13e^{-4}$$

(b) Let X denote the amount of time (in weeks) until the next earthquake. Because X will be greater than t if and only if no events occur within the next t units of time, we have from Equation (7.5) that

$$P\{X > t\} = P\{N(t) = 0\} = e^{-\lambda t}$$

so the probability distribution function F of the random variable X is given by

$$F(t) = P\{X \leq t\} = 1 - P\{X > t\} = 1 - e^{-\lambda t}$$
$$= 1 - e^{-2t}$$ ■

4.7.1 Computing the Poisson Distribution Function

If X is Poisson with parameter λ, then

$$\frac{P\{X = i + 1\}}{P\{X = i\}} = \frac{e^{-\lambda}\lambda^{i+1}/(i + 1)!}{e^{-\lambda}\lambda^i/i!} = \frac{\lambda}{i + 1} \tag{7.6}$$

Starting with $P\{X = 0\} = e^{-\lambda}$, we can use (7.6) to compute successively

$$P\{X = 1\} = \lambda P\{X = 0\}$$

$$P\{X = 2\} = \frac{\lambda}{2}P\{X = 1\}$$

$$\vdots$$

$$P\{X = i + 1\} = \frac{\lambda}{i + 1}P\{X = i\}$$

The website includes a program that uses Equation (7.6) to compute Poisson probabilities.

Example 7e (a) Determine $P\{X \leq 90\}$ when X is Poisson with mean 100.
(b) Determine $P\{Y \leq 1075\}$ when Y is Poisson with mean 1000.

Solution From the website we obtain the solutions

(a) $P\{X \leq 100\} \approx .1714$;
(b) $P\{Y \leq 1075\} \approx .9894$. ■

4.8 OTHER DISCRETE PROBABILITY DISTRIBUTIONS

4.8.1 The Geometric Random Variable

Suppose that independent trials, each having a probability $p, 0 < p < 1$, of being a success, are performed until a success occurs. If we let X equal the number of trials required, then

$$P\{X = n\} = (1 - p)^{n-1}p \qquad n = 1, 2, \ldots \tag{8.1}$$

Equation (8.1) follows because in order for X to equal n, it is necessary and sufficient that the first $n - 1$ trials are failures and the nth trial is a success. Equation (8.1) then follows, since the outcomes of the successive trials are assumed to be independent.

Since

$$\sum_{n=1}^{\infty} P\{X = n\} = p \sum_{n=1}^{\infty} (1 - p)^{n-1} = \frac{p}{1 - (1 - p)} = 1$$

it follows that with probability 1, a success will eventually occur. Any random variable X whose probability mass function is given by Equation (8.1) is said to be a *geometric* random variable with parameter p.

Example 8a. An urn contains N white and M black balls. Balls are randomly selected, one at a time, until a black one is obtained. If we assume that each selected ball is replaced before the next one is drawn, what is the probability that

(a) exactly n draws are needed;
(b) at least k draws are needed?

Solution If we let X denote the number of draws needed to select a black ball, then X satisfies Equation (8.1) with $p = M/(M + N)$. Hence

$$\text{(a)}\ P\{X = n\} = \left(\frac{N}{M + N}\right)^{n-1} \frac{M}{M + N} = \frac{MN^{n-1}}{(M + N)^n}$$

$$\begin{aligned}\text{(b)}\ P\{X \geq k\} &= \frac{M}{M + N} \sum_{n=k}^{\infty} \left(\frac{N}{M + N}\right)^{n-1} \\ &= \left(\frac{M}{M + N}\right)\left(\frac{N}{M + N}\right)^{k-1} \Big/ \left[1 - \frac{N}{M + N}\right] \\ &= \left(\frac{N}{M + N}\right)^{k-1}\end{aligned}$$

Of course, part (b) could have been obtained directly, since the probability that at least k trials are necessary to obtain a success is equal to the probability that the first $k - 1$ trials are all failures. That is, for a geometric random variable

$$P\{X \geq k\} = (1 - p)^{k-1}$$

■

Example 8b. Find the expected value of a geometric random variable.

Solution With $q = 1 - p$ we have that

$$\begin{aligned} E[X] &= \sum_{n=1}^{\infty} nq^{n-1}p \\ &= p\sum_{n=0}^{\infty} \frac{d}{dq}(q^n) \\ &= p\frac{d}{dq}\left(\sum_{n=0}^{\infty} q^n\right) \\ &= p\frac{d}{dq}\left(\frac{1}{1-q}\right) \\ &= \frac{p}{(1-q)^2} \\ &= \frac{1}{p} \end{aligned}$$

In other words, if independent trials, having a common probability p of being successful, are performed until the first success occurs, then the expected number of required trials equals $1/p$. For instance, the expected number of rolls of a fair die that it takes to obtain the value 1 is 6. ■

Example 8c. Find the variance of a geometric random variable.

Solution To determine $\text{Var}(X)$ let us first compute $E[X^2]$. With $q = 1 - p$,

$$\begin{aligned} E[X^2] &= \sum_{n=1}^{\infty} n^2q^{n-1}p \\ &= p\sum_{n=1}^{\infty} \frac{d}{dq}(nq^n) \\ &= p\frac{d}{dq}\left(\sum_{n=1}^{\infty} nq^n\right) \\ &= p\frac{d}{dq}\left(\frac{q}{1-q}E[X]\right) \\ &= p\frac{d}{dq}[q(1-q)^{-2}] \\ &= p\left[\frac{1}{p^2} + \frac{2(1-p)}{p^3}\right] \\ &= \frac{2}{p^2} - \frac{1}{p} \end{aligned}$$

Hence, since $E[X] = 1/p$,

$$\text{Var}(X) = \frac{1-p}{p^2}$$ ■

4.8.2 The Negative Binomial Random Variable

Suppose that independent trials, each having probability p, $0 < p < 1$, of being a success are performed until a total of r successes is accumulated. If we let X equal the number of trials required, then

$$P\{X = n\} = \binom{n-1}{r-1} p^r (1-p)^{n-r} \qquad n = r, r+1, \ldots \tag{8.2}$$

Equation (8.2) follows because, in order for the rth success to occur at the nth trial, there must be $r - 1$ successes in the first $n - 1$ trials, and the nth trial must be a success. The probability of the first event is

$$\binom{n-1}{r-1} p^{r-1} (1-p)^{n-r}$$

and the probability of the second is p; thus, by independence, Equation (8.2) is established. To verify that a total of r successes must eventually be accumulated, we can either analytically prove that

$$\sum_{n=r}^{\infty} P\{X = n\} = \sum_{n=r}^{\infty} \binom{n-1}{r-1} p^r (1-p)^{n-r} = 1 \tag{8.3}$$

or we can give a probabilistic argument as follows: The number of trials required to obtain r successes can be expressed as $Y_1 + Y_2 + \cdots + Y_r$, where Y_1 equals the number of trials required for the first success, Y_2 the number of additional trials after the first success until the second success occurs, Y_3 the number of additional trials until the third success, and so on. As the trials are independent and all have the same probability of a success, it follows that $Y_1, Y_2, \ldots, Y_r$ are all geometric random variables. Hence each is finite with probability 1, so $\sum_{i=1}^{r} Y_i$ must also be finite, establishing Equation (8.3).

Any random variable X whose probability mass function is given by Equation (8.2) is said to be a *negative binomial* random variable with parameters (r, p). Note that a geometric random variable is just a negative binomial with parameter $(1, p)$.

In the next example we use the negative binomial to obtain another solution of the problem of the points.

Example 8d. If independent trials, each resulting in a success with probability p, are performed, what is the probability of r successes occurring before m failures?

Solution The solution will be arrived at by noting that r successes will occur before m failures if and only if the rth success occurs no later than the $r + m - 1$ trial. This follows because if the rth success occurs before or at the $r + m - 1$ trial, then it must have occurred before the mth failure, and conversely. Hence, from Equation (8.2), the desired probability is

$$\sum_{n=r}^{r+m-1} \binom{n-1}{r-1} p^r (1-p)^{n-r}$$ ■

Example 8e. ***The Banach match problem.*** A pipe-smoking mathematician carries, at all times, 2 matchboxes, 1 in his left-hand pocket and 1 in his right-hand pocket. Each time he needs a match he is equally likely to take it from either pocket. Consider the moment when the mathematician first discovers that one of his matchboxes is empty. If it is assumed that both matchboxes initially contained N matches, what is the probability that there are exactly k matches in the other box, $k = 0, 1, \ldots, N$?

Solution Let E denote the event that the mathematician first discovers that the right-hand matchbox is empty and there are k matches in the left-hand box at the time. Now, this event will occur if and only if the $(N + 1)$th choice of the right-hand matchbox is made at the $N + 1 + N - k$ trial. Hence, from Equation (8.2) (with $p = \frac{1}{2}, r = N + 1, n = 2N - k + 1$), we see

$$P(E) = \binom{2N - k}{N}\left(\frac{1}{2}\right)^{2N-k+1}$$

As there is an equal probability that it is the left-hand box that is first discovered to be empty and there are k matches in the right-hand box at that time, the desired result is

$$2P(E) = \binom{2N - k}{N}\left(\frac{1}{2}\right)^{2N-k}$$ ■

Example 8f. Compute the expected value and the variance of a negative binomial random variable with parameters r and p.

Solution

$$\begin{aligned}
E[X^k] &= \sum_{n=r}^{\infty} n^k \binom{n-1}{r-1} p^r (1-p)^{n-r} \\
&= \frac{r}{p} \sum_{n=r}^{\infty} n^{k-1} \binom{n}{r} p^{r+1} (1-p)^{n-r} \qquad \text{since} \quad n\binom{n-1}{r-1} = r\binom{n}{r} \\
&= \frac{r}{p} \sum_{m=r+1}^{\infty} (m-1)^{k-1} \binom{m-1}{r} p^{r+1} (1-p)^{m-(r+1)} \qquad \text{by setting } m = n+1 \\
&= \frac{r}{p} E[(Y-1)^{k-1}]
\end{aligned}$$

where Y is a negative binomial random variable with parameters $r + 1, p$. Setting $k = 1$ in the preceding equation yields

$$E[X] = \frac{r}{p}$$

Setting $k = 2$ in the preceding equation, and using the formula for the expected value of a negative binomial random variable, gives that

$$E[X^2] = \frac{r}{p}E[Y - 1]$$

$$= \frac{r}{p}\left(\frac{r + 1}{p} - 1\right)$$

Therefore,

$$\text{Var}(X) = \frac{r}{p}\left(\frac{r + 1}{p} - 1\right) - \left(\frac{r}{p}\right)^2$$

$$= \frac{r(1 - p)}{p^2}$$ ■

Thus we see from Example 8f that if independent trials, each of which is a success with probability p, are performed, then the expected value and variance of the number of trials that it takes to amass r successes is r/p and $r(1 - p)/p^2$, respectively.

Since a geometric random variable is just a negative binomial with parameter $r = 1$, it follows from the preceding example that the variance of a geometric random variable with parameter p is equal to $(1 - p)/p^2$, which checks with the result of Example 8c.

Example 8g. Find the expected value and the variance of the number of times one must throw a die until the outcome 1 has occurred 4 times.

Solution Since the random variable of interest is a negative binomial with parameter $r = 4$ and $p = \frac{1}{6}$, we see that

$$E[X] = 24$$

$$\text{Var}(X) = \frac{4\left(\frac{5}{6}\right)}{\left(\frac{1}{6}\right)^2} = 120$$ ■

4.8.3 The Hypergeometric Random Variable

Suppose that a sample of size n is to be chosen randomly (without replacement) from an urn containing N balls, of which m are white and $N - m$ are black. If we let X denote the number of white balls selected, then

$$P\{X = i\} = \frac{\binom{m}{i}\binom{N - m}{n - i}}{\binom{N}{n}} \quad i = 0, 1, \ldots, n \tag{8.4}$$

A random variable X, whose probability mass function is given by Equation (8.4) for some values of n, N, m is said to be a *hypergeometric* random variable.

REMARK. Although we have written the hypergeometric probability mass function with i going from 0 to n, $P\{X = i\}$ will actually be 0 unless i satisfies the inequalities $n - (N - m) \leq i \leq \min(n, m)$. However, Equation (8.4) is always valid because of our convention that $\binom{r}{k}$ is equal to 0 when either $k < 0$ or $r < k$. ∎

Example 8h. An unknown number, say N, of animals inhabit a certain region. To obtain some information about the population size, ecologists often perform the following experiment: They first catch a number, say m, of these animals, mark them in some manner, and release them. After allowing the marked animals time to disperse throughout the region, a new catch of size, say n, is made. Let X denote the number of marked animals in this second capture. If we assume that the population of animals in the region remained fixed between the time of the two catches and that each time an animal was caught it was equally likely to be any of the remaining uncaught animals, it follows that X is a hypergeometric random variable such that

$$P\{X = i\} = \frac{\binom{m}{i}\binom{N-m}{n-i}}{\binom{N}{n}} \equiv P_i(N)$$

Suppose now that X is observed to equal i. Then, as $P_i(N)$ represents the probability of the observed event when there are actually N animals present in the region, it would appear that a reasonable estimate of N would be the value of N that maximizes $P_i(N)$. Such an estimate is called a *maximum likelihood* estimate. (See Theoretical Exercises 13 and 18 for other examples of this type of estimation procedure.)

The maximization of $P_i(N)$ can most simply be done by first noting that

$$\frac{P_i(N)}{P_i(N-1)} = \frac{(N-m)(N-n)}{N(N-m-n+i)}$$

Now, the preceding ratio is greater than 1 if and only if

$$(N - m)(N - n) \geq N(N - m - n + i)$$

or, equivalently, if and only if

$$N \leq \frac{mn}{i}$$

Thus $P_i(N)$ is first increasing, and then decreasing, and reaches its maximum value at the largest integral value not exceeding mn/i. This value is thus the maximum likelihood estimate of N. For example, suppose that the initial catch consisted of $m = 50$ animals, which are marked and then released. If a subsequent catch consists of $n = 40$ animals of which $i = 4$ are marked, then we would estimate that there are some 500 animals in the region. (It should be noted that the preceding estimate could also have been obtained by assuming that the proportion of marked animals in the region, m/N, is approximately equal to the proportion of marked animals in our second catch, i/n.) ∎

Example 8i. A purchaser of electrical components buys them in lots of size 10. It is his policy to inspect 3 components randomly from a lot and to accept the lot only if all 3 are nondefective. If 30 percent of the lots have 4 defective components and 70 percent have only 1, what proportion of lots does the purchaser reject?

Solution Let A denote the event that the purchaser accepts a lot. Now,

$$P(A) = P(A|\text{lot has 4 defectives})\frac{3}{10} + P(A|\text{lot has 1 defective})\frac{7}{10}$$

$$= \frac{\binom{4}{0}\binom{6}{3}}{\binom{10}{3}}\left(\frac{3}{10}\right) + \frac{\binom{1}{0}\binom{9}{3}}{\binom{10}{3}}\left(\frac{7}{10}\right)$$

$$= \frac{54}{100}$$

Hence 46 percent of the lots are rejected. ■

If n balls are randomly chosen without replacement from a set of N balls, of which the fraction $p = m/N$ is white, then the number of white balls selected is hypergeometric. Now, it would seem that when m and N are large in relation to n, it shouldn't make much difference whether the selection is being done with or without replacement. Because no matter which balls have previously been selected, each additional selection will, when m and N are large, be white with a probability approximately equal to p. In other words, it seems intuitive when m and N are large in relation to n that the probability mass function of X should approximately be that of a binomial random variable with parameters n and p. To verify this intuition note that if X is hypergeometric then, for $i \leq n$,

$$P\{X = i\} = \frac{\binom{m}{i}\binom{N-m}{n-i}}{\binom{N}{n}}$$

$$= \frac{m!}{(m-i)!\,i!}\,\frac{(N-m)!}{(N-m-n+i)!\,(n-i)!}\,\frac{(N-n)!\,n!}{N!}$$

$$= \binom{n}{i}\frac{m}{N}\frac{m-1}{N-1}\cdots\frac{m-i+1}{N-i+1}\frac{N-m}{N-i}\frac{N-m-1}{N-i-1}$$

$$\cdots\frac{N-m-(n-i-1)}{N-i-(n-i-1)}$$

$$\approx \binom{n}{i}p^i(1-p)^{n-i} \qquad \text{when } p = m/N \text{ and } m \text{ and } N \text{ are large in relation to } n \text{ and } i$$

Example 8j. Determine the expected value and the variance of X, a hypergeometric random variable with parameters n, N, m.

Solution

$$E[X^k] = \sum_{i=0}^{n} i^k P\{X = i\}$$
$$= \sum_{i=1}^{n} i^k \binom{m}{i}\binom{N-m}{n-i}\bigg/\binom{N}{n}$$

Using the identities

$$i\binom{m}{i} = m\binom{m-1}{i-1} \quad \text{and} \quad n\binom{N}{n} = N\binom{N-1}{n-1}$$

we obtain that

$$E[X^k] = \frac{nm}{N}\sum_{i=1}^{n} i^{k-1}\binom{m-1}{i-1}\binom{N-m}{n-i}\bigg/\binom{N-1}{n-1}$$
$$= \frac{nm}{N}\sum_{j=0}^{n-1} (j+1)^{k-1}\binom{m-1}{j}\binom{N-m}{n-1-j}\bigg/\binom{N-1}{n-1}$$
$$= \frac{nm}{N}E[(Y+1)^{k-1}]$$

where Y is a hypergeometric random variable with parameters $n-1$, $N-1$, $m-1$. Hence, upon setting $k = 1$ we see that

$$E[X] = \frac{nm}{N}$$

In words, if n balls are randomly selected from a set of N balls, of which m are white, then the expected number of white balls selected is nm/N.

Upon setting $k = 2$ in the equation for $E[X^k]$ we obtain that

$$E[X^2] = \frac{nm}{N}E[Y+1]$$
$$= \frac{nm}{N}\left[\frac{(n-1)(m-1)}{N-1} + 1\right]$$

where the final equality uses our preceding result to compute the expected value of the hypergeometric random variable Y.

As $E[X] = nm/N$ we can conclude that

$$\text{Var}(X) = \frac{nm}{N}\left[\frac{(n-1)(m-1)}{N-1} + 1 - \frac{nm}{N}\right] \tag{8.5}$$

If we let $p = m/N$ denote the fraction of balls that are white, then it follows from Equation (8.5), after a little algebra, that

$$\text{Var}(X) = \frac{N-n}{N-1}np(1-p) \tag{8.6}$$

■

REMARK. We have shown in Example 8j that if n balls are randomly selected without replacement from a set of N balls, of which the fraction p are white, then the expected number of white balls chosen is np. In addition, if N is large in relation to n [and so $(N - n)/(N - 1)$ is approximately equal to 1], then

$$\text{Var}(X) \approx np(1 - p)$$

In other words, $E[X]$ is the same as when the selection of the balls is done with replacement (so the number of white balls is binomial with parameters n and p), and if the total collection of balls is large, then $\text{Var}(X)$ is approximately equal to what it would be if the selection were done with replacement. This is, of course, exactly what we would have guessed given our earlier result that when the number of balls in the urn is large, the number of white balls chosen approximately has the mass function of a binomial random variable. ■

4.8.4 The Zeta (or Zipf) Distribution

A random variable is said to have a zeta (sometimes called the Zipf) distribution if its probability mass function is given by

$$P\{X = k\} = \frac{C}{k^{\alpha+1}} \qquad k = 1, 2, \ldots$$

for some value of $\alpha > 0$. Since the sum of the foregoing probabilities must equal 1, it follows that

$$C = \left[\sum_{k=1}^{\infty} \left(\frac{1}{k}\right)^{\alpha+1}\right]^{-1}$$

The zeta distribution owes its name to the fact that the function

$$\zeta(s) = 1 + \left(\frac{1}{2}\right)^s + \left(\frac{1}{3}\right)^s + \cdots + \left(\frac{1}{k}\right)^s + \cdots$$

is known in mathematical disciplines as the Riemann zeta function (after the German mathematician G. F. B. Riemann).

The zeta distribution was used by the Italian economist Pareto to describe the distribution of family incomes in a given country. However, it was G. K. Zipf who applied these distributions in a wide variety of different areas and, in doing so, popularized their use.

4.9 PROPERTIES OF THE CUMULATIVE DISTRIBUTION FUNCTION

Recall that for the distribution function F of X, $F(b)$ denotes the probability that the random variable X takes on a value that is less than or equal to b. Some properties of the cumulative distribution function (c.d.f.) F are

1. F is a nondecreasing function; that is, if $a < b$, then $F(a) \leq F(b)$.

2. $\lim_{b\to\infty} F(b) = 1$.

3. $\lim_{b \to -\infty} F(b) = 0$.

4. F is right continuous. That is, for any b and any decreasing sequence $b_n, n \geq 1$, that converges to b, $\lim_{n \to \infty} F(b_n) = F(b)$.

Property 1 follows, as was noted in Section 4.1, because for $a < b$ the event $\{X \leq a\}$ is contained in the event $\{X \leq b\}$ and so cannot have a larger probability. Properties 2, 3, and 4 all follow from the continuity property of probabilities (Section 2.6). For instance, to prove property 2 we note that if b_n increases to ∞, then the events $\{X \leq b_n\}, n \geq 1$, are increasing events whose union is the event $\{X < \infty\}$. Hence, by the continuity property of probabilities,

$$\lim_{n \to \infty} P\{X \leq b_n\} = P\{X < \infty\} = 1$$

which proves property 2.

The proof of property 3 is similar and is left as an exercise. To prove property 4, we note that if b_n decrease to b, then $\{X \leq b_n\}, n \geq 1$ are decreasing events whose intersection is $\{X \leq b\}$. Hence the continuity property yields that

$$\lim_{n} P\{X \leq b_n\} = P\{X \leq b\}$$

which verifies property 4.

All probability questions about X can be answered in terms of the c.d.f. F. For example,

$$P\{a < X \leq b\} = F(b) - F(a) \qquad \text{for all } a < b \tag{9.1}$$

This can best be seen by writing the event $\{X \leq b\}$ as the union of the mutually exclusive events $\{X \leq a\}$ and $\{a < X \leq b\}$. That is,

$$\{X \leq b\} = \{X \leq a\} \cup \{a < X \leq b\}$$

so

$$P\{X \leq b\} = P\{X \leq a\} + P\{a < X \leq b\}$$

which establishes Equation (9.1).

If we want to compute the probability that X is strictly less than b, we can again apply the continuity property to obtain

$$\begin{aligned} P\{X < b\} &= P\left(\lim_{n \to \infty} \left\{X \leq b - \frac{1}{n}\right\}\right) \\ &= \lim_{n \to \infty} P\left(X \leq b - \frac{1}{n}\right) \\ &= \lim_{n \to \infty} F\left(b - \frac{1}{n}\right) \end{aligned}$$

Note that $P\{X < b\}$ does not necessarily equal $F(b)$, since $F(b)$ also includes the probability that X equals b.

Example 9a. The distribution function of the random variable X is given by

$$F(x) = \begin{cases} 0 & x < 0 \\ \dfrac{x}{2} & 0 \le x < 1 \\ \dfrac{2}{3} & 1 \le x < 2 \\ \dfrac{11}{12} & 2 \le x < 3 \\ 1 & 3 \le x \end{cases}$$

A graph of $F(x)$ is presented in Figure 4.7. Compute (a) $P\{X < 3\}$, (b) $P\{X = 1\}$, (c) $P\{X > \frac{1}{2}\}$, and (d) $P\{2 < X \le 4\}$.

Solution (a) $P\{X < 3\} = \lim_n P\left\{X \le 3 - \frac{1}{n}\right\} = \lim_n F\left(3 - \frac{1}{n}\right) = \frac{11}{12}$

(b) $P\{X = 1\} = P\{X \le 1\} - P\{X < 1\}$

$$= F(1) - \lim_n F\left(1 - \frac{1}{n}\right) = \frac{2}{3} - \frac{1}{2} = \frac{1}{6}$$

(c) $P\left\{X > \frac{1}{2}\right\} = 1 - P\left\{X \le \frac{1}{2}\right\}$

$$= 1 - F\left(\frac{1}{2}\right) = \frac{3}{4}$$

(d) $P\{2 < X \le 4\} = F(4) - F(2)$

$$= \frac{1}{12}$$

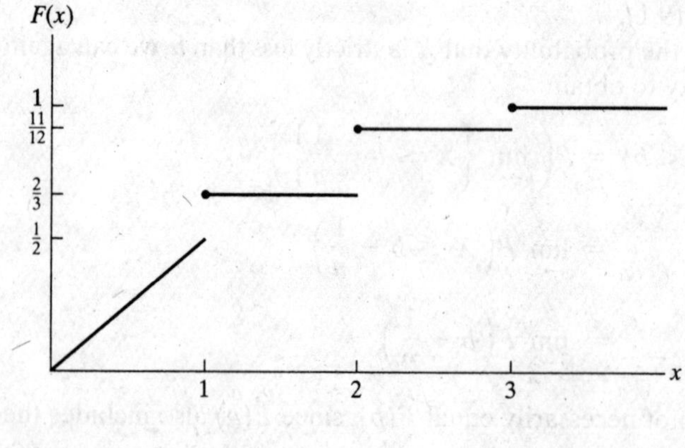

Figure 4.7 Graph of $F(x)$. ■

SUMMARY

A real-valued function defined on the outcome of a probability experiment is called a *random variable.*

If X is a random variable, then the function $F(x)$, defined by

$$F(x) = P\{X \leq x\}$$

is called the *distribution function* of X. All probabilities concerning X can be stated in terms of F.

A random variable whose set of possible values is either finite or countably infinite is called *discrete*. If X is a discrete random variable, then the function

$$p(x) = P\{X = x\}$$

is called the *probability mass function* of X. Also, the quantity $E[X]$, defined by

$$E[X] = \sum_{x:p(x)>0} xp(x)$$

is called the *expected value* of X. $E[X]$ is also commonly called the *mean* or the *expectation* of X.

A useful identity states that for a function g,

$$E[g(X)] = \sum_{x:p(x)>0} g(x)p(x)$$

The *variance* of a random variable X, denoted by $\text{Var}(X)$, is defined by

$$\text{Var}(X) = E[(X - E[X])^2]$$

The variance, which is equal to the expected square of the difference between X and its expected value, is a measure of the spread of the possible values of X. A useful identity is that

$$\text{Var}(X) = E[X^2] - (E[X])^2$$

The quantity $\sqrt{\text{Var}(X)}$ is called the *standard deviation* of X.

We now note some common types of discrete random variables.

The random variable X whose probability mass function is given by

$$p(i) = \binom{n}{i} p^i(1 - p)^{n-i} \qquad i = 0, \ldots, n$$

is said to be a binomial random variable with parameters n and p. Such a random variable can be interpreted as being the number of successes that occur when n independent trials, each of which results in a success with probability p, are performed. Its mean and variance are given by

$$E[X] = np \qquad \text{Var}(X) = np(1 - p)$$

The random variable X whose probability mass function is given by

$$p(i) = \frac{e^{-\lambda}\lambda^i}{i!} \qquad i \geq 0$$

is said to be a *Poisson* random variable with parameter λ. If a large number of (approximately) independent trials are performed, each having a small probability of being successful, then the number of successful trials that result will have a distribution that is approximately that of a Poisson random variable. The mean and variance of a Poisson random variable are both equal to its parameter λ. That is.

$$E[X] = \text{Var}(X) = \lambda$$

The random variable X whose probability mass function is given by

$$p(i) = p(1-p)^{i-1} \qquad i = 1, 2, \ldots$$

is said to be a *geometric* random variable with parameter p. Such a random variable represents the trial number of the first success when each trial is independently a success with probability p. Its mean and variance are given by

$$E[X] = \frac{1}{p} \qquad \text{Var}(X) = \frac{1-p}{p^2}$$

The random variable X whose probability mass function is given by

$$p(i) = \binom{i-1}{r-1} p^r (1-p)^{i-r} \qquad i \geq r$$

is said to be a *negative binomial* random variable with parameters r and p. Such a random variable represents the trial number of the rth success when each trial is independently a success with probability p. Its mean and variance are given by

$$E[X] = \frac{r}{p} \qquad \text{Var}(X) = \frac{r(1-p)}{p^2}$$

A *hypergeometric* random variable with parameters n, N, m represents the number of white balls selected when n balls are randomly chosen from an urn that contains N balls, of which m are white. Its probability mass function is given by

$$p(i) = \frac{\binom{m}{i}\binom{N-m}{n-i}}{\binom{N}{n}} \qquad i = 0, \ldots, m$$

With $p = m/N$, its mean and variance are

$$E[X] = np \qquad \text{Var}(X) = \frac{N-n}{N-1} np(1-p)$$

PROBLEMS

1. Two balls are chosen randomly from an urn containing 8 white, 4 black, and 2 orange balls. Suppose that we win \$2 for each black ball selected and we lose \$1 for each white ball selected. Let X denote our winnings. What are the possible values of X, and what are the probabilities associated with each value?

2. Two fair dice are rolled. Let X equal the product of the 2 dice. Compute $P\{X = i\}$ for $i = 1, 2, \ldots$

3. Three dice are rolled. By assuming that each of the $6^3 = 216$ possible outcomes is equally likely, find the probabilities attached to the possible values that X can take on, where X is the sum of the 3 dice.

4. Five men and 5 women are ranked according to their scores on an examination. Assume that no two scores are alike and all 10! possible rankings are equally likely. Let X denote the highest ranking achieved by a woman (for instance, $X = 1$ if the top-ranked person is female). Find $P\{X = i\}, i = 1, 2, 3, \ldots, 8, 9, 10$.

5. Let X represent the difference between the number of heads and the number of tails obtained when a coin is tossed n times. What are the possible values of X?

6. In Problem 5, if the coin is assumed fair, for $n = 3$ what are the probabilities associated with the values that X can take on?

7. Suppose that a die is rolled twice. What are the possible values that the following random variables can take on

(a) the maximum value to appear in the two rolls;
(b) the minimum value to appear on the two rolls;
(c) the sum of the two rolls;
(d) the value of the first roll minus the value of the second roll?

8. If the die in Problem 7 is assumed fair, calculate the probabilities associated with the random variables in parts (a) through (d).

9. Repeat Example 1b when the balls are selected with replacement.

10. In Example 1d compute the conditional probability that we win i dollars, given that we win something; compute it for $i = 1, 2, 3$.

11. (a) An integer N is to be selected at random from $\{1, 2, \ldots, (10)^3\}$ in the sense that each integer has the same probability of being selected. What is the probability that N will be divisible by 3? by 5? by 7? by 15? by 105? How would your answer change if $(10)^3$ is replaced by $(10)^k$ as k became larger and larger?

(b) An important function in number theory—one whose properties can be shown to be related to what is probably the most important unsolved problem of mathematics, the Riemann hypothesis—is the Möbius function $\mu(n)$, defined for all positive integral values n as follows: Factor n into its prime factors. If there is a repeated prime factor, as in $12 = 2 \cdot 2 \cdot 3$ or $49 = 7 \cdot 7$, then $\mu(n)$ is defined to equal 0. Now let N be chosen at random from $\{1, 2, \ldots (10)^k\}$, where k is large. Determine $P\{\mu(N) = 0\}$ as $k \to \infty$.

HINT: To compute $P\{\mu(N) \neq 0\}$, use the identity

$$\prod_{i=1}^{\infty} \frac{P_i^2 - 1}{P_i^2} = \left(\frac{3}{4}\right)\left(\frac{8}{9}\right)\left(\frac{24}{25}\right)\left(\frac{48}{49}\right)\cdots = \frac{6}{\pi^2}$$

where P_i is the ith smallest prime. (We do not include 1 as a prime.)

12. In the game of Two-Finger Morra, 2 players show 1 or 2 fingers and simultaneously guess the number of fingers their opponent will show. If only one of the players guesses correctly, he wins an amount (in dollars) equal to the sum of the fingers shown by him and his opponent. If both players guess correctly or if neither guesses correctly, then no money is exchanged. Consider a specified player and denote by X the amount of money he wins in a single game of Two-Finger Morra.

(a) If each player acts independently of the other, and if each player makes his choice of the number of fingers he will hold up and the number he will guess that his opponent will hold up in such a way that each of the 4 possibilities is equally likely, what are the possible values of X and what are their associated probabilities?

(b) Suppose that each player acts independently of the other. If each player decides to hold up the same number of fingers that he guesses his opponent will hold up, and if each player is equally likely to holu up 1 or 2 fingers, what are the possible values of X and their associated probabilities?

13. A salesman has scheduled two appointments to sell encyclopedias. His first appointment will lead to a sale with probability .3, and his second will lead independently to a sale with probability .6. Any sale made is equally likely to be either for the deluxe model, which costs \$1000, or the standard model, which costs \$500. Determine the probability mass function of X, the total dollar value of all sales.

14. Five distinct numbers are randomly distributed to players numbered 1 through 5. Whenever two players compare their numbers, the one with the higher one is declared the winner. Initially, players 1 and 2 compare their numbers; the winner then compares with player 3, and so on. Let X denote the number of times player 1 is a winner. Find $P\{X = i\}, i = 0, 1, 2, 3, 4$.

15. The National Basketball Association (NBA) draft lottery involves the 11 teams that had the worst won–lost records during the year. A total of 66 balls are placed in an urn. Each of these balls is inscribed with the name of a team: 11 have the name of the team with the worst record, 10 have the name of the team with the second worst record, 9 have the name of the team with the third worst record, and so on (with 1 ball having the name of the team with the eleventh worst record). A ball is then chosen at random and the team whose name is on the ball is given the first pick in the draft of players about to enter the league. Another ball is then chosen and if it "belongs" to a different team than the one that received the first draft pick, then the team to which it belongs receives the second draft pick. (If the ball belongs to the team receiving the first pick, then it is discarded and another one is chosen; this continues until the ball of another team is chosen.) Finally, another ball is chosen and the team named on the ball (provided that it is different from the previous two teams) receives the third draft pick. The remaining draft picks 4 through 11 are then awarded to the 8 teams that did not "win the lottery" in inverse order of their won–lost records. For instance, if the team with the worst record did not receive any of the 3 lottery picks, then that team would receive the fourth draft pick. Let X denote the draft pick of the team with the worst record. Find the probability mass function of X.

16. In Problem 15, let team number 1 be the team with the worst record, let team number 2 be the team with the second worst record, and so on. Let Y_i denote the team that gets draft pick number i. Thus $Y_1 = 3$ if the first ball chosen belongs to team number 3. Find the probability mass function of (a) Y_1, (b) Y_2, and (c) Y_3.

17. Suppose that the distribution function of X is given by

$$F(b) = \begin{cases} 0 & b < 0 \\ \dfrac{b}{4} & 0 \le b < 1 \\ \dfrac{1}{2} + \dfrac{b-1}{4} & 1 \le b < 2 \\ \dfrac{11}{12} & 2 \le b < 3 \\ 1 & 3 \le b \end{cases}$$

(a) Find $P\{X = i\}, i = 1, 2, 3$. **(b)** Find $P\{\frac{1}{2} < X < \frac{3}{2}\}$.

18. Four independent flips of a fair coin are made. Let X denote the number of heads obtained. Plot the probability mass function of the random variable $X - 2$.

19. If the distribution function of X is given by

$$F(b) = \begin{cases} 0 & b < 0 \\ \frac{1}{2} & 0 \le b < 1 \\ \frac{3}{5} & 1 \le b < 2 \\ \frac{4}{5} & 2 \le b < 3 \\ \frac{9}{10} & 3 \le b < 3.5 \\ 1 & b \ge 3.5 \end{cases}$$

calculate the probability mass function of X.

20. A gambling book recommends the following "winning strategy" for the game of roulette. It recommends that a gambler bet \$1 on red. If red appears (which has probability $\frac{18}{38}$), then the gambler should take her \$1 profit and quit. If the gambler loses this bet (which has probability $\frac{20}{38}$ of occurring), she should make additional \$1 bets on red on each of the next two spins of the roulette wheel and then quit. Let X denote the gambler's winnings when she quits.

(a) Find $P\{X > 0\}$.

(b) Are you convinced that the strategy is indeed a "winning" strategy? Explain your answer!

(c) Find $E[X]$.

21. A total of 4 buses carrying 148 students from the same school arrives at a football stadium. The buses carry, respectively, 40, 33, 25, and 50 students. One of the students is randomly selected. Let X denote the number of students that were on the bus carrying this randomly selected student. One of the 4 bus drivers is also randomly selected. Let Y denote the number of students on her bus.

(a) Which of $E[X]$ or $E[Y]$ do you think is larger? Why?

(b) Compute $E[X]$ and $E[Y]$.

22. Suppose that two teams play a series of games that ends when one of them has won i games. Suppose that each game played is, independently, won by team A with probability p. Find the expected number of games that are played when (a) $i = 2$ and (b) $i = 3$. Also show in both cases that this number is maximized when $p = \frac{1}{2}$.

23. You have \$1000 and a certain commodity presently sells for \$2 per ounce. Suppose that after one week the commodity will sell for either \$1 or \$4 an ounce, with these two possibilities being equally likely.

(a) If your objective is to maximize the expected amount of money that you possess at the end of the week, what strategy should you employ?

(b) If your objective is to maximize the expected amount of the commodity that you possess at the end of the week, what strategy should you employ?

24. A and B play the following game: A writes down either number 1 or number 2 and B must guess which one. If the number that A has written down is i and B has guessed correctly, B receives i units from A. If B makes a wrong guess. B pays $\frac{3}{4}$ unit to A. If B randomizes his decision by guessing 1 with probability p and 2 with probability $1 - p$, determine his expected gain if (a) A has written down number 1 and (b) A has written down number 2.

What value of p maximizes the minimum possible value of B's expected gain and what is this maximin value? (Note that B's expected gain depends not only on p but also on what A does.)

Consider now player A. Suppose that she also randomizes her decision, writing down number 1 with probability q. What is A's expected loss if (c) B chooses number 1 and (d) B chooses number 2?

What value of q minimizes A's maximum expected loss? Show that the minimum of A's maximum expected loss is equal to the maximum of B's minimum expected gain. This result, known as the minimax theorem, was first established in generality by the mathematician John von Neumann and is the fundamental result in the mathematical discipline known as the theory of games. The common value is called the value of the game to player B.

25. A typical slot machine has 3 dials, each with 20 symbols (cherries, lemons, plums, oranges, bells, and bars). A typical set of dials is set up as follows:

	Dial 1	Dial 2	Dial 3
Cherries	7	7	0
Oranges	3	7	6
Lemons	3	0	4
Plums	4	1	6
Bells	2	2	3
Bars	1	3	1
	20	20	20

According to this table, of the 20 slots on dial 1, 7 are cherries, 3 are oranges, and so on. A typical payoff on a 1-unit bet is as shown in the following table.

Dial 1	Dial 2	Dial 3	Payoff
Bar	Bar	Bar	60
Bell	Bell	Bell	20
Bell	Bell	Bar	18
Plum	Plum	Plum	14
Orange	Orange	Orange	10
Orange	Orange	Bar	8
Cherry	Cherry	Anything	2
Cherry	No cherry	Anything	0
	Anything else		−1

Compute the player's expected winnings on a single play of the slot machine. Assume that each dial acts independently.

26. One of the numbers 1 through 10 is randomly chosen. You are to try to guess the number chosen by asking questions with "yes–no" answers. Compute the expected number of questions you will need to ask in each of the two cases:

(a) Your ith question is to be "Is it i?", $i = 1, 2, 3, 4, 5, 6, 7, 8, 9, 10$.

(b) With each question you try to eliminate one-half of the remaining numbers, as nearly as possible.

27. An insurance company writes a policy to the effect that an amount of money A must be paid if some event E occurs within a year. If the company estimates that E will occur within a year with probability p, what should it charge the customer in order that its expected profit will be 10 percent of A?

28. A sample of 3 items is selected at random from a box containing 20 items of which 4 are defective. Find the expected number of defective items in the sample.

29. There are two possible causes for a breakdown of a machine. To check the first possibility would cost C_1 dollars, and, if that were the cause of the breakdown, the trouble could be repaired at a cost of R_1 dollars. Similarly, there are costs C_2 and R_2 associated with the second possibility. Let p and $1 - p$ denote, respectively, the probabilities that the breakdown is caused by the first and second possibilities. Under what conditions on p, C_i, R_i, $i = 1, 2$, should we check the first possible cause of breakdown and then the second, as opposed to reversing the checking order, so as to minimize the expected cost involved in returning the machine to working order?

NOTE: If the first check is negative, we must still check the other possibility.

30. A person tosses a fair coin until a tail appears for the first time. If the tail appears on the nth flip, the person wins 2^n dollars. Let X denote the player's winnings. Show that $E[X] = +\infty$. This problem is known as the St. Petersburg paradox.
(a) Would you be willing to pay \$1 million to play this game once?
(b) Would you be willing to pay \$1 million for each game if you could play for as long as you liked and only had to settle up when you stopped playing?

31. Each night different meteorologists give us the probability that it will rain the next day. To judge how well these people predict, we will score each of them as follows: If a meteorologist says that it will rain with probability p, then he or she will receive a score of

$$\begin{array}{ll} 1 - (1 - p)^2 & \text{if it does rain} \\ 1 - p^2 & \text{if it does not rain} \end{array}$$

We will then keep track of scores over a certain time span and conclude that the meteorologist with the highest average score is the best predictor of weather. Suppose now that a given meteorologist is aware of this and so wants to maximize his or her expected score. If this person truly believes that it will rain tomorrow with probability p^*, what value of p should he or she assert so as to maximize the expected score?

32. To determine whether or not they have a certain disease, 100 people are to have their blood tested. However, rather than testing each individual separately, it has been decided first to group the people in groups of 10. The blood samples of the 10 people in each group will be pooled and analyzed together. If the test is negative, one test will suffice for the 10 people; whereas, if the test is positive each of the 10 people will also be individually tested and, in all, 11 tests will be made on this group. Assume the probability that a person has the disease is .1 for all people, independently of each other, and compute the expected number of tests necessary for each group. (Note that we are assuming that the pooled test will be positive if at least one person in the pool has the disease.)

33. A newsboy purchases papers at 10 cents and sells them at 15 cents. However, he is not allowed to return unsold papers. If his daily demand is a binomial random variable with $n = 10, p = \frac{1}{3}$, approximately how many papers should he purchase so as to maximize his expected profit?

34. In Example 4b, suppose that the department store incurs an additional cost of c for each unit of unmet demand. (This is often referred to as a goodwill cost because the store loses the goodwill of those customers whose demands it cannot meet.) Compute the expected profit when the store stocks s units, and determine the value of s that maximizes the expected profit.

35. A box contains 5 red and 5 blue marbles. Two marbles are withdrawn randomly. If they are the same color, then you win \$1.10; if they are different colors, then you win $-$ \$1.00 (that is, you lose \$1.00). Calculate
(a) the expected value of the amount you win;
(b) the variance of the amount you win.

36. Consider Problem 22 with $i = 2$. Find the variance of the number of games played and show that this number is maximized when $p = \frac{1}{2}$.

37. Find $\text{Var}(X)$ and $\text{Var}(Y)$ for X and Y as given in Problem 21.

38. If $E[X] = 1$ and $\text{Var}(X) = 5$, find

(a) $E[(2 + X)^2]$; **(b)** $\text{Var}(4 + 3X)$.

39. A ball is drawn from an urn containing 3 white and 3 black balls. After the ball is drawn, it is then replaced and another ball is drawn. This goes on indefinitely. What is the probability that of the first 4 balls drawn, exactly 2 are white?

40. On a multiple-choice exam with 3 possible answers for each of the 5 questions, what is the probability that a student would get 4 or more correct answers just by guessing?

41. A man claims to have extrasensory perception. As a test, a fair coin is flipped 10 times, and the man is asked to predict the outcome in advance. He gets 7 out of 10 correct. What is the probability that he would have done at least this well if he had no ESP?

42. Suppose that when in flight, airplane engines will fail with probability $1 - p$ independently from engine to engine. If an airplane needs a majority of its engines operative to make a successful flight, for what values of p is a 5-engine plane preferable to a 3-engine plane?

43. A communications channel transmits the digits 0 and 1. However, due to static, the digit transmitted is incorrectly received with probability .2. Suppose that we want to transmit an important message consisting of one binary digit. To reduce the chance of error, we transmit 00000 instead of 0 and 11111 instead of 1. If the receiver of the message uses "majority" decoding, what is the probability that the message will be wrong when decoded? What independence assumptions are you making?

44. A satellite system consists of n components and functions on any given day if at least k of the n components function on that day. On a rainy day each of the components independently functions with probability p_1, whereas on a dry day they each independently function with probability p_2. If the probability of rain tomorrow is α, what is the probability that the satellite system will function?

45. A student is getting ready to take an important oral examination and is concerned about the possibility of having an "on" day or an "off" day. He figures that if he has an on day, then each of his examiners will pass him independently of each other, with probability .8, whereas, if he has an off day, this probability will be reduced to .4. Suppose that the student will pass the examination if a majority of the examiners pass him. If the student feels that he is twice as likely to have an off day as he is to have an on day, should he request an examination with 3 examiners or with 5 examiners?

46. Suppose that it takes at least 9 votes from a 12-member jury to convict a defendant. Suppose that the probability that a juror votes a guilty person innocent is .2, whereas the probability that the juror votes an innocent person guilty is .1. If each juror acts independently and if 65 percent of the defendants are guilty, find the probability that the jury renders a correct decision. What percentage of defendants is convicted?

47. In some military courts, 9 judges are appointed. However, both the prosecution and the defense attorneys are entitled to a peremptory challenge of any judge, in which case that judge is removed from the case and is not replaced. A defendant is declared guilty if the majority of judges cast votes of guilty, and he or she is declared innocent otherwise. Suppose that when the defendant is, in fact, guilty, each judge will (independently) vote guilty with probability .7, whereas when the defendant is, in fact, innocent, this probability drops to .3.

(a) What is the probability that a guilty defendant is declared guilty when there are (i) 9, (ii) 8, and (iii) 7 judges?

(b) Repeat part (a) for an innocent defendant.

(c) If the prosecution attorney does not exercise the right to a peremptory challenge of a judge and if the defense is limited to at most two such challenges, how many challenges should the defense attorney make if he or she is 60 percent certain that the client is guilty?

48. It is known that diskettes produced by a certain company will be defective with probability .01, independently of each other. The company sells the diskettes in packages of size 10 and offers a money-back guarantee that at most 1 of the 10 diskettes in the package will be defective. If someone buys 3 packages, what is the probability that he or she will return exactly 1 of them?

49. Suppose that 10 percent of the chips produced by a computer hardware manufacturer are defective. If we order 100 such chips, will the number of defective ones we receive be a binomial random variable?

50. Suppose that a biased coin that lands on heads with probability p is flipped 10 times. Given that a total of 6 heads result, find the conditional probability that the first 3 outcomes are
(a) H, T, T (meaning that the first flip is heads, the second is tails, and the third is tails);
(b) T, H, T.

51. The expected number of typographical errors on a page of a certain magazine is .2. What is the probability that the next page you read contains (a) 0 and (b) 2 or more typographical errors? Explain your reasoning!

52. The monthly worldwide average number of airplane crashes of commercial airlines is 3.5. What is the probability that there will be
(a) at least 2 such accidents in the next month;
(b) at most 1 accident in the next month?
Explain your reasoning!

53. Approximately 80,000 marriages took place in the state of New York last year. Estimate the probability that for at least one of these couples
(a) both partners were born on April 30;
(b) both partners celebrated their birthday on the same day of the year.
State your assumptions.

54. Suppose that the average number of cars abandoned weekly on a certain highway is 2.2. Approximate the probability that there will be
(a) no abandoned cars in the next week;
(b) at least 2 abandoned cars in the next week.

55. A certain typing agency employs 2 typists. The average number of errors per article is 3 when typed by the first typist and 4.2 when typed by the second. If your article is equally likely to be typed by either typist, approximate the probability that it will have no errors.

56. How many people are needed so that the probability that at least one of them has the same birthday as you is greater than $\frac{1}{2}$?

57. Suppose that the number of accidents occurring on a highway each day is a Poisson random variable with parameter $\lambda = 3$.
(a) Find the probability that 3 or more accidents occur today.
(b) Repeat part (a) under the assumption that at least 1 accident occurs today.

58. Compare the Poisson approximation with the correct binomial probability for the following cases:
(a) $P\{X = 2\}$ when $n = 8, p = .1$; **(b)** $P\{X = 9\}$ when $n = 10, p = .95$;
(c) $P\{X = 0\}$ when $n = 10, p = .1$; **(d)** $P\{X = 4\}$ when $n = 9, p = .2$.

59. If you buy a lottery ticket in 50 lotteries, in each of which your chance of winning a prize is $\frac{1}{100}$, what is the (approximate) probability that you will win a prize
(a) at least once; **(b)** exactly once; **(c)** at least twice?

60. The number of times that a person contracts a cold in a given year is a Poisson random variable with parameter $\lambda = 5$. Suppose that a new wonder drug (based on large quantities of vitamin C) has just been marketed that reduces the Poisson parameter to $\lambda = 3$ for 75 percent of the population. For the other 25 percent of the population the drug has no appreciable effect on colds. If an individual tries the drug for a year and has 2 colds in that time, how likely is it that the drug is beneficial for him or her?

61. The probability of being dealt a full house in a hand of poker is approximately .0014. Find an approximation for the probability that in 1000 hands of poker you will be dealt at least 2 full houses.

62. If n married couples are seated at random at a round table, approximately what is the probability that no wife sits next to her husband? When $n = 10$ compare your approximation with the exact answer given in Example 5n of Chapter 2.

63. People enter a gambling casino at a rate of 1 for every 2 minutes.

(a) What is the probability that no one enters between 12:00 and 12:05?

(b) What is the probability that at least 4 people enter the casino during that time?

64. The suicide rate in a certain state is 1 suicide per 100,000 inhabitants per month.

(a) Find the probability that in a city of 400,000 inhabitants within this state, there will be 8 or more suicides in a given month.

(b) What is the probability that there will be at least 2 months during the year that will have 8 or more suicides?

(c) Counting the present month as month number 1, what is the probability that the first month to have 8 or more suicides will be month number $i, i \geq 1$?

What assumptions are you making?

65. Each of 500 soldiers in an army company independently has a certain disease with probability $1/10^3$. This disease will show up in a blood test, and to facilitate matters blood samples from all 500 are pooled and tested.

(a) What is the (approximate) probability that the blood test will be positive (and so at least one person has the disease)?

Suppose now that the blood test yields a positive result.

(b) What is the probability, under this circumstance, that more than one person has the disease?

One of the 500 people is Jones, who knows that he has the disease.

(c) What does Jones think is the probability that more than one person has the disease?

As the pooled test was positive, the authorities have decided to test each individual separately. The first $i - 1$ of these tests were negative, and the ith one—which was on Jones—was positive.

(d) Given the preceding, as a function of i, what is the probability that any of the remaining people have the disease?

66. Consider a roulette wheel consisting of 38 numbers—1 through 36, 0, and double 0. If Smith always bets that the outcome will be one of the numbers 1 through 12, what is the probability that

(a) Smith will lose his first 5 bets;

(b) his first win will occur on his fourth bet?

67. Two athletic teams play a series of games; the first team to win 4 games is declared the overall winner. Suppose that one of the teams is stronger than the other and wins each game with probability .6, independent of the outcomes of the other games. Find the probability that the stronger team wins the series in exactly i games. Do it for $i = 4, 5, 6, 7$. Compare the probability that the stronger team wins with the probability that it would win a 2-out-of-3 series.

68. Suppose in Problem 67 that the two teams are evenly matched and each has probability $\frac{1}{2}$ of winning each game. Find the expected number of games played.

69. An interviewer is given a list of potential people she can interview. If the interviewer needs to interview 5 people and if each person (independently) agrees to be interviewed with probability $\frac{2}{3}$, what is the probability that her list of potential people will enable her to obtain her necessary number of interviews if the list consists of (a) 5 people and (b) 8 people? For part (b) what is the probability that the interviewer will speak to exactly (c) 6 people and (d) 7 people on the list?

70. A fair coin is continually flipped until heads appears for the tenth time. Let X denote the number of tails that occur. Compute the probability mass function of X.

71. Solve the Banach match problem (Example 8e) when the left-hand matchbox originally contained N_1 matches and the right-hand box contained N_2 matches.

72. In Banach's matchbox problem find the probability that at the moment when the first box is emptied (as opposed to being found empty), the other box contains exactly k matches.

73. An urn contains 4 white and 4 black balls. We randomly choose 4 balls. If 2 of them are white and 2 are black, we stop. If not, we replace the balls in the urn and again randomly select 4 balls. This continues until exactly 2 of the 4 chosen are white. What is the probability that we shall make exactly n selections?

74. Suppose that a batch of 100 items contains 6 that are defective and 94 that are nondefective. If X is the number of defective items in a randomly drawn sample of 10 items from the batch, find (a) $P\{X = 0\}$ and (b) $P\{X > 2\}$.

75. A game popular in Nevada gambling casinos is Keno, which is played as follows: Twenty numbers are selected at random by the casino from the set of numbers 1 through 80. A player can select from 1 to 15 numbers; a win occurs if some fraction of the player's chosen subset matches with any of the 20 numbers drawn by the house. The payoff is a function of the number of elements in the player's selection and the number of matches. For instance, if the player selects only 1 number, then he or she wins if this number is among the set of 20, and the payoff is \$2.2 won for every dollar bet. (As the player's probability of winning in this case is $\frac{1}{4}$, it is clear that the "fair" payoff should be \$3 won for every \$1 bet.) When the player selects 2 numbers, a payoff (of odds) of \$12 won for every \$1 bet is made when both numbers are among the 20,

(a) What would be the fair payoff in this case?
Let $P_{n,k}$ denote the probability that exactly k of the n numbers chosen by the player are among the 20 selected by the house.

(b) Compute $P_{n,k}$

(c) The most typical wager at Keno consists of selecting 10 numbers. For such a bet the casino pays off as shown in the following table. Compute the expected payoff:

Keno Payoffs in 10 Number Bets

Number of matches	Dollars won for each \$1 bet
0–4	−1
5	1
6	17
7	179
8	1,299
9	2,599
10	24,999

76. In Example 8i, what percentage of i defective lots does the purchaser reject? Find it for $i = 1, 4$. Given that a lot is rejected, what is the conditional probability that it contained 4 defective components?

77. A purchaser of transistors buys them in lots of 20. It is his policy to randomly inspect 4 components from a lot and to accept the lot only if all 4 are nondefective. If each component in a lot is, independently, defective with probability .1, what proportion of lots is rejected?

THEORETICAL EXERCISES

1. There are N distinct types of coupons, and each time one is obtained it will, independently of past choices, be of type i with probability P_i, $i = 1, \ldots, N$. Let T denote the number one need select to obtain at least one of each type. Compute $P\{T = n\}$.

HINT: Use an argument similar to the one used in Example 1e.

2. Use the continuity property of probabilities to find $\lim_{b \to -\infty} P\{X \le b\}$.

3. Express $P\{X \ge a\}$ in terms of the distribution function of X.

4. If X has distribution function F, what is the distribution function of e^X?

5. If X has distribution function F, what is the distribution function of the random variable $\alpha X + \beta$, where α and β are constants, $\alpha \ne 0$?

6. For a nonnegative integer-valued random variable N, show that

$$E[N] = \sum_{i=1}^{\infty} P\{N \ge i\}$$

HINT: $\sum_{i=1}^{\infty} P\{N \ge i\} = \sum_{i=1}^{\infty} \sum_{k=i}^{\infty} P\{N = k\}$. Now interchange the order of summation.

7. For a nonnegative integer-valued random variable N, show that

$$\sum_{i=0}^{\infty} iP\{N > i\} = \tfrac{1}{2}(E[N^2] - E[N])$$

HINT: $\sum_{i=0}^{\infty} iP\{N > i\} = \sum_{i=0}^{\infty} i \sum_{k=i+1}^{\infty} P\{N = k\}$. Now interchange the order of summation.

8. Let X be such that

$$P\{X = 1\} = p = 1 - P\{X = -1\}$$

Find $c \ne 1$ such that $E[c^X] = 1$.

9. Let X be a random variable having expected value μ and variance σ^2. Find the expected value and variance of

$$Y = \frac{X - \mu}{\sigma}$$

10. Let X be a binomial random variable with parameters n and p. Show that

$$E\left[\frac{1}{X + 1}\right] = \frac{1 - (1 - p)^{n+1}}{(n + 1)p}$$

11. Consider n independent sequential trials, each of which is successful with probability p. If there is a total of k successes, show that each of the $n!/[k!(n-k)!]$ possible arrangements of the k successes and $n-k$ failures is equally likely.

12. There are n components lined up in a linear arrangement. Suppose that each component independently functions with probability p. What is the probability that no 2 neighboring components are both nonfunctional?

HINT: Condition on the number of defective components and use the results of Example 4c of Chapter 1.

13. Let X be a binomial random variable with parameters (n, p). What value of p maximizes $P\{X = k\}, k = 0, 1, \ldots, n$? This is an example of a statistical method used to estimate p when a binomial (n, p) random variable is observed to equal k. If we assume that n is known, then we estimate p by choosing that value of p that maximizes $P\{X = k\}$. This is known as the method of maximum likelihood estimation.

14. A family has n children with probability $\alpha p^n, n \geq 1$, where $\alpha \leq (1-p)/p$.

(a) What proportion of families has no children?

(b) If each child is equally likely to be a boy or a girl (independently of each other), what proportion of families consists of k boys (and any number of girls)?

15. Suppose that n independent tosses of a coin having probability p of coming up heads are made. Show that the probability that an even number of heads results is $\frac{1}{2}[1 + (q-p)^n]$, where $q = 1 - p$. Do this by proving and then utilizing the identity

$$\sum_{i=0}^{[n/2]} \binom{n}{2i} p^{2i} q^{n-2i} = \frac{1}{2}[(p+q)^n + (q-p)^n]$$

where $[n/2]$ is the largest integer less than or equal to $n/2$. Compare this exercise with Theoretical Exercise 15 of Chapter 3.

16. Let X be a Poisson random variable with parameter λ. Show that $P\{X = i\}$ increases monotonically and then decreases monotonically as i increases, reaching its maximum when i is the largest integer not exceeding λ.

HINT: Consider $P\{X = i\}/P\{X = i-1\}$.

17. Let X be a Poisson random variable with parameter λ.

(a) Show that

$$P\{X \text{ is even}\} = \tfrac{1}{2}[1 + e^{-2\lambda}]$$

by using the results of Theoretical Exercise 15 and the relationship between Poisson and binomial random variables.

(b) Verify the above directly by making use of the expansion of $e^{-\lambda} + e^{\lambda}$.

18. Let X be a Poisson random variable with parameter λ. What value of λ maximizes $P\{X = k\}, k \geq 0$?

19. If X is a Poisson random variable with parameter λ, show that

$$E[X^n] = \lambda E[(X+1)^{n-1}]$$

Now use this result to compute $E[X^3]$.

20. Consider n coins, each of which independently comes up heads with probability p. Suppose that n is large, and p is small, and let $\lambda = np$. Suppose that all n coins are tossed; if at least one comes up heads the experiment ends, if not we again toss all n coins, and so on. That is, we stop the first time that at least one of the n coins come up heads. Let X denote the total number of heads that appear. Which of the following reasonings concerned with approximating $P\{X = 1\}$ is correct? In all cases Y is a Poisson random variable with parameter λ.

(a) Because the total number of heads that occur when all n coins are rolled is approximately a Poisson random variable with parameter λ,

$$P\{X = 1\} \approx P\{Y = 1\} = \lambda e^{-\lambda}$$

(b) Because the total number of heads that occur when all n coins are rolled is approximately a Poisson random variable with parameter λ, and because we only stop when this number is positive,

$$P\{X = 1\} \approx P\{Y = 1|Y > 0\} = \frac{\lambda e^{-\lambda}}{1 - e^{-\lambda}}$$

(c) Because at least one coin comes up heads, X will equal 1 if none of the other $n - 1$ come up heads. Because the number of heads resulting from these $n - 1$ coins is approximately Poisson with mean $(n - 1)p \approx \lambda$,

$$P\{X = 1\} \approx P\{Y = 0\} = e^{-\lambda}$$

21. From a set of n randomly chosen people let E_{ij} denote the event that persons i and j have the same birthday. Assume that each person is equally likely to have any of the 365 days of the year as his or her birthday. Find

(a) $P(E_{3,4}|E_{1,2})$; **(b)** $P(E_{1,3}|E_{1,2})$; **(c)** $P(E_{2,3}|E_{1,2} \cap E_{1,3})$.

What can you conclude from the above about the independence of the $\binom{n}{2}$ events E_{ij}?

22. An urn contains $2n$ balls, of which 2 are numbered 1, 2 are numbered 2, ..., and 2 are numbered n. Balls are successively withdrawn 2 at a time without replacement. Let T denote the first selection in which the balls withdrawn have the same number (and let it equal infinity if none of the pairs withdrawn has the same number). For $0 < \alpha < 1$ we want to show that

$$\lim_n P\{T > \alpha n\} = e^{-\alpha/2}$$

To verify the preceding, let M_k denote the number of pairs withdrawn in the first k selections, $k = 1, \ldots, n$.

(a) Argue that when n is large, M_k can be regarded as the number of successes in k (approximately) independent trials.

(b) When n is large, approximate $P\{M_k = 0\}$.

(c) Write the event $\{T > \alpha n\}$ in terms of the value of one of the variables M_k.

(d) Verify the limiting probability given for $P\{T > \alpha n\}$.

23. Suppose that the number of events that occur in a specified time is a Poisson random variable with parameter λ. If each event is counted with probability p, independently of every other event, show that the number of events that are counted is a Poisson random variable with parameter λp. Also, give an intuitive argument as to why this should be so.

As an application of the preceding paragraph, suppose that the number of distinct uranium deposits in a given area is a Poisson random variable with parameter $\lambda = 10$. If, in a fixed period of time, each deposit is discovered independently with probability $\frac{1}{50}$, find the probability that (a) exactly 1, (b) at least 1, and (c) at most 1 deposit is discovered during that time.

24. Prove

$$\sum_{i=0}^{n} e^{-\lambda}\frac{\lambda^i}{i!} = \frac{1}{n!}\int_{\lambda}^{\infty} e^{-x}x^n\,dx$$

HINT: Use integration by parts.

25. If X is a geometric random variable, show analytically that

$$P\{X = n + k | X > n\} = P\{X = k\}$$

Give a verbal argument using the interpretation of a geometric random variable as to why the preceding equation is true.

26. Let X be a negative binomial random variable with parameters r and p, and let Y be a binomial random variable with parameters n and p. Show that

$$P\{X > n\} = P\{Y < r\}$$

HINT: One could either attempt an analytical proof of the preceding, which is equivalent to proving the identity

$$\sum_{i=n+1}^{\infty} \binom{i-1}{r-1} p^r (1-p)^{i-r} = \sum_{i=0}^{r-1} \binom{n}{i} p^i (1-p)^{n-i}$$

or one could attempt a proof that uses the probabilistic interpretation of these random variables. That is, in the latter case start by considering a sequence of independent trials having a common success probability p. Then try to express the events $\{X > n\}$ and $\{Y < r\}$ in terms of the outcomes of this sequence.

27. For a hypergeometric random variable, determine

$$P\{X = k + 1\}/P\{X = k\}$$

28. Balls numbered 1 through N are in an urn. Suppose that $n, n \le N$, of them are randomly selected without replacement. Let Y denote the largest number selected.

(a) Find the probability mass function of Y.

(b) Derive an expression for $E[Y]$ and then use Fermat's combinatorial identity (see Theoretical Exercise 11 of Chapter 1) to simplify.

29. A jar contains $m + n$ chips, numbered $1, 2, \ldots, n + m$. A set of size n is drawn. If we let X denote the number of chips drawn having numbers that exceed all the numbers of those remaining, compute the probability mass function of X.

30. A jar contains n chips. Suppose that a boy successively draws a chip from the jar, each time replacing the one drawn before drawing another. This continues until the boy draws a chip that he has previously drawn before. Let X denote the number of draws, and compute its probability mass function.

31. Show that Equation (8.6) follows from (8.5).

32. From a set of n elements a nonempty subset is chosen at random in the sense that all of the nonempty subsets are equally likely to be selected. Let X denote the number of elements in the chosen subset. Using the identities given in Theoretical Exercise 12 of Chapter 1, show that

$$E[X] = \frac{n}{2 - \left(\frac{1}{2}\right)^{n-1}}$$

$$\text{Var}(X) = \frac{n \cdot 2^{2n-2} - n(n+1)2^{n-2}}{(2^n - 1)^2}$$

Show also that for n large,

$$\text{Var}(X) \sim \frac{n}{4}$$

in the sense that the ratio of the preceding approaches 1 as n approaches ∞. Compare this with the limiting form of $\text{Var}(Y)$ when $P\{Y = i\} = 1/n, i = 1, \ldots, n$.

33. An urn initially contains one red and one blue ball. At each stage a ball is randomly chosen and then replaced along with another of the same color. Let X denote the selection number of the first chosen ball that is blue. For instance, if the first selection is red and the second blue, then X is equal to 2.

(a) Find $P\{X > i\}, i \geq 1$.

(b) Show that with probability 1, a blue ball is eventually chosen. (That is, show that $P\{X < \infty\} = 1$.)

(c) Find $E[X]$.

SELF-TEST PROBLEMS AND EXERCISES

1. Suppose that the random variable X is equal to the number of hits obtained by a certain baseball player in his next 3 at bats. If $P\{X = 1\} = .3, P\{X = 2\} = .2$, and $P\{X = 0\} = 3P\{X = 3\}$, find $E[X]$.

2. Suppose that X takes on one of the values 0, 1, 2. If for some constant c, $P\{X = i\} = cP\{X = i - 1\}, i = 1, 2$, find $E[X]$.

3. A coin that when flipped comes up heads with probability p is flipped until either heads or tails has occurred twice. Find the expected number of flips.

4. A certain community is composed of m families, n_i of which have i children, $\sum_{i=1}^{r} n_i = m$. If one of the families is randomly chosen, let X denote the number of children in that family. If one of the $\sum_{i=1}^{r} in_i$ children is randomly chosen, let Y denote the total number of children in the family of that child. Show that $E[Y] \geq E[X]$.

5. Suppose that $P\{X = 0\} = 1 - P\{X = 1\}$. If $E[X] = 3\text{Var}(X)$, find $P\{X = 0\}$.

6. There are 2 coins in a bin. When one of them is flipped it lands on heads with probability .6, and when the other is flipped it lands on heads with probability .3. One of these coins is to be randomly chosen and then flipped. Without knowing which coin is chosen, you can bet any amount up to 10 dollars and you then either win that amount if the coin comes up heads or lose it if it comes up tails. Suppose, however, that an insider is willing to sell you, for an amount C, the information as to which coin was selected. What is your expected payoff if you buy this information? Note that if you buy it and then bet x, then you will end up either winning $x - C$ or $-x - C$ (that is, losing $x + C$ in the latter case). Also, for what values of C does it pay to purchase the information?

7. A philanthropist writes a positive number x on a piece of red paper, shows it to an impartial observer, and then turns it face down on the table. The observer then flips a fair coin. If it shows heads, she writes the value $2x$, and, if tails, the value $x/2$, on a piece of blue paper which she then turns face down on the table. Without knowing either the value x or the result of the coin flip, you have the option of turning over either the red or the blue piece of paper. After doing so, and observing the number written on that paper, you may elect to receive as a reward either that amount or the (unknown) amount written on the other piece of paper. For instance, if you elect to turn over the blue paper and observe the value 100, then you can elect either to accept 100 as your reward or to take the amount (either 200 or 50) on the red paper. Suppose that you would like your expected reward to be large.

(a) Argue that there is no reason to turn over the red paper first because if you do so, then no matter what value you observe, it is always better to switch to the blue paper.
(b) Let y be a fixed nonnegative value, and consider the following strategy. Turn over the blue paper and if its value is at least y, then accept that amount. If it is less than y, then switch to the red paper. Let $R_y(x)$ denote the reward obtained if the philanthropist writes the amount x and you employ this strategy. Find $E[R_y(x)]$. Note that $E[R_0(x)]$ is the expected reward if the philanthropist writes the amount x when you employ the strategy of always choosing the blue paper.

8. Let $B(n, p)$ represent a binomial random variable with parameters n and p. Argue that

$$P\{B(n, p) \leq i\} = 1 - P\{B(n, 1 - p) \leq n - i - 1\}$$

HINT: The number of successes less than or equal to i is equivalent to what statement about the number of failures?

9. If X is a binomial random variable with expected value 6 and variance 2.4, find $P\{X = 5\}$.

10. An urn contains n balls, numbered 1 through n. If m balls are randomly withdrawn in sequence, each time replacing the ball selected previously, find $P\{X = k\}, k = 1, \ldots, m$, where X is the maximum of the m chosen numbers.

HINT: First find $P\{X \leq k\}$.

11. Teams A and B play a series of games with the first team to win 3 games being declared the winner of the series. Suppose that team A independently wins each game with probability p. Find the conditional probability that team A wins
(a) the series given that it wins the first game;
(b) the first game given that it wins the series.

12. A local soccer team has 5 more games that it will play. If it wins its game this weekend, then it will play its final 4 games in the upper bracket of its league, and if it loses, then it will play its final games in the lower bracket. If it plays in the upper bracket, then it will independently win each of its games in this bracket with probability .4, and if it plays in the lower bracket, then it will independently win each of its games with probability .7. If the probability that it wins its game this weekend is .5, what is the probability that it wins at least 3 of its final 4 games?

13. On average, 5.2 hurricanes hit a certain region in a year. What is the probability that there will be 3 or fewer hurricanes hitting this year?

14. The number of eggs laid on a tree leaf by an insect of a certain type is a Poisson random variable with parameter λ. However, such a random variable can only be observed if it is positive, since if it is 0 then we cannot know that such an insect was on the leaf. If we let Y denote the observed number of eggs, then

$$P\{Y = i\} = P\{X = i | X > 0\}$$

where X is Poisson with parameter λ. Find $E[Y]$.

15. A casino patron will continue to make \$5 bets on red in roulette until she has won 4 of these bets.
(a) What is the probability that she places a total of 9 bets?
(b) What is her expected winnings when she stops?

REMARK: On each bet she will either win \$5 with probability $\frac{18}{38}$ or lose \$5 with probability $\frac{20}{38}$.

16. When three friends go for coffee, they decide who will pay the check by each flipping a coin and then letting the "odd person" pay. If all three flips are the same (so there is no odd person), then they make a second round of flips, and continue to do so until there is an odd person. What is the probability that

(a) exactly 3 rounds of flips are made;

(b) more than 4 rounds are needed?

17. If X is a geometric random variable with parameter p, show that

$$E[1/X] = \frac{-p\log(p)}{1-p}$$

HINT: You will need to evaluate an expression of the form $\sum_{i=1}^{\infty} a^i/i$. To do so, write $a^i/i = \int_0^a x^{i-1}\,dx$, and then interchange the sum and the integral.

18. Suppose that

$$P\{X = a\} = p, \qquad P\{X = b\} = 1 - p$$

(a) Show that $\dfrac{X-b}{a-b}$ is a Bernoulli random variable.

(b) Find $\text{Var}(X)$.

19. Each game you play is a win with probability p. You plan to play 5 games, but if you win the fifth game then you will keep on playing until you lose.

(a) Find the expected number of games that you play.

(b) Find the expected number of games that you lose.

20. Balls are randomly withdrawn, one at a time without replacement, from an urn that initially has N white and M black balls. Find the probability that n white balls are drawn before m black balls, $n \le N$, $m \le M$.

CHAPTER 5

Continuous Random Variables

5.1 INTRODUCTION

In Chapter 4 we considered discrete random variables, that is, random variables whose set of possible values is either finite or countably infinite. However, there also exist random variables whose set of possible values is uncountable. Two examples would be the time that a train arrives at a specified stop and the lifetime of a transistor. Let X be such a random variable. We say that X is a *continuous*[†] random variable if there exists a nonnegative function f, defined for all real $x \in (-\infty, \infty)$, having the property that for any set B of real numbers[‡]

$$P\{X \in B\} = \int_B f(x)\,dx \tag{1.1}$$

The function f is called the *probability density function* of the random variable X (see Figure 5.1).

In words, Equation (1.1) states that the probability that X will be in B may be obtained by integrating the probability density function over the set B. Since X must assume some value, f must satisfy

$$1 = P\{X \in (-\infty, \infty)\} = \int_{-\infty}^{\infty} f(x)\,dx$$

All probability statements about X can be answered in terms of f. For instance, letting $B = [a, b]$, we obtain from Equation (1.1) that

$$P\{a \le X \le b\} = \int_a^b f(x)\,dx \tag{1.2}$$

If we let $a = b$ in Equation (1.2), we obtain

$$P\{X = a\} = \int_a^a f(x)\,dx = 0$$

[†] Sometimes called *absolutely continuous*.

[‡] Actually, for technical reasons Equation (1.1) is true only for the measurable sets B, which, fortunately, includes all sets of practical interest.

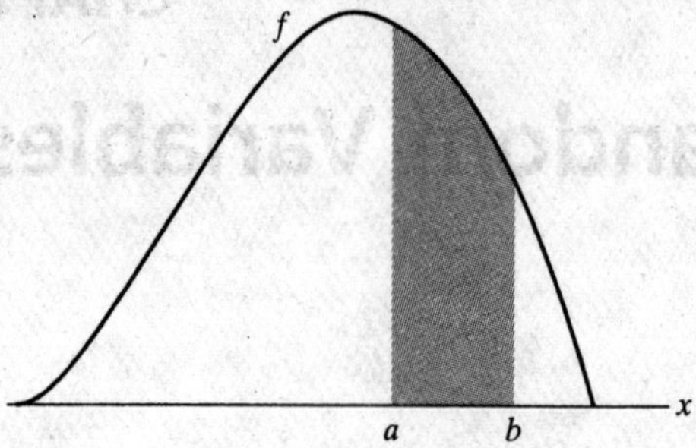

$P(a \le X \le b)$ = area of shaded region

Figure 5.1 Probability density function f.

In words, this equation states that the probability that a continuous random variable will assume any fixed value is zero. Hence, for a continuous random variable,

$$P\{X < a\} = P\{X \le a\} = F(a) = \int_{-\infty}^{a} f(x)\,dx$$

Example 1a. Suppose that X is a continuous random variable whose probability density function is given by

$$f(x) = \begin{cases} C(4x - 2x^2) & 0 < x < 2 \\ 0 & \text{otherwise} \end{cases}$$

(a) What is the value of C?
(b) Find $P\{X > 1\}$.

Solution (a) Since f is a probability density function, we must have that $\int_{-\infty}^{\infty} f(x)\,dx = 1$, implying that

$$C \int_{0}^{2} (4x - 2x^2)\,dx = 1$$

or

$$C\left[2x^2 - \frac{2x^3}{3}\right]\Bigg|_{x=0}^{x=2} = 1$$

or

$$C = \frac{3}{8}$$

Hence

(b) $$P\{X > 1\} = \int_{1}^{\infty} f(x)\,dx = \tfrac{3}{8}\int_{1}^{2} (4x - 2x^2)\,dx = \tfrac{1}{2}$$ ■

Example 1b. The amount of time, in hours, that a computer functions before breaking down is a continuous random variable with probability density function given by

$$f(x) = \begin{cases} \lambda e^{-x/100} & x \geq 0 \\ 0 & x < 0 \end{cases}$$

What is the probability that

(a) a computer will function between 50 and 150 hours before breaking down;
(b) it will function less than 100 hours?

Solution (a) Since

$$1 = \int_{-\infty}^{\infty} f(x)\,dx = \lambda \int_0^{\infty} e^{-x/100}\,dx$$

we obtain

$$1 = -\lambda(100)e^{-x/100}\Big|_0^{\infty} = 100\lambda \qquad \text{or} \qquad \lambda = \frac{1}{100}$$

Hence the probability that a computer will function between 50 and 150 hours before breaking down is given by

$$P\{50 < X < 150\} = \int_{50}^{150} \frac{1}{100} e^{-x/100}\,dx = -e^{-x/100}\Big|_{50}^{150}$$

$$= e^{-1/2} - e^{-3/2} \approx .384$$

(b) Similarly,

$$P\{X < 100\} = \int_0^{100} \frac{1}{100} e^{-x/100}\,dx = -e^{-x/100}\Big|_0^{100} = 1 - e^{-1} \approx .633$$

In other words, approximately 63.3 percent of the time a computer will fail before registering 100 hours of use. ■

Example 1c. The lifetime in hours of a certain kind of radio tube is a random variable having a probability density function given by

$$f(x) = \begin{cases} 0 & x \leq 100 \\ \dfrac{100}{x^2} & x > 100 \end{cases}$$

What is the probability that exactly 2 of 5 such tubes in a radio set will have to be replaced within the first 150 hours of operation? Assume that the events E_i, $i = 1, 2, 3, 4, 5$, that the ith such tube will have to be replaced within this time, are independent.

Solution Now,

$$P(E_i) = \int_0^{150} f(x)\,dx$$

$$= 100 \int_{100}^{150} x^{-2}\,dx$$

$$= \frac{1}{3}$$

Hence, from the independence of the events E_i, it follows that the desired probability is

$$\binom{5}{2}\left(\frac{1}{3}\right)^2\left(\frac{2}{3}\right)^3 = \frac{80}{243}$$ ■

The relationship between the cumulative distribution F and the probability density f is expressed by

$$F(a) = P\{X \in (-\infty, a]\} = \int_{-\infty}^{a} f(x)\,dx$$

Differentiating both sides of the preceding equation yields

$$\frac{d}{da}F(a) = f(a)$$

That is, the density is the derivative of the cumulative distribution function. A somewhat more intuitive interpretation of the density function may be obtained from Equation (1.2) as follows:

$$P\left\{a - \frac{\varepsilon}{2} \le X \le a + \frac{\varepsilon}{2}\right\} = \int_{a-\varepsilon/2}^{a+\varepsilon/2} f(x)\,dx \approx \varepsilon f(a)$$

when ε is small and when $f(\cdot)$ is continuous at $x = a$. In other words, the probability that X will be contained in an interval of length ε around the point a is approximately $\varepsilon f(a)$. From this, we see that $f(a)$ is a measure of how likely it is that the random variable will be near a.

5.2 EXPECTATION AND VARIANCE OF CONTINUOUS RANDOM VARIABLES

In Chapter 4 we defined the expected value of a discrete random variable X by

$$E[X] = \sum_x xP\{X = x\}$$

If X is a continuous random variable having probability density function $f(x)$, then as

$$f(x)\,dx \approx P\{x \le X \le x + dx\} \qquad \text{for } dx \text{ small}$$

it is easy to see that the analogous definition is to define the expected value of X by

$$E[X] = \int_{-\infty}^{\infty} x f(x)\, dx$$

Example 2a. Find $E[X]$ when the density function of X is

$$f(x) = \begin{cases} 2x & \text{if } 0 \le x \le 1 \\ 0 & \text{otherwise} \end{cases}$$

Solution

$$\begin{aligned} E[X] &= \int x f(x)\, dx \\ &= \int_0^1 2x^2\, dx \\ &= \tfrac{2}{3} \end{aligned}$$

■

Example 2b. The density function of X is given by

$$f(x) = \begin{cases} 1 & \text{if } 0 \le x \le 1 \\ 0 & \text{otherwise} \end{cases}$$

Find $E[e^X]$.

Solution Let $Y = e^X$. We start by determining F_Y, the probability distribution function of Y. Now, for $1 \le x \le e$,

$$\begin{aligned} F_Y(x) &= P\{Y \le x\} \\ &= P\{e^X \le x\} \\ &= P\{X \le \log(x)\} \\ &= \int_0^{\log(x)} f(y)\, dy \\ &= \log(x) \end{aligned}$$

By differentiating $F_Y(x)$, we obtain that the probability density function of Y is given by

$$f_Y(x) = \frac{1}{x} \qquad 1 \le x \le e$$

Hence

$$\begin{aligned} E[e^X] = E[Y] &= \int_{-\infty}^{\infty} x f_Y(x)\, dx \\ &= \int_1^e dx \\ &= e - 1 \end{aligned}$$

■

Although the method employed in Example 2b to compute the expected value of a function of X is always applicable, there is, as in the discrete case, an alternative way of proceeding. The following is a direct analog of Proposition 5.1 of Chapter 4.

Proposition 2.1

If X is a continuous random variable with probability density function $f(x)$, then for any real-valued function g,

$$E[g(X)] = \int_{-\infty}^{\infty} g(x)f(x)\,dx$$

An application of Proposition 2.1 to Example 2b yields that

$$E[e^X] = \int_0^1 e^x\,dx \qquad \text{since } f(x) = 1, \quad 0 < x < 1$$
$$= e - 1$$

which is in accord with the result of that example.

The proof of Proposition 2.1 is more involved than its discrete random variable analog, and we will present one under the provision that the random variable $g(X)$ is nonnegative. (The general proof, which follows the argument in the case we present, is indicated in Theoretical Exercises 2 and 3.) We will need the following lemma, which is of independent interest.

Lemma 2.1

For a nonnegative random variable Y,

$$E[Y] = \int_0^{\infty} P\{Y > y\}\,dy$$

Proof: We present a proof when Y is a continuous random variable with probability density function f_Y. We have

$$\int_0^{\infty} P\{Y > y\}\,dy = \int_0^{\infty}\int_y^{\infty} f_Y(x)\,dx\,dy$$

where we have used the fact that $P\{Y > y\} = \int_y^{\infty} f_Y(x)\,dx$. Interchanging the order of integration in the preceding equation yields

$$\int_0^{\infty} P\{Y > y\}\,dy = \int_0^{\infty}\left(\int_0^{x} dy\right)f_Y(x)\,dx$$
$$= \int_0^{\infty} x f_Y(x)\,dx$$
$$= E[Y]$$

✦

Proof of Proposition 2.1: For any function g for which $g(x) \geq 0$, we have from Lemma 2.1 that

$$E[g(X)] = \int_0^\infty P\{g(X) > y\}\,dy$$

$$= \int_0^\infty \int_{x:g(x)>y} f(x)\,dx\,dy$$

$$= \int_{x:g(x)>0} \int_0^{g(x)} dy\,f(x)\,dx$$

$$= \int_{x:g(x)>0} g(x)f(x)\,dx$$

which completes the proof. ✦

Example 2c. A stick of length 1 is split at a point U that is uniformly distributed over $(0, 1)$. Determine the expected length of the piece that contains the point $p, 0 \leq p \leq 1$.

Solution Let $L_p(U)$ denote the length of the substick that contains the point p, and note (see Figure 5.2) that

$$L_p(U) = \begin{cases} 1 - U & U < p \\ U & U > p \end{cases}$$

Hence from Proposition 2.1 we have that

$$E[L_p(U)] = \int_0^1 L_p(u)\,du$$

$$= \int_0^p (1 - u)\,du + \int_p^1 u\,du$$

$$= \frac{1}{2} - \frac{(1-p)^2}{2} + \frac{1}{2} - \frac{p^2}{2}$$

$$= \frac{1}{2} + p(1 - p)$$

Since $p(1 - p)$ is maximized when $p = \frac{1}{2}$, it is interesting to note that the expected length of the substick containing the point p is maximized when p is the midpoint of the original stick. ■

Figure 5.2 Substick containing point p: (a) $U < p$; (b) $U > p$.

Example 2d. Suppose that if you are s minutes early for an appointment, then you incur the cost cs, and if you are s minutes late, then you incur the cost ks. Suppose that the travel time from where you presently are to the location of your appointment is a continuous random variable having probability density function f. Determine the time at which you should depart if you want to minimize your expected cost.

Solution Let X denote the travel time. If you leave t minutes before your appointment, then your cost, call it $C_t(X)$, is given by

$$C_t(X) = \begin{cases} c(t - X) & \text{if } X \le t \\ k(X - t) & \text{if } X \ge t \end{cases}$$

Therefore,

$$\begin{aligned} E[C_t(X)] &= \int_0^\infty C_t(x)f(x)\,dx \\ &= \int_0^t c(t-x)f(x)\,dx + \int_t^\infty k(x-t)f(x)\,dx \\ &= ct\int_0^t f(x)\,dx - c\int_0^t xf(x)\,dx + k\int_t^\infty xf(x)\,dx - kt\int_t^\infty f(x)\,dx \end{aligned}$$

The value of t that minimizes $E[C_t(X)]$ can now be obtained by calculus. Differentiation yields

$$\begin{aligned} \frac{d}{dt}E[C_t(X)] &= ct\,f(t) + cF(t) - ct\,f(t) - kt\,f(t) + kt\,f(t) - k[1 - F(t)] \\ &= (k + c)F(t) - k \end{aligned}$$

Equating to zero shows that the minimal expected cost is obtained when you leave t^* minutes before your appointment, where t^* satisfies

$$F(t^*) = \frac{k}{k + c}$$ ■

As in Chapter 4, we can use Proposition 2.1 to show the following.

Corollary 2.1

If a and b are constants, then

$$E[aX + b] = aE[X] + b$$

The proof of Corollary 2.1 for a continuous random variable X is the same as the one given for a discrete random variable. The only modification is that the sum is replaced by an integral and the probability mass function by a probability density function.

The variance of a continuous random variable is defined exactly as it is for a discrete one. Namely, if X is a random variable with expected value μ, then the variance of X is defined (for any type of random variable) by

$$\text{Var}(X) = E[(X - \mu)^2]$$

The alternative formula,

$$\text{Var}(X) = E[X^2] - (E[X])^2$$

is established in a similar manner as in the discrete case.

Example 2e. Find $\text{Var}(X)$ for X as given in Example 2a.

Solution We first compute $E[X^2]$.

$$\begin{aligned} E[X^2] &= \int_{-\infty}^{\infty} x^2 f(x)\,dx \\ &= \int_0^1 2x^3\,dx \\ &= \tfrac{1}{2} \end{aligned}$$

Hence, since $E[X] = \frac{2}{3}$ we obtain that

$$\text{Var}(X) = \tfrac{1}{2} - \left(\tfrac{2}{3}\right)^2 = \tfrac{1}{18}$$ ■

It can be shown, with the proof mimicking the one given for discrete random variables, that for constants a and b

$$\text{Var}(aX + b) = a^2\,\text{Var}(X)$$

There are several important classes of continuous random variables that appear frequently in applications of probability; the next few sections are devoted to a study of some of them.

5.3 THE UNIFORM RANDOM VARIABLE

A random variable is said to be *uniformly* distributed over the interval (0, 1) if its probability density function is given by

$$f(x) = \begin{cases} 1 & 0 < x < 1 \\ 0 & \text{otherwise} \end{cases} \tag{3.1}$$

Note that Equation (3.1) is a density function, since $f(x) \geq 0$ and $\int_{-\infty}^{\infty} f(x)\,dx = \int_0^1 dx = 1$. Because $f(x) > 0$ only when $x \in (0, 1)$, it follows that X must assume a value in $(0, 1)$. Also, since $f(x)$ is constant for $x \in (0, 1)$, X is just as likely to be near any value in $(0, 1)$ as any other value. To check this, note that for any $0 < a < b < 1$,

$$P\{a \leq X \leq b\} = \int_a^b f(x)\,dx = b - a$$

In other words, the probability that X is in any particular subinterval of $(0, 1)$ equals the length of that subinterval.

In general, we say that X is a uniform random variable on the interval (α, β) if its probability density function is given by

$$f(x) = \begin{cases} \dfrac{1}{\beta - \alpha} & \text{if } \alpha < x < \beta \\ 0 & \text{otherwise} \end{cases} \tag{3.2}$$

Since $F(a) = \int_{-\infty}^{a} f(x)\,dx$, we obtain from Equation (3.2) that the distribution function of a uniform random variable on the interval (α, β) is given by

$$F(a) = \begin{cases} 0 & a \le \alpha \\ \dfrac{a - \alpha}{\beta - \alpha} & \alpha < a < \beta \\ 1 & a \ge \beta \end{cases}$$

Figure 5.3 presents a graph of $f(a)$ and $F(a)$.

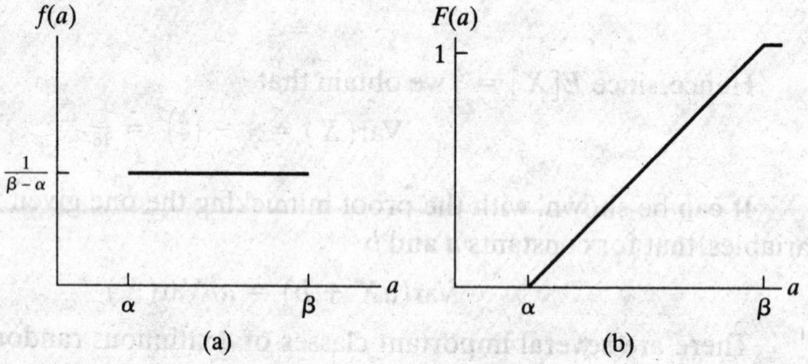

Figure 5.3 Graph of (a) $f(a)$ and (b) $F(a)$ for a uniform (α, β) random variable.

Example 3a. Let X be uniformly distributed over (α, β). Find (a) $E[X]$ and (b) $\text{Var}(X)$.

Solution

$$\text{(a) } E[X] = \int_{-\infty}^{\infty} x f(x)\,dx$$

$$= \int_{\alpha}^{\beta} \frac{x}{\beta - \alpha}\,dx$$

$$= \frac{\beta^2 - \alpha^2}{2(\beta - \alpha)}$$

$$= \frac{\beta + \alpha}{2}$$

In words, the expected value of a random variable uniformly distributed over some interval is equal to the midpoint of that interval.

(b) To find $\text{Var}(X)$, we first calculate $E[X^2]$.

$$E[X^2] = \int_\alpha^\beta \frac{1}{\beta - \alpha} x^2\,dx$$
$$= \frac{\beta^3 - \alpha^3}{3(\beta - \alpha)}$$
$$= \frac{\beta^2 + \alpha\beta + \alpha^2}{3}$$

Hence

$$\text{Var}(X) = \frac{\beta^2 + \alpha\beta + \alpha^2}{3} - \frac{(\alpha + \beta)^2}{4}$$
$$= \frac{(\beta - \alpha)^2}{12}$$

Therefore, the variance of a random variable that is uniformly distributed over some interval is the square of the length of that interval divided by 12. ■

Example 3b. If X is uniformly distributed over $(0, 10)$, calculate the probability that (a) $X < 3$, (b) $X > 6$, and (c) $3 < X < 8$.

Solution

$$\text{(a) } P\{X < 3\} = \int_0^3 \tfrac{1}{10}\,dx = \tfrac{3}{10}$$

$$\text{(b) } P\{X > 6\} = \int_6^{10} \tfrac{1}{10}\,dx = \tfrac{4}{10}$$

$$\text{(c) } P\{3 < X < 8\} = \int_3^8 \tfrac{1}{10}\,dx = \tfrac{1}{2}$$ ■

Example 3c. Buses arrive at a specified stop at 15-minute intervals starting at 7 A.M. That is, they arrive at 7, 7:15, 7:30, 7:45, and so on. If a passenger arrives at the stop at a time that is uniformly distributed between 7 and 7:30, find the probability that he waits

(a) less than 5 minutes for a bus;
(b) more than 10 minutes for a bus.

Solution Let X denote the number of minutes past 7 that the passenger arrives at the stop. Since X is a uniform random variable over the interval $(0, 30)$, it follows that the passenger will have to wait less than 5 minutes if (and only if) he arrives between 7:10 and 7:15 or between 7:25 and 7:30. Hence the desired probability for part (a) is

$$P\{10 < X < 15\} + P\{25 < X < 30\} = \int_{10}^{15} \tfrac{1}{30}\, dx + \int_{25}^{30} \tfrac{1}{30}\, dx = \tfrac{1}{3}$$

Similarly, he would have to wait more than 10 minutes if he arrives between 7 and 7:05 or between 7:15 and 7:20, and so the probability for part (b) is

$$P\{0 < X < 5\} + P\{15 < X < 20\} = \tfrac{1}{3}$$ ■

The next example was first considered by the French mathematician L. F. Bertrand in 1889 and is often referred to as *Bertrand's paradox*. It represents our initial introduction to a subject commonly referred to as geometrical probability.

Example 3d. Consider a random chord of a circle. What is the probability that the length of the chord will be greater than the side of the equilateral triangle inscribed in that circle?

Solution The problem as stated is incapable of solution because it is not clear what is meant by a random chord. To give meaning to this phrase, we shall reformulate the problem in two distinct ways.

The first formulation is as follows: The position of the chord can be determined by its distance from the center of the circle. This distance can vary between 0 and r, the radius of the circle. Now, the length of the chord will be greater than the side of the equilateral triangle inscribed in the circle if its distance from the center is less than $r/2$. Hence, by assuming that a random chord is one whose distance D from the center is uniformly distributed between 0 and r, we see that the probability that it is greater than the side of an inscribed equilateral triangle is

$$P\left\{D < \frac{r}{2}\right\} = \frac{r/2}{r} = \frac{1}{2}$$

For our second formulation of the problem consider an arbitrary chord of the circle; through one end of the chord draw a tangent. The angle θ between the chord and the tangent, which can vary from 0° to 180°, determines the position of the chord (see Figure 5.4). Furthermore, the length of the chord will be greater than the side of the inscribed equilateral triangle if the angle θ is between 60° and 120°. Hence, assuming that a random chord is one whose angle

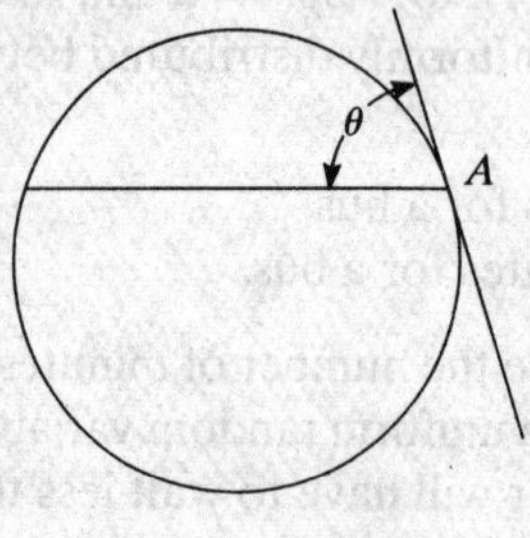

Figure 5.4

θ is uniformly distributed between 0° and 180°, we see that the desired answer in this formulation is

$$P\{60 < \theta < 120\} = \frac{120 - 60}{180} = \frac{1}{3}$$

It should be noted that random experiments could be performed in such a way that $\frac{1}{2}$ or $\frac{1}{3}$ would be the correct probability. For instance, if a circular disk of radius r is thrown on a table ruled with parallel lines a distance $2r$ apart, then one and only one of these lines would cross the disk and form a chord. All distances from this chord to the center of the disk would be equally likely so that the desired probability that the chord's length will be greater than the side of an inscribed equilateral triangle is $\frac{1}{2}$. On the other hand, if the experiment consisted of rotating a needle freely about a point A on the edge (see Figure 5.4) of the circle, the desired answer would be $\frac{1}{3}$. ■

5.4 NORMAL RANDOM VARIABLES

We say that X is a normal random variable, or simply that X is normally distributed, with parameters μ and σ^2 if the density of X is given by

$$f(x) = \frac{1}{\sqrt{2\pi}\,\sigma} e^{-(x-\mu)^2/2\sigma^2} \qquad -\infty < x < \infty$$

This density function is a bell-shaped curve that is symmetric about μ (see Figure 5.5).

The normal distribution was introduced by the French mathematician Abraham DeMoivre in 1733 and was used by him to approximate probabilities associated

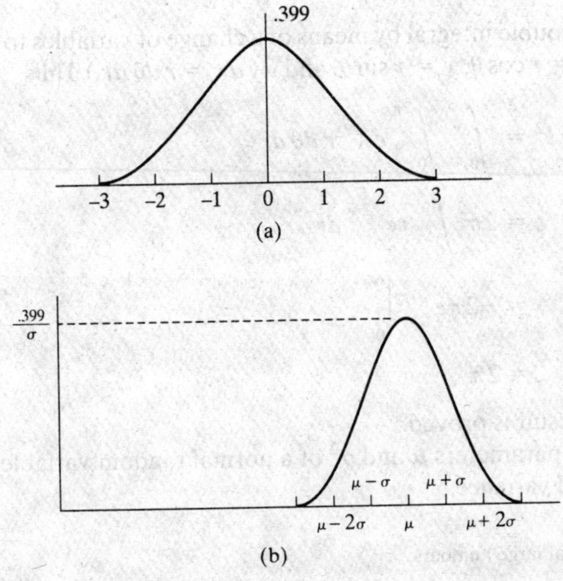

Figure 5.5 Normal density function: (a) $\mu = 0, \sigma = 1$; (b) arbitrary μ, σ^2.

with binomial random variables when the binomial parameter n is large. This result was later extended by Laplace and others and is now encompassed in a probability theorem known as the central limit theorem, which is discussed in Chapter 8. The central limit theorem, one of the two most important results in probability theory,† gives a theoretical base to the often noted empirical observation that, in practice, many random phenomena obey, at least approximately, a normal probability distribution. Some examples of this behavior are the height of a man, the velocity in any direction of a molecule in gas, and the error made in measuring a physical quantity.

To prove that $f(x)$ is indeed a probability density function, we need to show that

$$\frac{1}{\sqrt{2\pi}\,\sigma}\int_{-\infty}^{\infty} e^{-(x-\mu)^2/2\sigma^2}\,dx = 1$$

By making the substitution $y = (x - \mu)/\sigma$, we see that

$$\frac{1}{\sqrt{2\pi}\,\sigma}\int_{-\infty}^{\infty} e^{-(x-\mu)^2/2\sigma^2}\,dx = \frac{1}{\sqrt{2\pi}}\int_{-\infty}^{\infty} e^{-y^2/2}\,dy$$

and hence we must show that

$$\int_{-\infty}^{\infty} e^{-y^2/2}\,dy = \sqrt{2\pi}$$

Toward this end, let $I = \int_{-\infty}^{\infty} e^{-y^2/2}\,dy$. Then

$$\begin{aligned}
I^2 &= \int_{-\infty}^{\infty} e^{-y^2/2}\,dy \int_{-\infty}^{\infty} e^{-x^2/2}\,dx \\
&= \int_{-\infty}^{\infty}\int_{-\infty}^{\infty} e^{-(y^2+x^2)/2}\,dy\,dx
\end{aligned}$$

We now evaluate the double integral by means of a change of variables to polar coordinates. (That is, let $x = r\cos\theta$, $y = r\sin\theta$, and $dy\,dx = r\,d\theta\,dr$.) Thus

$$\begin{aligned}
I^2 &= \int_0^{\infty}\int_0^{2\pi} e^{-r^2/2} r\,d\theta\,dr \\
&= 2\pi \int_0^{\infty} re^{-r^2/2}\,dr \\
&= -2\pi e^{-r^2/2}\Big|_0^{\infty} \\
&= 2\pi
\end{aligned}$$

Hence $I = \sqrt{2\pi}$, and the result is proved.

We now show that the parameters μ and σ^2 of a normal random variable represent its expected value and variance.

† The other is the strong law of large numbers.

Example 4a. Find (a) $E[X]$ and (b) $\text{Var}(X)$ when X is a normal random variable with parameters μ and σ^2.

Solution (a) $E[X] = \dfrac{1}{\sqrt{2\pi}\,\sigma}\displaystyle\int_{-\infty}^{\infty} xe^{-(x-\mu)^2/2\sigma^2}\,dx$

Writing x as $(x - \mu) + \mu$ yields

$$E[X] = \frac{1}{\sqrt{2\pi}\,\sigma}\int_{-\infty}^{\infty} (x-\mu)e^{-(x-\mu)^2/2\sigma^2}\,dx + \mu\frac{1}{\sqrt{2\pi}\,\sigma}\int_{-\infty}^{\infty} e^{-(x-\mu)^2/2\sigma^2}\,dx$$

Letting $y = x - \mu$ in the first integral yields

$$E[X] = \frac{1}{\sqrt{2\pi}\,\sigma}\int_{-\infty}^{\infty} ye^{-y^2/2\sigma^2}\,dy + \mu\int_{-\infty}^{\infty} f(x)\,dx$$

where $f(x)$ is the normal density. By symmetry, the first integral must be 0, so

$$E[X] = \mu\int_{-\infty}^{\infty} f(x)\,dx = \mu$$

(b) Since $E[X] = \mu$, we have that

$$\begin{aligned}\text{Var}(X) &= E[(X-\mu)^2] \\ &= \frac{1}{\sqrt{2\pi}\,\sigma}\int_{-\infty}^{\infty} (x-\mu)^2 e^{-(x-\mu)^2/2\sigma^2}\,dx \qquad (4.1)\end{aligned}$$

Substituting $y = (x - \mu)/\sigma$ in Equation (4.1) yields

$$\begin{aligned}\text{Var}(X) &= \frac{\sigma^2}{\sqrt{2\pi}}\int_{-\infty}^{\infty} y^2 e^{-y^2/2}\,dy \\ &= \frac{\sigma^2}{\sqrt{2\pi}}\left[-ye^{-y^2/2}\Big|_{-\infty}^{\infty} + \int_{-\infty}^{\infty} e^{-y^2/2}\,dy\right] \text{ by integration by parts} \\ &= \sigma^2\frac{1}{\sqrt{2\pi}}\int_{-\infty}^{\infty} e^{-y^2/2}\,dy \\ &= \sigma^2\end{aligned}$$

■

An important fact about normal random variables is that if X is normally distributed with parameters μ and σ^2, then $Y = \alpha X + \beta$ is normally distributed with parameters $\alpha\mu + \beta$ and $\alpha^2\sigma^2$. To show this, suppose that $\alpha > 0$. (The verification when $\alpha < 0$ is similar.) Now, F_Y,[†] the cumulative distribution function of the random variable Y, is given by

[†]When there is more than one random variable under consideration, we shall denote the cumulative distribution function of a random variable Z by F_Z. Similarly, we shall denote the density of Z by f_Z.

$$\begin{aligned} F_Y(a) &= P\{\alpha X + \beta \le a\} \\ &= P\left\{X \le \frac{a-\beta}{\alpha}\right\} \\ &= F_X\left(\frac{a-\beta}{\alpha}\right) \end{aligned}$$

Differentiation yields that the density function of Y is

$$\begin{aligned} f_Y(a) &= \frac{1}{\alpha} f_X\left(\frac{a-\beta}{\alpha}\right) \\ &= \frac{1}{\sqrt{2\pi}\,\alpha\sigma} \exp\left\{-\left(\frac{a-\beta}{\alpha} - \mu\right)^2 \Big/ 2\sigma^2\right\} \\ &= \frac{1}{\sqrt{2\pi}\,\alpha\sigma} \exp\{-(a - \beta - \alpha\mu)^2/2(\alpha\sigma)^2\} \end{aligned} \tag{4.2}$$

which shows that Y is normal with mean $\alpha\mu + \beta$ and variance $\alpha^2\sigma^2$.

An important implication of the preceding result is that if X is normally distributed with parameters μ and σ^2, then $Z = (X - \mu)/\sigma$ is normally distributed with parameters 0 and 1. Such a random variable Z is said to have the *standard*, or *unit*, normal distribution.

It is traditional to denote the cumulative distribution function of a standard normal random variable by $\Phi(x)$. That is,

$$\Phi(x) = \frac{1}{\sqrt{2\pi}} \int_{-\infty}^{x} e^{-y^2/2}\, dy$$

The values of $\Phi(x)$ for nonnegative x are given in Table 5.1. For negative values of x, $\Phi(x)$ can be obtained from Equation (4.3):

$$\Phi(-x) = 1 - \Phi(x) \qquad -\infty < x < \infty \tag{4.3}$$

[The values of $\Phi(x)$ can also be obtained from the website.] The proof of Equation (4.3), which follows from the symmetry of the standard normal density, is left as an exercise. This equation states that if Z is a standard normal random variable, then

$$P\{Z \le -x\} = P\{Z > x\} \qquad -\infty < x < \infty$$

Since $Z = (X - \mu)/\sigma$ is a standard normal random variable whenever X is normally distributed with parameters μ and σ^2, it follows that the distribution function of X can be expressed as

$$\begin{aligned} F_X(a) &= P\{X \le a\} \\ &= P\left(\frac{X-\mu}{\sigma} \le \frac{a-\mu}{\sigma}\right) \\ &= \Phi\left(\frac{a-\mu}{\sigma}\right) \end{aligned}$$

TABLE 5.1 AREA $\Phi(x)$ UNDER THE STANDARD NORMAL CURVE TO THE LEFT OF x

x	.00	.01	.02	.03	.04	.05	.06	.07	.08	.09
.0	.5000	.5040	.5080	.5120	.5160	.5199	.5239	.5279	.5319	.5359
.1	.5398	.5438	.5478	.5517	.5557	.5596	.5636	.5675	.5714	.5753
.2	.5793	.5832	.5871	.5910	.5948	.5987	.6026	.6064	.6103	.6141
.3	.6179	.6217	.6255	.6293	.6331	.6368	.6406	.6443	.6480	.6517
.4	.6554	.6591	.6628	.6664	.6700	.6736	.6772	.6808	.6844	.6879
.5	.6915	.6950	.6985	.7019	.7054	.7088	.7123	.7157	.7190	.7224
.6	.7257	.7291	.7324	.7357	.7389	.7422	.7454	.7486	.7517	.7549
.7	.7580	.7611	.7642	.7673	.7704	.7734	.7764	.7794	.7823	.7852
.8	.7881	.7910	.7939	.7967	.7995	.8023	.8051	.8078	.8106	.8133
.9	.8159	.8186	.8212	.8238	.8264	.8289	.8315	.8340	.8365	.8389
1.0	.8413	.8438	.8461	.8485	.8508	.8531	.8554	.8577	.8599	.8621
1.1	.8643	.8665	.8686	.8708	.8729	.8749	.8770	.8790	.8810	.8830
1.2	.8849	.8869	.8888	.8907	.8925	.8944	.8962	.8980	.8997	.9015
1.3	.9032	.9049	.9066	.9082	.9099	.9115	.9131	.9147	.9162	.9177
1.4	.9192	.9207	.9222	.9236	.9251	.9265	.9279	.9292	.9306	.9319
1.5	.9332	.9345	.9357	.9370	.9382	.9394	.9406	.9418	.9429	.9441
1.6	.9452	.9463	.9474	.9484	.9495	.9505	.9515	.9525	.9535	.9545
1.7	.9554	.9564	.9573	.9582	.9591	.9599	.9608	.9616	.9625	.9633
1.8	.9641	.9649	.9656	.9664	.9671	.9678	.9686	.9693	.9699	.9706
1.9	.9713	.9719	.9726	.9732	.9738	.9744	.9750	.9756	.9761	.9767
2.0	.9772	.9778	.9783	.9788	.9793	.9798	.9803	.9808	.9812	.9817
2.1	.9821	.9826	.9830	.9834	.9838	.9842	.9846	.9850	.9854	.9857
2.2	.9861	.9864	.9868	.9871	.9875	.9878	.9881	.9884	.9887	.9890
2.3	.9893	.9896	.9898	.9901	.9904	.9906	.9909	.9911	.9913	.9916
2.4	.9918	.9920	.9922	.9925	.9927	.9929	.9931	.9932	.9934	.9936
2.5	.9938	.9940	.9941	.9943	.9945	.9946	.9948	.9949	.9951	.9952
2.6	.9953	.9955	.9956	.9957	.9959	.9960	.9961	.9962	.9963	.9964
2.7	.9965	.9966	.9967	.9968	.9969	.9970	.9971	.9972	.9973	.9974
2.8	.9974	.9975	.9976	.9977	.9977	.9978	.9979	.9979	.9980	.9981
2.9	.9981	.9982	.9982	.9983	.9984	.9984	.9985	.9985	.9986	.9986
3.0	.9987	.9987	.9987	.9988	.9988	.9989	.9989	.9989	.9990	.9990
3.1	.9990	.9991	.9991	.9991	.9992	.9992	.9992	.9992	.9993	.9993
3.2	.9993	.9993	.9994	.9994	.9994	.9994	.9994	.9995	.9995	.9995
3.3	.9995	.9995	.9995	.9996	.9996	.9996	.9996	.9996	.9996	.9997
3.4	.9997	.9997	.9997	.9997	.9997	.9997	.9997	.9997	.9997	.9998

Example 4b. If X is a normal random variable with parameters $\mu = 3$ and $\sigma^2 = 9$, find (a) $P\{2 < X < 5\}$; (b) $P\{X > 0\}$; (c) $P\{|X - 3| > 6\}$.

Solution

$$
\begin{aligned}
\text{(a)}\quad P\{2 < X < 5\} &= P\left\{\frac{2-3}{3} < \frac{X-3}{3} < \frac{5-3}{3}\right\} \\
&= P\left\{-\frac{1}{3} < Z < \frac{2}{3}\right\} \\
&= \Phi\left(\frac{2}{3}\right) - \Phi\left(-\frac{1}{3}\right) \\
&= \Phi\left(\frac{2}{3}\right) - \left[1 - \Phi\left(\frac{1}{3}\right)\right] \approx .3779
\end{aligned}
$$

$$\text{(b) } P\{X > 0\} = P\left\{\frac{X-3}{3} > \frac{0-3}{3}\right\} = P\{Z > -1\}$$
$$= 1 - \Phi(-1)$$
$$= \Phi(1)$$
$$\approx .8413$$

$$\text{(c) } P\{|X-3| > 6\} = P\{X > 9\} + P\{X < -3\}$$
$$= P\left\{\frac{X-3}{3} > \frac{9-3}{3}\right\} + P\left\{\frac{X-3}{3} < \frac{-3-3}{3}\right\}$$
$$= P\{Z > 2\} + P\{Z < -2\}$$
$$= 1 - \Phi(2) + \Phi(-2)$$
$$= 2[1 - \Phi(2)]$$
$$\approx .0456$$

■

Example 4c. An examination is often regarded as being good (in the sense of determining a valid grade spread for those taking it) if the test scores of those taking the examination can be approximated by a normal density function. (In other words, a graph of the frequency of grade scores should have approximately the bell-shaped form of the normal density.) The instructor often uses the test scores to estimate the normal parameters μ and σ^2 and then assigns the letter grade A to those whose test score is greater than $\mu + \sigma$, B to those whose score is between μ and $\mu + \sigma$, C to those whose score is between $\mu - \sigma$ and μ, D to those whose score is between $\mu - 2\sigma$ and $\mu - \sigma$, and F to those getting a score below $\mu - 2\sigma$. (This is sometimes referred to as grading "on the curve.") Since

$$P\{X > \mu + \sigma\} = P\left\{\frac{X-\mu}{\sigma} > 1\right\} = 1 - \Phi(1) \approx .1587$$

$$P\{\mu < X < \mu + \sigma\} = P\left\{0 < \frac{X-\mu}{\sigma} < 1\right\} = \Phi(1) - \Phi(0) \approx .3413$$

$$P\{\mu - \sigma < X < \mu\} = P\left\{-1 < \frac{X-\mu}{\sigma} < 0\right\}$$
$$= \Phi(0) - \Phi(-1) \approx .3413$$

$$P\{\mu - 2\sigma < X < \mu - \sigma\} = P\left\{-2 < \frac{X-\mu}{\sigma} < -1\right\}$$
$$= \Phi(2) - \Phi(1) \approx .1359$$

$$P\{X < \mu - 2\sigma\} = P\left\{\frac{X-\mu}{\sigma} < -2\right\} = \Phi(-2) \approx .0228$$

it follows that approximately 16 percent of the class will receive an A grade on the examination, 34 percent a B grade, 34 percent a C grade, and 14 percent a D grade; 2 percent will fail. ■

Example 4d. An expert witness in a paternity suit testifies that the length (in days) of pregnancy (that is, the time from impregnation to the delivery of the child) is approximately normally distributed with parameters $\mu = 270$ and $\sigma^2 = 100$. The defendant in the suit is able to prove that he was out of the country during a period that began 290 days before the birth of the child and ended 240 days before the birth. If the defendant was, in fact, the father of the child, what is the probability that the mother could have had the very long or very short pregnancy indicated by the testimony?

Solution Let X denote the length of the pregnancy and assume that the defendant is the father. Then the probability that the birth could occur within the indicated period is

$$\begin{aligned} P\{X > 290 \text{ or } X < 240\} &= P\{X > 290\} + P\{X < 240\} \\ &= P\left\{\frac{X - 270}{10} > 2\right\} + P\left\{\frac{X - 270}{10} < -3\right\} \\ &= 1 - \Phi(2) + 1 - \Phi(3) \\ &\approx .0241 \end{aligned}$$

■

Example 4e. Suppose that a binary message—either 0 or 1—must be transmitted by wire from location A to location B. However, the data sent over the wire are subject to a channel noise disturbance, so to reduce the possibility of error, the value 2 is sent over the wire when the message is 1 and the value -2 is sent when the message is 0. If $x, x = \pm 2$, is the value sent at location A, then R, the value received at location B, is given by $R = x + N$, where N is the channel noise disturbance. When the message is received at location B the receiver decodes it according to the following rule:

If $R \geq .5$, then 1 is concluded.
If $R < .5$, then 0 is concluded.

As the channel noise is often normally distributed, we will determine the error probabilities when N is a unit normal random variable.

There are two types of errors that can occur: One is that the message 1 can be incorrectly concluded to be 0, and the other that 0 is concluded to be 1. The first type of error will occur if the message is 1 and $2 + N < .5$, whereas the second will occur if the message is 0 and $-2 + N \geq .5$. Hence

$$\begin{aligned} P\{\text{error} \mid \text{message is } 1\} &= P\{N < -1.5\} \\ &= 1 - \Phi(1.5) \approx .0668 \end{aligned}$$

and

$$\begin{aligned} P\{\text{error} \mid \text{message is } 0\} &= P\{N \geq 2.5\} \\ &= 1 - \Phi(2.5) \approx .0062 \end{aligned}$$

■

5.4.1 The Normal Approximation to the Binomial Distribution

An important result in probability theory, known as the DeMoivre–Laplace limit theorem, states that when n is large, a binomial random variable with parameters n and p will have approximately the same distribution as a normal random variable with the same mean and variance as the binomial. This result was proved originally for the special case $p = \frac{1}{2}$ by DeMoivre in 1733 and was then extended to general p by Laplace in 1812. It formally states that if we "standardize" the binomial by first subtracting its mean np and then dividing the result by its standard deviation $\sqrt{np(1-p)}$, then the distribution function of this standardized random variable (which has mean 0 and variance 1) will converge to the standard normal distribution function as $n \to \infty$.

The DeMoivre–Laplace limit theorem

If S_n denotes the number of successes that occur when n independent trials, each resulting in a success with probability p, are performed then, for any $a < b$,

$$P\left\{a \le \frac{S_n - np}{\sqrt{np(1-p)}} \le b\right\} \to \Phi(b) - \Phi(a)$$

as $n \to \infty$.

As the preceding theorem is only a special case of the central limit theorem, which is presented in Chapter 8, we shall not present a proof.

It should be noted that we now have two possible approximations to binomial probabilities: the Poisson approximation, which yields a good approximation when n is large and p is small, and the normal approximation, which can be shown to be quite good when $np(1-p)$ is large (see Figure 5.6). [The normal approximation will, in general, be quite good for values of n satisfying $np(1-p) \ge 10$.]

Example 4f. Let X be the number of times that a fair coin, flipped 40 times, lands heads. Find the probability that $X = 20$. Use the normal approximation and then compare it to the exact solution.

Solution To employ the normal approximation, note that because the binomial is a discrete integer valued random variable whereas the normal is a continuous random variable, it is best to write $P\{X = i\}$ as $P\{i - 1/2 < X < i + 1/2\}$ before applying the normal approximation (this is called the *continuity correction*). This gives

$$\begin{aligned} P\{X = 20\} &= P\{19.5 \le X < 20.5\} \\ &= P\left\{\frac{19.5 - 20}{\sqrt{10}} < \frac{X - 20}{\sqrt{10}} < \frac{20.5 - 20}{\sqrt{10}}\right\} \\ &\approx P\left\{-.16 < \frac{X - 20}{\sqrt{10}} < .16\right\} \\ &\approx \Phi(.16) - \Phi(-.16) \approx .1272 \end{aligned}$$

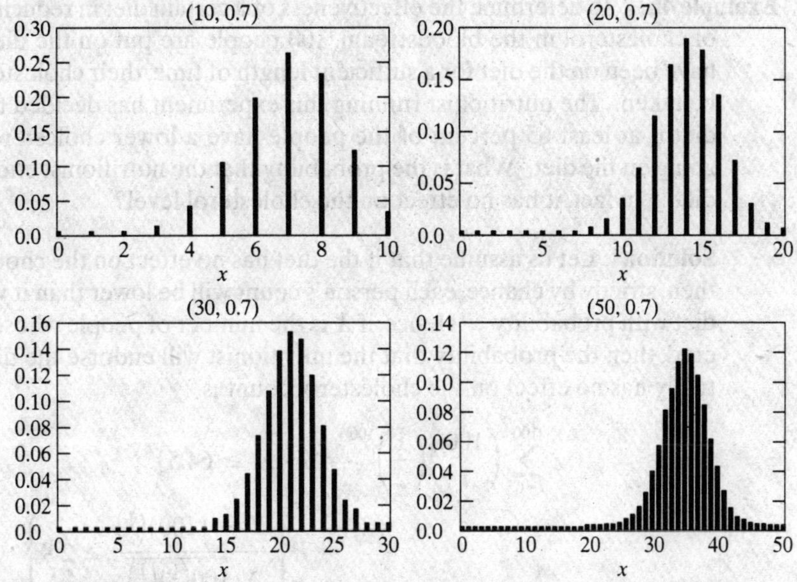

Figure 5.6 The probability mass function of a binomial (n, p) random variable becomes more and more "normal" as n becomes larger and larger.

The exact result is

$$P\{X = 20\} = \binom{40}{20}\left(\frac{1}{2}\right)^{40} \approx .1254 \qquad \blacksquare$$

Example 4g. The ideal size of a first-year class at a particular college is 150 students. The college, knowing from past experience that on the average only 30 percent of those accepted for admission will actually attend, uses a policy of approving the applications of 450 students. Compute the probability that more than 150 first-year students attend this college.

Solution If X denotes the number of students that attend, then X is a binomial random variable with parameters $n = 450$ and $p = .3$. Using the continuity correction, the normal approximation yields that

$$\begin{aligned} P\{X \geq 150.5\} &= P\left\{\frac{X - (450)(.3)}{\sqrt{450(.3)(.7)}} \geq \frac{150.5 - (450)(.3)}{\sqrt{450(.3)(.7)}}\right\} \\ &\approx 1 - \Phi(1.59) \\ &\approx .0559 \end{aligned}$$

Hence less than 6 percent of the time do more than 150 of the first 450 accepted actually attend. (What independence assumptions have we made?) ■

Example 4h. To determine the effectiveness of a certain diet in reducing the amount of cholesterol in the bloodstream, 100 people are put on the diet. After they have been on the diet for a sufficient length of time, their cholesterol count will be taken. The nutritionist running this experiment has decided to endorse the diet if at least 65 percent of the people have a lower cholesterol count after going on the diet. What is the probability that the nutritionist endorses the new diet if, in fact, it has no effect on the cholesterol level?

Solution Let us assume that if the diet has no effect on the cholesterol count, then, strictly by chance, each person's count will be lower than it was before the diet with probability $\frac{1}{2}$. Hence, if X is the number of people whose count is lowered, then the probability that the nutritionist will endorse the diet when it actually has no effect on the cholesterol count is

$$\begin{aligned}\sum_{i=65}^{100}\binom{100}{i}\left(\frac{1}{2}\right)^{100} &= P\{X \geq 64.5\}\\ &= P\left\{\frac{X-(100)\left(\frac{1}{2}\right)}{\sqrt{100\left(\frac{1}{2}\right)\left(\frac{1}{2}\right)}} \geq 2.9\right\}\\ &\approx 1-\Phi(2.9)\\ &\approx .0019\end{aligned}$$

■

Example 4i. Fifty-two percent of the residents of New York City are in favor of outlawing cigarette smoking in publicly owned areas. Approximate the probability that greater than 50 percent of a random sample of n people from New York are in favor of this prohibition when

(a) $n = 11$ (b) $n = 101$ (c) $n = 1001$

How large would n have to be to make this probability exceed .95?

Solution Let N denote the number of residents of New York City. To answer the preceding question, we must first understand that a random sample of size n means that the n people were chosen in such a manner that each of the $\binom{N}{n}$ subsets of n people had the same chance of being the chosen subset. Consequently, S_n, the number of people in the sample who are in favor of the smoking prohibition, is a hypergeometric random variable. That is, S_n has the same distribution as the number of white balls obtained when n balls are chosen from an urn of N balls, of which $.52N$ are white. But because N and $.52N$ are both large in comparison with the sample size n, it follows from the binomial approximation to the hypergeometric (see Section 4.9.3) that the distribution of S_n is very closely approximated by a binomial distribution with parameters n and $p = .52$. The normal approximation to the binomial distribution then shows that

$$P\{S_n > .5n\} = P\left\{\frac{S_n - .52n}{\sqrt{n(.52)(.48)}} > \frac{.5n - .52n}{\sqrt{n(.52)(.48)}}\right\}$$

$$= P\left\{\frac{S_n - .52n}{\sqrt{n(.52)(.48)}} > -.04\sqrt{n}\right\}$$

$$\approx \Phi(.04\sqrt{n})$$

Thus,

$$P\{S_n > .5n\} \approx \begin{cases} \Phi(.1328) = .5528, & \text{if } n = 11 \\ \Phi(.4020) = .6562, & \text{if } n = 101 \\ \Phi(1.2665) = .8973, & \text{if } n = 1001 \end{cases}$$

In order for this probability to be at least .95, we would need that $\Phi(.04\sqrt{n}) > .95$. Because $\Phi(x)$ is an increasing function and $\Phi(1.645) = .95$, this means that

$$.04\sqrt{n} > 1.645$$

or

$$n \geq 1691.266$$

That is, the sample size would have to be at least 1692. ■

Historical Notes Concerning the Normal Distribution

The normal distribution was introduced by the French mathematician Abraham DeMoivre in 1733. DeMoivre, who used this distribution to approximate probabilities connected with coin tossing, called it the exponential bell-shaped curve. Its usefulness, however, became truly apparent only in 1809, when the famous German mathematician K. F. Gauss used it as an integral part of his approach to predicting the location of astronomical entities. As a result, it became common after this time to call it the *Gaussian distribution*.

During the mid to late nineteenth century, however, most statisticians started to believe that the majority of data sets would have histograms conforming to the Gaussian bell-shaped form. Indeed, it came to be accepted that it was "normal" for any well-behaved data set to follow this curve. As a result, following the lead of the British statistician Karl Pearson, people began referring to the Gaussian curve by calling it simply the *normal* curve. (A partial explanation as to why so many data sets conform to the normal curve is provided by the central limit theorem, which is presented in Chapter 8.)

(continues on next page)

Abraham DeMoivre (1667–1754)

Today there is no shortage of statistical consultants, many of whom ply their trade in the most elegant of settings. However, the first of their breed worked, in the early years of the eighteenth century, out of a dark, grubby betting shop in Long Acres, London, known as Slaughter's Coffee House. He was Abraham DeMoivre, a Protestant refugee from Catholic France, and, for a price, he would compute the probability of gambling bets in all types of games of chance.

Although DeMoivre, the discoverer of the normal curve, made his living at the coffee shop, he was a mathematician of recognized abilities. Indeed, he was a member of the Royal Society and was reported to be an intimate of Isaac Newton.

Listen to Karl Pearson imagining DeMoivre at work at Slaughter's Coffee House. "*I picture DeMoivre working at a dirty table in the coffee house with a broken-down gambler beside him and Isaac Newton walking through the crowd to his corner to fetch out his friend. It would make a great picture for an inspired artist.*"

Karl Friedrich Gauss

Karl Friedrich Gauss (1777–1855), one of the earliest users of the normal curve, was one of the greatest mathematicians of all time. Listen to the words of the well-known mathematical historian E. T. Bell, as expressed in his 1954 book *Men of Mathematics*. In a chapter entitled "The Prince of Mathematicians," he writes, "*Archimedes, Newton, and Gauss; these three are in a class by themselves among the great mathematicians, and it is not for ordinary mortals to attempt to rank them in order of merit. All three started tidal waves in both pure and applied mathematics. Archimedes esteemed his pure mathematics more highly than its applications; Newton appears to have found the chief justification for his mathematical inventions in the scientific uses to which he put them; while Gauss declared it was all one to him whether he worked on the pure or on the applied side.*"

5.5 EXPONENTIAL RANDOM VARIABLES

A continuous random variable whose probability density function is given, for some $\lambda > 0$, by

$$f(x) = \begin{cases} \lambda e^{-\lambda x} & \text{if } x \geq 0 \\ 0 & \text{if } x < 0 \end{cases}$$

is said to be an *exponential* random variable (or, more simply, is said to be exponentially distributed) with parameter λ. The cumulative distribution function $F(a)$ of an exponential random variable is given by

$$\begin{aligned} F(a) &= P\{X \le a\} \\ &= \int_0^a \lambda e^{-\lambda x}\,dx \\ &= -e^{-\lambda x}\Big|_0^a \\ &= 1 - e^{-\lambda a} \qquad a \ge 0 \end{aligned}$$

Note that $F(\infty) = \int_0^\infty \lambda e^{-\lambda x}\,dx = 1$, as, of course, it must. The parameter λ will now be shown to equal the reciprocal of the expected value.

Example 5a. Let X be an exponential random variable with parameter λ. Calculate (a) $E[X]$ and (b) $\text{Var}(X)$.

Solution (a) Since the density function is given by

$$f(x) = \begin{cases} \lambda e^{-\lambda x} & x \ge 0 \\ 0 & x < 0 \end{cases}$$

we obtain

$$E[X] = \int_0^\infty x\lambda e^{-\lambda x}\,dx$$

Integrating by parts ($\lambda e^{-\lambda x}\,dx = dv, u = x$) yields

$$\begin{aligned} E[X] &= -xe^{-\lambda x}\Big|_0^\infty + \int_0^\infty e^{-\lambda x}\,dx \\ &= 0 - \frac{e^{-\lambda x}}{\lambda}\Big|_0^\infty \\ &= \frac{1}{\lambda} \end{aligned}$$

(b) To obtain the variance of X, we first find $E[X^2]$.

$$E[X^2] = \int_0^\infty x^2\lambda e^{-\lambda x}\,dx$$

Integrating by parts ($\lambda e^{-\lambda x}dx = dv, u = x^2$) gives

$$\begin{aligned} E[X^2] &= -x^2e^{-\lambda x}\Big|_0^\infty + \int_0^\infty 2xe^{-\lambda x}\,dx \\ &= 0 + \frac{2}{\lambda}E[X] \\ &= \frac{2}{\lambda^2} \end{aligned}$$

Hence

$$\text{Var}(X) = \frac{2}{\lambda^2} - \left(\frac{1}{\lambda}\right)^2$$
$$= \frac{1}{\lambda^2}$$

Thus the mean of the exponentiaı is the reciprocal of its parameter λ and the variance is the mean squared. ■

The exponential distribution often arises, in practice, as being the distribution of the amount of time until some specific event occurs. For instance, the amount of time (starting from now) until an earthquake occurs, or until a new war breaks out, or until a telephone call you receive turns out to be a wrong number are all random variables that tend in practice to have exponential distributions. (For a theoretical explanation for this the reader should consult Section 4.8, in particular, Example 8d.)

Example 5b. Suppose that the length of a phone call in minutes is an exponential random variable with parameter $\lambda = \frac{1}{10}$. If someone arrives immediately ahead of you at a public telephone booth, find the probability that you will have to wait

(a) more than 10 minutes;
(b) between 10 and 20 minutes.

Solution Letting X denote the length of the call made by the person in the booth, we have that the desired probabilities are

$$\text{(a) } P\{X > 10\} = 1 - F(10)$$
$$= e^{-1} \approx .368$$
$$\text{(b) } P\{10 < X < 20\} = F(20) - F(10)$$
$$= e^{-1} - e^{-2} \approx .233$$

■

We say that a nonnegative random variable X is *memoryless* if

$$P\{X > s + t | X > t\} = P\{X > s\} \qquad \text{for all } s, t \geq 0 \tag{5.1}$$

If we think of X as being the lifetime of some instrument, Equation (5.1) states that the probability that the instrument survives for at least $s + t$ hours, given that it has survived t hours, is the same as the initial probability that it survives for at least s hours. In other words, if the instrument is alive at age t, the distribution of the remaining amount of time that it survives is the same as the original lifetime distribution (that is, it is as if the instrument does not remember that it has already been in use for a time t).

The condition (5.1) is equivalent to

$$\frac{P\{X > s + t, X > t\}}{P\{X > t\}} = P\{X > s\}$$

or

$$P\{X > s + t\} = P\{X > s\}P\{X > t\} \tag{5.2}$$

Since Equation (5.2) is satisfied when X is exponentially distributed (for $e^{-\lambda(s+t)} = e^{-\lambda s}e^{-\lambda t}$), it follows that exponentially distributed random variables are memoryless.

Example 5c. Consider a post office that is staffed by two clerks. Suppose that when Mr. Smith enters the system, he discovers that Ms. Jones is being served by one of the clerks and Mr. Brown by the other. Suppose also that Mr. Smith is told that his service will begin as soon as either Jones or Brown leaves. If the amount of time that a clerk spends with a customer is exponentially distributed with parameter λ, what is the probability that, of the three customers, Mr. Smith is the last to leave the post office?

Solution The answer is obtained by reasoning as follows: Consider the time at which Mr. Smith first finds a free clerk. At this point either Ms. Jones or Mr. Brown would have just left and the other one would still be in service. However, by the lack of memory of the exponential, it follows that the additional amount of time that this other person (either Jones or Brown) would still have to spend in the post office is exponentially distributed with parameter λ. That is, it is the same as if service for this person were just starting at this point. Hence, by symmetry, the probability that the remaining person finishes before Smith must equal $\frac{1}{2}$. ■

It turns out that not only is the exponential distribution memoryless, but it is also the unique distribution possessing this property. To see this, suppose that X is memoryless and let $\bar{F}(x) = P\{X > x\}$. Then, by Equation (5.2), it follows that

$$\bar{F}(s + t) = \bar{F}(s)\bar{F}(t)$$

That is, $\bar{F}(\cdot)$ satisfies the functional equation

$$g(s + t) = g(s)g(t)$$

However, it turns out that the only right continuous solution of this functional equation is†

$$g(x) = e^{-\lambda x} \tag{5.3}$$

† One can prove equation (5.3) as follows: If $g(s + t) = g(s)g(t)$, then

$$g\left(\frac{2}{n}\right) = g\left(\frac{1}{n} + \frac{1}{n}\right) = g^2\left(\frac{1}{n}\right)$$

and repeating this yields $g(m/n) = g^m(1/n)$. Also,

$$g(1) = g\left(\frac{1}{n} + \frac{1}{n} + \cdots + \frac{1}{n}\right) = g^n\left(\frac{1}{n}\right) \quad \text{or} \quad g\left(\frac{1}{n}\right) = (g(1))^{1/n}$$

Hence $g(m/n) = (g(1))^{m/n}$, which since g is right continuous, implies that $g(x) = (g(1))^x$. Since $g(1) = \left(g\left(\frac{1}{2}\right)\right)^2 \geq 0$, we obtain $g(x) = e^{-\lambda x}$, where $\lambda = -\log(g(1))$.

and, since a distribution function is always right continuous, we must have

$$\bar{F}(x) = e^{-\lambda x} \qquad \text{or} \qquad F(x) = P\{X \le x\} = 1 - e^{-\lambda x}$$

which shows that X is exponentially distributed.

Example 5d. Suppose that the number of miles that a car can run before its battery wears out is exponentially distributed with an average value of 10,000 miles. If a person desires to take a 5000-mile trip, what is the probability that he or she will be able to complete the trip without having to replace the car battery? What can be said when the distribution is not exponential?

Solution It follows by the memoryless property of the exponential distribution that the remaining lifetime (in thousands of miles) of the battery is exponential with parameter $\lambda = \frac{1}{10}$. Hence the desired probability is

$$P\{\text{remaining lifetime} > 5\} = 1 - F(5) = e^{-5\lambda} = e^{-1/2} \approx .604$$

However, if the lifetime distribution F is not exponential, then the relevant probability is

$$P\{\text{lifetime} > t + 5 | \text{lifetime} > t\} = \frac{1 - F(t + 5)}{1 - F(t)}$$

where t is the number of miles that the battery had been in use prior to the start of the trip. Therefore, if the distribution is not exponential, additional information is needed (namely, t) before the desired probability can be calculated. ■

A variation of the exponential distribution is the distribution of a random variable that is equally likely to be either positive or negative and whose absolute value is exponentially distributed with parameter λ, $\lambda \ge 0$. Such a random variable is said to have a *Laplace* distribution[‡] and its density is given by

$$f(x) = \tfrac{1}{2}\lambda e^{-\lambda|x|} \qquad -\infty < x < \infty$$

Its distribution function is given by

$$F(x) = \begin{cases} \frac{1}{2}\displaystyle\int_{-\infty}^{x} \lambda e^{\lambda x}\,dx & x < 0 \\ \frac{1}{2}\displaystyle\int_{-\infty}^{0} \lambda e^{\lambda x}\,dx + \frac{1}{2}\int_{0}^{x} \lambda e^{-\lambda x}\,dx & x > 0 \end{cases}$$

$$= \begin{cases} \frac{1}{2}e^{\lambda x} & x < 0 \\ 1 - \frac{1}{2}e^{-\lambda x} & x > 0 \end{cases}$$

Example 5e. Let us reconsider Example 4e, which supposes that a binary message is to be transmitted from A to B, with the value 2 being sent when the message is 1 and −2 when it is 0. However, suppose now that rather than being a

[‡] It also is sometimes called the double exponential random variable.

standard normal random variable, the channel noise N is a Laplacian random variable with parameter $\lambda = 1$. Again suppose that if R is the value received at location B, then the message is decoded as follows:

If $R \geq .5$, then 1 is concluded.
If $R < .5$, then 0 is concluded.

In this case, where the noise is Laplacian with parameter $\lambda = 1$, the 2 types of errors will have probabilities given by

$$\begin{aligned} P\{\text{error}|\text{message 1 is sent}\} &= P\{N < -1.5\} \\ &= \tfrac{1}{2}e^{-1.5} \\ &\approx .1116 \\ P\{\text{error}|\text{message 0 is sent}\} &= P\{N \geq 2.5\} \\ &= \tfrac{1}{2}e^{-2.5} \\ &\approx .041 \end{aligned}$$

On comparing this with the results of Example 4e, we see that the error probabilities are higher when the noise is Laplacian with $\lambda = 1$ than when it is a standard normal variable. ■

5.5.1 Hazard Rate Functions

Consider a positive continuous random variable X that we interpret as being the lifetime of some item, having distribution function F and density f. The *hazard rate* (sometimes called the *failure rate*) function $\lambda(t)$ of F is defined by

$$\lambda(t) = \frac{f(t)}{\bar{F}(t)} \qquad \bar{F} = 1 - F$$

To interpret $\lambda(t)$, suppose that the item has survived for a time t and we desire the probability that it will not survive for an additional time dt. That is, consider $P\{X \in (t, t + dt)|X > t\}$. Now

$$\begin{aligned} P\{X \in (t, t + dt)|X > t\} &= \frac{P\{X \in (t, t + dt), X > t\}}{P\{X > t\}} \\ &= \frac{P\{X \in (t, t + dt)\}}{P\{X > t\}} \\ &\approx \frac{f(t)}{\bar{F}(t)}\,dt \end{aligned}$$

Thus, $\lambda(t)$ represents the conditional probability intensity that a t-unit-old item will fail.

Suppose now that the lifetime distribution is exponential. Then, by the memoryless property, it follows that the distribution of remaining life for a t-year-old item is the same as for a new item. Hence $\lambda(t)$ should be constant. This checks out, since

$$\lambda(t) = \frac{f(t)}{\bar{F}(t)} = \frac{\lambda e^{-\lambda t}}{e^{-\lambda t}} = \lambda$$

Thus the failure rate function for the exponential distribution is constant. The parameter λ is often referred to as the *rate* of the distribution.

It turns out that the failure rate function $\lambda(t)$ uniquely determines the distribution F. To prove this, note that by definition

$$\lambda(t) = \frac{\frac{d}{dt}F(t)}{1 - F(t)}$$

Integrating both sides yields

$$\log(1 - F(t)) = -\int_0^t \lambda(t)\,dt + k$$

or

$$1 - F(t) = e^k \exp\left\{-\int_0^t \lambda(t)\,dt\right\}$$

Letting $t = 0$ shows that $k = 0$ and thus

$$F(t) = 1 - \exp\left\{-\int_0^t \lambda(t)\,dt\right\} \tag{5.4}$$

Hence a distribution function of a positive continuous random variable can be specified by giving its hazard rate function. For instance, if a random variable has a linear hazard rate function—that is, if

$$\lambda(t) = a + bt$$

then its distribution function is given by

$$F(t) = 1 - e^{-at - bt^2/2}$$

and differentiation yields that its density is

$$f(t) = (a + bt)e^{-(at + bt^2/2)} \qquad t \geq 0$$

When $a = 0$, the preceding is known as the Rayleigh density function.

Example 5f. One often hears that the death rate of a person who smokes is, at each age, twice that of a nonsmoker. What does this mean? Does it mean that a nonsmoker has twice the probability of surviving a given number of years as does a smoker of the same age?

Solution If $\lambda_s(t)$ denotes the hazard rate of a smoker of age t and $\lambda_n(t)$ that of a nonsmoker of age t, then the preceding is equivalent to the statement that

$$\lambda_s(t) = 2\lambda_n(t)$$

The probability that an A-year-old nonsmoker will survive until age B, $A < B$, is

$$\begin{aligned}
P\{A\text{-year-old nonsmoker reaches age } B\} \\
&= P\{\text{nonsmoker's lifetime} > B \mid \text{nonsmoker's lifetime} > A\} \\
&= \frac{1 - F_{\text{non}}(B)}{1 - F_{\text{non}}(A)} \\
&= \frac{\exp\left\{-\int_0^B \lambda_n(t)\,dt\right\}}{\exp\left\{-\int_0^A \lambda_n(t)\,dt\right\}} \qquad \text{from (5.4)} \\
&= \exp\left\{-\int_A^B \lambda_n(t)\,dt\right\}
\end{aligned}$$

whereas the corresponding probability for a smoker is, by the same reasoning,

$$\begin{aligned}
P\{A\text{-year-old smoker reaches age } B\} &= \exp\left\{-\int_A^B \lambda_s(t)\,dt\right\} \\
&= \exp\left\{-2\int_A^B \lambda_n(t)\,dt\right\} \\
&= \left[\exp\left\{-\int_A^B \lambda_n(t)\,dt\right\}\right]^2
\end{aligned}$$

In other words, of two people of the same age, one of whom is a smoker and the other a nonsmoker, the probability that the smoker survives to any given age is the *square* (not one-half) of the corresponding probability for a nonsmoker. For instance, if $\lambda_n(t) = \frac{1}{30}$, $50 \le t \le 60$, then the probability that a 50-year-old nonsmoker reaches age 60 is $e^{-1/3} \approx .7165$, whereas the corresponding probability for a smoker is $e^{-2/3} \approx .5134$. ■

5.6 OTHER CONTINUOUS DISTRIBUTIONS

5.6.1 The Gamma Distribution

A random variable is said to have a gamma distribution with parameters (α, λ), $\lambda > 0$, and $\alpha > 0$ if its density function is given by

$$f(x) = \begin{cases} \dfrac{\lambda e^{-\lambda x}(\lambda x)^{\alpha-1}}{\Gamma(\alpha)} & x \ge 0 \\ 0 & x < 0 \end{cases}$$

where $\Gamma(\alpha)$, called the gamma function, is defined as

$$\Gamma(\alpha) = \int_0^\infty e^{-y} y^{\alpha-1}\, dy$$

The integration by parts of $\Gamma(\alpha)$ yields that

$$\begin{aligned}\Gamma(\alpha) &= -e^{-y} y^{\alpha-1}\Big|_0^\infty + \int_0^\infty e^{-y}(\alpha-1)y^{\alpha-2}\, dy \\ &= (\alpha-1)\int_0^\infty e^{-y} y^{\alpha-2}\, dy \qquad (6.1) \\ &= (\alpha-1)\Gamma(\alpha-1)\end{aligned}$$

For integral values of α, say $\alpha = n$, we obtain by applying Equation (6.1) repeatedly that

$$\begin{aligned}\Gamma(n) &= (n-1)\Gamma(n-1) \\ &= (n-1)(n-2)\Gamma(n-2) \\ &= \cdots \\ &= (n-1)(n-2)\cdots 3\cdot 2\Gamma(1)\end{aligned}$$

Since $\Gamma(1) = \int_0^\infty e^{-x}\, dx = 1$, it follows that for integral values of n.

$$\Gamma(n) = (n-1)!$$

When α is a positive integer, say $\alpha = n$, the gamma distribution with parameters (α, λ) often arises, in practice, as the distribution of the amount of time one has to wait until a total of n events has occurred. More specifically, if events are occurring randomly in time and in accordance with the three axioms of Section 4.8, then it turns out that the amount of time one has to wait until a total of n events has occurred will be a gamma random variable with parameters (n, λ). To prove this, let T_n denote the time at which the nth event occurs, and note that T_n is less than or equal to t if and only if the number of events that have occurred by time t is at least n. That is, with $N(t)$ equal to the number of events in $[0, t]$,

$$\begin{aligned}P\{T_n \le t\} &= P\{N(t) \ge n\} \\ &= \sum_{j=n}^{\infty} P\{N(t) = j\} \\ &= \sum_{j=n}^{\infty} \frac{e^{-\lambda t}(\lambda t)^j}{j!}\end{aligned}$$

where the final identity follows because the number of events in $[0, t]$ has a Poisson distribution with parameter λt. Differentiation of the above yields that the density function of T_n is as follows:

$$f(t) = \sum_{j=n}^{\infty} \frac{e^{-\lambda t} j(\lambda t)^{j-1}\lambda}{j!} - \sum_{j=n}^{\infty} \frac{\lambda e^{-\lambda t}(\lambda t)^j}{j!}$$

$$= \sum_{j=n}^{\infty} \frac{\lambda e^{-\lambda t}(\lambda t)^{j-1}}{(j-1)!} - \sum_{j=n}^{\infty} \frac{\lambda e^{-\lambda t}(\lambda t)^j}{j!}$$

$$= \frac{\lambda e^{-\lambda t}(\lambda t)^{n-1}}{(n-1)!}$$

Hence T_n is the gamma distribution with parameters (n, λ). (This distribution is often referred to in the literature as the n-Erlang distribution.) Note that when $n = 1$, this distribution reduces to the exponential.

The gamma distribution with $\lambda = \frac{1}{2}$ and $\alpha = n/2$ (n being a positive integer) is called the χ_n^2 (read "chi-squared") distribution with n degrees of freedom. The chi-squared distribution often arises in practice as being the distribution of the error involved in attempting to hit a target in n dimensional space when each coordinate error is normally distributed. This distribution will be studied in Chapter 6, where its relation to the normal distribution is detailed.

Example 6a. Let X be a gamma random variable with parameters α and λ. Calculate (a) $E[X]$ and (b) $\text{Var}(X)$.

Solution

(a) $$E[X] = \frac{1}{\Gamma(\alpha)} \int_0^{\infty} \lambda x e^{-\lambda x}(\lambda x)^{\alpha-1}\, dx$$

$$= \frac{1}{\lambda\Gamma(\alpha)} \int_0^{\infty} \lambda e^{-\lambda x}(\lambda x)^{\alpha}\, dx$$

$$= \frac{\Gamma(\alpha+1)}{\lambda\Gamma(\alpha)}$$

$$= \frac{\alpha}{\lambda} \quad \text{by Equation (6.1)}$$

(b) By first calculating $E[X^2]$, we can show that

$$\text{Var}(X) = \frac{\alpha}{\lambda^2}$$

The details are left as an exercise. ■

5.6.2 The Weibull Distribution

The Weibull distribution is widely used in engineering practice due to its versatility. It was originally proposed for the interpretation of fatigue data, but now its use has extended to many other engineering problems. In particular, it is widely used, in the field of life phenomena, as the distribution of the lifetime of some object, particularly when the "weakest link" model is appropriate for the object. That is, consider an object consisting of many parts and suppose that the object experiences death (failure) when any of its parts fail. Under these conditions, it has been shown (both theoretically and empirically) that a Weibull distribution provides a close approximation to the distribution of the lifetime of the item.

The Weibull distribution function has the form

$$F(x) = \begin{cases} 0 & x \leq \nu \\ 1 - \exp\left\{-\left(\dfrac{x-\nu}{\alpha}\right)^{\beta}\right\} & x > \nu \end{cases} \tag{6.2}$$

A random variable whose cumulative distribution function is given by Equation (6.2) is said to be a Weibull random variable with parameters ν, α, and β. Differentiation yields that the density is

$$f(x) = \begin{cases} 0 & x \leq \nu \\ \dfrac{\beta}{\alpha}\left(\dfrac{x-\nu}{\alpha}\right)^{\beta-1} \exp\left\{-\left(\dfrac{x-\nu}{\alpha}\right)^{\beta}\right\} & x > \nu \end{cases}$$

5.6.3 The Cauchy Distribution

A random variable is said to have a Cauchy distribution with parameter $\theta, -\infty < \theta < \infty$, if its density is given by

$$f(x) = \frac{1}{\pi}\frac{1}{1+(x-\theta)^2} \qquad -\infty < x < \infty$$

Example 6b. Suppose that a narrow beam flashlight is spun around its center, which is located a unit distance from the x-axis (see Figure 5.7). When the flashlight has stopped spinning, consider the point X at which the beam intersects the x-axis. (If the beam is not pointing toward the x-axis, repeat the experiment.)

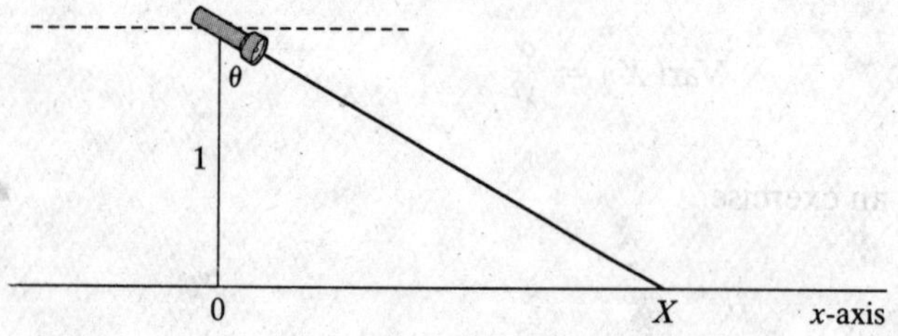

Figure 5.7

As indicated in Figure 5.7, the point X is determined by the angle θ between the flashlight and the y-axis, which from the physical situation appears to be uniformly distributed between $-\pi/2$ and $\pi/2$. The distribution function of X is thus given by

$$\begin{aligned} F(x) &= P\{X \le x\} \\ &= P\{\tan\theta \le x\} \\ &= P\{\theta \le \tan^{-1} x\} \\ &= \frac{1}{2} + \frac{1}{\pi}\tan^{-1} x \end{aligned}$$

where the last equality follows since θ, being uniform over $(-\pi/2, \pi/2)$, has distribution

$$P\{\theta \le a\} = \frac{a - (-\pi/2)}{\pi} = \frac{1}{2} + \frac{a}{\pi} \qquad -\frac{\pi}{2} < a < \frac{\pi}{2}$$

Hence the density function of X is given by

$$f(x) = \frac{d}{dx}F(x) = \frac{1}{\pi(1 + x^2)} \qquad -\infty < x < \infty$$

and we see that X has the Cauchy distribution.† ■

5.6.4 The Beta Distribution

A random variable is said to have a beta distribution if its density is given by

$$f(x) = \begin{cases} \dfrac{1}{B(a,b)} x^{a-1}(1-x)^{b-1} & 0 < x < 1 \\ 0 & \text{otherwise} \end{cases}$$

where

$$B(a,b) = \int_0^1 x^{a-1}(1-x)^{b-1}\,dx$$

The beta distribution can be used to model a random phenomenon whose set of possible values is some finite interval $[c, d]$—which by letting c denote the origin and taking $d - c$ as a unit measurement can be transformed into the interval $[0, 1]$.

†That $\frac{d}{dx}(\tan^{-1} x) = 1/(1 + x^2)$ can be seen as follows: If $y = \tan^{-1} x$, then $\tan y = x$, so

$$\begin{aligned} 1 = \frac{d}{dx}(\tan y) &= \frac{d}{dy}(\tan y)\frac{dy}{dx} = \frac{d}{dy}\left(\frac{\sin y}{\cos y}\right)\frac{dy}{dx} \\ &= \left(\frac{\cos^2 y + \sin^2 y}{\cos^2}\right)\frac{dy}{dx} \end{aligned}$$

or

$$\frac{dy}{dx} = \frac{\cos^2 y}{\sin^2 y + \cos^2 y} = \frac{1}{\tan^2 y + 1} = \frac{1}{x^2 + 1}$$

When $a = b$, the beta density is symmetric about $\frac{1}{2}$, giving more and more weight to regions about $\frac{1}{2}$ as the common value a increases (see Figure 5.8). When $b > a$, the density is skewed to the left (in the sense that smaller values become more likely); and it is skewed to the right when $a > b$ (see Figure 5.9).

The following relationship can be shown to exist between

$$B(a, b) = \int_0^1 x^{a-1}(1 - x)^{b-1}\,dx$$

and the gamma function:

$$B(a, b) = \frac{\Gamma(a)\Gamma(b)}{\Gamma(a + b)} \tag{6.3}$$

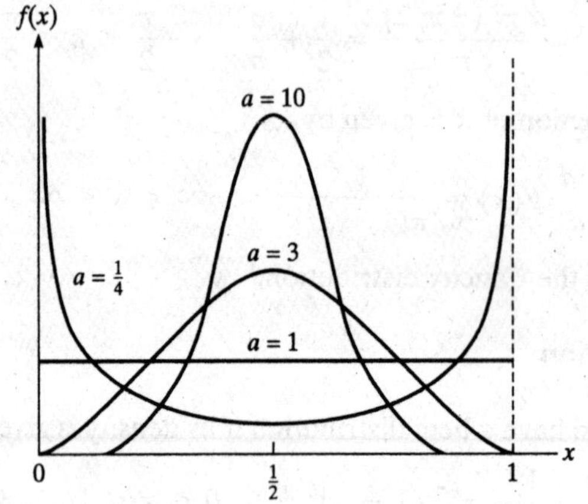

Figure 5.8 Beta densities with parameters (a, b) when $a = b$.

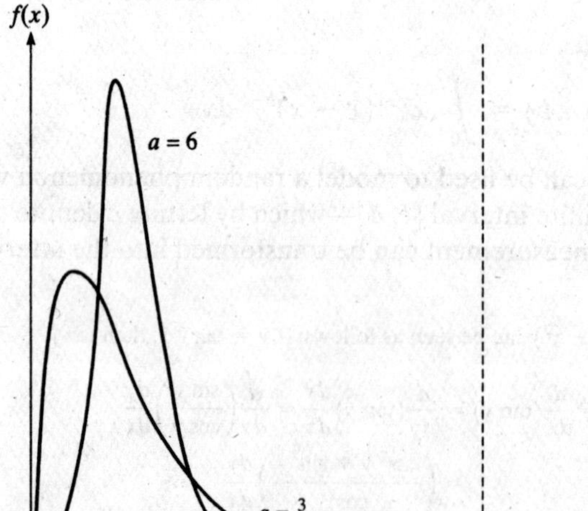

Figure 5.9 Beta densities with parameters (a, b) when $a/(a + b) = 1/20$.

Upon using Equation (6.1) along with the identity (6.3), it is an easy matter to show that if X is a beta random variable with parameters a and b, then

$$E[X] = \frac{a}{a+b}$$

$$\text{Var}(X) = \frac{ab}{(a+b)^2(a+b+1)}$$

REMARK. A verification of Equation (6.3) appears in Example 7c of Chapter 6. ■

5.7 THE DISTRIBUTION OF A FUNCTION OF A RANDOM VARIABLE

It is often the case that we know the probability distribution of a random variable and are interested in determining the distribution of some function of it. For instance, suppose that we know the distribution of X and want to find the distribution of $g(X)$. To do so, it is necessary to express the event that $g(X) \le y$ in terms of X being in some set. We illustrate by the following examples.

Example 7a. Let X be uniformly distributed over $(0, 1)$. We obtain the distribution of the random variable Y, defined by $Y = X^n$, as follows: For $0 \le y \le 1$,

$$\begin{aligned} F_Y(y) &= P\{Y \le y\} \\ &= P\{X^n \le y\} \\ &= P\{X \le y^{1/n}\} \\ &= F_X(y^{1/n}) \\ &= y^{1/n} \end{aligned}$$

For instance, the density function of Y is given by

$$f_Y(y) = \begin{cases} \dfrac{1}{n} y^{1/n-1} & 0 \le y \le 1 \\ 0 & \text{otherwise} \end{cases}$$

■

Example 7b. If X is a continuous random variable with probability density f_X, then the distribution of $Y = X^2$ is obtained as follows: For $y \ge 0$,

$$\begin{aligned} F_Y(y) &= P\{Y \le y\} \\ &= P\{X^2 \le y\} \\ &= P\{-\sqrt{y} \le X \le \sqrt{y}\} \\ &= F_X(\sqrt{y}) - F_X(-\sqrt{y}) \end{aligned}$$

Differentiation yields

$$f_Y(y) = \frac{1}{2\sqrt{y}}[f_X(\sqrt{y}) + f_X(-\sqrt{y})] \qquad \blacksquare$$

Example 7c. If X has a probability density f_X, then $Y = |X|$ has a density function that is obtained as follows: For $y \ge 0$,

$$\begin{aligned} F_Y(y) &= P\{Y \le y\} \\ &= P\{|X| \le y\} \\ &= P\{-y \le X \le y\} \\ &= F_X(y) - F_X(-y) \end{aligned}$$

Hence, on differentiation, we obtain

$$f_Y(y) = f_X(y) + f_X(-y) \qquad y \ge 0 \qquad \blacksquare$$

The method employed in Examples 7a through 7c can be used to prove Theorem 7.1.

Theorem 7.1

Let X be a continuous random variable having probability density function f_X. Suppose that $g(x)$ is a strictly monotone (increasing or decreasing), differentiable (and thus continuous) function of x. Then the random variable Y defined by $Y = g(X)$ has a probability density function given by

$$f_Y(y) = \begin{cases} f_X[g^{-1}(y)]\left|\dfrac{d}{dy}g^{-1}(y)\right| & \text{if } y = g(x) \text{ for some } x \\ 0 & \text{if } y \ne g(x) \text{ for all } x \end{cases}$$

where $g^{-1}(y)$ is defined to equal that value of x such that $g(x) = y$.

We shall prove Theorem 7.1 when $g(x)$ is an increasing function.

Proof: Suppose that $y = g(x)$ for some x. Then, with $Y = g(X)$,

$$\begin{aligned} F_Y(y) &= P\{g(X) \le y\} \\ &= P\{X \le g^{-1}(y)\} \\ &= F_X(g^{-1}(y)) \end{aligned}$$

Differentiation gives that

$$f_Y(y) = f_X(g^{-1}(y))\frac{d}{dy}g^{-1}(y)$$

which agrees with Theorem 7.1 since $g^{-1}(y)$ is nondecreasing, so its derivative is nonnegative.

When $y \neq g(x)$ for any x, then $F_Y(y)$ is either 0 or 1, and in either case $f_Y(y) = 0$. ✦

Example 7d. Let X be a continuous nonnegative random variable with density function f, and let $Y = X^n$. Find f_Y, the probability density function of Y.

Solution If $g(x) = x^n$, then

$$g^{-1}(y) = y^{1/n}$$

and

$$\frac{d}{dy}\{g^{-1}(y)\} = \frac{1}{n}y^{1/n-1}$$

Hence, from Theorem 7.1, we obtain that

$$f_Y(y) = \frac{1}{n}y^{1/n-1}f(y^{1/n})$$

If $n = 2$, this gives

$$f_Y(y) = \frac{1}{2\sqrt{y}}f(\sqrt{y})$$

which (since $X \geq 0$) is in agreement with the result of Example 7b. ■

SUMMARY

A random variable X is called *continuous* if there is a nonnegative function f, called the *probability density function* of X, such that for any set B

$$P\{X \in B\} = \int_B f(x)\,dx$$

If X is continuous, then its distribution function F will be differentiable and

$$\frac{d}{dx}F(x) = f(x)$$

The expected value of a continuous random variable X is defined by

$$E[X] = \int_{-\infty}^{\infty} xf(x)\,dx$$

A useful identity is that for any function g,

$$E[g(X)] = \int_{-\infty}^{\infty} g(x)f(x)\,dx$$

As in the case of a discrete random variable, the variance of X is defined by

$$\text{Var}(X) = E[(X - E[X])^2]$$

A random variable X is said to be *uniform* over the interval (a, b) if its probability density function is given by

$$f(x) = \begin{cases} \dfrac{1}{b-a} & a \leq x \leq b \\ 0 & \text{otherwise} \end{cases}$$

Its expected value and variance are

$$E[X] = \frac{a+b}{2} \qquad \text{Var}(X) = \frac{(b-a)^2}{12}$$

A random variable X is said to be *normal* with parameters μ and σ^2 if its probability density function is given by

$$f(x) = \frac{1}{\sqrt{2\pi}\,\sigma} e^{-(x-\mu)^2/2\sigma^2} \qquad -\infty < x < \infty$$

It can be shown that

$$\mu = E[X] \qquad \sigma^2 = \text{Var}(X)$$

If X is normal with mean μ and variance σ^2, then Z, defined by

$$Z = \frac{X - \mu}{\sigma}$$

is normal with mean 0 and variance 1. Such a random variable is said to be a *standard* normal random variable. Probabilities about X can be expressed in terms of probabilities about the standard normal variable Z, whose probability distribution function can be obtained either from Table 5.1 or from the website.

The probability distribution function of a binomial random variable with parameters n and p can, when n is large, be approximated by that of a normal random variable having mean np and variance $np(1 - p)$.

A random variable whose probability density function is of the form

$$f(x) = \begin{cases} \lambda e^{-\lambda x} & x \geq 0 \\ 0 & \text{otherwise} \end{cases}$$

is said to be an *exponential* random variable with parameter λ. Its expected value and variance are

$$E[X] = \frac{1}{\lambda} \qquad \text{Var}(X) = \frac{1}{\lambda^2}$$

A key property, possessed only by exponential random variables, is that they are *memoryless* in the sense that for positive s and t,

$$P\{X > s + t | X > t\} = P\{X > s\}$$

If X represents the life of an item, then the memoryless property states that for any t, the remaining life of a t-year-old item has the same probability distribution as the life of a new item. Thus one need not remember the age of an item to know its distribution of remaining life.

Let X be a nonnegative continuous random variable with distribution function F and density function f. The function

$$\lambda(t) = \frac{f(t)}{1 - F(t)} \qquad t \geq 0$$

is called the *hazard rate*, or *failure rate*, function of F. If we interpret X as being the life of an item, then for small values of dt, $\lambda(t)\,dt$ is approximately the probability that a t-unit-old item will fail within an additional time dt. If F is the exponential distribution with parameter λ, then

$$\lambda(t) = \lambda \qquad t \geq 0$$

In addition, the exponential is the unique distribution having a constant failure rate.

A random variable is said to have a *gamma* distribution with parameters α and λ if its probability density function is equal to

$$f(x) = \frac{\lambda e^{-\lambda x}(\lambda x)^{\alpha-1}}{\Gamma(\alpha)} \qquad x \geq 0$$

and is 0 otherwise. The quantity $\Gamma(\alpha)$ is called the gamma function and is defined by

$$\Gamma(\alpha) = \int_0^\infty e^{-x}x^{\alpha-1}\,dx$$

The expected value and variance of a gamma random variable are

$$E[X] = \frac{\alpha}{\lambda} \qquad \text{Var}(X) = \frac{\alpha}{\lambda^2}$$

A random variable is said to have a *beta* distribution with parameters (a, b) if its probability density function is equal to

$$f(x) = \frac{1}{B(a, b)}x^{a-1}(1 - x)^{b-1} \qquad 0 \leq x \leq 1$$

and is equal to 0 otherwise. The constant $B(a, b)$ is given by

$$B(a, b) = \int_0^1 x^{a-1}(1 - x)^{b-1}\,dx$$

The mean and variance of such a random variable are

$$E[X] = \frac{a}{a + b} \qquad \text{Var}(X) = \frac{ab}{(a + b)^2(a + b + 1)}$$

PROBLEMS

1. Let X be a random variable with probability density function

$$f(x) = \begin{cases} c(1 - x^2) & -1 < x < 1 \\ 0 & \text{otherwise} \end{cases}$$

(a) What is the value of c? **(b)** What is the cumulative distribution function of X?

2. A system consisting of one original unit plus a spare can function for a random amount of time X. If the density of X is given (in units of months) by

$$f(x) = \begin{cases} Cxe^{-x/2} & x > 0 \\ 0 & x \leq 0 \end{cases}$$

what is the probability that the system functions for at least 5 months?

3. Consider the function

$$f(x) = \begin{cases} C(2x - x^3) & 0 < x < \frac{5}{2} \\ 0 & \text{otherwise} \end{cases}$$

Could f be a probability density function? If so, determine C. Repeat if $f(x)$ were given by

$$f(x) = \begin{cases} C(2x - x^2) & 0 < x < \frac{5}{2} \\ 0 & \text{otherwise} \end{cases}$$

4. The probability density function of X, the lifetime of a certain type of electronic device (measured in hours), is given by

$$f(x) = \begin{cases} \dfrac{10}{x^2} & x > 10 \\ 0 & x \leq 10 \end{cases}$$

(a) Find $P\{X > 20\}$.

(b) What is the cumulative distribution function of X?

(c) What is the probability that of 6 such types of devices at least 3 will function for at least 15 hours? What assumptions are you making?

5. A filling station is supplied with gasoline once a week. If its weekly volume of sales in thousands of gallons is a random variable with probability density function

$$f(x) = \begin{cases} 5(1 - x)^4 & 0 < x < 1 \\ 0 & \text{otherwise} \end{cases}$$

what need the capacity of the tank be so that the probability of the supply's being exhausted in a given week is .01?

6. Compute $E[X]$ if X has a density function given by

(a) $f(x) = \begin{cases} \frac{1}{4}xe^{-x/2} & x > 0 \\ 0 & \text{otherwise} \end{cases}$; **(b)** $f(x) = \begin{cases} c(1 - x^2) & -1 < x < 1 \\ 0 & \text{otherwise} \end{cases}$;

(c) $f(x) = \begin{cases} \dfrac{5}{x^2} & x > 5 \\ 0 & x \leq 5 \end{cases}$

7. The density function of X is given by

$$f(x) = \begin{cases} a + bx^2 & 0 \le x \le 1 \\ 0 & \text{otherwise} \end{cases}$$

If $E[X] = \frac{3}{5}$, find a and b.

8. The lifetime in hours of an electronic tube is a random variable having a probability density function given by

$$f(x) = xe^{-x} \quad x \ge 0$$

Compute the expected lifetime of such a tube.

9. Consider Example 5b of Chapter 4, but now suppose that the seasonal demand is a continuous random variable having probability density function f. Show that the optimal amount to stock is the value s^* that satisfies

$$F(s^*) = \frac{b}{b + \ell}$$

where b is net profit per unit sale, ℓ is the net loss per unit unsold, and F is the cumulative distribution function of the seasonal demand.

10. Trains headed for destination A arrive at the train station at 15-minute intervals starting at 7 A.M., whereas trains headed for destination B arrive at 15-minute intervals starting at 7:05 A.M.
    (a) If a certain passenger arrives at the station at a time uniformly distributed between 7 and 8 A.M. and then gets on the first train that arrives, what proportion of time does he or she go to destination A?
    (b) What if the passenger arrives at a time uniformly distributed between 7:10 and 8:10 A.M.?

11. A point is chosen at random on a line segment of length L. Interpret this statement and find the probability that the ratio of the shorter to the longer segment is less than $\frac{1}{4}$.

12. A bus travels between the two cities A and B, which are 100 miles apart. If the bus has a breakdown, the distance from the breakdown to city A has a uniform distribution over $(0, 100)$. There is a bus service station in city A, in B, and in the center of the route between A and B. It is suggested that it would be more efficient to have the three stations located 25, 50, and 75 miles, respectively, from A. Do you agree? Why?

13. You arrive at a bus stop at 10 o'clock, knowing that the bus will arrive at some time uniformly distributed between 10 and 10:30.
    (a) What is the probability that you will have to wait longer than 10 minutes?
    (b) If at 10:15 the bus has not yet arrived, what is the probability that you will have to wait at least an additional 10 minutes?

14. Let X be a uniform $(0, 1)$ random variable. Compute $E[X^n]$ by using Proposition 2.1 and then check the result by using the definition of expectation.

15. If X is a normal random variable with parameters $\mu = 10$ and $\sigma^2 = 36$, compute
    (a) $P\{X > 5\}$; **(b)** $P\{4 < X < 16\}$; **(c)** $P\{X < 8\}$;
    (d) $P\{X < 20\}$; **(e)** $P\{X > 16\}$.

16. The annual rainfall (in inches) in a certain region is normally distributed with $\mu = 40$ and $\sigma = 4$. What is the probability that starting with this year, it will take over 10 years before a year occurs having a rainfall of over 50 inches? What assumptions are you making?

17. A man aiming at a target receives 10 points if his shot is within 1 inch of the target, 5 points if it is between 1 and 3 inches of the target, and 3 points if it is between 3 and 5 inches of the target. Find the expected number of points scored if the distance from the shot to the target is uniformly distributed between 0 and 10.

18. Suppose that X is a normal random variable with mean 5. If $P\{X > 9\} = .2$, approximately what is $\text{Var}(X)$?

19. Let X be a normal random variable with mean 12 and variance 4. Find the value of c such that $P\{X > c\} = .10$.

20. If 65 percent of the population of a large community is in favor of a proposed rise in school taxes, approximate the probability that a random sample of 100 people will contain

(a) at least 50 who are in favor of the proposition;

(b) between 60 and 70 inclusive who are in favor;

(c) fewer than 75 in favor.

21. Suppose that the height, in inches, of a 25-year-old man is a normal random variable with parameters $\mu = 71$ and $\sigma^2 = 6.25$. What percentage of 25-year-old men are over 6 feet 2 inches tall? What percentage of men in the 6-footer club are over 6 foot 5 inches?

22. The width of a slot of a duralumin forging is (in inches) normally distributed with $\mu = .9000$ and $\sigma = .0030$. The specification limits were given as $.9000 \pm .0050$.

(a) What percentage of forgings will be defective?

(b) What is the maximum allowable value of σ that will permit no more than 1 in 100 defectives when the widths are normally distributed with $\mu = .9000$ and σ?

23. One thousand independent rolls of a fair die will be made. Compute an approximation to the probability that number 6 will appear between 150 and 200 times inclusively. If number 6 appears exactly 200 times, find the probability that number 5 will appear less than 150 times.

24. The lifetimes of interactive computer chips produced by a certain semiconductor manufacturer are normally distributed with parameters $\mu = 1.4 \times 10^6$ hours and $\sigma = 3 \times 10^5$ hours. What is the approximate probability that a batch of 100 chips will contain at least 20 whose lifetimes are less than 1.8×10^6?

25. Each item produced by a certain manufacturer is, independently, of acceptable quality with probability .95. Approximate the probability that at most 10 of the next 150 items produced are unacceptable.

26. Two types of coins are produced at a factory: a fair coin and a biased one that comes up heads 55 percent of the time. We have one of these coins but do not know whether it is a fair coin or a biased one. In order to ascertain which type of coin we have, we shall perform the following statistical test: We shall toss the coin 1000 times. If the coin lands on heads 525 or more times, then we shall conclude that it is a biased coin, whereas, if it lands heads less than 525 times, then we shall conclude that it is the fair coin. If the coin is actually fair, what is the probability that we shall reach a false conclusion? What would it be if the coin were biased?

27. In 10,000 independent tosses of a coin, the coin landed heads 5800 times. Is it reasonable to assume that the coin is not fair? Explain.

28. Twelve percent of the population is lefthanded. Approximate the probability that there are at least 20 lefthanders in a school of 200 students. State your assumptions.

29. A model for the movement of a stock supposes that if the present price of the stock is s, then after one time period it will be either us with probability p, or ds with probability $1 - p$. Assuming that successive movements are independent, approximate the probability that the stock's price will be up at least 30 percent after the next 1000 time periods if $u = 1.012$, $d = 0.990$, and $p = .52$.

30. An image is partitioned into 2 regions—one white and the other black. A reading taken from a randomly chosen point in the white section will give a reading that is normally distributed with $\mu = 4$ and $\sigma^2 = 4$, whereas one taken from a randomly chosen point in the black region will have a normally distributed reading with parameters (6, 9). A point is randomly chosen on the image and has a reading of 5. If the fraction of the image that is black is α, for what value of α would the probability of making an error be the same whether one concluded the point was in the black region or in the white region?

31. **(a)** A fire station is to be located along a road of length A, $A < \infty$. If fires will occur at points uniformly chosen on $(0, A)$, where should the station be located so as to minimize the expected distance from the fire? That is, choose a so as to

$$\text{minimize } E[|X - a|]$$

when X is uniformly distributed over $(0, A)$.

(b) Now suppose that the road is of infinite length—stretching from point 0 outward to ∞. If the distance of a fire from point 0 is exponentially distributed with rate λ, where should the fire station now be located? That is, we want to minimize $E[|X - a|]$, where X is now exponential with rate λ.

32. The time (in hours) required to repair a machine is an exponentially distributed random variable with parameter $\lambda = \frac{1}{2}$. What is

(a) the probability that a repair time exceeds 2 hours;

(b) the conditional probability that a repair takes at least 10 hours, given that its duration exceeds 9 hours?

33. The number of years a radio functions is exponentially distributed with parameter $\lambda = \frac{1}{8}$. If Jones buys a used radio, what is the probability that it will be working after an additional 8 years?

34. Jones figures that the total number of thousands of miles that an auto can be driven before it would need to be junked is an exponential random variable with parameter $\frac{1}{20}$. Smith has a used car that he claims has been driven only 10,000 miles. If Jones purchases the car, what is the probability that she would get at least 20,000 additional miles out of it? Repeat under the assumption that the lifetime mileage of the car is not exponentially distributed but rather is (in thousands of miles) uniformly distributed over (0, 40).

35. The lung cancer hazard rate of a t-year-old male smoker, $\lambda(t)$, is such that

$$\lambda(t) = .027 + .00025(t - 40)^2 \qquad t \geq 40$$

Assuming that a 40-year-old male smoker survives all other hazards, what is the probability that he survives to (a) age 50 and (b) age 60 without contracting lung cancer?

36. Suppose that the life distribution of an item has hazard rate function $\lambda(t) = t^3, t > 0$. What is the probability that

(a) the item survives to age 2;
(b) the item's lifetime is between .4 and 1.4;
(c) a 1-year-old item will survive to age 2?

37. If X is uniformly distributed over $(-1, 1)$, find

(a) $P\{|X| > \frac{1}{2}\}$;
(b) the density function of the random variable $|X|$.

38. If Y is uniformly distributed over (0, 5), what is the probability that the roots of the equation $4x^2 + 4xY + Y + 2 = 0$ are both real?

39. If X is an exponential random variable with parameter $\lambda = 1$, compute the probability density function of the random variable Y defined by $Y = \log X$.

40. If X is uniformly distributed over (0, 1), find the density function of $Y = e^X$.

41. Find the distribution of $R = A \sin \theta$, where A is a fixed constant and θ is uniformly distributed on $(-\pi/2, \pi/2)$. Such a random variable R arises in the theory of ballistics. If a projectile is fired from the origin at an angle α from the earth with a speed v, then the point R at which it returns to the earth can be expressed as $R = (v^2/g) \sin 2\alpha$, where g is the gravitational constant, equal to 980 centimeters per second squared.

THEORETICAL EXERCISES

1. The speed of a molecule in a uniform gas at equilibrium is a random variable whose probability density function is given by

$$f(x) = \begin{cases} ax^2e^{-bx^2} & x \geq 0 \\ 0 & x < 0 \end{cases}$$

where $b = m/2kT$ and k, T, and m denote, respectively, Boltzmann's constant, the absolute temperature, and the mass of the molecule. Evaluate a in terms of b.

2. Show that

$$E[Y] = \int_0^\infty P\{Y > y\}\,dy - \int_0^\infty P\{Y < -y\}\,dy$$

HINT: Show that

$$\int_0^\infty P\{Y < -y\}\,dy = -\int_{-\infty}^0 xf_Y(x)\,dx$$

$$\int_0^\infty P\{Y > y\}\,dy = \int_0^\infty xf_Y(x)\,dx$$

3. If X has density function f, show that

$$E[g(X)] = \int_{-\infty}^\infty g(x)f(x)\,dx$$

HINT: Using Theoretical Exercise 2, start with

$$E[g(X)] = \int_0^\infty P\{g(X) > y\}\,dy - \int_0^\infty P\{g(X) < -y\}\,dy$$

and then proceed as in the proof given in the text when $g(X) \geq 0$.

4. Prove Corollary 2.1.

5. Use the result that for a nonnegative random variable Y,

$$E[Y] = \int_0^\infty P\{Y > t\}\,dt$$

to show that for a nonnegative random variable X,

$$E[X^n] = \int_0^\infty nx^{n-1}P\{X > x\}\,dx$$

HINT: Start with

$$E[X^n] = \int_0^\infty P\{X^n > t\}\,dt$$

and make the change of variables $t = x^n$.

6. Define a collection of events $E_a, 0 < a < 1$, having the property that $P(E_a) = 1$ for all a, but $P\left(\bigcap_a E_a\right) = 0$.

HINT: Let X be uniform over $(0, 1)$ and define each E_a in terms of X.

7. The standard deviation of X, denoted $SD(X)$, is given by

$$SD(X) = \sqrt{Var(X)}$$

Find $SD(aX + b)$ if X has variance σ^2.

8. Let X be a random variable that takes on values between 0 and c. That is, $P\{0 \le X \le c\} = 1$. Show that

$$Var(X) \le \frac{c^2}{4}$$

HINT: One approach is to first argue that

$$E[X^2] \le cE[X]$$

Then use this to show that

$$Var(X) \le c^2[\alpha(1 - \alpha)] \quad \text{where} \quad \alpha = \frac{E[X]}{c}$$

9. If Z is a standard normal random variable, show that for $x > 0$,
(a) $P\{Z > x\} = P\{Z < -x\}$; **(b)** $P\{|Z| > x\} = 2P\{Z > x\}$;
(c) $P\{|Z| < x\} = 2P\{Z < x\} - 1$.

10. Let $f(x)$ denote the probability density function of a normal random variable with mean μ and variance σ^2. Show that $\mu - \sigma$ and $\mu + \sigma$ are points of inflection of this function. That is, show that $f''(x) = 0$ when $x = \mu - \sigma$ or $x = \mu + \sigma$.

11. Use the identity of Theoretical Exercise 5 to derive $E[X^2]$ when X is an exponential random variable with parameter λ.

12. The median of a continuous random variable having distribution function F is that value m such that $F(m) = \frac{1}{2}$. That is, a random variable is just as likely to be larger than its median as it is to be smaller. Find the median of X if X is
(a) uniformly distributed over (a, b); **(b)** normal with parameters μ, σ^2;
(c) exponential with rate λ.

13. The mode of a continuous random variable having density f is the value of x for which $f(x)$ attains its maximum. Compute the mode of X in cases (a), (b), and (c) of Theoretical Exercise 12.

14. If X is an exponential random variable with parameter λ, and $c > 0$, show that cX is exponential with parameter λ/c.

15. Compute the hazard rate function of X when X is uniformly distributed over $(0, a)$.

16. If X has hazard rate function $\lambda_X(t)$, compute the hazard rate function of aX where a is a positive constant.

17. Verify that the gamma density function integrates to 1.

18. If X is an exponential random variable with mean $1/\lambda$, show that

$$E[X^k] = \frac{k!}{\lambda^k} \quad k = 1, 2, \ldots$$

HINT: Make use of the gamma density function to evaluate the preceding.

19. Verify that

$$Var(X) = \frac{\alpha}{\lambda^2}$$

when X is a gamma random variable with parameters α and λ.

20. Show that $\Gamma\left(\frac{1}{2}\right) = \sqrt{\pi}$.

HINT: $\Gamma\left(\frac{1}{2}\right) = \int_0^\infty e^{-x}x^{-1/2}\,dx$. Make the change of variables $y = \sqrt{2x}$ and then relate the resulting expression to the normal distribution.

21. Compute the hazard rate function of a gamma random variable with parameters (α, λ) and show it is increasing when $\alpha \geq 1$ and decreasing when $\alpha \leq 1$.

22. Compute the hazard rate function of a Weibull random variable and show it is increasing when $\beta \geq 1$ and decreasing when $\beta \leq 1$.

23. Show that a plot of $\log(\log(1 - F(x))^{-1})$ against $\log x$ will be a straight line with slope β when $F(\cdot)$ is a Weibull distribution function. Show also that approximately 63.2 percent of all observations from such a distribution will be less than α. Assume that $v = 0$.

24. Let

$$Y = \left(\frac{X - \nu}{\alpha}\right)^\beta$$

Show that if X is a Weibull random variable with parameters ν, α, and β, then Y is an exponential random variable with parameter $\lambda = 1$ and vice versa.

25. If X is a beta random variable with parameters a and b, show that

$$E[X] = \frac{a}{a+b} \qquad \text{Var}(X) = \frac{ab}{(a+b)^2(a+b+1)}$$

26. If X is uniformly distributed over (a, b), what random variable, having a linear relation with X, is uniformly distributed over $(0, 1)$?

27. Consider the beta distribution with parameters (a, b). Show that

(a) when $a > 1$ and $b > 1$, the density is unimodal (that is, it has a unique mode) with mode equal to $(a - 1)/(a + b - 2)$;

(b) when $a \leq 1, b \leq 1$, and $a + b < 2$, the density is either unimodal with mode at 0 or 1 or U-shaped with modes at both 0 and 1;

(c) when $a = 1 = b$, all points in $[0, 1]$ are modes.

28. Let X be a continuous random variable having cumulative distribution function F. Define the random variable Y by $Y = F(X)$. Show that Y is uniformly distributed over $(0, 1)$.

29. Let X have probability density f_X. Find the probability density function of the random variable Y, defined by $Y = aX + b$.

30. Find the probability density function of $Y = e^X$ when X is normally distributed with parameters μ and σ^2. The random variable Y is said to have a lognormal distribution (since $\log Y$ has a normal distribution) with parameters μ and σ^2.

31. Let X and Y be independent random variables that are both equally likely to be either $1, 2, \ldots, (10)^N$, where N is very large. Let D denote the greatest common divisor of X and Y, and let $Q_k = P\{D = k\}$.

(a) Give a heuristic argument that $Q_k = \dfrac{1}{k^2}Q_1$.

HINT: Note that in order for D to equal k, k must divide both X and Y and also X/k and Y/k must be relatively prime. (That is, they must have a greatest common divisor equal to 1.)

(b) Use part (a) to show that

$$Q_1 = P\{X \text{ and } Y \text{ are relatively prime}\} = \frac{1}{\sum_{k=1}^{\infty} 1/k^2}$$

It is a well-known identity that $\sum_{1}^{\infty} 1/k^2 = \pi^2/6$, so $Q_1 = 6/\pi^2$ (In number theory this is known as the Legendre theorem.)

(c) Now argue that

$$Q_1 = \prod_{i=1}^{\infty}\left(\frac{P_i^2 - 1}{P_i^2}\right)$$

where P_i is the ith smallest prime greater than 1.

HINT: X and Y will be relatively prime if they have no common prime factors. Hence, from part (b), we see that

$$\prod_{i=1}^{\infty}\left(\frac{P_i^2 - 1}{P_i^2}\right) = \frac{6}{\pi^2}$$

which was noted without explanation in Problem 11 of Chapter 4. (The relationship between this problem and Problem 11 of Chapter 4 is that X and Y are relatively prime if XY has no multiple prime factors.)

32. Prove Theorem 7.1 when $g(x)$ is a decreasing function.

SELF-TEST PROBLEMS AND EXERCISES

1. The number of minutes of playing time of a certain high school basketball player in a randomly chosen game is a random variable whose probability density function is given in the following figure.

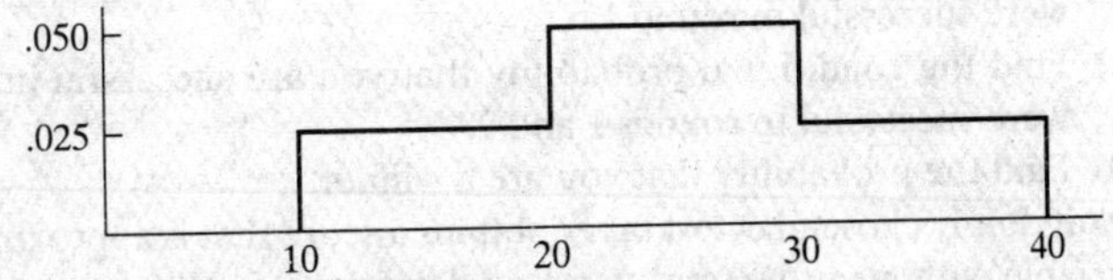

Find the probability that the player plays

(a) over 15 minutes;
(b) between 20 and 35 minutes;
(c) less than 30 minutes;
(d) more than 36 minutes.

2. For some constant c, the random variable X has probability density function

$$f(x) = \begin{cases} cx^n & 0 < x < 1 \\ 0 & \text{otherwise} \end{cases}$$

Find (a) c and (b) $P\{X > x\}, 0 < x < 1$.

3. For some constant c, the random variable X has probability density function

$$f(x) = \begin{cases} cx^4 & 0 < x < 2 \\ 0 & \text{otherwise} \end{cases}$$

Find (a) $E[X]$ and (b) $\text{Var}(X)$.

4. The random variable X has probability density function

$$f(x) = \begin{cases} ax + bx^2 & 0 < x < 1 \\ 0 & \text{otherwise} \end{cases}$$

If $E[X] = .6$, find (a) $P\{X < \frac{1}{2}\}$ and (b) $\text{Var}(X)$.

5. The random variable X is said to be a discrete uniform random variable on the integers $1, 2, \ldots, n$ if

$$P\{X = i\} = \frac{1}{n} \qquad i = 1, 2, \ldots, n$$

For any nonnegative real number x, let $\text{Int}(x)$ (sometimes written as $[x]$) be the largest integer that is less than or equal to x. If U is a uniform random variable on $(0, 1)$, show that $X = \text{Int}(nU) + 1$ is a discrete uniform random variable on $1, \ldots, n$.

6. Your company must make a sealed bid for a construction project. If you succeed in winning the contract (by having the lowest bid), then you plan to pay another firm 100 thousand dollars to do the work. If you believe that the maximum bid (in thousands of dollars) of the other participating companies can be modeled as being the value of a random variable that is uniformly distributed on (70, 140), how much should you bid to maximize your expected profit?

7. To be a winner in the following game, you must be successful in three successive rounds. The game depends on the value of U, a uniform random variable on $(0, 1)$. If $U > .1$, then you are successful in round 1; if $U > .2$, then you are successful in round 2; and if $U > .3$, then you are successful in round 3.

(a) Find the probability that you are successful in round 1.

(b) Find the conditional probability that you are successful in round 2 given that you were successful in round 1.

(c) Find the conditional probability that you are successful in round 3 given that you were successful in rounds 1 and 2.

(d) Find the probability that you are a winner.

8. A randomly chosen IQ test taker obtains a score that is approximately a normal random variable with mean 100 and standard deviation 15. What is the probability that the test score of such a person is (a) above 125; (b) between 90 and 110?

9. Suppose that the travel time from your home to your office is normally distributed with mean 40 minutes and standard deviation 7 minutes. If you want to be 95 percent certain that you will not be late for an office appointment at 1 P.M., what is the latest time that you should leave home?

10. The life of a certain type of automobile tire is normally distributed with mean 34,000 miles and standard deviation 4000 miles.

(a) What is the probability that such a tire lasts over 40,000 miles?

(b) What is the probability that it lasts between 30,000 and 35,000 miles?

(c) Given that it has survived 30,000 miles, what is the conditional probability that it survives another 10,000 miles?

11. The annual rainfall in Cleveland, Ohio is approximately a normal random variable with mean 40.2 inches and standard deviation 8.4 inches. What is the probability that
(a) next year's rainfall will exceed 44 inches;
(b) the yearly rainfalls in exactly three of the next seven years will exceed 44 inches? Assume that if A_i is the event that the rainfall exceeds 44 inches in year i (from now), then the events $A_i, i \geq 1$, are independent.

12. The following table uses 1992 data concerning the percentages of male and female full-time workers whose annual salaries fall in different ranges.

Earnings range	Percentage of females	Percentage of males
≤9999	8.6	4.4
10,000–19,999	38.0	21.1
20,000–24,999	19.4	15.8
25,000–49,999	29.2	41.5
≥50,000	4.8	17.2

Suppose that random samples of 200 male and 200 female fulltime workers are chosen. Approximate the probability that
(a) at least 70 of the women earn $25,000 or more;
(b) at most 60 percent of the men earn $25,000 or more;
(c) at least three-fourths of the men and at least half the women earn $20,000 or more.

13. At a certain bank, the amount of time that a customer spends being served by a teller is an exponential random variable with mean 5 minutes. If there is a customer in service when you enter the bank, what is the probability that he or she will still be with the teller after an additional 4 minutes?

14. Suppose that the cumulative distribution function of the random variable X is given by

$$F(x) = 1 - e^{-x^2} \qquad x > 0$$

Evaluate (a) $P\{X > 2\}$; (b) $P\{1 < X < 3\}$; (c) the hazard rate function of F; (d) $E[X]$; (e) $\text{Var}(X)$.

HINT: For parts (d) and (e) you might want to make use of the results of Theoretical Exercise 5.

15. The number of years that a washing machine functions is a random variable whose hazard rate function is given by

$$\lambda(t) = \begin{cases} .2 & 0 < t < 2 \\ .2 + .3(t - 2) & 2 \leq t < 5 \\ 1.1 & t > 5 \end{cases}$$

(a) What is the probability that the machine will still be working six years after being purchased?
(b) If it is still working six years after being purchased, what is the conditional probability that it will fail within the succeeding two years?

16. A standard Cauchy random variable has density function

$$f(x) = \frac{1}{\pi(1 + x^2)} \qquad -\infty < x < \infty$$

If X is a standard Cauchy random variable, show that $1/X$ is also a standard Cauchy random variable.

17. A roulette wheel has 38 slots, numbered 0, 00, and 1 through 36. If you bet 1 on a specified number then you either win 35 if the roulette ball lands on that number or lose 1 if it does not. If you continually make such bets, approximate the probability that
(a) you are winning after 34 bets; **(b)** you are winning after 1000 bets;
(c) you are winning after 100,000 bets.
Assume that each roll of the roulette ball is equally likely to land on any of the 38 numbers.

18. There are two types of batteries in a bin. When in use, type i batteries last (in hours) an exponentially distributed time with rate λ_i, $i = 1, 2$. A battery that is randomly chosen from the bin will be a type i battery with probability p_i, $\sum_{i=1}^{2} p_i = 1$. If a randomly chosen battery is still operating after t hours of use, what is the probability it will still be operating after an additional s hours?

CHAPTER 6

Jointly Distributed Random Variables

6.1 JOINT DISTRIBUTION FUNCTIONS

Thus far, we have only concerned ourselves with probability distributions for single random variables. However, we are often interested in probability statements concerning two or more random variables. In order to deal with such probabilities, we define, for any two random variables X and Y, the *joint cumulative probability distribution function* of X and Y by

$$F(a, b) = P\{X \le a, Y \le b\} \qquad -\infty < a, b < \infty$$

The distribution of X can be obtained from the joint distribution of X and Y as follows:

$$\begin{aligned} F_X(a) &= P\{X \le a\} \\ &= P\{X \le a, Y < \infty\} \\ &= P\Big(\lim_{b\to\infty} \{X \le a, Y \le b\}\Big) \\ &= \lim_{b\to\infty} P\{X \le a, Y \le b\} \\ &= \lim_{b\to\infty} F(a, b) \\ &\equiv F(a, \infty) \end{aligned}$$

The reader should note that we have, in the preceding set of equalities, once again made use of the fact that probability is a continuous set (that is, event) function. Similarly, the cumulative distribution function of Y is given by

$$\begin{aligned} F_Y(b) &= P\{Y \le b\} \\ &= \lim_{a\to\infty} F(a, b) \\ &\equiv F(\infty, b) \end{aligned}$$

The distribution functions F_X and F_Y are sometimes referred to as the *marginal* distributions of X and Y.

All joint probability statements about X and Y can, in theory, be answered in terms of their joint distribution function. For instance, suppose we wanted to compute the joint probability that X is greater than a and Y is greater than b. This could be done as follows:

$$\begin{aligned} P\{X > a, Y > b\} &= 1 - P(\{X > a, Y > b\}^c) \\ &= 1 - P(\{X > a\}^c \cup \{Y > b\}^c) \\ &= 1 - P(\{X \le a\} \cup \{Y \le b\}) \\ &= 1 - [P\{X \le a\} + P\{Y \le b\} - P\{X \le a, Y \le b\}] \\ &= 1 - F_X(a) - F_Y(b) + F(a, b) \end{aligned} \tag{1.1}$$

Equation (1.1) is a special case of Equation (1.2), whose verification is left as an exercise.

$$\begin{aligned} &P\{a_1 < X \le a_2, b_1 < Y \le b_2\} \\ &\quad = F(a_2, b_2) + F(a_1, b_1) - F(a_1, b_2) - F(a_2, b_1) \end{aligned} \tag{1.2}$$

whenever $a_1 < a_2, b_1 < b_2$.

In the case when X and Y are both discrete random variables, it is convenient to define the *joint probability mass function* of X and Y by

$$p(x, y) = P\{X = x, Y = y\}$$

The probability mass function of X can be obtained from $p(x, y)$ by

$$\begin{aligned} p_X(x) &= P\{X = x\} \\ &= \sum_{y:p(x,y)>0} p(x, y) \end{aligned}$$

Similarly,

$$p_Y(y) = \sum_{x:p(x,y)>0} p(x, y)$$

Example 1a. Suppose that 3 balls are randomly selected from an urn containing 3 red, 4 white, and 5 blue balls. If we let X and Y denote, respectively, the number of red and white balls chosen, then the joint probability mass function of X and Y, $p(i, j) = P\{X = i, Y = j\}$, is given by

$$p(0, 0) = \binom{5}{3} \bigg/ \binom{12}{3} = \frac{10}{220}$$

$$p(0, 1) = \binom{4}{1}\binom{5}{2} \bigg/ \binom{12}{3} = \frac{40}{220}$$

$$p(0, 2) = \binom{4}{2}\binom{5}{1} \bigg/ \binom{12}{3} = \frac{30}{220}$$

$$p(0, 3) = \binom{4}{3} \bigg/ \binom{12}{3} = \frac{4}{220}$$

$$p(1, 0) = \binom{3}{1}\binom{5}{2} \bigg/ \binom{12}{3} = \frac{30}{220}$$

$$p(1,1) = \binom{3}{1}\binom{4}{1}\binom{5}{1}\Big/\binom{12}{3} = \frac{60}{220}$$

$$p(1,2) = \binom{3}{1}\binom{4}{2}\Big/\binom{12}{3} = \frac{18}{220}$$

$$p(2,0) = \binom{3}{2}\binom{5}{1}\Big/\binom{12}{3} = \frac{15}{220}$$

$$p(2,1) = \binom{3}{2}\binom{4}{1}\Big/\binom{12}{3} = \frac{12}{220}$$

$$p(3,0) = \binom{3}{3}\Big/\binom{12}{3} = \frac{1}{220}$$

These probabilities can most easily be expressed in tabular form as in Table 6.1. The reader should note that the probability mass function of X is obtained by computing the row sums, whereas the probability mass function of Y is obtained by computing the column sums. As the individual probability mass functions of X and Y thus appear in the margin of such a table, they are often referred to as being the marginal probability mass functions of X and Y, respectively. ■

TABLE 6.1 $P\{X = i, Y = j\}$

i \ j	0	1	2	3	Row sum $= P\{X = i\}$
0	$\frac{10}{220}$	$\frac{40}{220}$	$\frac{30}{220}$	$\frac{4}{220}$	$\frac{84}{220}$
1	$\frac{30}{220}$	$\frac{60}{220}$	$\frac{18}{220}$	0	$\frac{108}{220}$
2	$\frac{15}{220}$	$\frac{12}{220}$	0	0	$\frac{27}{220}$
3	$\frac{1}{220}$	0	0	0	$\frac{1}{220}$
Column sum $= P\{Y = j\}$	$\frac{56}{220}$	$\frac{112}{220}$	$\frac{48}{220}$	$\frac{4}{220}$	

Example 1b. Suppose that 15 percent of the families in a certain community have no children, 20 percent have 1, 35 percent have 2, and 30 percent have 3; and suppose, further, that in each family, each child is equally likely (independently) to be a boy or a girl. If a family is chosen at random from this community, then B, the number of boys, and G, the number of girls, in this family will have the joint probability mass function shown in Table 6.2.

TABLE 6.2 $P\{B = i, G = j\}$

i \ j	0	1	2	3	Row sum = $P\{B = i\}$
0	.15	.10	.0875	.0375	.3750
1	.10	.175	.1125	0	.3875
2	.0875	.1125	0	0	.2000
3	.0375	0	0	0	.0375
Column sum = $P\{G = j\}$	.3750	.3875	.2000	.0375	

These probabilities are obtained as follows:

$$
\begin{aligned}
P\{B = 0, G = 0\} &= P\{\text{no children}\} = .15 \\
P\{B = 0, G = 1\} &= P\{\text{1 girl and total of 1 child}\} \\
&= P\{\text{1 child}\}P\{\text{1 girl}|\text{1 child}\} = (.20)\left(\tfrac{1}{2}\right) \\
P\{B = 0, G = 2\} &= P\{\text{2 girls and total of 2 children}\} \\
&= P\{\text{2 children}\}P\{\text{2 girls}|\text{2 children}\} = (.35)\left(\tfrac{1}{2}\right)^2
\end{aligned}
$$

We leave the verification of the remaining probabilities in Table 6.2 to the reader. ■

We say that X and Y are *jointly continuous* if there exists a function $f(x, y)$ defined for all real x and y, having the property that for every set C of pairs of real numbers (that is, C is a set in the two-dimensional plane)

$$P\{(X, Y) \in C\} = \iint_{(x,y)\in C} f(x, y)\,dx\,dy \tag{1.3}$$

The function $f(x, y)$ is called the *joint probability density function* of X and Y. If A and B are any sets of real numbers, then by defining $C = \{(x, y): x \in A, y \in B\}$, we see from Equation (1.3) that

$$P\{X \in A, Y \in B\} = \int_B \int_A f(x, y)\,dx\,dy \tag{1.4}$$

Because

$$
\begin{aligned}
F(a, b) &= P\{X \in (-\infty, a], Y \in (-\infty, b]\} \\
&= \int_{-\infty}^{b} \int_{-\infty}^{a} f(x, y)\,dx\,dy
\end{aligned}
$$

it follows, upon differentiation, that

$$f(a, b) = \frac{\partial^2}{\partial a\,\partial b} F(a, b)$$

wherever the partial derivatives are defined. Another interpretation of the joint density function is obtained from Equation (1.4) as follows:

$$P\{a < X < a + da, b < Y < b + db\} = \int_b^{d+db} \int_a^{a+da} f(x, y)\, dx\, dy$$
$$\approx f(a, b)\, da\, db$$

when da and db are small and $f(x, y)$ is continuous at a, b. Hence $f(a, b)$ is a measure of how likely it is that the random vector (X, Y) will be near (a, b).

If X and Y are jointly continuous, they are individually continuous, and their probability density functions can be obtained as follows:

$$P\{X \in A\} = P\{X \in A, Y \in (-\infty, \infty)\}$$
$$= \int_A \int_{-\infty}^{\infty} f(x, y)\, dy\, dx$$
$$= \int_A f_X(x)\, dx$$

where

$$f_X(x) = \int_{-\infty}^{\infty} f(x, y)\, dy$$

is thus the probability density function of X. Similarly, the probability density function of Y is given by

$$f_Y(y) = \int_{-\infty}^{\infty} f(x, y)\, dx$$

Example 1c. The joint density function of X and Y is given by

$$f(x, y) = \begin{cases} 2e^{-x}e^{-2y} & 0 < x < \infty, 0 < y < \infty \\ 0 & \text{otherwise} \end{cases}$$

Compute (a) $P\{X > 1, Y < 1\}$, (b) $P\{X < Y\}$, and (c) $P\{X < a\}$.

Solution

$$\text{(a) } P\{X > 1, Y < 1\} = \int_0^1 \int_1^{\infty} 2e^{-x}e^{-2y}\, dx\, dy$$
$$= \int_0^1 2e^{-2y}\left(-e^{-x}\Big|_1^{\infty}\right) dy$$
$$= e^{-1} \int_0^1 2e^{-2y}\, dy$$
$$= e^{-1}(1 - e^{-2})$$

$$\text{(b) } P\{X < Y\} = \iint\limits_{(x,y):\, x<y} 2e^{-x}e^{-2y}\,dx\,dy$$

$$= \int_0^{\infty}\int_0^{y} 2e^{-x}e^{-2y}\,dx\,dy$$

$$= \int_0^{\infty} 2e^{-2y}(1 - e^{-y})\,dy$$

$$= \int_0^{\infty} 2e^{-2y}\,dy - \int_0^{\infty} 2e^{-3y}\,dy$$

$$= 1 - \tfrac{2}{3}$$

$$= \tfrac{1}{3}$$

$$\text{(c) } P\{X < a\} = \int_0^{a}\int_0^{\infty} 2e^{-2y}e^{-x}\,dy\,dx$$

$$= \int_0^{a} e^{-x}\,dx$$

$$= 1 - e^{-a}$$

■

Example 1d. Consider a circle of radius R and suppose that a point within the circle is randomly chosen in such a manner that all regions within the circle of equal area are equally likely to contain the point. (In other words, the point is uniformly distributed within the circle.) If we let the center of the circle denote the origin and define X and Y to be the coordinates of the point chosen (Figure 6.1), it follows, since (X, Y) is equally likely to be near each point in the circle, that the joint density function of X and Y is given by

$$f(x, y) = \begin{cases} c & \text{if } x^2 + y^2 \le R^2 \\ 0 & \text{if } x^2 + y^2 > R^2 \end{cases}$$

for some value of c.

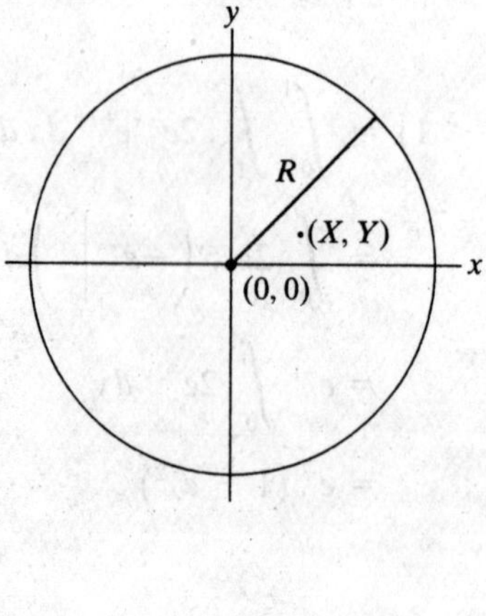

Figure 6.1 Joint probability distribution.

(a) Determine c.
(b) Find the marginal density functions of X and Y.
(c) Compute the probability that D, the distance from the origin of the point selected, is less than or equal to a.
(d) Find $E[D]$.

Solution (a) Because

$$\int_{-\infty}^{\infty}\int_{-\infty}^{\infty} f(x, y)\, dy\, dx = 1$$

it follows that

$$c \iint\limits_{x^2+y^2\le R^2} dy\, dx = 1$$

We can evaluate $\iint_{x^2+y^2\le R^2} dy\, dx$ either by using polar coordinates, or more simply, by noting that it represents the area of the circle and is thus equal to πR^2. Hence

$$c = \frac{1}{\pi R^2}$$

$$\begin{aligned}
\text{(b) } f_X(x) &= \int_{-\infty}^{\infty} f(x, y)\, dy \\
&= \frac{1}{\pi R^2}\int_{x^2+y^2\le R^2} dy \\
&= \frac{1}{\pi R^2}\int_{-c}^{c} dy \qquad c = \sqrt{R^2 - x^2} \\
&= \frac{2}{\pi R^2}\sqrt{R^2 - x^2} \qquad x^2 \le R^2
\end{aligned}$$

and it equals 0 when $x^2 > R^2$. By symmetry the marginal density of Y is given by

$$\begin{aligned}
f_Y(y) &= \frac{2}{\pi R^2}\sqrt{R^2 - y^2} \qquad y^2 \le R^2 \\
&= 0 \qquad y^2 > R^2
\end{aligned}$$

(c) The distribution function of $D = \sqrt{X^2 + Y^2}$, the distance from the origin, is obtained as follows: for $0 \le a \le R$,

$$\begin{aligned}
F_D(a) &= P\{\sqrt{X^2 + Y^2} \le a\} \\
&= P\{X^2 + Y^2 \le a^2\} \\
&= \iint\limits_{x^2+y^2\le a^2} f(x, y)\, dy\, dx
\end{aligned}$$

$$= \frac{1}{\pi R^2} \iint\limits_{x^2+y^2 \le a} dy\,dx$$

$$= \frac{\pi a^2}{\pi R^2}$$

$$= \frac{a^2}{R^2}$$

where we have used the fact that $\iint_{x^2+y^2\le a^2} dy\,dx$ is the area of a circle of radius a and thus is equal to πa^2.

(d) From part (c) we obtain that the density function of D is

$$f_D(a) = \frac{2a}{R^2} \qquad 0 \le a \le R$$

Hence

$$E[D] = \frac{2}{R^2}\int_0^R a^2\,da = \frac{2R}{3}$$ ■

Example 1e. The joint density of X and Y is given by

$$f(x, y) = \begin{cases} e^{-(x+y)} & 0 < x < \infty, 0 < y < \infty \\ 0 & \text{otherwise} \end{cases}$$

Find the density function of the random variable X/Y.

Solution We start by computing the distribution function of X/Y. For $a > 0$,

$$F_{X/Y}(a) = P\left\{\frac{X}{Y} \le a\right\}$$

$$= \iint\limits_{x/y \le a} e^{-(x+y)}\,dx\,dy$$

$$= \int_0^\infty \int_0^{ay} e^{-(x+y)}\,dx\,dy$$

$$= \int_0^\infty (1 - e^{-ay})e^{-y}\,dy$$

$$= \left\{-e^{-y} + \frac{e^{-(a+1)y}}{a+1}\right\}\Bigg|_0^\infty$$

$$= 1 - \frac{1}{a+1}$$

Differentiation yields that the density function of X/Y is given by $f_{X/Y}(a) = 1/(a+1)^2, 0 < a < \infty$. ■

We can also define joint probability distributions for n random variables in exactly the same manner as we did for $n = 2$. For instance, the joint cumulative probability distribution function $F(a_1, a_2, \dots, a_n)$ of the n random variables $X_1, X_2, \dots, X_n$ is defined by

$$F(a_1, a_2, \dots, a_n) = P\{X_1 \le a_1, X_2 \le a_2, \dots, X_n \le a_n\}$$

Further, the n random variables are said to be jointly continuous if there exists a function $f(x_1, x_2, \dots, x_n)$, called the joint probability density function, such that for any set C in n-space

$$P\{(X_1, X_2, \dots, X_n) \in C\} = \iint_{(x_1, \dots, x_n) \in C} \cdots \int f(x_1, \dots, x_n)\, dx_1\, dx_2 \cdots dx_n$$

In particular, for any n sets of real numbers $A_1, A_2, \dots, A_n$,

$$P\{X_1 \in A_1, X_2 \in A_2, \dots, X_n \in A_n\}$$
$$= \int_{A_n} \int_{A_{n-1}} \cdots \int_{A_1} f(x_1, \dots, x_n)\, dx_1\, dx_2 \cdots dx_n$$

Example 1f. ***The multinomial distribution.*** One of the most important joint distributions is the multinomial, which arises when a sequence of n independent and identical experiments is performed. Suppose that each experiment can result in any one of r possible outcomes, with respective probabilities $p_1, p_2, \dots, p_r$, $\sum_{i=1}^{r} p_i = 1$. If we let X_i denote the number of the n experiments that result in outcome number i, then

$$P\{X_1 = n_1, X_2 = n_2, \dots, X_r = n_r\} = \frac{n!}{n_1!\, n_2! \cdots n_r!} p_1^{n_1} p_2^{n_2} \cdots p_r^{n_r} \qquad (1.5)$$

whenever $\sum_{i=1}^{r} n_i = n$.

Equation (1.5) is verified by noting that any sequence of outcomes for the n experiments that leads to outcome i occurring n_i times for $i = 1, 2, \dots, r$, will, by the assumed independence of experiments, have probability $p_1^{n_1} p_2^{n_2} \cdots p_r^{n_r}$ of occurring. As there are $n!/(n_1!\, n_2! \dots n_r!)$ such sequences of outcomes (there are $n!/n_1! \dots n_r!$ different permutations of n things of which n_1 are alike, n_2 are alike, $\dots$, n_r are alike). Equation (1.5) is established. The joint distribution whose joint probability mass function is specified by Equation (1.5) is called the multinomial distribution. The reader should note that when $r = 2$, the multinomial reduces to the binomial distribution.

As an application of the multinomial, suppose that a fair die is rolled 9 times. The probability that 1 appears three times, 2 and 3 twice each, 4 and 5 once each, and 6 not at all is

$$\frac{9!}{3!\,2!\,2!\,1!\,1!\,0!}\left(\frac{1}{6}\right)^3\left(\frac{1}{6}\right)^2\left(\frac{1}{6}\right)^2\left(\frac{1}{6}\right)^1\left(\frac{1}{6}\right)^1\left(\frac{1}{6}\right)^0 = \frac{9!}{3!\,2!\,2!}\left(\frac{1}{6}\right)^9 \qquad \blacksquare$$

6.2 INDEPENDENT RANDOM VARIABLES

The random variables X and Y are said to be *independent* if for any two sets of real numbers A and B,

$$P\{X \in A, Y \in B\} = P\{X \in A\}P\{Y \in B\} \tag{2.1}$$

In other words, X and Y are independent if, for all A and B, the events $E_A = \{X \in A\}$ and $F_B = \{Y \in B\}$ are independent.

It can be shown by using the three axioms of probability that Equation (2.1) will follow if and only if for all a, b,

$$P\{X \le a, Y \le b\} = P\{X \le a\}P\{Y \le b\}$$

Hence, in terms of the joint distribution function F of X and Y, we have that X and Y are independent if

$$F(a, b) = F_X(a)F_Y(b) \qquad \text{for all } a, b$$

When X and Y are discrete random variables, the condition of independence (2.1) is equivalent to

$$p(x, y) = p_X(x)p_Y(y) \qquad \text{for all } x, y \tag{2.2}$$

The equivalence follows because, if (2.1) is satisfied, then we obtain Equation (2.2) by letting A and B be, respectively, the one point sets $A = \{x\}$, $B = \{y\}$. Furthermore, if Equation (2.2) is valid, then for any sets A, B,

$$\begin{aligned}
P\{X \in A, Y \in B\} &= \sum_{y \in B}\sum_{x \in A} p(x, y) \\
&= \sum_{y \in B}\sum_{x \in A} p_X(x)p_Y(y) \\
&= \sum_{y \in B} p_Y(y) \sum_{x \in A} p_X(x) \\
&= P\{Y \in B\}P\{X \in A\}
\end{aligned}$$

and thus Equation (2.1) is established.

In the jointly continuous case the condition of independence is equivalent to

$$f(x, y) = f_X(x)f_Y(y) \qquad \text{for all } x, y$$

Thus, loosely speaking, X and Y are independent if knowing the value of one does not change the distribution of the other. Random variables that are not independent are said to be *dependent*.

Example 2a. Suppose that $n + m$ independent trials, having a common success probability p, are performed. If X is the number of successes in the first n trials, and Y is the number of successes in the final m trials, then X and Y are independent, since knowing the number of successes in the first n trials does not affect the distribution of the number of successes in the final m trials (by the assumption of independent trials). In fact, for integral x and y,

$$P\{X = x, Y = y\} = \binom{n}{x} p^x (1-p)^{n-x} \binom{m}{y} p^y (1-p)^{m-y} \qquad \begin{array}{l} 0 \le x \le n, \\ 0 \le y \le m \end{array}$$

$$= P\{X = x\}P\{Y = y\}$$

On the other hand, X and Z will be dependent, where Z is the total number of successes in the $n + m$ trials. (Why is this?) ■

Example 2b. Suppose that the number of people that enter a post office on a given day is a Poisson random variable with parameter λ. Show that if each person that enters the post office is a male with probability p and a female with probability $1 - p$, then the number of males and females entering the post office are independent Poisson random variables with respective parameters λp and $\lambda(1 - p)$.

Solution Let X and Y denote, respectively, the number of males and females that enter the post office. We shall show the independence of X and Y by establishing Equation (2.2). To obtain an expression for $P\{X = i, Y = j\}$, we condition on $X + Y$ as follows:

$$P\{X = i, Y = j\} = P\{X = i, Y = j | X + Y = i + j\}P\{X + Y = i + j\}$$
$$+ P\{X = i, Y = j | X + Y \ne i + j\}P\{X + Y \ne i + j\}$$

[The reader should note that this equation is merely a special case of the formula $P(E) = P(E|F)P(F) + P(E|F^c)P(F^c)$.]

As $P\{X = i, Y = j | X + Y \ne i + j\}$ is clearly 0, we obtain

$$P\{X = i, Y = j\} = P\{X = i, Y = j | X + Y = i + j\}P\{X + Y = i + j\} \tag{2.3}$$

Now, as $X + Y$ is the total number that enter the post office, it follows, by assumption, that

$$P\{X + Y = i + j\} = e^{-\lambda} \frac{\lambda^{i+j}}{(i + j)!} \tag{2.4}$$

Furthermore, given that $i + j$ people do enter the post office, since each person entering will be male with probability p, it follows that the probability that exactly i of them will be male (and thus j of them female) is just the binomial probability $\binom{i+j}{i} p^i (1 - p)^j$. That is,

$$P\{X = i, Y = j | X + Y = i + j\} = \binom{i+j}{i} p^i (1-p)^j \tag{2.5}$$

Substituting Equations (2.4) and (2.5) into Equation (2.3) yields

$$\begin{aligned} P\{X = i, Y = j\} &= \binom{i+j}{i} p^i (1-p)^j e^{-\lambda} \frac{\lambda^{i+j}}{(i+j)!} \\ &= e^{-\lambda} \frac{(\lambda p)^i}{i!\, j!} [\lambda(1-p)]^j \\ &= \frac{e^{-\lambda p} (\lambda p)^i}{i!} e^{-\lambda(1-p)} \frac{[\lambda(1-p)]^j}{j!} \end{aligned} \tag{2.6}$$

Hence

$$P\{X = i\} = e^{-\lambda p} \frac{(\lambda p)^i}{i!} \sum_j e^{-\lambda(1-p)} \frac{[\lambda(1-p)]^j}{j!} = e^{-\lambda p} \frac{(\lambda p)^i}{i!} \tag{2.7}$$

and similarly,

$$P\{Y = j\} = e^{-\lambda(1-p)} \frac{[\lambda(1-p)]^j}{j!} \tag{2.8}$$

Equations (2.6), (2.7), and (2.8) establish the desired result. ■

Example 2c. A man and a woman decide to meet at a certain location. If each person independently arrives at a time uniformly distributed between 12 noon and 1 P.M., find the probability that the first to arrive has to wait longer than 10 minutes.

Solution If we let X and Y denote, respectively, the time past 12 that the man and the woman arrive, then X and Y are independent random variables, each of which is uniformly distributed over $(0, 60)$. The desired probability, $P\{X + 10 < Y\} + P\{Y + 10 < X\}$, which by symmetry equals $2P\{X + 10 < Y\}$, is obtained as follows:

$$\begin{aligned} 2P\{X + 10 < Y\} &= 2 \iint_{x+10<y} f(x, y)\, dx\, dy \\ &= 2 \iint_{x+10<y} f_X(x) f_Y(y)\, dx\, dy \\ &= 2 \int_{10}^{60} \int_{0}^{y-10} \left(\frac{1}{60}\right)^2 dx\, dy \\ &= \frac{2}{(60)^2} \int_{10}^{60} (y - 10)\, dy \\ &= \frac{25}{36} \end{aligned}$$

■

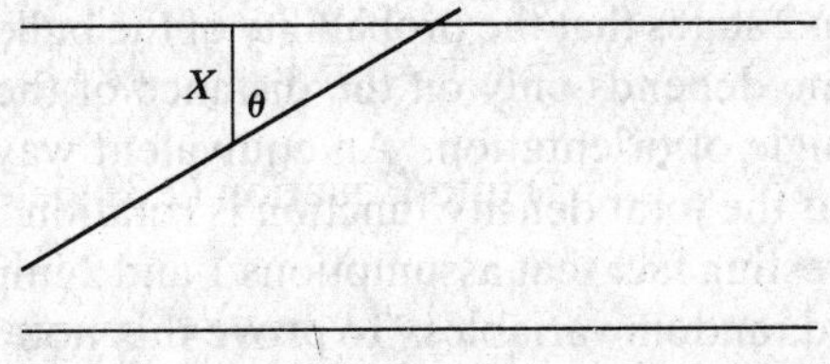

Figure 6.2

Our next example presents the oldest problem dealing with geometrical probabilities. It was first considered and solved by Buffon, a French naturalist of the eighteenth century, and is usually referred to as Buffon's needle problem.

Example 2d. ***Buffon's needle problem.*** A table is ruled with equidistant parallel lines a distance D apart. A needle of length L, where $L \leq D$, is randomly thrown on the table. What is the probability that the needle will intersect one of the lines (the other possibility being that the needle will be completely contained in the strip between two lines)?

Solution Let us determine the position of the needle by specifying the distance X from the middle point of the needle to the nearest parallel line, and the angle θ between the needle and the projected line of length X (see Figure 6.2). The needle will intersect a line if the hypotenuse of the right triangle in Figure 6.2 is less than $L/2$, that is, if

$$\frac{X}{\cos\theta} < \frac{L}{2} \quad \text{or} \quad X < \frac{L}{2}\cos\theta$$

As X varies between 0 and $D/2$ and θ between 0 and $\pi/2$, it is reasonable to assume that they are independent, uniformly distributed random variables over these respective ranges. Hence

$$\begin{aligned} P\left\{X < \frac{L}{2}\cos\theta\right\} &= \iint\limits_{x<L/2\cos y} f_X(x) f_\theta(y)\,dx\,dy \\ &= \frac{4}{\pi D}\int_0^{\pi/2}\int_0^{L/2\cos y} dx\,dy \\ &= \frac{4}{\pi D}\int_0^{\pi/2}\frac{L}{2}\cos y\,dy \\ &= \frac{2L}{\pi D} \end{aligned}$$

■

Example 2e.** ***Characterization of the normal distribution. Let X and Y denote the horizontal and vertical miss distances when a bullet is fired at a target, and assume that

1. X and Y are independent continuous random variables having differentiable density functions.
2. The joint density $f(x, y) = f_X(x)f_Y(y)$ of X and Y depends on (x, y) only through $x^2 + y^2$.

Loosely put, assumption 2 states that the probability of the bullet landing on any point of the $x - y$ plane depends only on the distance of the point from the target and not on its angle of orientation. An equivalent way of phrasing assumption 2 is to say that the joint density function is rotation invariant.

It is a rather interesting fact that assumptions 1 and 2 imply that X and Y are normally distributed random variables. To prove this, note first that the assumptions yield the relation

$$f(x, y) = f_X(x)f_Y(y) = g(x^2 + y^2) \tag{2.9}$$

for some function g. Differentiating Equation (2.9) with respect to x yields

$$f'_X(x)f_Y(y) = 2xg'(x^2 + y^2) \tag{2.10}$$

Dividing Equation (2.10) by Equation (2.9) gives

$$\frac{f'_X(x)}{f_X(x)} = \frac{2xg'(x^2 + y^2)}{g(x^2 + y^2)}$$

or

$$\frac{f'_X(x)}{2xf_X(x)} = \frac{g'(x^2 + y^2)}{g(x^2 + y^2)} \tag{2.11}$$

As the value of the left-hand side of Equation (2.11) depends only on x, whereas the value of the right-hand side depends on $x^2 + y^2$, it follows that the left-hand side must be the same for all x. To see this, consider any x_1, x_2 and let y_1, y_2 be such that $x_1^2 + y_1^2 = x_2^2 + y_2^2$. Then, from Equation (2.11), we obtain

$$\frac{f'_X(x_1)}{2x_1f_X(x_1)} = \frac{g'(x_1^2 + y_1^2)}{g(x_1^2 + y_1^2)} = \frac{g'(x_2^2 + y_2^2)}{g(x_2^2 + y_2^2)} = \frac{f'_X(x_2)}{2x_2f_X(x_2)}$$

Hence

$$\frac{f'_X(x)}{xf_X(x)} = c \qquad \text{or} \qquad \frac{d}{dx}(\log f_X(x)) = cx$$

which implies, upon integration of both sides, that

$$\log f_X(x) = a + \frac{cx^2}{2} \qquad \text{or} \qquad f_X(x) = ke^{cx^2/2}$$

Since $\int_{-\infty}^{\infty} f_X(x)dx = 1$, it follows that c is necessarily negative, and we may write $c = -1/\sigma^2$. Hence

$$f_X(x) = ke^{-x^2/2\sigma^2}$$

That is, X is a normal random variable with parameters $\mu = 0$ and σ^2. A similar argument can be applied to $f_Y(y)$ to show that

$$f_Y(y) = \frac{1}{\sqrt{2\pi}\,\bar{\sigma}}e^{-y^2/2\bar{\sigma}^2}$$

Furthermore, it follows from assumption 2 that $\sigma^2 = \bar{\sigma}^2$, and thus X and Y are independent, identically distributed normal random variables with parameters $\mu = 0$ and σ^2 ■

A necessary and sufficient condition for the random variables X and Y to be independent is for their joint probability density function (or joint probability mass function in the discrete case) $f(x, y)$ to factor into two terms, one depending only on x and the other depending only on y.

Proposition 2.1

The continuous (discrete) random variables X and Y are independent if and only if their joint probability density (mass) function can be expressed as

$$f_{X,Y}(x, y) = h(x)g(y) \qquad -\infty < x < \infty, -\infty < y < \infty$$

Proof: Let us give the proof in the continuous case. First note that independence implies that the joint density is the product of the marginal densities of X and Y, so the preceding factorization will hold when the random variables are independent. Now, suppose that

$$f_{X,Y}(x, y) = h(x)g(y)$$

Then

$$\begin{aligned} 1 &= \int_{-\infty}^{\infty}\int_{-\infty}^{\infty} f_{X,Y}(x, y)\,dx\,dy \\ &= \int_{-\infty}^{\infty} h(x)\,dx \int_{-\infty}^{\infty} g(y)\,dy \\ &= C_1C_2 \end{aligned}$$

where $C_1 = \int_{-\infty}^{\infty} h(x)\,dx$ and $C_2 = \int_{-\infty}^{\infty} g(y)\,dy$. Also,

$$f_X(x) = \int_{-\infty}^{\infty} f_{X,Y}(x, y)\,dy = C_2h(x)$$

$$f_Y(y) = \int_{-\infty}^{\infty} f_{X,Y}(x, y)\,dx = C_1g(y)$$

Since $C_1C_2 = 1$, we thus see that

$$f_{X,Y}(x, y) = f_X(x)f_Y(y)$$

and the proof is complete. ✦

Example 2f. If the joint density function of X and Y is

$$f(x, y) = 6e^{-2x}e^{-3y} \qquad 0 < x < \infty, 0 < y < \infty$$

and is equal to 0 outside this region, are the random variables independent? What if the joint density function is

$$f(x, y) = 24xy \qquad 0 < x < 1, 0 < y < 1, 0 < x + y < 1$$

and is equal to 0 otherwise?

Solution In the first instance, the joint density function factors, and thus the random variables, are independent (with one being exponential with rate 2 and the other exponential with rate 3). In the second instance, because the region in which the joint density is nonzero cannot be expressed in the form $x \in A, y \in B$, the joint density does not factor, so the random variables are not independent. This can be seen clearly by letting

$$I(x, y) = \begin{cases} 1 & \text{if } 0 < x < 1, 0 < y < 1, 0 < x + y < 1 \\ 0 & \text{otherwise} \end{cases}$$

and writing

$$f(x, y) = 24xy\, I(x, y)$$

which clearly does not factor into a part depending only on x and another depending only on y. ■

The concept of independence may, of course, be defined for more than two random variables. In general, the n random variables $X_1, X_2, \ldots, X_n$ are said to be independent if, for all sets of real numbers $A_1, A_2, \ldots, A_n$,

$$P\{X_1 \in A_1, X_2 \in A_2, \ldots, X_n \in A_n\} = \prod_{i=1}^{n} P\{X_i \in A_i\}$$

As before, it can be shown that this condition is equivalent to

$$\begin{aligned} &P\{X_1 \le a_1, X_2 \le a_2, \ldots, X_n \le a_n\} \\ &\quad = \prod_{i=1}^{n} P\{X_i \le a_i\} \qquad \text{for all } a_1, a_2, \ldots, a_n \end{aligned}$$

Finally, we say that an infinite collection of random variables is independent if every finite subcollection of them is independent.

Example 2g. ***How can a computer choose a random subset?*** Most computers are able to generate the value of, or *simulate*, a uniform (0, 1) random variable by means of a built-in subroutine that (to a high degree of approximation) produces such "random numbers." As a result, it is quite easy for the computer to simulate an indicator (that is, a Bernoulli) random variable. Suppose I is an indicator variable such that

$$P\{I = 1\} = p = 1 - P\{I = 0\}$$

The computer can simulate I by choosing a uniform $(0, 1)$ random number U and then letting

$$I = \begin{cases} 1 & \text{if } U < p \\ 0 & \text{if } U \geq p \end{cases}$$

Suppose that we are interested in having the computer select $k, k \leq n$, of the numbers $1, 2, \ldots, n$ in such a way that each of the $\binom{n}{k}$ subsets of size k is equally likely to be chosen. We now present a method that will enable the computer to solve this task. To generate such a subset, we will first simulate, in sequence, n indicator variables $I_1, I_2, \ldots, I_n$, of which exactly k will equal 1. Those i for which $I_i = 1$ will then constitute the desired subset.

To generate the random variables $I_1, \ldots, I_n$, start by simulating n independent uniform $(0, 1)$ random variables $U_1, U_2, \ldots, U_n$. Now define

$$I_1 = \begin{cases} 1 & \text{if } U_1 < \dfrac{k}{n} \\ 0 & \text{otherwise} \end{cases}$$

and then recursively, once $I_1, \ldots, I_i$ are determined, set

$$I_{i+1} = \begin{cases} 1 & \text{if } U_{i+1} < \dfrac{k - (I_1 + \cdots + I_i)}{n - i} \\ 0 & \text{otherwise} \end{cases}$$

In words, at the $i + 1$ stage we set I_{i+1} equal to 1 (and thus put $i + 1$ into the desired subset) with a probability equal to the remaining number of places in the subset $\left(\text{namely, } k - \sum_{j=1}^{i} I_j\right)$ divided by the remaining number of possibilities (namely, $n - i$). Hence the joint distribution of $I_1, I_2, \ldots, I_n$ is determined from

$$P\{I_1 = 1\} = \frac{k}{n}$$

$$P\{I_{i+1} = 1 | I_1, \ldots, I_i\} = \frac{k - \sum_{j=1}^{i} I_j}{n - i} \qquad 1 < i < n$$

The proof that the preceding results in all subsets of size k being equally likely to be chosen is by induction on $k + n$. It is immediate when $k + n = 2$ (that is, when $k = 1, n = 1$), and so assume it to be true whenever $k + n \leq l$. Now suppose that $k + n = l + 1$ and consider any subset of size k—say $i_1 \leq i_2 \leq \cdots \leq i_k$—and consider the following two cases.

Case 1: $i_1 = 1$

$$P\{I_1 = I_{i_2} = \cdots = I_{i_k} = 1, I_j = 0 \text{ otherwise}\}$$
$$= P\{I_1 = 1\}P\{I_{i_2} = \cdots = I_{i_k} = 1, I_j = 0 \text{ otherwise} | I_1 = 1\}$$

Now given that $I_1 = 1$, the remaining elements of the subset are chosen as if a subset of size $k - 1$ were to be chosen from the $n - 1$ elements $2, 3, \ldots, n$. Hence, by the induction hypothesis, the conditional probability that this will result in a given subset of size $k - 1$ being selected is $1\Big/\binom{n-1}{k-1}$. Hence

$$P\{I_1 = I_{i_2} = \cdots = I_{i_k} = 1, I_j = 0 \text{ otherwise}\}$$
$$= \frac{k}{n} \frac{1}{\binom{n-1}{k-1}} = \frac{1}{\binom{n}{k}}$$

Case 2: $i_1 \neq 1$

$$P\{I_{i_1} = I_{i_2} = \cdots = I_{i_k} = 1, I_j = 0 \text{ otherwise}\}$$
$$= P\{I_{i_1} = \cdots = I_{i_k} = 1, I_j = 0 \text{ otherwise} | I_1 = 0\}P\{I_1 = 0\}$$
$$= \frac{1}{\binom{n-1}{k}}\left(1 - \frac{k}{n}\right) = \frac{1}{\binom{n}{k}}$$

where the induction hypothesis was used to evaluate the preceding conditional probability.

Hence in all cases the probability that a given subset of size k will be the subset chosen is $1\Big/\binom{n}{k}$. ■

REMARK. The foregoing method for generating a random subset has a very low memory requirement. A faster algorithm that requires somewhat more memory is presented in Section 10.1. (The latter algorithm uses the last k elements of a random permutation of $1, 2, \ldots, n$.) ■

Example 2h. Let X, Y, Z be independent and uniformly distributed over $(0, 1)$. Compute $P\{X \geq YZ\}$.

Solution Since

$$f_{X,Y,Z}(x, y, z) = f_X(x)f_Y(y)f_Z(z) = 1 \qquad 0 \leq x \leq 1, 0 \leq y \leq 1, 0 \leq z \leq 1$$

we have

$$
\begin{aligned}
P\{X \geq YZ\} &= \iiint\limits_{x \geq yz} f_{X,Y,Z}(x, y, z)\, dx\, dy\, dz \\
&= \int_0^1 \int_0^1 \int_{yz}^1 dx\, dy\, dz \\
&= \int_0^1 \int_0^1 (1 - yz)\, dy\, dz \\
&= \int_0^1 \left(1 - \frac{z}{2}\right) dz \\
&= \frac{3}{4}
\end{aligned}
$$

■

Example 2i. ***Probabilistic interpretation of half-life.*** Let $N(t)$ denote the number of nuclei contained in a radioactive mass of material at time t. The concept of half-life is often defined in a deterministic fashion by stating that it is an empirical fact that for some value h, called the half-life,

$$N(t) = 2^{-t/h}N(0) \qquad t > 0$$

[Note that $N(h) = N(0)/2$.] Since the preceding implies that for any nonnegative s and t,

$$N(t + s) = 2^{-(s+t)/h}N(0) = 2^{-t/h}N(s)$$

it follows that no matter how much time s has already elapsed, in an additional time t the number of existing nuclei will decrease by the factor $2^{-t/h}$.

Since the deterministic relationship given results from observations of radioactive masses containing huge numbers of nuclei, it would seem that it might be consistent with a probabilistic interpretation. The clue to deriving the appropriate probability model for half-life resides in the empirical observation that the proportion of decay in any time interval depends neither on the total number of nuclei at the beginning at the interval nor on the location of this interval (since $N(t + s)/N(s)$ depends neither on $N(s)$ nor on s). Thus it appears that the individual nuclei act independently and with a memoryless life distribution. Thus, since the unique life distribution which is memoryless is the exponential distribution, and since exactly one-half of a given amount of mass decays every h time units, we propose the following probabilistic model for radioactive extinction.

Probabilistic interpretation of the half-life h: The lifetimes of the individual nuclei are independent random variables having a life distribution that is exponential with median equal to h. That is, if L represents the lifetime of a given nucleus then

$$P\{L < t\} = 1 - 2^{-t/h}$$

(As $P\{L < h\} = \frac{1}{2}$ and the preceding can be written as

$$P\{L < t\} = 1 - \exp\left\{-t\frac{\log 2}{h}\right\}$$

we see that L indeed has an exponential distribution with median h.)

It should be noted that under the probabilistic interpretation of half-life given previously, if one starts with $N(0)$ nuclei at time 0 then $N(t)$, the number that remain at time t, will have a binomial distribution with parameters $n = N(0)$ and $p = 2^{-t/h}$. Results of Chapter 8 will show that this interpretation of half-life is consistent with the deterministic model when considering the proportion of a large number of nuclei that decay over a given time frame. However, the difference between the deterministic and probabilistic interpretation becomes apparent when one considers the actual number of decayed nuclei. We will now indicate this with regard to the question of whether protons decay.

There appears to be some controversy over whether or not protons decay. Indeed, one theory appears to predict that protons should decay with a half-life of about $h = 10^{30}$ years. To check this empirically, it has been suggested that one follow a large number of protons for, say, one or two years and determine whether any of them decay within this time period. (Clearly, it would not be feasible to follow a mass of protons for 10^{30} years to see whether one-half of it decays.) Let us suppose that we are able to keep track of $N(0) = 10^{30}$ protons for c years. The number of decays predicted by the deterministic model would then be given by

$$\begin{aligned}
N(0) - N(c) &= h(1 - 2^{-c/h}) \\
&= \frac{1 - 2^{-c/h}}{1/h} \\
&\approx \lim_{x\to 0} \frac{1 - 2^{-cx}}{x} \qquad \text{since } \frac{1}{h} = 10^{-30} \approx 0 \\
&= \lim_{x\to 0} (c2^{-cx} \log 2) \qquad \text{by L'Hospital's rule} \\
&= c \log 2 \approx .6931c
\end{aligned}$$

For instance, in 2 years the deterministic model predicts that there should be 1.3863 decays, and it would thus appear to be a serious blow to the hypothesis that protons decay with a half-life of 10^{30} years if no decays are observed over these 2 years.

Let us now contrast the conclusions above with those obtained from the probabilistic model. Again let us consider the hypothesis that the half-life of protons is $h = 10^{30}$ years, and suppose that we follow h protons for c years. Since there is a huge number of independent protons, each of which will have a very small probability of decaying within this time period, it follows that the number of protons that decay will have (to a very strong approximation) a Poisson distribution with parameter equal to $h(1 - 2^{-c/h}) \approx c \log 2$. Thus

$$P\{0 \text{ decays}\} = e^{-c\log 2}$$

$$= e^{-\log(2^c)} = \frac{1}{2^c}$$

and, in general,

$$P\{n \text{ decays}\} = \frac{2^{-c}[c\log 2]^n}{n!} \qquad n \geq 0$$

Thus we see that even though the average number of decays over 2 years is (as predicted by the deterministic model) 1.3863, there is one chance in 4 that there will not be any decays, thereby indicating that such a result in no way invalidates the original hypothesis of proton decay. ■

REMARK. *Independence is a symmetric relation.* The random variables X and Y are independent if their joint density function (or mass function in the discrete case) is the product of their individual density (or mass) functions. Therefore, to say that X is independent of Y is equivalent to saying that Y is independent of X, or just that X and Y are independent. As a result, in considering whether X is independent of Y in situations where it is not at all intuitive that knowing the value of Y will not change the probabilities concerning X, it can be beneficial to interchange the roles of X and Y and ask instead whether Y is independent of X. The following example illustrates this point. ■

Example 2j. If the initial throw of the dice in the game of craps results in the sum of the dice equaling 4, then the player will continue to throw the dice until the sum is either 4 or 7. If this sum is 4, then the player wins, and if it is 7, then the player loses. Let N denote the number of throws needed until either 4 or 7 appears, and let X denote the value (either 4 or 7) of the final throw. Is N independent of X? That is, does knowing which of 4 or 7 occurs first affect the distribution of the number of throws needed until that number appears? Most people do not find the answer to this question to be intuitively obvious. However, suppose that we turn it around and ask whether X is independent of N. That is, does knowing how many throws it takes to obtain a sum of either 4 or 7 affect the probability that that sum is equal to 4? For instance, suppose we know that it takes n throws of the dice to obtain a sum either 4 or 7. Does this affect the probability distribution of the final sum? Clearly not, since all that is important is that its value is either 4 or 7, and the fact that none of the first $n - 1$ throws were either 4 or 7 does not change the probabilities for the nth throw. Thus we can conclude that X is independent of N, or equivalently, that N is independent of X.

For another example, let $X_1, X_2, \ldots$ be a sequence of independent and identically distributed continuous random variables, and suppose that we observe these random variables in sequence. If $X_n > X_i$ for each $i = 1, \ldots, n - 1$, then we say that X_n is a *record value*. That is, each random variable that is larger than all those preceding it is called a record value. Let A_n denote the event that X_n is a record value. Is A_{n+1} independent of A_n? That is, does knowing that the nth random variable is the largest of the first n change the probability that the $(n+1)$st random variable is the largest of the first $n+1$? While it is true that A_{n+1} is independent of A_n, this may not be intuitively obvious. However, if we turn the question around and ask whether A_n is independent of A_{n+1}, then the result is more easily understood. For knowing that the $(n+1)$st value is larger than $X_1, \ldots, X_n$ clearly gives us no information about the relative size of X_n among the first n random variables. Indeed, by symmetry it is clear that each of these n random variables is equally likely to be the largest of this set, so $P(A_n|A_{n+1}) = P(A_n) = 1/n$. Hence we can conclude that A_n and A_{n+1} are independent events. ■

REMARK: It follows from the identity

$$\begin{aligned} &P\{X_1 \le a_1, \ldots, X_n \le a_n\} \\ &\quad = P\{X_1 \le a_1\}P\{X_2 \le a_2|X_1 \le a_1\} \cdots P\{X_n \le a_n|X_1 \le a_1, \ldots, X_{n-1} \le a_{n-1}\} \end{aligned}$$

that the independence of $X_1, \ldots, X_n$ can be established sequentially. That is, we can show that these random variables are independent by showing that

$$\begin{array}{lll} X_2 & \text{is independent of} & X_1 \\ X_3 & \text{is independent of} & X_1, X_2 \\ X_4 & \text{is independent of} & X_1, X_2, X_3 \\ & \vdots & \\ X_n & \text{is independent of} & X_1, \ldots, X_{n-1} \end{array}$$

6.3 SUMS OF INDEPENDENT RANDOM VARIABLES

It is often important to be able to calculate the distribution of $X + Y$ from the distributions of X and Y when X and Y are independent. Suppose that X and Y are independent, continuous random variables having probability density functions f_X and f_Y. The cumulative distribution function of $X + Y$ is obtained as follows:

$$\begin{aligned} F_{X+Y}(a) &= P\{X + Y \le a\} \\ &= \iint\limits_{x+y \le a} f_X(x) f_Y(y)\, dx\, dy \\ &= \int_{-\infty}^{\infty} \int_{-\infty}^{a-y} f_X(x) f_Y(y)\, dx\, dy \qquad (3.1) \\ &= \int_{-\infty}^{\infty} \int_{-\infty}^{a-y} f_X(x)\, dx\, f_Y(y)\, dy \\ &= \int_{-\infty}^{\infty} F_X(a - y) f_Y(y)\, dy \end{aligned}$$

The cumulative distribution function F_{X+Y} is called the *convolution* of the distributions F_X and F_Y (the cumulative distribution functions of X and Y, respectively).

By differentiating Equation (3.1), we obtain that the probability density function f_{X+Y} of $X + Y$ is given by

$$\begin{aligned} f_{X+Y}(a) &= \frac{d}{da}\int_{-\infty}^{\infty} F_X(a-y)f_Y(y)\,dy \\ &= \int_{-\infty}^{\infty} \frac{d}{da}F_X(a-y)f_Y(y)\,dy \qquad (3.2) \\ &= \int_{-\infty}^{\infty} f_X(a-y)f_Y(y)\,dy \end{aligned}$$

Example 3a. ***Sum of two independent uniform random variables.*** If X and Y are independent random variables, both uniformly distributed on $(0, 1)$, calculate the probability density of $X + Y$.

Solution From Equation (3.2), since

$$f_X(a) = f_Y(a) = \begin{cases} 1 & 0 < a < 1 \\ 0 & \text{otherwise} \end{cases}$$

we obtain

$$f_{X+Y}(a) = \int_0^1 f_X(a-y)\,dy$$

For $0 \le a \le 1$, this yields

$$f_{X+Y}(a) = \int_0^a dy = a$$

For $1 < a < 2$, we get

$$f_{X+Y}(a) = \int_{a-1}^1 dy = 2 - a$$

Hence

$$f_{X+Y}(a) = \begin{cases} a & 0 \le a \le 1 \\ 2-a & 1 < a < 2 \\ 0 & \text{otherwise} \end{cases}$$

Because of the shape of its density function (see Figure 6.3), the random variable $X + Y$ is said to have a *triangular* distribution. ■

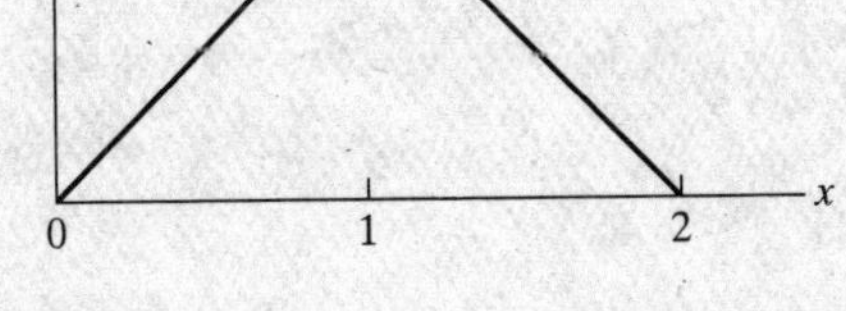

Figure 6.3 Triangular density function.

Recall that a gamma random variable has a density of the form

$$f(y) = \frac{\lambda e^{-\lambda y}(\lambda y)^{t-1}}{\Gamma(t)} \qquad 0 < y < \infty$$

An important property of this family of distributions is that for a fixed value of λ, it is closed under convolutions.

Proposition 3.1

If X and Y are independent gamma random variables with respective parameters (s, λ) and (t, λ), then $X + Y$ is a gamma random variable with parameters $(s + t, \lambda)$.

Proof: Using Equation (3.2), we obtain

$$\begin{aligned}
f_{X+Y}(a) &= \frac{1}{\Gamma(s)\Gamma(t)} \int_0^a \lambda e^{-\lambda(a-y)}[\lambda(a-y)]^{s-1}\lambda e^{-\lambda y}(\lambda y)^{t-1}\,dy \\
&= Ke^{-\lambda a}\int_0^a (a-y)^{s-1}y^{t-1}\,dy \\
&= Ke^{-\lambda a}a^{s+t-1}\int_0^1 (1-x)^{s-1}x^{t-1}\,dx \qquad \text{by letting } x = \frac{y}{a} \\
&= Ce^{-\lambda a}a^{s+t-1}
\end{aligned}$$

where C is a constant that does not depend on a. But as the above is a density function and thus must integrate to 1, the value of C is determined, and we have

$$f_{X+Y}(a) = \frac{\lambda e^{-\lambda a}(\lambda a)^{s+t-1}}{\Gamma(s+t)}$$

Hence the result is proved. ✦

It is now a simple matter to establish, by using Proposition 3.1 and induction, that if $X_i, i = 1, \ldots, n$ are independent gamma random variables with respective parameters $(t_i, \lambda), i = 1, \ldots, n$, then $\sum_{i=1}^{n} X_i$ is gamma with parameters $\left(\sum_{i=1}^{n} t_i, \lambda\right)$. We leave the proof of this as an exercise.

Example 3b. Let $X_1, X_2, \ldots, X_n$ be n independent exponential random variables each having parameter λ. Then, as an exponential random variable with parameter λ is the same as a gamma random variable with parameters $(1, \lambda)$, we see from Proposition 3.1 that $X_1 + X_2 + \cdots + X_n$ is a gamma random variable with parameters(n, λ). ■

If $Z_1, Z_2, \ldots, Z_n$ are independent standard normal random variables, then $Y \equiv \sum_{i=1}^{n} Z_i^2$ is said to have the *chi-squared* (sometimes seen as χ^2) distribution with n degrees of freedom. Let us compute its density function. When $n = 1, Y = Z_1^2$, and from Example 7b of Chapter 5 we see that its probability density function is given by

$$\begin{aligned} f_{Z^2}(y) &= \frac{1}{2\sqrt{y}}[f_Z(\sqrt{y}) + f_Z(-\sqrt{y})] \\ &= \frac{1}{2\sqrt{y}} \frac{2}{\sqrt{2\pi}} e^{-y/2} \\ &= \frac{\frac{1}{2}e^{-y/2}(y/2)^{1/2-1}}{\sqrt{\pi}} \end{aligned}$$

But we recognize the above as the gamma distribution with parameters $\left(\frac{1}{2}, \frac{1}{2}\right)$. [A by-product of this analysis is that $\Gamma\left(\frac{1}{2}\right) = \sqrt{\pi}$.] But as each Z_i^2 is gamma $\left(\frac{1}{2}, \frac{1}{2}\right)$, we obtain from Proposition 3.1 that the χ^2 distribution with n degrees of freedom is just the gamma distribution with parameters $\left(n/2, \frac{1}{2}\right)$ and hence has a probability density function given by

$$\begin{aligned} f_{\chi^2}(y) &= \frac{\frac{1}{2}e^{-y/2}\left(\frac{y}{2}\right)^{n/2-1}}{\Gamma\left(\frac{n}{2}\right)} \qquad y > 0 \\ &= \frac{e^{-y/2}y^{n/2-1}}{2^{n/2}\Gamma\left(\frac{n}{2}\right)} \qquad y > 0 \end{aligned}$$

When n is an even integer, $\Gamma(n/2) = [(n/2) - 1]!$, whereas when n is odd, $\Gamma(n/2)$ can be obtained from iterating the relationship $\Gamma(t) = (t - 1)\Gamma(t - 1)$ and then using the result obtained previously that $\Gamma\left(\frac{1}{2}\right) = \sqrt{\pi}$. [For instance, $\Gamma\left(\frac{5}{2}\right) = \frac{3}{2}\Gamma\left(\frac{3}{2}\right) = \frac{3}{2}\frac{1}{2}\Gamma\left(\frac{1}{2}\right) = \frac{3}{4}\sqrt{\pi}$.]

The chi-squared distribution often arises in practice as being the distribution of the square of the error involved when one attempts to hit a target in n-dimensional space when the coordinate errors are taken to be independent unit normal random variables. It is also important in statistical analysis.

We can also use Equation (3.2) to prove the following important result about normal random variables.

Proposition 3.2

If $X_i, i = 1, \ldots, n$, are independent random variables that are normally distributed with respective parameters $\mu_i, \sigma_i^2, i = 1, \ldots, n$, then $\sum_{i=1}^{n} X_i$ is normally distributed with parameters $\sum_{i=1}^{n} \mu_i$ and $\sum_{i=1}^{n} \sigma_i^2$.

Proof of Proposition 3.2: To begin, let X and Y be independent normal random variables, with X having mean 0 and variance σ^2, and Y having mean 0 and variance 1. We will determine the density function of $X + Y$ by utilizing Equation (3.2). Now, with

$$c = \frac{1}{2\sigma^2} + \frac{1}{2} = \frac{1 + \sigma^2}{2\sigma^2}$$

we have

$$\begin{aligned} f_X(a - y)f_Y(y) &= \frac{1}{\sqrt{2\pi}\,\sigma}\exp\left\{-\frac{(a - y)^2}{2\sigma^2}\right\}\frac{1}{\sqrt{2\pi}}\exp\left\{-\frac{y^2}{2}\right\} \\ &= \frac{1}{2\pi\sigma}\exp\left\{-\frac{a^2}{2\sigma^2}\right\}\exp\left\{-c\left(y^2 - 2y\frac{a}{1 + \sigma^2}\right)\right\} \end{aligned}$$

Hence, from Equation (3.2),

$$\begin{aligned} f_{X+Y}(a) &= \frac{1}{2\pi\sigma}\exp\left\{-\frac{a^2}{2\sigma^2}\right\}\exp\left\{\frac{a^2}{2\sigma^2(1 + \sigma^2)}\right\}\int_{-\infty}^{\infty}\exp\left\{-c\left(y - \frac{a}{1 + \sigma^2}\right)^2\right\}dy \\ &= \frac{1}{2\pi\sigma}\exp\left\{-\frac{a^2}{2(1 + \sigma^2)}\right\}\int_{-\infty}^{\infty}\exp\{-cx^2\}\,dx \\ &= C\exp\left\{-\frac{a^2}{2(1 + \sigma^2)}\right\} \end{aligned}$$

where C doesn't depend on a. But this implies that $X + Y$ is normal with mean 0 and variance $1 + \sigma^2$.

Now, suppose that X_1 and X_2 are independent normal random variables, with X_i having mean μ_i and variance $\sigma_i^2, i = 1, 2$. Then

$$X_1 + X_2 = \sigma_2\left(\frac{X_1 - \mu_1}{\sigma_2} + \frac{X_2 - \mu_2}{\sigma_2}\right) + \mu_1 + \mu_2$$

But since $(X_1 - \mu_1)/\sigma_2$ is normal with mean 0 and variance σ_1^2/σ_2^2, and $(X_2 - \mu_2)/\sigma_2$ is normal with mean 0 and variance 1, it follows from our previous result that $(X_1 - \mu_1)/\sigma_2 + (X_2 - \mu_2)/\sigma_2$ is normal with mean 0 and variance $1 + \sigma_1^2/\sigma_2^2$, implying that $X_1 + X_2$ is normal with mean $\mu_1 + \mu_2$ and variance $\sigma_2^2(1 + \sigma_1^2/\sigma_2^2) = \sigma_1^2 + \sigma_2^2$.

Thus, Proposition 3.2 is established when $n = 2$. The general case now follows by induction. That is, assume that it is true when there are $n - 1$ random variables. Now consider the case of n, and write

$$\sum_{i=1}^{n} X_i = \sum_{i=1}^{n-1} X_i + X_n$$

By the induction hypothesis, $\sum_{i=1}^{n-1} X_i$ is normal with mean $\sum_{i=1}^{n-1} \mu_i$ and variance $\sum_{i=1}^{n-1} \sigma_i^2$. Therefore, by the result for $n = 2$, we can conclude that $\sum_{i=1}^{n} X_i$ is normal with mean $\sum_{i=1}^{n} \mu_i$ and variance $\sum_{i=1}^{n} \sigma_i^2$. ✦

Example 3c. A club basketball team will play a 44-game season. Twenty-six of these games are against class A teams and 18 are against class B teams. Suppose that the team will win each game against a class A team with probability .4, and will win each game against a class B team with probability .7. Assume also that the results of the different games are independent. Approximate the probability that

(a) the team wins 25 games or more;
(b) the team wins more games against class A teams than it does against class B teams.

Solution (a) Let X_A and X_B denote, respectively, the number of games the team wins against class A and against class B teams. Note that X_A and X_B are independent binomial random variables, and

$$E[X_A] = 26(.4) = 10.4 \qquad \text{Var}(X_A) = 26(.4)(.6) = 6.24$$
$$E[X_B] = 18(.7) = 12.6 \qquad \text{Var}(X_B) = 18(.7)(.3) = 3.78$$

By the normal approximation to the binomial it follows that X_A and X_B will approximately have the same distribution as would independent normal random variables with expected values and variances as given in the preceding. Hence, by Proposition 3.2, $X_A + X_B$ will approximately have a normal distribution with mean 23 and variance 10.02. Therefore, letting Z denote a standard normal random variable, we have

$$\begin{aligned} P\{X_A + X_B \geq 25\} &= P\{X_A + X_B \geq 24.5\} \\ &= P\left\{\frac{X_A + X_B - 23}{\sqrt{10.02}} \geq \frac{24.5 - 23}{\sqrt{10.02}}\right\} \\ &\approx P\left\{Z \geq \frac{1.5}{\sqrt{10.02}}\right\} \\ &\approx 1 - P\{Z < .4739\} \\ &\approx .3178 \end{aligned}$$

(b) We note that $X_A - X_B$ will approximately have a normal distribution with mean -2.2 and variance 10.02. Hence

$$\begin{aligned} P\{X_A - X_B \geq 1\} &= P\{X_A - X_B \geq .5\} \\ &= P\left\{\frac{X_A - X_B + 2.2}{\sqrt{10.02}} \geq \frac{.5 + 2.2}{\sqrt{10.02}}\right\} \\ &\approx P\left\{Z \geq \frac{2.7}{\sqrt{10.02}}\right\} \\ &\approx 1 - P\{Z < .8530\} \\ &\approx .1968 \end{aligned}$$

Therefore, there is approximately a 31.78 percent chance that the team will win at least 25 games, and approximately a 19.68 percent chance that it will win more games against class A teams than against class B teams. ■

The random variable Y is said to be a *lognormal* random variable with parameters μ and σ if $\log(Y)$ is a normal random variable with mean μ and variance σ^2. That is, Y is lognormal if it can be expressed as

$$Y = e^X$$

where X is a normal random variable.

Example 3d. Starting at some fixed time, let $S(n)$ denote the price of a certain security at the end of n additional weeks, $n \geq 1$. A popular model for the evolution of these prices assumes that the price ratios $S(n)/S(n-1)$, $n \geq 1$, are independent and identically distributed lognormal random variables. Assuming this model, with parameters $\mu = .0165$, $\sigma = .0730$, what is the probability that

(a) the price of the security increases over each of the next two weeks;
(b) the price at the end of two weeks is higher than it is today?

Solution Let Z be a standard normal random variable. To solve part (a), we use the fact that $\log(x)$ increases in x to conclude that $x > 1$ if and only if $\log(x) > \log(1) = 0$. As a result, we have

$$\begin{aligned} P\left\{\frac{S(1)}{S(0)} > 1\right\} &= P\left\{\log\left(\frac{S(1)}{S(0)}\right) > 0\right\} \\ &= P\left\{Z > \frac{-.0165}{.0730}\right\} \\ &= P\{Z < .2260\} \\ &= .5894 \end{aligned}$$

Therefore, the probability that the price is up after one week is .5894. Since the successive price ratios are independent, the probability that the price increases over each of the next two weeks is $(.5894)^2 = .3474$.

To solve part (b), reason as follows:

$$P\left\{\frac{S(2)}{S(0)} > 1\right\} = P\left\{\frac{S(2)}{S(1)}\frac{S(1)}{S(0)} > 1\right\}$$
$$= P\left\{\log\left(\frac{S(2)}{S(1)}\right) + \log\left(\frac{S(1)}{S(0)}\right) > 0\right\}$$

However, $\log\left(\frac{S(2)}{S(1)}\right) + \log\left(\frac{S(1)}{S(0)}\right)$, being the sum of two independent normal random variables with a common mean .0165 and a common standard deviation .0730, is a normal random variable with mean .0330 and variance $2(.0730)^2$. Consequently,

$$P\left\{\frac{S(2)}{S(0)} > 1\right\} = P\left\{Z > \frac{-.0330}{.0730\sqrt{2}}\right\}$$
$$= P\{Z < .31965\}$$
$$= .6254$$ ■

Rather than attempt to derive a general expression for the distribution of $X + Y$ in the discrete case, we shall consider some examples.

Example 3e. ***Sums of independent Poisson random variables.*** If X and Y are independent Poisson random variables with respective parameters λ_1 and λ_2, compute the distribution of $X + Y$.

Solution Because the event $\{X + Y = n\}$ may be written as the union of the disjoint events $\{X = k, Y = n - k\}, 0 \le k \le n$, we have

$$\begin{aligned}
P\{X + Y = n\} &= \sum_{k=0}^{n} P\{X = k, Y = n - k\} \\
&= \sum_{k=0}^{n} P\{X = k\}P\{Y = n - k\} \\
&= \sum_{k=0}^{n} e^{-\lambda_1}\frac{\lambda_1^k}{k!}e^{-\lambda_2}\frac{\lambda_2^{n-k}}{(n-k)!} \\
&= e^{-(\lambda_1+\lambda_2)}\sum_{k=0}^{n}\frac{\lambda_1^k\lambda_2^{n-k}}{k!\,(n-k)!} \\
&= \frac{e^{-(\lambda_1+\lambda_2)}}{n!}\sum_{k=0}^{n}\frac{n!}{k!\,(n-k)!}\lambda_1^k\lambda_2^{n-k} \\
&= \frac{e^{-(\lambda_1+\lambda_2)}}{n!}(\lambda_1 + \lambda_2)^n
\end{aligned}$$

In words, $X_1 + X_2$ has a Poisson distribution with parameter $\lambda_1 + \lambda_2$. ■

Example 3f. ***Sums of independent binomial random variables.*** Let X and Y be independent binomial random variables with respective parameters (n, p) and (m, p). Calculate the distribution of $X + Y$.

Solution Without any computation at all we can immediately conclude, by recalling the interpretation of a binomial random variable, that $X + Y$ is binomial with parameters $(n + m, p)$. This follows because X represents the number of successes in n independent trials, each of which results in a success with probability p; similarly, Y represents the number of successes in m independent trials, each trial being a success with probability p. Hence, as X and Y are assumed independent, it follows that $X + Y$ represents the number of successes in $n + m$ independent trials when each trial has a probability p of being a success. Therefore, $X + Y$ is a binomial random variable with parameters $(n + m, p)$. To check this result analytically, note that

$$\begin{aligned} P\{X + Y = k\} &= \sum_{i=0}^{n} P\{X = i, Y = k - i\} \\ &= \sum_{i=0}^{n} P\{X = i\}P\{Y = k - i\} \\ &= \sum_{i=0}^{n} \binom{n}{i} p^i q^{n-i} \binom{m}{k-i} p^{k-i} q^{m-k+i} \end{aligned}$$

where $q = 1 - p$ and where $\binom{r}{j} = 0$ when $j > r$. Hence

$$P\{X + Y = k\} = p^k q^{n+m-k} \sum_{i=0}^{n} \binom{n}{i}\binom{m}{k-i}$$

and the result follows upon application of the combinatorial identity

$$\binom{n+m}{k} = \sum_{i=0}^{n} \binom{n}{i}\binom{m}{k-i}$$

■

6.4 CONDITIONAL DISTRIBUTIONS: DISCRETE CASE

Recall that for any two events E and F, the conditional probability of E given F is defined, provided that $P(F) > 0$, by

$$P(E|F) = \frac{P(EF)}{P(F)}$$

Hence, if X and Y are discrete random variables, it is natural to define the conditional probability mass function of X given that $Y = y$, by

$$\begin{aligned} p_{X|Y}(x|y) &= P\{X = x|Y = y\} \\ &= \frac{P\{X = x, Y = y\}}{P\{Y = y\}} \\ &= \frac{p(x, y)}{p_Y(y)} \end{aligned}$$

for all values of y such that $p_Y(y) > 0$. Similarly, the conditional probability distribution function of X given that $Y = y$ is defined, for all y such that $p_Y(y) > 0$, by

$$F_{X|Y}(x|y) = P\{X \le x|Y = y\}$$
$$= \sum_{a \le x} p_{X|Y}(a|y)$$

In other words, the definitions are exactly the same as in the unconditional case except that everything is now conditional on the event that $Y = y$. If X is independent of Y, then the conditional mass function and distribution function are the same as the unconditional ones. This follows because if X is independent of Y, then

$$\begin{aligned} p_{X|Y}(x|y) &= P\{X = x|Y = y\} \\ &= \frac{P\{X = x, Y = y\}}{P\{Y = y\}} \\ &= \frac{P\{X = x\}P\{Y = y\}}{P\{Y = y\}} \\ &= P\{X = x\} \end{aligned}$$

Example 4a. Suppose that $p(x, y)$, the joint probability mass function of X and Y, is given by

$$p(0,0) = .4 \qquad p(0,1) = .2 \qquad p(1,0) = .1 \qquad p(1,1) = .3$$

Calculate the conditional probability mass function of X, given that $Y = 1$.

Solution We first note that

$$p_Y(1) = \sum_x p(x,1) = p(0,1) + p(1,1) = .5$$

Hence

$$p_{X|Y}(0|1) = \frac{p(0,1)}{p_Y(1)} = \frac{2}{5}$$

and

$$p_{X|Y}(1|1) = \frac{p(1,1)}{p_Y(1)} = \frac{3}{5}$$ ■

Example 4b. If X and Y are independent Poisson random variables with respective parameters λ_1 and λ_2, calculate the conditional distribution of X, given that $X + Y = n$.

Solution We calculate the conditional probability mass function of X given that $X + Y = n$ as follows:

$$P\{X = k | X + Y = n\} = \frac{P\{X = k, X + Y = n\}}{P\{X + Y = n\}}$$
$$= \frac{P\{X = k, Y = n - k\}}{P\{X + Y = n\}}$$
$$= \frac{P\{X = k\}P\{Y = n - k\}}{P\{X + Y = n\}}$$

where the last equality follows from the assumed independence of X and Y. Recalling (Example 3d) that $X + Y$ has a Poisson distribution with parameter $\lambda_1 + \lambda_2$, we see that the preceding equals

$$P\{X = k | X + Y = n\} = \frac{e^{-\lambda_1}\lambda_1^k}{k!} \frac{e^{-\lambda_2}\lambda_2^{n-k}}{(n-k)!} \left[\frac{e^{-(\lambda_1+\lambda_2)}(\lambda_1 + \lambda_2)^n}{n!}\right]^{-1}$$
$$= \frac{n!}{(n-k)!\,k!} \frac{\lambda_1^k \lambda_2^{n-k}}{(\lambda_1 + \lambda_2)^n}$$
$$= \binom{n}{k}\left(\frac{\lambda_1}{\lambda_1 + \lambda_2}\right)^k \left(\frac{\lambda_2}{\lambda_1 + \lambda_2}\right)^{n-k}$$

In other words, the conditional distribution of X, given that $X + Y = n$, is the binomial distribution with parameters n and $\lambda_1/(\lambda_1 + \lambda_2)$. ■

6.5 CONDITIONAL DISTRIBUTIONS: CONTINUOUS CASE

If X and Y have a joint probability density function $f(x, y)$, then the conditional probability density function of X, given that $Y = y$, is defined for all values of y such that $f_Y(y) > 0$, by

$$f_{X|Y}(x|y) = \frac{f(x, y)}{f_Y(y)}$$

To motivate this definition, multiply the left-hand side by dx and the right-hand side by $(dx\,dy)/dy$ to obtain

$$f_{X|Y}(x|y)\,dx = \frac{f(x, y)\,dx\,dy}{f_Y(y)\,dy}$$
$$\approx \frac{P\{x \le X \le x + dx, y \le Y \le y + dy\}}{P\{y \le Y \le y + dy\}}$$
$$= P\{x \le X \le x + dx | y \le Y \le y + dy\}$$

In other words, for small values of dx and dy, $f_{X|Y}(x|y)dx$ represents the conditional probability that X is between x and $x + dx$, given that Y is between y and $y + dy$.

The use of conditional densities allows us to define conditional probabilities of events associated with one random variable when we are given the value of a second random variable. That is, if X and Y are jointly continuous, then for any set A,

$$P\{X \in A|Y = y\} = \int_A f_{X|Y}(x|y)\,dx$$

In particular, by letting $A = (-\infty, a]$, we can define the conditional cumulative distribution function of X, given that $Y = y$, by

$$F_{X|Y}(a|y) \equiv P\{X \leq a|Y = y\} = \int_{-\infty}^{a} f_{X|Y}(x|y)\,dx$$

The reader should note that, by using the ideas presented in the preceding discussion, we have been able to give workable expressions for conditional probabilities, even though the event on which we are conditioning (namely, the event $\{Y = y\}$) has probability 0.

Example 5a. The joint density of X and Y is given by

$$f(x, y) = \begin{cases} \frac{15}{2}x(2 - x - y) & 0 < x < 1, 0 < y < 1 \\ 0 & \text{otherwise} \end{cases}$$

Compute the conditional density of X, given that $Y = y$, where $0 < y < 1$.

Solution For $0 < x < 1, 0 < y < 1$, we have

$$\begin{aligned} f_{X|Y}(x|y) &= \frac{f(x, y)}{f_Y(y)} \\ &= \frac{f(x, y)}{\int_{-\infty}^{\infty} f(x, y)\,dx} \\ &= \frac{x(2 - x - y)}{\int_0^1 x(2 - x - y)\,dx} \\ &= \frac{x(2 - x - y)}{\frac{2}{3} - y/2} \\ &= \frac{6x(2 - x - y)}{4 - 3y} \end{aligned}$$

■

Example 5b. Suppose that the joint density of X and Y is given by

$$f(x, y) = \begin{cases} \dfrac{e^{-x/y}e^{-y}}{y} & 0 < x < \infty, 0 < y < \infty \\ 0 & \text{otherwise} \end{cases}$$

Find $P\{X > 1|Y = y\}$.

Solution We first obtain the conditional density of X, given that $Y = y$.

$$\begin{aligned}
f_{X|Y}(x|y) &= \frac{f(x, y)}{f_Y(y)} \\
&= \frac{e^{-x/y}e^{-y}/y}{e^{-y}\int_0^\infty (1/y)e^{-x/y}\,dx} \\
&= \frac{1}{y}e^{-x/y}
\end{aligned}$$

Hence

$$\begin{aligned}
P\{X > 1|Y = y\} &= \int_1^\infty \frac{1}{y}e^{-x/y}\,dx \\
&= -e^{-x/y}\Big|_1^\infty \\
&= e^{-1/y}
\end{aligned}$$

■

If X and Y are independent continuous random variables, the conditional density of X, given $Y = y$, is just the unconditional density of X. This is so because, in the independent case,

$$f_{X|Y}(x|y) = \frac{f(x, y)}{f_Y(y)} = \frac{f_X(x)f_Y(y)}{f_Y(y)} = f_X(x)$$

We can also talk about conditional distributions when the random variables are neither jointly continuous nor jointly discrete. For example, suppose that X is a continuous random variable having probability density function f and N is a discrete random variable, and consider the conditional distribution of X given that $N = n$. Then

$$\frac{P\{x < X < x + dx|N = n\}}{dx} = \frac{P\{N = n|x < X < x + dx\}}{P\{N = n\}}\frac{P\{x < X < x + dx\}}{dx}$$

and letting dx approach 0 gives

$$\lim_{dx\to 0}\frac{P\{x < X < x + dx|N = n\}}{dx} = \frac{P\{N = n|X = x\}}{P\{N = n\}}f(x)$$

thus showing that the conditional density of X given that $N = n$ is given by

$$f_{X|N}(x|n) = \frac{P\{N = n|X = x\}}{P\{N = n\}}f(x)$$

Example 5c. Consider $n + m$ trials having a common probability of success. Suppose, however, that this success probability is not fixed in advance but is chosen from a uniform $(0, 1)$ population. What is the conditional distribution of the success probability given that the $n + m$ trials result in n successes?

Solution If we let X denote the trial success probability, then X is a uniform $(0,1)$ random variable. Also, given that $X = x$, the $n + m$ trials are independent with common success probability x, and so N, the number of successes, is a binomial random variable with parameters $(n + m, x)$. Hence the conditional density of X given that $N = n$ is as follows:

$$
\begin{aligned}
f_{X|N}(x|n) &= \frac{P\{N = n|X = x\}f_X(x)}{P\{N = n\}} \\
&= \frac{\binom{n+m}{n}x^n(1-x)^m}{P\{N = n\}} \qquad 0 < x < 1 \\
&= cx^n(1-x)^m
\end{aligned}
$$

where c does not depend on x. Hence the conditional density is that of a beta random variable with parameters $n + 1, m + 1$.

The preceding result is quite interesting, for it states that if the original or *prior* (to the collection of data) distribution of a trial success probability is uniformly distributed over $(0,1)$ [or, equivalently, is beta with parameters $(1,1)$] then the posterior (or conditional) distribution given a total of n successes in $n + m$ trials is beta with parameters $(1 + n, 1 + m)$. This is valuable, for it enhances our intuition as to what it means to assume that a random variable has a beta distribution. ■

*6.6 ORDER STATISTICS

Let $X_1, X_2, \ldots, X_n$ be n independent and identically distributed, continuous random variables having a common density f and distribution function F. Define

$$
\begin{aligned}
X_{(1)} &= \text{smallest of } X_1, X_2, \ldots, X_n \\
X_{(2)} &= \text{second smallest of } X_1, X_2, \ldots, X_n \\
&\vdots \\
X_{(j)} &= j\text{th smallest of } X_1, X_2, \ldots, X_n \\
&\vdots \\
X_{(n)} &= \text{largest of } X_1, X_2, \ldots, X_n
\end{aligned}
$$

The ordered values $X_{(1)} \le X_{(2)} \le \cdots \le X_{(n)}$ are known as the *order statistics* corresponding to the random variables $X_1, X_2, \ldots, X_n$. In other words, $X_{(1)}, \ldots, X_{(n)}$ are the ordered values of $X_1, \ldots, X_n$.

The joint density function of the order statistics is obtained by noting that the order statistics $X_{(1)}, \ldots, X_{(n)}$ will take on the values $x_1 \le x_2 \le \cdots \le x_n$ if and only if for some permutation $(i_1, i_2, \ldots, i_n)$ of $(1, 2, \ldots, n)$

$$X_1 = x_{i_1}, X_2 = x_{i_2}, \ldots, X_n = x_{i_n}$$

Since, for any permutation $(i_1, \ldots, i_n)$ of $(1, 2, \ldots, n)$,

$$P\left\{x_{i_1} - \frac{\varepsilon}{2} < X_1 < x_{i_1} + \frac{\varepsilon}{2}, \ldots, x_{i_n} - \frac{\varepsilon}{2} < X_n < x_{i_n} + \frac{\varepsilon}{2}\right\}$$
$$\approx \varepsilon^n f_{X_1,\ldots,X_n}(x_{i_1}, \ldots, x_{i_n})$$
$$= \varepsilon^n f(x_{i_1}) \cdots f(x_{i_n})$$
$$= \varepsilon^n f(x_1) \cdots f(x_n)$$

we see that for $x_1 < x_2 < \cdots < x_n$,

$$P\left\{x_1 - \frac{\varepsilon}{2} < X_{(1)} < x_1 + \frac{\varepsilon}{2}, \ldots, x_n - \frac{\varepsilon}{2} < X_{(n)} < x_n + \frac{\varepsilon}{2}\right\}$$
$$\approx n!\, \varepsilon^n f(x_1) \cdots f(x_n)$$

Dividing by ε^n and letting $\varepsilon \to 0$ yields

$$f_{X_{(1)},\ldots,X_{(n)}}(x_1, x_2, \ldots, x_n) = n!\, f(x_1) \cdots f(x_n) \qquad x_1 < x_2 < \cdots < x_n \tag{6.1}$$

Equation (6.1) is most simply explained by arguing that in order for the vector $\langle X_{(1)}, \ldots, X_{(n)} \rangle$ to equal $\langle x_1, \ldots, x_n \rangle$, it is necessary and sufficient for $\langle X_1, \ldots, X_n \rangle$ to equal one of the $n!$ permutations of $\langle x_1, \ldots, x_n \rangle$. As the probability (density) that $\langle X_1, \ldots, X_n \rangle$ equals any given permutation of $\langle x_1, \ldots, x_n \rangle$ is just $f(x_1) \cdots f(x_n)$, Equation (6.1) follows.

Example 6a. Along a road 1 mile long are 3 people "distributed at random." Find the probability that no 2 people are less than a distance of d miles apart, when $d \leq \frac{1}{2}$.

Solution Let us assume that "distributed at random" means that the positions of the 3 people are independent and uniformly distributed over the road. If X_i denotes the position of the ith person, the desired probability is $P\{X_{(i)} > X_{(i-1)} + d, i = 2, 3\}$. As

$$f_{X_{(1)},X_{(2)},X_{(3)}}(x_1, x_2, x_3) = 3! \qquad 0 < x_1 < x_2 < x_3 < 1$$

it follows that

$$P\{X_{(i)} > X_{(i-1)} + d, i = 2, 3\} = \iiint\limits_{\substack{x_i > x_{i-1}+d \\ i=2,3}} f_{X_{(1)},X_{(2)},X_{(3)}}(x_1, x_2, x_3)\, dx_1\, dx_2\, dx_3$$
$$= 3! \int_0^{1-2d} \int_{x_1+d}^{1-d} \int_{x_2+d}^{1} dx_3\, dx_2\, dx_1$$
$$= 6 \int_0^{1-2d} \int_{x_1+d}^{1-d} (1 - d - x_2)\, dx_2\, dx_1$$
$$= 6 \int_0^{1-2d} \int_0^{1-2d-x_1} y_2\, dy_2\, dx_1$$

where we have made the change of variables $y_2 = 1 - d - x_2$. Hence continuing the string of equalities yields

$$= 3 \int_0^{1-2d} (1 - 2d - x_1)^2\, dx_1$$

$$= 3 \int_0^{1-2d} y_1^2\, dy_1$$

$$= (1 - 2d)^3$$

Hence the desired probability that no 2 people are within a distance d of each other when 3 people are uniformly and independently distributed over an interval of size 1 is $(1 - 2d)^3$ when $d \le \frac{1}{2}$. In fact, the same method can be used to prove that when there are n people distributed at random over the unit interval the desired probability is

$$[1 - (n-1)d]^n \qquad \text{when } d \le \frac{1}{n-1}$$

The proof is left as an exercise. ■

The density function of the jth-order statistic $X_{(j)}$ can be obtained either by integrating the joint density function (6.1) or by direct reasoning as follows: In order for $X_{(j)}$ to equal x, it is necessary for $j - 1$ of the n values $X_1, \ldots, X_n$ to be less than x, $n - j$ of them to be greater than x, and 1 of them to equal x. Now, the probability density that any given set of $j - 1$ of the X_i's are less than x, another given set of $n - j$ are all greater than x, and the remaining value is equal to x, equals

$$[F(x)]^{j-1}[1 - F(x)]^{n-j} f(x)$$

Hence, as there are

$$\binom{n}{j-1,\, n-j,\, 1} = \frac{n!}{(n-j)!\,(j-1)!}$$

different partitions of the n random variables $X_1, \ldots, X_n$ into the three groups, we see that the density function of $X_{(j)}$ is given by

$$f_{X_{(j)}}(x) = \frac{n!}{(n-j)!\,(j-1)!}[F(x)]^{j-1}[1 - F(x)]^{n-j} f(x) \tag{6.2}$$

Example 6b. When a sample of $2n + 1$ random variables (that is, when $2n + 1$ independent and identically distributed random variables) are observed, the $(n + 1)$st smallest is called the sample median. If a sample of size 3 from a uniform distribution over $(0, 1)$ is observed, find the probability that the sample median is between $\frac{1}{4}$ and $\frac{3}{4}$.

Solution From Equation (6.2) the density of $X_{(2)}$ is given by

$$f_{X_{(2)}}(x) = \frac{3!}{1!\,1!} x(1 - x) \qquad 0 < x < 1$$

Hence

$$P\{\tfrac{1}{4} < X_{(2)} < \tfrac{3}{4}\} = 6\int_{1/4}^{3/4} x(1-x)\,dx$$

$$= 6\left\{\frac{x^2}{2} - \frac{x^3}{3}\right\}\Bigg|_{x=1/4}^{x=3/4} = \frac{11}{16}$$ ■

The cumulative distribution function of $X_{(j)}$ can be obtained by integrating Equation (6.2). That is,

$$F_{X_{(j)}}(y) = \frac{n!}{(n-j)!\,(j-1)!}\int_{-\infty}^{y}[F(x)]^{j-1}[1-F(x)]^{n-j}f(x)\,dx \tag{6.3}$$

However, $F_{X_{(j)}}(y)$ could also have been derived directly by noting that the jth order statistic is less than or equal to y if and only if there are j or more of the X_i's that are less than or equal to y. Hence, as the number of the X_i's that are less than or equal to y is a binomial random variable with parameters $[n, p = F(y)]$, it follows that

$$\begin{aligned} F_{X_{(j)}}(y) &= P\{X_{(j)} \le y\} = P\{j \text{ or more of the } X_i\text{'s are } \le y\} \\ &= \sum_{k=j}^{n}\binom{n}{k}[F(y)]^k[1-F(y)]^{n-k} \end{aligned} \tag{6.4}$$

If, in Equations (6.3) and (6.4), we take F to be the uniform $(0, 1)$ distribution [that is, $f(x) = 1, 0 < x < 1$], then we obtain the interesting analytical identity

$$\sum_{k=j}^{n}\binom{n}{k}y^k(1-y)^{n-k} = \frac{n!}{(n-j)!\,(j-1)!}\int_0^y x^{j-1}(1-x)^{n-j}\,dx \qquad 0 \le y \le 1 \tag{6.5}$$

By employing the same type of argument that we used in establishing Equation (6.2), we can show that the joint density function of the order statistics $X_{(i)}$ and $X_{(j)}$, when $i < j$, is

$$\begin{aligned} f_{X_{(i)},X_{(j)}}(x_i, x_j) = &\frac{n!}{(i-1)!\,(j-i-1)!\,(n-j)!}[F(x_i)]^{i-1} \\ &\times [F(x_j) - F(x_i)]^{j-i-1}[1-F(x_j)]^{n-j}f(x_i)f(x_j) \end{aligned} \tag{6.6}$$

for all $x_i < x_j$.

Example 6c. ***Distribution of the range of a random sample.*** Suppose that n independent and identically distributed random variables $X_1, X_2, \ldots, X_n$ are observed. The random variable R, defined by $R = X_{(n)} - X_{(1)}$, is called the *range* of the observed random variables. If the random variables X_i have distribution function F and density function f, then the distribution of R can be obtained from Equation (6.6) as follows: for $a \ge 0$,

$$P\{R \le a\} = P\{X_{(n)} - X_{(1)} \le a\}$$
$$= \iint_{x_n - x_1 \le a} f_{X_{(1)}, X_{(n)}}(x_1, x_n)\, dx_1\, dx_n$$
$$= \int_{-\infty}^{\infty} \int_{x_1}^{x_1+a} \frac{n!}{(n-2)!} [F(x_n) - F(x_1)]^{n-2} f(x_1) f(x_n)\, dx_n\, dx_1$$

Making the change of variable $y = F(x_n) - F(x_1)$, $dy = f(x_n)\, dx_n$ yields

$$\int_{x_1}^{x_1+a} [F(x_n) - F(x_1)]^{n-2} f(x_n)\, dx_n = \int_0^{F(x_1+a)-F(x_1)} y^{n-2}\, dy$$
$$= \frac{1}{n-1} [F(x_1 + a) - F(x_1)]^{n-1}$$

and thus

$$P\{R \le a\} = n \int_{-\infty}^{\infty} [F(x_1 + a) - F(x_1)]^{n-1} f(x_1)\, dx_1 \qquad (6.7)$$

Equation (6.7) can be explicitly evaluated only in a few special cases. One such case is when the X_i's are all uniformly distributed on $(0, 1)$. In this case we obtain from Equation (6.7) that for $0 < a < 1$,

$$P\{R < a\} = n \int_0^1 [F(x_1 + a) - F(x_1)]^{n-1} f(x_1)\, dx_1$$
$$= n \int_0^{1-a} a^{n-1}\, dx_1 + n \int_{1-a}^1 (1 - x_1)^{n-1}\, dx_1$$
$$= n(1 - a)a^{n-1} + a^n$$

Differentiation yields that the density function of the range is given, in this case, by

$$f_R(a) = \begin{cases} n(n-1)a^{n-2}(1-a) & 0 \le a \le 1 \\ 0 & \text{otherwise} \end{cases}$$

That is, the range of n independent uniform $(0, 1)$ random variables is a beta random variable with parameters $n - 1, 2$. ∎

6.7 JOINT PROBABILITY DISTRIBUTION OF FUNCTIONS OF RANDOM VARIABLES

Let X_1 and X_2 be jointly continuous random variables with joint probability density function f_{X_1, X_2}. It is sometimes necessary to obtain the joint distribution of the random variables Y_1 and Y_2, which arise as functions of X_1 and X_2. Specifically, suppose that $Y_1 = g_1(X_1, X_2)$ and $Y_2 = g_2(X_1, X_2)$ for some functions g_1 and g_2.

Assume that the functions g_1 and g_2 satisfy the following conditions:

1. The equations $y_1 = g_1(x_1, x_2)$ and $y_2 = g_2(x_1, x_2)$ can be uniquely solved for x_1 and x_2 in terms of y_1 and y_2 with solutions given by, say, $x_1 = h_1(y_1, y_2)$, $x_2 = h_2(y_1, y_2)$.
2. The functions g_1 and g_2 have continuous partial derivatives at all points (x_1, x_2) and are such that the following 2×2 determinant

$$J(x_1, x_2) = \begin{vmatrix} \dfrac{\partial g_1}{\partial x_1} & \dfrac{\partial g_1}{\partial x_2} \\ \dfrac{\partial g_2}{\partial x_1} & \dfrac{\partial g_2}{\partial x_2} \end{vmatrix} \equiv \frac{\partial g_1}{\partial x_1}\frac{\partial g_2}{\partial x_2} - \frac{\partial g_1}{\partial x_2}\frac{\partial g_2}{\partial x_1} \neq 0$$

at all points (x_1, x_2).

Under these two conditions it can be shown that the random variables Y_1 and Y_2 are jointly continuous with joint density function given by

$$f_{Y_1Y_2}(y_1, y_2) = f_{X_1,X_2}(x_1, x_2)|J(x_1, x_2)|^{-1} \tag{7.1}$$

where $x_1 = h_1(y_1, y_2)$, $x_2 = h_2(y_1, y_2)$.

A proof of Equation (7.1) would proceed along the following lines:

$$P\{Y_1 \le y_1, Y_2 \le y_2\} = \iint\limits_{\substack{(x_1,x_2): \\ g_1(x_1,x_2)\le y_1 \\ g_2(x_1,x_2)\le y_2}} f_{X_1,X_2}(x_1, x_2)\,dx_1\,dx_2 \tag{7.2}$$

The joint density function can now be obtained by differentiating Equation (7.2) with respect to y_1 and y_2. That the result of this differentiation will be equal to the right-hand side of Equation (7.1) is an exercise in advanced calculus whose proof will not be presented in this book.

Example 7a. Let X_1 and X_2 be jointly continuous random variables with probability density function f_{X_1,X_2}. Let $Y_1 = X_1 + X_2$, $Y_2 = X_1 - X_2$. Find the joint density function of Y_1 and Y_2 in terms of f_{X_1,X_2}.

Solution Let $g_1(x_1, x_2) = x_1 + x_2$ and $g_2(x_1, x_2) = x_1 - x_2$. Then

$$J(x_1, x_2) = \begin{vmatrix} 1 & 1 \\ 1 & -1 \end{vmatrix} = -2$$

Also, as the equations $y_1 = x_1 + x_2$ and $y_2 = x_1 - x_2$ have as their solution $x_1 = (y_1 + y_2)/2$, $x_2 = (y_1 - y_2)/2$, it follows from Equation (7.1) that the desired density is

$$f_{Y_1,Y_2}(y_1, y_2) = \tfrac{1}{2} f_{X_1,X_2}\left(\frac{y_1 + y_2}{2}, \frac{y_1 - y_2}{2}\right)$$

For instance, if X_1 and X_2 are independent, uniform $(0, 1)$ random variables, then

$$f_{Y_1,Y_2}(y_1, y_2) = \begin{cases} \frac{1}{2} & 0 \le y_1 + y_2 \le 2, 0 \le y_1 - y_2 \le 2 \\ 0 & \text{otherwise} \end{cases}$$

or if X_1 and X_2 were independent, exponential random variables with respective parameters λ_1 and λ_2, then

$$f_{Y_1,Y_2}(y_1, y_2) = \begin{cases} \dfrac{\lambda_1\lambda_2}{2}\exp\left\{-\lambda_1\left(\dfrac{y_1 + y_2}{2}\right) - \lambda_2\left(\dfrac{y_1 - y_2}{2}\right)\right\} & y_1 + y_2 \ge 0, y_1 - y_2 \ge 0 \\ 0 & \text{otherwise} \end{cases}$$

Finally, if X_1 and X_2 are independent standard normal random variables,

$$\begin{aligned} f_{Y_1,Y_2}(y_1, y_2) &= \frac{1}{4\pi}e^{-[(y_1+y_2)^2/8+(y_1-y_2)^2/8]} \\ &= \frac{1}{4\pi}e^{-(y_1^2+y_2^2)/4} \\ &= \frac{1}{\sqrt{4\pi}}e^{-y_1^2/4}\frac{1}{\sqrt{4\pi}}e^{-y_2^2/4} \end{aligned}$$

Thus, not only do we obtain (in agreement with Proposition 3.2) that both $X_1 + X_2$ and $X_1 - X_2$ are normal with mean 0 and variance 2, we also obtain the interesting result that these two random variables are independent. (In fact, it can be shown that if X_1 and X_2 are independent random variables having a common distribution function F, then $X_1 + X_2$ will be independent of $X_1 - X_2$ if and only if F is a normal distribution function.) ■

Example 7b. Let (X, Y) denote a random point in the plane and assume that the rectangular coordinates X and Y are independent unit normal random variables. We are interested in the joint distribution of R, Θ, the polar coordinate representation of this point (see Figure 6.4).

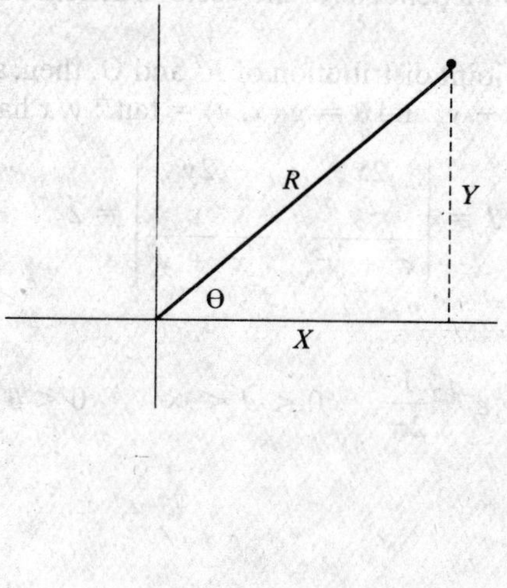

Figure 6.4 • = Random point. $(X, Y) = R, \Theta$.

Letting $r = g_1(x, y) = \sqrt{x^2 + y^2}$ and $\theta = g_2(x, y) = \tan^{-1} y/x$, we see that

$$\frac{\partial g_1}{\partial x} = \frac{x}{\sqrt{x^2 + y^2}} \qquad \frac{\partial g_1}{\partial y} = \frac{y}{\sqrt{x^2 + y^2}}$$

$$\frac{\partial g_2}{\partial x} = \frac{1}{1 + (y/x)^2}\left(\frac{-y}{x^2}\right) = \frac{-y}{x^2 + y^2} \qquad \frac{\partial g_2}{\partial y} = \frac{1}{x[1 + (y/x)^2]} = \frac{x}{x^2 + y^2}$$

Hence

$$J(x, y) = \frac{x^2}{(x^2 + y^2)^{3/2}} + \frac{y^2}{(x^2 + y^2)^{3/2}} = \frac{1}{\sqrt{x^2 + y^2}} = \frac{1}{r}$$

As the joint density function of X and Y is

$$f(x, y) = \frac{1}{2\pi} e^{-(x^2+y^2)/2}$$

we see that the joint density function of $R = \sqrt{x^2 + y^2}$, $\Theta = \tan^{-1} y/x$, is given by

$$f(r, \theta) = \frac{1}{2\pi} r e^{-r^2/2} \qquad 0 < \theta < 2\pi, \qquad 0 < r < \infty$$

As this joint density factors into the marginal densities for R and Θ, we obtain that R and Θ are independent random variables, with Θ being uniformly distributed over $(0, 2\pi)$ and R having the Rayleigh distribution with density

$$f(r) = re^{-r^2/2} \qquad 0 < r < \infty$$

(Thus, for instance, when one is aiming at a target in the plane, if the horizontal and vertical miss distances are independent unit normals, then the absolute value of the error has the above Rayleigh distribution.)

The preceding result is quite interesting, for it certainly is not evident a priori that a random vector whose coordinates are independent unit normal random variables will have an angle of orientation that is not only uniformly distributed, but is also independent of the vector's distance from the origin.

If we wanted the joint distribution of R^2 and Θ, then, as the transformation $d = g_1(x, y) = x^2 + y^2$ and $\theta = g_2(x, y) = \tan^{-1} y/x$ has a Jacobian

$$J = \begin{vmatrix} 2x & 2y \\ \dfrac{-y}{x^2 + y^2} & \dfrac{x}{x^2 + y^2} \end{vmatrix} = 2$$

we see that

$$f(d, \theta) = \tfrac{1}{2} e^{-d/2} \frac{1}{2\pi} \qquad 0 < d < \infty, \qquad 0 < \theta < 2\pi$$

Therefore, R^2 and Θ are independent, with R^2 having an exponential distribution with parameter $\frac{1}{2}$. But as $R^2 = X^2 + Y^2$, it follows, by definition, that R^2 has a chi-squared distribution with 2 degrees of freedom. Hence we have a verification of the result that the exponential distribution with parameter $\frac{1}{2}$ is the same as the chi-squared distribution with 2 degrees of freedom.

The preceding result can be used to simulate (or generate) normal random variables by making a suitable transformation on uniform random variables. Let U_1 and U_2 be independent random variables each uniformly distributed over $(0, 1)$. We will transform U_1, U_2 into two independent unit normal random variables X_1 and X_2 by first considering the polar coordinate representation (R, Θ) of the random vector (X_1, X_2). From the above, R^2 and Θ will be independent, and, in addition, $R^2 = X_1^2 + X_2^2$ will have an exponential distribution with parameter $\lambda = \frac{1}{2}$. But $-2 \log U_1$ has such a distribution since, for $x > 0$,

$$\begin{aligned} P\{-2 \log U_1 < x\} &= P\left\{\log U_1 > -\frac{x}{2}\right\} \\ &= P\{U_1 > e^{-x/2}\} \\ &= 1 - e^{-x/2} \end{aligned}$$

Also, as $2\pi U_2$ is a uniform $(0, 2\pi)$ random variable, we can use it to generate Θ. That is, if we let

$$\begin{aligned} R^2 &= -2 \log U_1 \\ \Theta &= 2\pi U_2 \end{aligned}$$

then R^2 can be taken to be the square of the distance from the origin and θ as the angle of orientation of (X_1, X_2). As $X_1 = R \cos \Theta$, $X_2 = R \sin \Theta$, we obtain that

$$X_1 = \sqrt{-2 \log U_1} \cos(2\pi U_2)$$

$$X_2 = \sqrt{-2 \log U_1} \sin(2\pi U_2)$$

are independent standard normal random variables. ■

Example 7c. If X and Y are independent gamma random variables with parameters (α, λ) and (β, λ), respectively, compute the joint density of $U = X + Y$ and $V = X/(X + Y)$.

Solution The joint density of X and Y is given by

$$\begin{aligned} f_{X,Y}(x, y) &= \frac{\lambda e^{-\lambda x}(\lambda x)^{\alpha-1}}{\Gamma(\alpha)} \frac{\lambda e^{-\lambda y}(\lambda y)^{\beta-1}}{\Gamma(\beta)} \\ &= \frac{\lambda^{\alpha+\beta}}{\Gamma(\alpha)\Gamma(\beta)} e^{-\lambda(x+y)} x^{\alpha-1} y^{\beta-1} \end{aligned}$$

Now, if $g_1(x, y) = x + y$, $g_2(x, y) = x/(x + y)$, then

$$\frac{\partial g_1}{\partial x} = \frac{\partial g_1}{\partial y} = 1 \qquad \frac{\partial g_2}{\partial x} = \frac{y}{(x+y)^2} \qquad \frac{\partial g_2}{\partial y} = -\frac{x}{(x+y)^2}$$

and so

$$J(x, y) = \begin{vmatrix} 1 & 1 \\ \dfrac{y}{(x+y)^2} & \dfrac{-x}{(x+y)^2} \end{vmatrix} = -\frac{1}{x+y}$$

Finally, as the equations $u = x + y$, $v = x/(x + y)$ have as their solutions $x = uv$, $y = u(1 - v)$, we see that

$$\begin{aligned} f_{U,V}(u, v) &= f_{X,Y}[uv, u(1-v)]u \\ &= \frac{\lambda e^{-\lambda u}(\lambda u)^{\alpha+\beta-1}}{\Gamma(\alpha+\beta)} \frac{v^{\alpha-1}(1-v)^{\beta-1}\Gamma(\alpha+\beta)}{\Gamma(\alpha)\Gamma(\beta)} \end{aligned}$$

Hence $X + Y$ and $X/(X + Y)$ are independent, with $X + Y$ having a gamma distribution with parameters $(\alpha + \beta, \lambda)$ and $X/(X + Y)$ having a beta distribution with parameters (α, β). The preceding also shows that $B(\alpha, \beta)$, the normalizing factor in the beta density, is such that

$$\begin{aligned} B(\alpha, \beta) &\equiv \int_0^1 v^{\alpha-1}(1-v)^{\beta-1}\, dv \\ &= \frac{\Gamma(\alpha)\Gamma(\beta)}{\Gamma(\alpha+\beta)} \end{aligned}$$

The preceding result is quite interesting. For suppose there are $n + m$ jobs to be performed, with each (independently) taking an exponential amount of time with rate λ for performance, and suppose that we have two workers to perform these jobs. Worker I will do jobs $1, 2, \ldots, n$, and worker II will do the remaining m jobs. If we let X and Y denote the total working times of workers I and II, respectively, then (either from the above result or from Example 3b) X and Y will be independent gamma random variables having parameters (n, λ) and (m, λ), respectively. Then the preceding result yields that independently of the working time needed to complete all $n + m$ jobs (that is, of $X + Y$), the proportion of this work that will be performed by worker I has a beta distribution with parameters (n, m). ■

When the joint density function of the n random variables $X_1, X_2, \ldots, X_n$ is given and we want to compute the joint density function of $Y_1, Y_2, \ldots, Y_n$, where

$$Y_1 = g_1(X_1, \ldots, X_n) \qquad Y_2 = g_2(X_1, \ldots, X_n), \ldots$$
$$Y_n = g_n(X_1, \ldots, X_n)$$

the approach is the same. Namely, we assume that the functions g_i have continuous partial derivatives and that the Jacobian determinant $J(x_1, \ldots, x_n) \neq 0$ at all points $(x_1, \ldots, x_n)$, where

$$J(x_1, \ldots, x_n) = \begin{vmatrix} \dfrac{\partial g_1}{\partial x_1} & \dfrac{\partial g_1}{\partial x_2} & \cdots & \dfrac{\partial g_1}{\partial x_n} \\ \dfrac{\partial g_2}{\partial x_1} & \dfrac{\partial g_2}{\partial x_2} & \cdots & \dfrac{\partial g_2}{\partial x_n} \\ \dfrac{\partial g_n}{\partial x_1} & \dfrac{\partial g_n}{\partial x_2} & \cdots & \dfrac{\partial g_n}{\partial x_n} \end{vmatrix}$$

Furthermore, we suppose that the equations $y_1 = g_1(x_1, \ldots, x_n)$, $y_2 = g_2(x_1, \ldots, x_n), \ldots$, $y_n = g_n(x_1, \ldots, x_n)$ have a unique solution, say, $x_1 = h_1(y_1, \ldots, y_n), \ldots, x_n = h_n(y_1, \ldots, y_n)$. Under these assumptions, the joint density function of the random variables Y_i is given by

$$f_{Y_1, \ldots, Y_n}(y_1, \ldots, y_n) = f_{X_1, \ldots, X_n}(x_1, \ldots, x_n)|J(x_1, \ldots, x_n)|^{-1} \tag{7.3}$$

where $x_i = h_i(y_1, \ldots, y_n)$, $i = 1, 2, \ldots, n$.

Example 7d. Let X_1, X_2, and X_3 be independent unit normal random variables. If $Y_1 = X_1 + X_2 + X_3$, $Y_2 = X_1 - X_2$, $Y_3 = X_1 - X_3$, compute the joint density function of Y_1, Y_2, Y_3.

Solution Letting $Y_1 = X_1 + X_2 + X_3$, $Y_2 = X_1 - X_2$, $Y_3 = X_1 - X_3$, the Jacobian of these transformations is given by

$$J = \begin{vmatrix} 1 & 1 & 1 \\ 1 & -1 & 0 \\ 1 & 0 & -1 \end{vmatrix} = 3$$

As the preceding transformations yield that

$$X_1 = \frac{Y_1 + Y_2 + Y_3}{3} \qquad X_2 = \frac{Y_1 - 2Y_2 + Y_3}{3} \qquad X_3 = \frac{Y_1 + Y_2 - 2Y_3}{3}$$

we see from Equation (7.3) that

$$f_{Y_1, Y_2, Y_3}(y_1, y_2, y_3) = \frac{1}{3} f_{X_1, X_2, X_3}\left(\frac{y_1 + y_2 + y_3}{3}, \frac{y_1 - 2y_2 + y_3}{3}, \frac{y_1 + y_2 - 2y_3}{3}\right)$$

Hence, as

$$f_{X_1, X_2, X_3}(x_1, x_2, x_3) = \frac{1}{(2\pi)^{3/2}} e^{-\sum_{i=1}^{3} x_i^2/2}$$

we see that

$$f_{Y_1, Y_2, Y_3}(y_1, y_2, y_3) = \frac{1}{3(2\pi)^{3/2}} e^{-Q(y_1, y_2, y_3)/2}$$

where

$$Q(y_1, y_2, y_3) = \left(\frac{y_1 + y_2 + y_3}{3}\right)^2 + \left(\frac{y_1 - 2y_2 + y_3}{3}\right)^2 + \left(\frac{y_1 + y_2 - 2y_3}{3}\right)^2$$

$$= \frac{y_1^2}{3} + \frac{2}{3}y_2^2 + \frac{2}{3}y_3^2 - \frac{2}{3}y_2y_3 \qquad \blacksquare$$

Example 7e. Let $X_1, X_2, \ldots, X_n$ be independent and identically distributed exponential random variables with rate λ. Let

$$Y_i = X_1 + \cdots + X_i \qquad i = 1, \ldots, n$$

(a) Find the joint density function of $Y_1, \ldots, Y_n$.
(b) Use the result of part (a) to find the density of Y_n.

Solution (a) The Jacobian of the transformations $Y_1 = X_1, Y_2 = X_1 + X_2, \ldots, Y_n = X_1 + \cdots + X_n$ is

$$J = \begin{vmatrix} 1 & 0 & 0 & 0 & \cdots & 0 \\ 1 & 1 & 0 & 0 & \cdots & 0 \\ 1 & 1 & 1 & 0 & \cdots & 0 \\ \cdots & & \cdots & & & \\ \cdots & & \cdots & & & \\ 1 & 1 & 1 & 1 & \cdots & 1 \end{vmatrix}$$

Since only the first term of the determinant will be nonzero, we have that $J = 1$. Now the joint density function of $X_1, \ldots, X_n$ is given by

$$f_{X_1,\ldots,X_n}(x_1, \ldots, x_n) = \prod_{i=1}^{n} \lambda e^{-\lambda x_i} \qquad 0 < x_i < \infty, i = 1, \ldots, n$$

Hence, as the preceding transformations yield that

$$X_1 = Y_1, X_2 = Y_2 - Y_1, \ldots, X_i = Y_i - Y_{i-1}, \ldots, X_n = Y_n - Y_{n-1}$$

we obtain from Equation (7.3) that the joint density function of $Y_1, \ldots, Y_n$ is

$$\begin{aligned} &f_{Y_1,\ldots,Y_n}(y_1, y_2, \ldots, y_n) \\ &\quad = f_{X_1,\ldots,X_n}(y_1, y_2 - y_1, \ldots, y_i - y_{i-1}, \ldots, y_n - y_{n-1}) \\ &\quad = \lambda^n \exp\left\{-\lambda\left[y_1 + \sum_{i=2}^{n} (y_i - y_{i-1})\right]\right\} \\ &\quad = \lambda^n e^{-\lambda y_n} \qquad 0 < y_1, 0 < y_i - y_{i-1}, i = 2, \ldots, n \\ &\quad = \lambda^n e^{-\lambda y_n} \qquad 0 < y_1 < y_2 < \cdots < y_n \end{aligned}$$

(b) To obtain the marginal density of Y_n, let us integrate out the other variables one at a time. This gives

$$f_{Y_2,\ldots,Y_n}(y_2,\ldots,y_n) = \int_0^{y_2} \lambda^n e^{-\lambda y_n}\,dy_1$$
$$= \lambda^n y_2 e^{-\lambda y_n} \qquad 0 < y_2 < y_3 < \cdots < y_n$$

Continuing gives that

$$f_{Y_3,\ldots,Y_n}(y_3,\ldots,y_n) = \int_0^{y_3} \lambda^n y_2 e^{-\lambda y_n}\,dy_2$$
$$= \lambda^n \frac{y_3^2}{2} e^{-\lambda y_n} \qquad 0 < y_3 < y_4 < \cdots < y_n$$

The next integration yields that

$$f_{Y_4,\ldots,Y_n}(y_4,\ldots,y_n) = \lambda^n \frac{y_4^3}{3!} e^{-\lambda y_n} \qquad 0 < y_4 < \cdots < y_n$$

Continuing in this fashion yields

$$f_{Y_n}(y_n) = \lambda^n \frac{y_n^{n-1}}{(n-1)!} e^{-\lambda y_n} \qquad 0 < y_n$$

which, in agreement with the result obtained in Example 3b, shows that $X_1 + \cdots + X_n$ is a gamma random variable with parameters n and λ. ■

*6.8 EXCHANGEABLE RANDOM VARIABLES

The random variables $X_1, X_2, \ldots, X_n$ are said to be *exchangeable* if for every permutation $i_1, \ldots, i_n$ of the integers $1, \ldots, n$

$$P\{X_{i_1} \le x_1, X_{i_2} \le x_2, \ldots, X_{i_n} \le x_n\} = P\{X_1 \le x_1, X_2 \le x_2, \ldots, X_n \le x_n\}$$

for all $x_1, \ldots, x_n$. That is, the n random variables are exchangeable if their joint distribution is the same no matter in which order they are observed.

Discrete random variables will be exchangeable if

$$P\{X_{i_1} = x_1, X_{i_2} = x_2, \ldots, X_{i_n} = x_n\} = P\{X_1 = x_1, X_2 = x_2, \ldots, X_n = x_n\}$$

for all permutations $i_1, \ldots, i_n$, and all values $x_1, \ldots, x_n$. This is equivalent to stating that $p(x_1, x_2, \ldots, x_n) = P\{X_1 = x_1, \ldots, X_n = x_n\}$ is a symmetric function of the vector $(x_1, \ldots, x_n)$, which means that its value does not change when the values of the vector are permuted.

Example 8a. Suppose that balls are withdrawn one at a time and without replacement from an urn that initially contains n balls, of which k are considered special, in such a manner that each withdrawal is equally likely to be any of the balls that remain in the urn at the time. Let $X_i = 1$ if the ith ball withdrawn is special and let it be 0 otherwise. We will show that the random variables

$X_1, \ldots, X_n$ are exchangeable. To do so, let $(x_1, \ldots, x_n)$ be a vector consisting of k ones and $n - k$ zeros. However, before considering the joint mass function evaluated at $(x_1, \ldots, x_n)$, let us try to gain some insight by considering a fixed such vector—for instance, consider the vector $(1, 1, 0, 1, 0, \ldots, 0, 1)$, which is assumed to have k ones and $n - k$ zeros. Then

$$p(1, 1, 0, 1, 0, \ldots, 0, 1) = \frac{k}{n}\frac{k-1}{n-1}\frac{n-k}{n-2}\frac{k-2}{n-3}\frac{n-k-1}{n-4}\cdots\frac{1}{2}\frac{1}{1}$$

which follows since the probability that the first ball is special is k/n, the conditional probability that the next one is special is $(k-1)/(n-1)$, the conditional probability that the next one is not special is $(n-k)/(n-2)$, and so on. By the same argument, it follows that $p(x_1, \ldots, x_n)$ can be expressed as the product of n fractions. The successive denominator terms of these fractions will go from n down to 1. The numerator term at the location where the vector $(x_1, \ldots, x_n)$ is 1 for the ith time is $k - (i - 1)$, and where it is 0 for the ith time it is $n - k - (i - 1)$. Hence, since the vector $(x_1, \ldots, x_n)$ consists of k ones and $n - k$ zeros, we obtain that

$$p(x_1, \ldots, x_n) = \frac{k!\,(n-k)!}{n!} \qquad x_i = 0, 1, \sum_{i=1}^{n} x_i = k$$

Since this is a symmetric function of $(x_1, \ldots, x_n)$, it follows that the random variables are exchangeable. ■

REMARK. Another way to obtain the preceding formula for the joint probability mass function is to regard all the n balls as distinguishable from each other. Then, as the outcome of the experiment is an ordering of these balls, it follows that there are $n!$ equally likely outcomes. As the number of outcomes having special and nonspecial balls in specified places is equal to the number of ways of permuting the special and the nonspecial balls among themselves, namely $k!\,(n-k)!$, we obtain the preceding density function. ■

If $X_1, X_2, \ldots, X_n$ are exchangeable, it easily follows that each X_i has the same probability distribution. For instance, if X and Y are exchangeable discrete random variables, then

$$P\{X = x\} = \sum_y P\{X = x, Y = y\} = \sum_y P\{X = y, Y = x\} = P\{Y = x\}$$

For instance, it follows from Example 8a that the ith ball withdrawn will be special with probability k/n, which is intuitively clear since each of the n balls is equally likely to be the ith one selected.

Example 8b. In Example 8a, let Y_i denote the selection number of the first special ball withdrawn, let Y_2 denote the additional number that are then withdrawn until the second special ball appears, and in general, let Y_i denote the additional number of balls withdrawn after the $(i-1)$st special ball is selected until the ith is selected, $i = 1, \ldots, k$. For instance, if $n = 4$, $k = 2$ and $X_1 = 1$, $X_2 = 0$, $X_3 = 0$, $X_4 = 1$, then $Y_1 = 1$, $Y_2 = 3$. Now, $Y_1 = i_1$, $Y_2 = i_2, \ldots,$ $Y_k = i_k \Leftrightarrow X_{i_1} = X_{i_1+i_2} = \cdots = X_{i_1+\cdots+i_k} = 1$, $X_j = 0$, otherwise; thus, we obtain from the joint mass function of the X_i that

$$P\{Y_1 = i_1, Y_2 = i_2, \ldots, Y_k = i_k\} = \frac{k!\,(n-k)!}{n!} \qquad i_1 + \cdots + i_k \le n$$

Hence we see that the random variables $Y_1, \ldots, Y_k$ are exchangeable. For instance, it follows from this that the number of cards one must select from a well-shuffled deck until an ace appears has the same distribution as the number of additional cards one must select after the first ace appears until the next one does, and so on. ■

Example 8c. The following is known as Polya's urn model. Suppose that an urn initially contains n red and m blue balls. At each stage a ball is randomly chosen, its color is noted, and it is then replaced along with another ball of the same color. Let $X_i = 1$ if the ith ball selected is red and let it equal 0 if the ith ball is blue, $i \ge 1$. To obtain a feeling for the joint probabilities of these X_i, note the following special cases.

$$P\{X_1 = 1, X_2 = 1, X_3 = 0, X_4 = 1, X_5 = 0\}$$

$$= \frac{n}{n+m}\,\frac{n+1}{n+m+1}\,\frac{m}{n+m+2}\,\frac{n+2}{n+m+3}\,\frac{m+1}{n+m+4}$$

$$= \frac{n(n+1)(n+2)m(m+1)}{(n+m)(n+m+1)(n+m+2)(n+m+3)(n+m+4)}$$

and

$$P\{X_1 = 0, X_2 = 1, X_3 = 0, X_4 = 1, X_5 = 1\}$$

$$= \frac{m}{n+m}\,\frac{n}{n+m+1}\,\frac{m+1}{n+m+2}\,\frac{n+1}{n+m+3}\,\frac{n+2}{n+m+4}$$

$$= \frac{n(n+1)(n+2)m(m+1)}{(n+m)(n+m+1)(n+m+2)(n+m+3)(n+m+4)}$$

By the same reasoning it follows that for any sequence $x_1, \ldots, x_k$ that contains r ones and $k - r$ zeros, we have

$$P\{X_1 = x_1, \ldots, X_k = x_k\}$$

$$= \frac{n(n+1)\cdots(n+r-1)m(m+1)\cdots(m+k-r-1)}{(n+m)\cdots(n+m+k-1)}$$

Therefore, we see that for any value of k, the random variables $X_1, \ldots, X_k$ are exchangeable. ■

Our final example deals with continuous random variables that are exchangeable.

Example 8d. Let $X_1, X_2, \ldots, X_n$ be independent uniform $(0, 1)$ random variables, and let $X_{(1)}, \ldots, X_{(n)}$ denote their order statistics. That is, $X_{(j)}$ is the jth smallest of $X_1, X_2, \ldots, X_n$. Also, let

$$Y_1 = X_{(1)},$$
$$Y_i = X_{(i)} - X_{(i-1)}, \qquad i = 2, \ldots n$$

Show that $Y_1, \ldots, Y_n$ are exchangeable.

Solution The transformations

$$y_1 = x_1, \ldots, y_i = x_i - x_{i-1} \qquad i = 2, \ldots, n$$

yield that

$$x_i = y_1 + \cdots + y_i \qquad i = 1, \ldots, n$$

As it is easy to see that the Jacobian of the preceding transformations is equal to 1, we obtain from Equation (7.3) that

$$f_{Y_1,\ldots,Y_n}(y_1, y_2, \ldots, y_n) = f(y_1, y_1 + y_2, \ldots, y_1 + \cdots + y_n)$$

where f is the joint density function of the order statistics. Hence from Equation (6.1) we obtain that

$$f_{Y_1,\ldots,Y_n}(y_1, y_2, \ldots, y_n) = n! \qquad 0 < y_1 < y_1 + y_2 < \cdots < y_1 + \cdots + y_n < 1$$

or, equivalently,

$$f_{Y_1,\ldots,Y_n}(y_1, y_2, \ldots, y_n) = n! \qquad 0 < y_i < 1, i = 1, \ldots, n, \qquad y_1 + \cdots + y_n < 1$$

As the preceding joint density is a symmetric function of $y_1, \ldots, y_n$, we see that the random variables $Y_1, \ldots, Y_n$ are exchangeable. ■

SUMMARY

The *joint cumulative probability distribution function* of the pair of random variables X and Y is defined by

$$F(x, y) = P\{X \le x, Y \le y\} \qquad -\infty < x, y < \infty$$

All probabilities regarding the pair can be obtained from F. To obtain the individual probability distribution functions of X and Y, use

$$F_X(x) = \lim_{y\to\infty} F(x, y) \qquad F_Y(y) = \lim_{x\to\infty} F(x, y)$$

If X and Y are both discrete random variables, then their *joint probability mass function* is defined by

$$p(i, j) = P\{X = i, Y = j\}$$

The individual mass functions are

$$P\{X = i\} = \sum_j p(i,j) \qquad P\{Y = j\} = \sum_i p(i,j)$$

The random variables X and Y are said to be *jointly continuous* if there is a function $f(x, y)$, called the *joint probability density function*, such that for any two-dimensional set C,

$$P\{(X,Y) \in C\} = \iint_C f(x,y)\,dx\,dy$$

It follows from the preceding that

$$P\{x < X < x + dx, y < Y < y + dy\} \approx f(x,y)\,dx\,dy$$

If X and Y are jointly continuous, then they are individually continuous with density functions

$$f_X(x) = \int_{-\infty}^{\infty} f(x,y)\,dy \qquad f_Y(y) = \int_{-\infty}^{\infty} f(x,y)\,dx$$

The random variables X and Y are *independent* if for all sets A and B

$$P\{X \in A, Y \in B\} = P\{X \in A\}P\{Y \in B\}$$

If the joint distribution function (or the joint probability mass function in the discrete case, or the joint density function in the continuous case) factors into a part depending only on x and a part depending only on y, then X and Y are independent.

In general, the random variables $X_1, \ldots, X_n$ are independent if for all sets of real numbers $A_1, \ldots, A_n$,

$$P\{X_1 \in A_1, \ldots, X_n \in A_n\} = P\{X_1 \in A_1\} \cdots P\{X_n \in A_n\}$$

If X and Y are independent continuous random variables, then the distribution function of their sum can be obtained from the identity

$$F_{X+Y}(a) = \int_{-\infty}^{\infty} F_X(a - y) f_Y(y)\,dy$$

If X_i, $i = 1, \ldots, n$, are independent normal random variables with respective parameters μ_i and σ_i^2, $i = 1, \ldots, n$, then $\sum_{i=1}^{n} X_i$ is normal with parameters $\sum_{i=1}^{n} \mu_i$ and $\sum_{i=1}^{n} \sigma_i^2$.

If X_i, $i = 1, \ldots, n$, are independent Poisson random variables with respective parameters λ_i, $i = 1, \ldots, n$, then $\sum_{i=1}^{n} X_i$ is Poisson with parameter $\sum_{i=1}^{n} \lambda_i$.

If X and Y are discrete random variables, then the *conditional probability mass function* of X given that $Y = y$ is defined by

$$P\{X = x | Y = y\} = \frac{p(x,y)}{p_Y(y)}$$

where p is their joint probability mass function. Also, if X and Y are jointly continuous with joint density function f, then the *conditional probability density function* of X given that $Y = y$ is given by

$$f_{X|Y}(x|y) = \frac{f(x, y)}{f_Y(y)}$$

The ordered values $X_{(1)} \leq X_{(2)} \leq \cdots \leq X_{(n)}$ of a set of independent and identically distributed random variables are called the *order statistics* of that set. If the random variables are continuous with density function f, then the joint density function of the order statistics is

$$f(x_1, \ldots, x_n) = n!\, f(x_1) \cdots f(x_n) \qquad x_1 \leq x_2 \leq \cdots \leq x_n$$

The random variables $X_1, \ldots, X_n$ are *exchangeable* if the joint distribution of $X_{i_1}, \ldots, X_{i_n}$ is the same for every permutation $i_1, \ldots, i_n$ of $1, \ldots, n$.

PROBLEMS

1. Two fair dice are rolled. Find the joint probability mass function of X and Y when
(a) X is the largest value obtained on any die and Y is the sum of the values;
(b) X is the value on the first die and Y is the larger of the two values;
(c) X is the smallest and Y is the largest value obtained on the dice.

2. Suppose that 3 balls are chosen without replacement from an urn consisting of 5 white and 8 red balls. Let X_i equal 1 if the ith ball selected is white, and let it equal 0 otherwise. Give the joint probability mass function of
(a) X_1, X_2; **(b)** X_1, X_2, X_3.

3. In Problem 2, suppose that the white balls are numbered, and let Y_i equal 1 if the ith white ball is selected and 0 otherwise. Find the joint probability mass function of
(a) Y_1, Y_2; **(b)** Y_1, Y_2, Y_3.

4. Repeat Problem 2 when the ball selected is replaced in the urn before the next selection.

5. Repeat Problem 3a when the ball selected is replaced in the urn before the next selection.

6. A bin of 5 transistors is known to contain 2 that are defective. The transistors are to be tested, one at a time, until the defective ones are identified. Denote by N_1 the number of tests made until the first defective is identified and by N_2 the number of additional tests until the second defective is identified; find the joint probability mass function of N_1 and N_2.

7. Consider a sequence of independent Bernoulli trials, each of which is a success with probability p. Let X_1 be the number of failures preceding the first success, and let X_2 be the number of failures between the first two successes. Find the joint mass function of X_1 and X_2.

8. The joint probability density function of X and Y is given by

$$f(x, y) = c(y^2 - x^2)e^{-y} \qquad -y \leq x \leq y, 0 < y < \infty$$

(a) Find c. **(b)** Find the marginal densities of X and Y.
(c) Find $E[X]$.

9. The joint probability density function of X and Y is given by

$$f(x, y) = \frac{6}{7}\left(x^2 + \frac{xy}{2}\right) \qquad 0 < x < 1, 0 < y < 2$$

(a) Verify that this is indeed a joint density function.
(b) Compute the density function of X.

(c) Find $P\{X > Y\}$. **(d)** Find $P\{Y > \frac{1}{2}|X < \frac{1}{2}\}$.
(e) Find $E[X]$. **(f)** Find $E[Y]$.

10. The joint probability density function of X and Y is given by

$$f(x, y) = e^{-(x+y)} \qquad 0 \le x < \infty, 0 \le y < \infty$$

Find (a) $P\{X < Y\}$ and (b) $P\{X < a\}$.

11. A television store owner figures that 45 percent of the customers entering his store will purchase an ordinary television set, 15 percent will purchase a plasma television set, and 40 percent will just be browsing. If 5 customers enter his store on a given day, what is the probability that he will sell exactly 2 ordinary sets and 1 plasma set on that day?

12. The number of people that enter a drugstore in a given hour is a Poisson random variable with parameter $\lambda = 10$. Compute the conditional probability that at most 3 men entered the drugstore, given that 10 women entered in that hour. What assumptions have you made?

13. A man and a woman agree to meet at a certain location about 12:30 P.M. If the man arrives at a time uniformly distributed between 12:15 and 12:45 and if the woman independently arrives at a time uniformly distributed between 12:00 and 1 P.M., find the probability that the first to arrive waits no longer than 5 minutes. What is the probability that the man arrives first?

14. An ambulance travels back and forth, at a constant speed, along a road of length L. At a certain moment of time an accident occurs at a point uniformly distributed on the road. [That is, its distance from one of the fixed ends of the road is uniformly distributed over $(0, L)$.] Assuming that the ambulance's location at the moment of the accident is also uniformly distributed, compute, assuming independence, the distribution of its distance from the accident.

15. The random vector (X, Y) is said to be uniformly distributed over a region R in the plane if, for some constant c, its joint density is

$$f(x, y) = \begin{cases} c & \text{if } (x, y) \in R \\ 0 & \text{otherwise} \end{cases}$$

(a) Show that $1/c$ = area of region R.
Suppose that (X, Y) is uniformly distributed over the square centered at $(0, 0)$, whose sides are of length 2.
(b) Show that X and Y are independent, with each being distributed uniformly over $(-1, 1)$.
(c) What is the probability that (X, Y) lies in the circle of radius 1 centered at the origin? That is, find $P\{X^2 + Y^2 \le 1\}$.

16. Suppose that n points are independently chosen at random on the perimeter of a circle, and we want the probability that they all lie in some semicircle. (That is, we want the probability that there is a line passing through the center of the circle such that all the points are on one side of that line.)

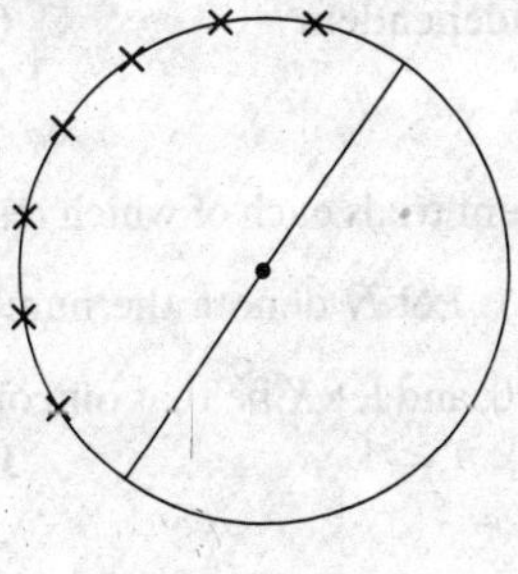

Let $P_1, \ldots, P_n$ denote the n points. Let A denote the event that all the points are contained in some semicircle, and let A_i be the event that all the points lie in the semicircle beginning at the point P_i and going clockwise for 180°, $i = 1, \ldots, n$.
(a) Express A in terms of the A_i. **(b)** Are the A_i mutually exclusive?
(c) Find $P(A)$.

17. Three points X_1, X_2, X_3 are selected at random on a line L. What is the probability that X_2 lies between X_1 and X_3?

18. Two points are selected randomly on a line of length L so as to be on opposite sides of the midpoint of the line. [In other words, the two points X and Y are independent random variables such that X is uniformly distributed over $(0, L/2)$ and Y is uniformly distributed over $(L/2, L)$.] Find the probability that the distance between the two points is greater than $L/3$.

19. In Problem 18 find the probability that the 3 line segments from 0 to X, from X to Y, and from Y to L could be made to form the three sides of a triangle. (Note that three line segments can be made to form a triangle if the length of each of them is less than the sum of the lengths of the others.)

20. The joint density of X and Y is given by

$$f(x, y) = \begin{cases} xe^{-(x+y)} & x > 0, y > 0 \\ 0 & \text{otherwise} \end{cases}$$

Are X and Y independent? What if $f(x, y)$ were given by

$$f(x, y) = \begin{cases} 2 & 0 < x < y, 0 < y < 1 \\ 0 & \text{otherwise} \end{cases}$$

21. Let

$$f(x, y) = 24xy \qquad 0 \le x \le 1, 0 \le y \le 1, 0 \le x + y \le 1$$

and let it equal 0 otherwise.
(a) Show that $f(x, y)$ is a joint probability density function.
(b) Find $E[X]$. **(c)** Find $E[Y]$.

22. The joint density function of X and Y is

$$f(x, y) = \begin{cases} x + y & 0 < x < 1, 0 < y < 1 \\ 0 & \text{otherwise} \end{cases}$$

(a) Are X and Y independent? **(b)** Find the density function of X.
(c) Find $P\{X + Y < 1\}$.

23. The random variables X and Y have joint density function.

$$f(x, y) = 12xy(1 - x) \qquad 0 < x < 1, 0 < y < 1$$

and equal to 0 otherwise.
(a) Are X and Y independent? **(b)** Find $E[X]$.
(c) Find $E[Y]$. **(d)** Find $\text{Var}(X)$.
(e) Find $\text{Var}(Y)$.

24. Consider independent trials each of which results in outcome $i, i = 0, 1, \ldots, k$ with probability p_i, $\sum_{i=0}^{k} p_i = 1$. Let N denote the number of trials needed to obtain an outcome that is not equal to 0, and let X be that outcome.
(a) Find $P\{N = n\}, n \ge 1$. **(b)** Find $P\{X = j\}, j = 1, \ldots, k$.

(c) Show that $P\{N = n, X = j\} = P\{N = n\}P\{X = j\}$.
(d) Is it intuitive to you that N is independent of X?
(e) Is it intuitive to you that X is independent of N?

25. Suppose that 10^6 people arrive at a service station at times that are independent random variables, each of which is uniformly distributed over $(0, 10^6)$. Let N denote the number that arrive in the first hour. Find an approximation for $P\{N = i\}$.

26. Suppose that A, B, C, are independent random variables, each being uniformly distributed over $(0, 1)$.
(a) What is the joint cumulative distribution function of A, B, C?
(b) What is the probability that all of the roots of the equation $Ax^2 + Bx + C = 0$ are real?

27. If X is uniformly distributed over $(0, 1)$ and Y is exponentially distributed with parameter $\lambda = 1$, find the distribution of (a) $Z = X + Y$ and (b) $Z = X/Y$. Assume independence.

28. If X_1 and X_2 are independent exponential random variables with respective parameters λ_1 and λ_2, find the distribution of $Z = X_1/X_2$. Also compute $P\{X_1 < X_2\}$.

29. When a current I (measured in amperes) flows through a resistance R (measured in ohms), the power generated is given by $W = I^2R$ (measured in watts). Suppose that I and R are independent random variables with densities

$$f_I(x) = 6x(1 - x) \qquad 0 \le x \le 1$$

$$f_R(x) = 2x \qquad 0 \le x \le 1$$

Determine the density function of W.

30. The expected number of typographical errors on a page of a certain magazine is .2. What is the probability that an article of 10 pages contains (a) 0, and (b) 2 or more typographical errors? Explain your reasoning!

31. The monthly worldwide average number of airplane crashes of commercial airlines is 2.2. What is the probability that there will be
(a) more than 2 such accidents in the next month;
(b) more than 4 such accidents in the next 2 months;
(c) more than 5 such accidents in the next 3 months?
Explain your reasoning!

32. The gross weekly sales at a certain restaurant is a normal random variable with mean \$2200 and standard deviation \$230. What is the probability that
(a) the total gross sales over the next 2 weeks exceeds \$5000;
(b) weekly sales exceed \$2000 in at least 2 of the next 3 weeks?
What independence assumptions have you made?

33. Jill's bowling scores are approximately normally distributed with mean 170 and standard deviation 20, while Jack's scores are approximately normally distributed with mean 160 and standard deviation 15. If Jack and Jill each bowl one game, then assuming that their scores are independent random variables, approximate the probability that
(a) Jack's score is higher;
(b) the total of their scores is above 350.

34. According to the U.S. National Center for Health Statistics, 25.2 percent of males and 23.6 percent of females never eat breakfast. Suppose that random samples of 200 men and 200 women are chosen. Approximate the probability that
(a) at least 110 of these 400 people never eat breakfast;
(b) the number of the women who never eat breakfast is at least as large as the number of the men who never eat breakfast.

35. In Problem 2, calculate the conditional probability mass function of X_1 given that
(a) $X_2 = 1$; **(b)** $X_2 = 0$.

36. In Problem 4, calculate the conditional probability mass function of X_1 given that
(a) $X_2 = 1$; **(b)** $X_2 = 0$.

37. In Problem 3, calculate the conditional probability mass function of Y_1 given that
(a) $Y_2 = 1$; **(b)** $Y_2 = 0$.

38. In Problem 5, calculate the conditional probability mass function of Y_1 given that
(a) $Y_2 = 1$; **(b)** $Y_2 = 0$.

39. Choose a number X at random from the set of numbers $\{1, 2, 3, 4, 5\}$. Now choose a number at random from the subset no larger than X, that is, from $\{1, \ldots, X\}$. Call this second number Y.
(a) Find the joint mass function of X and Y.
(b) Find the conditional mass function of X given that $Y = i$. Do it for $i = 1, 2, 3, 4, 5$.
(c) Are X and Y independent? Why?

40. Two dice are rolled. Let X and Y denote, respectively, the largest and smallest values obtained. Compute the conditional mass function of Y given $X = i$, for $i = 1, 2, \ldots, 6$. Are X and Y independent? Why?

41. The joint probability mass function of X and Y is given by

$$p(1,1) = \tfrac{1}{8} \qquad p(1,2) = \tfrac{1}{4}$$
$$p(2,1) = \tfrac{1}{8} \qquad p(2,2) = \tfrac{1}{2}$$

(a) Compute the conditional mass function of X given $Y = i, i = 1, 2$.
(b) Are X and Y independent?
(c) Compute $P\{XY \le 3\}, P\{X + Y > 2\}, P\{X/Y > 1\}$.

42. The joint density function of X and Y is given by

$$f(x, y) = xe^{-x(y+1)} \qquad x > 0, y > 0$$

(a) Find the conditional density of X, given $Y = y$, and that of Y, given $X = x$.
(b) Find the density function of $Z = XY$.

43. The joint density of X and Y is

$$f(x, y) = c(x^2 - y^2)e^{-x} \qquad 0 \le x < \infty, -x \le y \le x$$

Find the conditional distribution of Y, given $X = x$.

44. An insurance company supposes that each person has an accident parameter and that the yearly number of accidents of someone whose accident parameter is λ is Poisson distributed with mean λ. They also suppose that the parameter value of a newly insured person can be assumed to be the value of a gamma random variable with parameters s and α. If a newly insured person has n accidents in her first year, find the conditional density of her accident parameter. Also, determine the expected number of accidents that she will have in the following year.

45. If X_1, X_2, X_3 are independent random variables that are uniformly distributed over $(0, 1)$, compute the probability that the largest of the three is greater than the sum of the other two.

46. A complex machine is able to operate effectively as long as at least 3 of its 5 motors are functioning. If each motor independently functions for a random amount of time with density function $f(x) = xe^{-x}, x > 0$, compute the density function of the length of time that the machine functions.

47. If 3 trucks break down at points randomly distributed on a road of length L, find the probability that no 2 of the trucks are within a distance d of each other when $d \le L/2$.

48. Consider a sample of size 5 from a uniform distribution over $(0, 1)$. Compute the probability that the median is in the interval $\left(\frac{1}{4}, \frac{3}{4}\right)$.

49. If X_1, X_2, X_3, X_4, X_5 are independent and identically distributed exponential random variables with the parameter λ, compute

(a) $P\{\min(X_1, \ldots, X_5) \le a\}$; **(b)** $P\{\max(X_1, \ldots, X_5) \le a\}$.

50. Derive the distribution of the range of a sample of size 2 from a distribution having density function $f(x) = 2x, 0 < x < 1$.

51. Let X and Y denote the coordinates of a point uniformly chosen in the circle of radius 1 centered at the origin. That is, their joint density is

$$f(x, y) = \frac{1}{\pi} \qquad x^2 + y^2 \le 1$$

Find the joint density function of the polar coordinates $R = (X^2 + Y^2)^{1/2}$ and $\Theta = \tan^{-1} Y/X$.

52. If X and Y are independent random variables both uniformly distributed over $(0, 1)$, find the joint density function of $R = \sqrt{X^2 + Y^2}$, $\Theta = \tan^{-1} Y/X$.

53. If U is uniform on $(0, 2\pi)$ and Z, independent of U, is exponential with rate 1, show directly (without using the results of Example 7b) that X and Y defined by

$$X = \sqrt{2Z} \cos U$$

$$Y = \sqrt{2Z} \sin U$$

are independent unit normal random variables.

54. X and Y have joint density function

$$f(x, y) = \frac{1}{x^2 y^2} \qquad x \ge 1, y \ge 1$$

(a) Compute the joint density function of $U = XY, V = X/Y$.
(b) What are the marginal densities?

55. If X and Y are independent and identically distributed uniform random variables on $(0, 1)$, compute the joint density of

(a) $U = X + Y, V = X/Y$; **(b)** $U = X, V = X/Y$;
(c) $U = X + Y, V = X/(X + Y)$.

56. Repeat Problem 55 when X and Y are independent exponential random variables, each with parameter $\lambda = 1$.

57. If X_1 and X_2 are independent exponential random variables each having parameter λ, find the joint density function of $Y_1 = X_1 + X_2$ and $Y_2 = e^{X_1}$.

58. If X, Y, and Z are independent random variables having identical density functions $f(x) = e^{-x}, 0 < x < \infty$, derive the joint distribution of $U = X + Y$, $V = X + Z$, $W = Y + Z$.

59. In Example 8b, let $Y_{k+1} = n + 1 - \sum_{i=1}^{k} Y_i$. Show that $Y_1, \ldots, Y_k, Y_{k+1}$ are exchangeable. Note that Y_{k+1} is the number of balls one must observe to obtain a special ball if one considers the balls in their reverse order of withdrawal.

60. Consider an urn containing n balls, numbered $1, \ldots, n$, and suppose that k of them are randomly withdrawn. Let X_i equal 1 if ball numbered i is removed and let it be 0 otherwise. Show that $X_1, \ldots, X_n$ are exchangeable.

THEORETICAL EXERCISES

1. Verify Equation (1.2).
2. Suppose that the number of events that occur in a given time period is a Poisson random variable with parameter λ. If each event is classified as a type i event with probability p_i, $i = 1, \ldots, n$, $\sum p_i = 1$, independently of other events, show that the numbers of type i events that occur, $i = 1, \ldots, n$, are independent Poisson random variables with respective parameters λp_i, $i = 1, \ldots, n$.
3. Suggest a procedure for using Buffon's needle problem to estimate π. Surprisingly enough, this was once a common method of evaluating π.
4. Solve Buffon's needle problem when $L > D$.

   ANSWER: $\dfrac{2L}{\pi D}(1 - \sin\theta) + 2\theta/\pi$, where θ is such that $\cos\theta = D/L$.
5. If X and Y are independent continuous positive random variables, express the density function of (a) $Z = X/Y$ and (b) $Z = XY$ in terms of the density functions of X and Y. Evaluate these expressions in the special case where X and Y are both exponential random variables.
6. Show analytically (by induction) that $X_1 + \cdots + X_n$ has a negative binomial distribution when the X_i, $i = 1, \ldots, n$ are independent and identically distributed geometric random variables. Also, give a second argument that verifies the preceding without any need for computations.
7. **(a)** If X has a gamma distribution with parameters (t, λ), what is the distribution of cX, $c > 0$?

   (b) Show that
   $$\frac{1}{2\lambda}\chi^2_{2n}$$
   has a gamma distribution with parameters n, λ when n is a positive integer and χ^2_{2n} is a chi-squared random variable with $2n$ degrees of freedom.
8. Let X and Y be independent continuous random variables with respective hazard rate functions $\lambda_X(t)$ and $\lambda_Y(t)$, and set $W = \min(X, Y)$.

   (a) Determine the distribution function of W in terms of those of X and Y.

   (b) Show that $\lambda_W(t)$, the hazard rate function of W, is given by
   $$\lambda_W(t) = \lambda_X(t) + \lambda_Y(t)$$
9. Let $X_1, \ldots, X_n$ be independent exponential random variables having a common parameter λ. Determine the distribution of $\min(X_1, \ldots, X_n)$.
10. The lifetimes of batteries are independent exponential random variables, each having parameter λ. A flashlight needs 2 batteries to work. If one has a flashlight and a stockpile of n batteries, what is the distribution of time that the flashlight can operate?
11. Let X_1, X_2, X_3, X_4, X_5, be independent continuous random variables having a common distribution function F and density function f, and set
    $$I = P\{X_1 < X_2 < X_3 < X_4 < X_5\}$$
    (a) Show that I does not depend on F.

    HINT: Write I as a five-dimensional integral and make the change of variables $u_i = F(x_i)$, $i = 1, \ldots, 5$.

    (b) Evaluate I.

    (c) Give an intuitive explanation for your answer to (b).

12. Show that the jointly continuous (discrete) random variables $X_1, \ldots, X_n$ are independent if and only if their joint probability density (mass) function $f(x_1, \ldots, x_n)$ can be written as

$$f(x_1, \ldots, x_n) = \prod_{i=1}^{n} g_i(x_i)$$

for nonnegative functions $g_i(x), i = 1, \ldots, n$.

13. In Example 5c we computed the conditional density of a success probability for a sequence of trials when the first $n + m$ trials resulted in n successes. Would the conditional density change if we specified which n of these trials resulted in successes?

14. Suppose that X and Y are independent geometric random variables with the same parameter p.

(a) Without any computations, what do you think is the value of

$$P\{X = i | X + Y = n\}?$$

HINT: Imagine that you continually flip a coin having probability p of coming up heads. If the second head occurs on the nth flip, what is the probability mass function of the time of the first head?

(b) Verify your conjecture in part (a).

15. If X and Y are independent binomial random variables with identical parameters n and p, show analytically that the conditional distribution of X, given that $X + Y = m$, is the hypergeometric distribution. Also, give a second argument that yields the result without any computations.

HINT: Suppose that $2n$ coins are flipped. Let X denote the number of heads in the first n flips and Y the number in the second n flips. Argue that given a total of m heads, the number of heads in the first n flips has the same distribution as the number of white balls selected when a sample of size m is chosen from n white and n black balls.

16. Consider an experiment that results in one of three possible outcomes, outcome i occurring with probability $p_i, i = 1, 2, 3$. Suppose that n independent replications of this experiment are performed and let $X_i, i = 1, 2, 3$ denote the number of times that outcome i occurs. Determine the conditional probability mass function of X_1, given that $X_2 = m$.

17. Let X_1, X_2, X_3 be independent and identically distributed continuous random variables. Compute

(a) $P\{X_1 > X_2 | X_1 > X_3\}$;
(b) $P\{X_1 > X_2 | X_1 < X_3\}$;
(c) $P\{X_1 > X_2 | X_2 > X_3\}$;
(d) $P\{X_1 > X_2 | X_2 < X_3\}$.

18. Let U denote a random variable uniformly distributed over $(0, 1)$. Compute the conditional distribution of U given that

(a) $U > a$;
(b) $U < a$;

where $0 < a < 1$.

19. Suppose that W, the amount of moisture in the air on a given day, is a gamma random variable with parameters (t, β). That is, its density is $f(w) = \beta e^{-\beta w} (\beta w)^{t-1}/\Gamma(t), w > 0$. Suppose also that given that $W = w$, the number of accidents during that day—call it N—has a Poisson distribution with mean w. Show that the conditional distribution of W given that $N = n$ is the gamma distribution with parameters $(t + n, \beta + 1)$.

20. Let W be a gamma random variable with parameters (t, β), and suppose that conditional on $W = w, X_1, X_2, \ldots, X_n$ are independent exponential random variables with rate w. Show that the conditional distribution of W given that $X_1 = x_1, X_2 = x_2, \ldots, X_n = x_n$ is gamma with parameters $\left(t + n, \beta + \sum_{i=1}^{n} x_i\right)$.

21. A rectangular array of mn numbers arranged in n rows, each consisting of m columns, is said to contain a *saddlepoint* if there is a number that is both the minimum of its row and the maximum of its column. For instance, in the array

1	3	2
0	−2	6
.5	12	3

the number 1 in the first row, first column is a saddlepoint. The existence of a saddlepoint is of significance in the theory of games. Consider a rectangular array of numbers as described previously and suppose that there are two individuals—A and B—that are playing the following game: A is to choose one of the numbers $1, 2, \ldots, n$ and B one of the numbers $1, 2, \ldots, m$. These choices are announced simultaneously, and if A chose i and B chose j, then A wins from B the amount specified by the number in the ith row, jth column of the array. Now suppose that the array contains a saddlepoint—say the number in the row r and column k—call this number x_{rk}. Now if player A chooses row r, then that player can guarantee herself a win at least x_{rk} (since x_{rk} is the minimum number in the row r). On the other hand, if player B chooses column k, then he can guarantee that he will lose no more than x_{rk} (since x_{rk} is the maximum number in the column k). Hence, as A has a way of playing that guarantees her a win of x_{rk} and as B has a way of playing that guarantees he will lose no more than x_{rk}, it seems reasonable to take these two strategies as being optimal and declare that the value of the game to player A is x_{rk}.

If the nm numbers in the rectangular array described above are independently chosen from an arbitrary continuous distribution, what is the probability that the resulting array will contain a saddlepoint?

22. The random variables X and Y are said to have a bivariate normal distribution if their joint density function is given by

$$f(x, y) = \frac{1}{2\pi\sigma_x\sigma_y\sqrt{1-\rho^2}}$$
$$\times \exp\left\{-\frac{1}{2(1-\rho^2)}\left[\left(\frac{x-\mu_x}{\sigma_x}\right)^2 + \left(\frac{y-\mu_y}{\sigma_y}\right)^2 - 2\rho\frac{(x-\mu_x)(y-\mu_y)}{\sigma_x\sigma_y}\right]\right\}$$

(a) Show that the conditional density of X, given that $Y = y$, is the normal density with parameters

$$\mu_x + \rho\frac{\sigma_x}{\sigma_y}(y - \mu_y) \qquad \text{and} \qquad \sigma_x^2(1 - \rho^2)$$

(b) Show that X and Y are both normal random variables with respective parameters μ_x, σ_x^2 and μ_y, σ_y^2.

(c) Show that X and Y are independent when $\rho = 0$.

23. Suppose that $F(x)$ is a cumulative distribution function. Show that (a) $F^n(x)$ and (b) $1 - [1 - F(x)]^n$ are also cumulative distribution functions when n is a positive integer.

HINT: Let $X_1, \ldots, X_n$ be independent random variables having the common distribution function F. Define random variables Y and Z in terms of the X_i so that $P\{Y \le x\} = F^n(x)$, and $P\{Z \le x\} = 1 - [1 - F(x)]^n$.

24. Show that if n people are distributed at random along a road L miles long, then the probability that no 2 people are less than a distance of D miles apart is, when $D \le L/(n-1)$, $[1-(n-1)D/L]^n$. What if $D > L/(n-1)$?

25. Establish Equation (6.2) by differentiating Equation (6.4).

26. Show that the median of a sample of size $2n+1$ from a uniform distribution on $(0, 1)$ has a beta distribution with parameters $(n+1, n+1)$.

27. Verify Equation (6.6), which gives the joint density of $X_{(i)}$ and $X_{(j)}$.

28. Compute the density of the range of a sample of size n from a continuous distribution having density function f.

29. Let $X_{(1)} \le X_{(2)} \le \cdots \le X_{(n)}$ be the ordered values of n independent uniform $(0, 1)$ random variables. Prove that for $1 \le k \le n+1$,

$$P\{X_{(k)} - X_{(k-1)} > t\} = (1-t)^n$$

where $X_{(0)} \equiv 0$, $X_{(n+1)} \equiv t$.

30. Let $X_1, \ldots, X_n$ be a set of independent and identically distributed continuous random variables having distribution function F, and let $X_{(i)}$, $i = 1, \ldots, n$ denote their ordered values. If X, independent of the X_i, $i = 1, \ldots, n$, also has distribution F, determine

(a) $P\{X > X_{(n)}\}$; **(b)** $P\{X > X_{(1)}\}$;

(c) $P\{X_{(i)} < X < X_{(j)}\}$, $1 \le i < j \le n$.

31. Let $X_1, \ldots, X_n$ be independent and identically distributed random variables having distribution function F and density f. The quantity $M \equiv [X_{(1)} + X_{(n)}]/2$, defined to be the average of the smallest and largest value, is called the midrange. Show that its distribution function is

$$F_M(m) = n \int_{-\infty}^{m} [F(2m - x) - F(x)]^{n-1} f(x)\, dx$$

32. Let $X_1, \ldots, X_n$ be independent uniform $(0, 1)$ random variables. Let $R = X_{(n)} - X_{(1)}$ denote the range and $M = [X_{(n)} + X_{(1)}]/2$ the midrange. Compute the joint density function of R and M.

33. If X and Y are independent standard normal random variables, determine the joint density function of

$$U = X \qquad V = \frac{X}{Y}$$

Then use your result to show that X/Y has a Cauchy distribution.

SELF-TEST PROBLEMS AND EXERCISES

1. Each throw of an unfair die lands on each of the odd numbers 1, 3, 5 with probability C and on each of the even numbers with probability $2C$.

(a) Find C.

(b) Suppose that the die is tossed. Let X equal 1 if the result is an even number, and let it be 0 otherwise. Also, let Y equal 1 if the result is a number greater than three and let it be 0 otherwise. Find the joint probability mass function of X and Y. Suppose now that 12 independent tosses of the die are made.

(c) Find the probability that each of the six outcomes occurs exactly twice.

(d) Find the probability that 4 of the outcomes are either one or two, 4 are either three or four, and 4 are either five or six.

(e) Find the probability that at least 8 of the tosses land on even numbers.

2. The joint probability mass function of the random variables X, Y, Z is

$$p(1,2,3) = p(2,1,1) = p(2,2,1) = p(2,3,2) = \tfrac{1}{4}$$

Find (a) $E[XYZ]$, and (b) $E[XY + XZ + YZ]$.

3. The joint density of X and Y is given by

$$f(x,y) = C(y - x)e^{-y} \qquad -y < x < y, \qquad 0 < y < \infty$$

(a) Find C.
(b) Find the density function of X.
(c) Find the density function of Y.
(d) Find $E[X]$.
(e) Find $E[Y]$.

4. Suppose that X, Y, and Z are independent random variables that are each equally likely to be either 1 or 2. Find the probability mass function of (a) XYZ, (b) $XY + XZ + YZ$, and (c) $X^2 + YZ$.

5. Let X and Y be continuous random variables with joint density function

$$f(x) = \begin{cases} \dfrac{x}{5} + cy & 0 < x < 1, 1 < y < 5 \\ 0 & \text{otherwise} \end{cases}$$

where c is a constant.

(a) What is the value of c?
(b) Are X and Y independent?
(c) Find $P\{X + Y > 3\}$.

6. The joint density function of X and Y is

$$f(x,y) = \begin{cases} xy & 0 < x < 1, 0 < y < 2 \\ 0 & \text{otherwise} \end{cases}$$

(a) Are X and Y independent?
(b) Find the density function of X.
(c) Find the density function of Y.
(d) Find the joint distribution function.
(e) Find $E[Y]$.
(f) Find $P\{X + Y < 1\}$.

7. Consider two components and three types of shocks. A type 1 shock causes component 1 to fail, a type 2 shock causes component 2 to fail, and a type 3 shock causes both components 1 and 2 to fail. The times until shocks 1, 2, and 3 occur are independent exponential random variables with respective rates λ_1, λ_2, and λ_3. Let X_i denote the time at which component i fails, $i = 1, 2$. The random variables X_1, X_2 are said to have a joint bivariate exponential distribution. Find $P\{X_1 > s, X_2 > t\}$.

8. Consider a directory of classified advertisements that consists of m pages, where m is very large. Suppose that the number of advertisements per page varies and that your only method of finding out how many advertisements there are on a specified page is to count them. In addition, suppose that there are too many pages for it to be feasible to make a complete count of the total number of advertisements and that your objective is to choose a directory advertisement in such a way that each of them has an equal chance of being selected.

(a) If you randomly choose a page and then randomly choose an advertisement from that page, would that satisfy your objective? Why or why not?

Let $n(i)$ denote the number of advertisements on page i, $i = 1, \ldots, m$, and suppose that whereas these quantities are unknown, we can assume that they are all less than or equal to some specified value n. Consider the following algorithm for choosing an advertisement.

Step 1. Choose a page at random. Suppose it is page X. Determine $n(X)$ by counting the number of advertisements on page X.

Step 2. "Accept" page X with probability $n(X)/n$. If page X is accepted, go to step 3. Otherwise, return to step 1.

Step 3. Randomly choose one of the advertisements on page X.

Call each pass of the algorithm through step 1 an iteration. For instance, if the first randomly chosen page is rejected and the second accepted, than we would have needed 2 iterations of the algorithm to obtain an advertisement.

(b) What is the probability that a single iteration of the algorithm results in the acceptance of an advertisement on page i?

(c) What is the probability that a single iteration of the algorithm results in the acceptance of an advertisement?

(d) What is the probability that the algorithm goes through k iterations, accepting the jth advertisement on page i on the final iteration?

(e) What is the probability that the jth advertisement on page i is the advertisement obtained from the algorithm?

(f) What is the expected number of iterations taken by the algorithm?

9. The "random" parts of the algorithm in Self-Test Problem 8 can be written in terms of the generated values of a sequence of independent uniform $(0, 1)$ random variables, known as random numbers. With $[x]$ defined as the largest integer less than or equal to x, the first step can be written as follows:

Step 1. Generate a uniform $(0, 1)$ random variable U. Let $X = [mU] + 1$, and determine the value of $n(X)$.

(a) Explain why the above is equivalent to step 1 of Problem 8.

HINT: What is the probability mass function of X?

(b) Write the remaining steps of the algorithm in a similar style.

10. Let $X_1, X_2, \ldots$ be a sequence of independent uniform $(0, 1)$ random variables. For a fixed constant c, define the random variable N by

$$N = \min\{n : X_n > c\}$$

Is N independent of X_N? That is, does knowing the value of the first random variable that is greater than c affect the probability distribution of when this random variable occurs? Give an intuitive explanation for your answer.

11. The following dartboard is a square whose sides are of length 6. The three circles are all centered at the center of the board and are of radii 1, 2, and 3. Darts landing within the circle of radius 1 score 30 points, those landing outside this circle but within the circle of radius 2 are worth 20 points, and those landing outside the circle of radius 2 but within the circle of radius 3 are worth 10 points. Darts that do not land within the circle of radius

3 do not score any points. Assuming that each dart that you throw will, independent of what occurred on your previous throws, land on a point uniformly distributed in the square, find the probabilities of the following events.

(a) You score 20 on a throw of the dart.
(b) You score at least 20 on a throw of the dart.
(c) You score 0 on a throw of the dart.
(d) The expected value of your score on a throw of the dart.
(e) Both of your first two throws score at least 10.
(f) Your total score after two throws is 30.

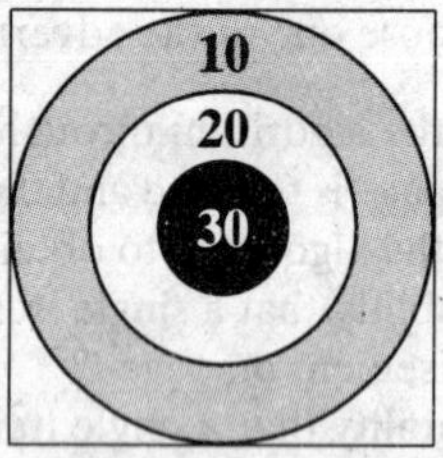

12. A model proposed for NBA basketball supposes that when two teams with roughly the same record play each other, then the number of points scored in a quarter by the home team minus the number scored by the visiting team is approximately a normal random variable with mean 1.5 and variance 6. In addition, the model supposes that the point differentials for the four quarters are independent. Assuming this model:
(a) What is the probability that the home team wins?
(b) What is the conditional probability that the home team wins given that it is behind by 5 points at halftime?
(c) What is the conditional probability that the home team wins given that it is ahead by 5 points at the end of the first quarter?

13. Let N be a geometric random variable with parameter p. Suppose that the conditional distribution of X given that $N = n$ is the gamma distribution with parameters n and λ. Find the conditional probability mass function of N given that $X = x$.

14. Let X and Y be independent uniform $(0, 1)$ random variables.
(a) Find the joint density of $U = X, V = X + Y$.
(b) Use the result obtained in part (a) to compute the density function of V.

15. You and 3 others are to place bids for an object, with the high bid winning. If you win, you plan to immediately sell the object for 10 thousand dollars. How much should you bid to maximize your expected profit if you believe that the bids of the others can be regarded as being independent and uniformly distributed between 7 and 11 thousand dollars?

16. Find the probability that $X_1, X_2, \ldots, X_n$ is a permutation of $1, 2, \ldots, n$, when $X_1, X_2, \ldots, X_n$ are independent, and
(a) each is equally likely to be any of the values $1, \ldots, n$;
(b) each has the probability mass function $P\{X_i = j\} = p_j, j = 1, \ldots, n$.

17. Let $X_1, \ldots, X_n$ and $Y_1, \ldots, Y_n$ be independent random vectors, with each vector being a random ordering of k ones and $n - k$ zeroes. That is, their joint probability mass functions are

$$P\{X_1 = i_1, \ldots, X_n = i_n\} = P\{Y_1 = i_1, \ldots, Y_n = i_n\} = \frac{1}{\binom{n}{k}}, \quad i_j = 0, 1, \quad \sum_{j=1}^{n} i_j = k$$

Let

$$N = \sum_{i=1}^{n} |X_i - Y_i|$$

denote the number of coordinates at which the two vectors have different values. Also, let M denote the number of values of i for which $X_i = 1, Y_i = 0$.

(a) Relate N and M.
(b) What is the distribution of M?
(c) Find $E[N]$.
(d) Find $\text{Var}(N)$.

***18.** Let $Z_1, Z_2, \ldots, Z_n$ be independent standard normal random variables, and let

$$S_j = \sum_{i=1}^{j} Z_i$$

(a) What is the conditional distribution of S_n given that $S_k = y$, for $k = 1, \ldots, n$?
(b) Show that, for $1 \le k \le n$, the conditional distribution of S_k given that $S_n = x$ is normal with mean xk/n and variance $k(n - k)/n$.

CHAPTER 7

Properties of Expectation

7.1 INTRODUCTION

In this chapter we develop and exploit additional properties of expected values. To begin, recall that the expected value of the random variable X is defined by

$$E[X] = \sum_x xp(x)$$

where X is a discrete random variable with probability mass function $p(x)$, and by

$$E[X] = \int_{-\infty}^{\infty} xf(x)\,dx$$

when X is a continuous random variable with probability density function $f(x)$.

Since $E[X]$ is a weighted average of the possible values of X, it follows that if X must lie between a and b, then so must its expected value. That is, if

$$P\{a \le X \le b\} = 1$$

then

$$a \le E[X] \le b$$

To verify the preceding, suppose that X is a discrete random variable for which $P\{a \le X \le b\} = 1$. Since this implies that $p(x) = 0$ for all x outside of the interval $[a, b]$, we see that

$$\begin{aligned}
E[X] &= \sum_{x:p(x)>0} xp(x) \\
&\ge \sum_{x:p(x)>0} ap(x) \\
&= a \sum_{x:p(x)>0} p(x) \\
&= a
\end{aligned}$$

In the same manner it can be shown that $E[X] \le b$, so the result follows for discrete random variables. As the proof in the continuous case is similar, the result follows.

7.2 EXPECTATION OF SUMS OF RANDOM VARIABLES

For a two-dimensional analog of Propositions 5.1 of Chapter 4 and 2.1 of Chapter 5, which give the computational formulas for the expected value of a function of a random variable, suppose that X and Y are random variables and g is a function of two variables. Then we have the following result.

Proposition 2.1

If X and Y have a joint probability mass function $p(x, y)$, then

$$E[g(X, Y)] = \sum_y \sum_x g(x, y)p(x, y)$$

If X and Y have a joint probability density function $f(x, y)$, then

$$E[g(X, Y)] = \int_{-\infty}^{\infty} \int_{-\infty}^{\infty} g(x, y)f(x, y)\, dx\, dy$$

Example 2a. An accident occurs at a point X that is uniformly distributed on a road of length L. At the time of the accident an ambulance is at a location Y that is also uniformly distributed on the road. Assuming that X and Y are independent, find the expected distance between the ambulance and the point of the accident.

Solution We need to compute $E[|X - Y|]$. Since the joint density function of X and Y is

$$f(x, y) = \frac{1}{L^2}, \qquad 0 < x < L, \quad 0 < y < L$$

we obtain from Proposition 2.1 that

$$E[|X - Y|] = \frac{1}{L^2} \int_0^L \int_0^L |x - y|\, dy\, dx$$

Now,

$$\begin{aligned}\int_0^L |x - y|\, dy &= \int_0^x (x - y)\, dy + \int_x^L (y - x)\, dy \\ &= \frac{x^2}{2} + \frac{L^2}{2} - \frac{x^2}{2} - x(L - x) \\ &= \frac{L^2}{2} + x^2 - xL\end{aligned}$$

Therefore,

$$E[|X - Y|] = \frac{1}{L^2}\int_0^L \left(\frac{L^2}{2} + x^2 - xL\right) dx$$
$$= \frac{L}{3}$$
■

For an important application of Proposition 2.1, suppose that $E[X]$ and $E[Y]$ are both finite and let $g(X,Y) = X + Y$. Then, in the continuous case,

$$\begin{aligned}
E[X + Y] &= \int_{-\infty}^{\infty}\int_{-\infty}^{\infty} (x + y)f(x,y)\,dx\,dy \\
&= \int_{-\infty}^{\infty}\int_{-\infty}^{\infty} xf(x,y)\,dy\,dx + \int_{-\infty}^{\infty}\int_{-\infty}^{\infty} yf(x,y)\,dx\,dy \\
&= \int_{-\infty}^{\infty} xf_X(x)\,dx + \int_{-\infty}^{\infty} yf_Y(y)\,dy \\
&= E[X] + E[Y]
\end{aligned}$$

The same result holds in general; thus, whenever $E[X]$ and $E[Y]$ are finite.

$$E[X + Y] = E[X] + E[Y] \tag{2.1}$$

Example 2b. Suppose that for random variables X and Y,

$$X \geq Y$$

That is, for any outcome of the probability experiment, the value of the random variable X is greater than or equal the value of the random variable Y. Since the preceding is equivalent to the inequality $X - Y \geq 0$, it follows that $E[X - Y] \geq 0$, or, equivalently,

$$E[X] \geq E[Y]$$
■

Using Equation (2.1), a simple induction proof shows that if $E[X_i]$ is finite for all $i = 1, \ldots, n$, then

$$E[X_1 + \cdots + X_n] = E[X_1] + \cdots + E[X_n] \tag{2.2}$$

Equation (2.2) is an extremely useful formula whose utility will now be illustrated by a series of examples.

Example 2c. ***The sample mean.*** Let $X_1, \ldots, X_n$ be independent and identically distributed random variables having distribution function F and expected value μ. Such a sequence of random variables is said to constitute a sample from the distribution F. The quantity $\bar{X}$, defined by

$$\bar{X} = \sum_{i=1}^{n} \frac{X_i}{n}$$

is called the *sample mean*. Compute $E[\bar{X}]$.

Solution

$$\begin{aligned} E[\bar{X}] &= E\left[\sum_{i=1}^{n} \frac{X_i}{n}\right] \\ &= \frac{1}{n} E\left[\sum_{i=1}^{n} X_i\right] \\ &= \frac{1}{n} \sum_{i=1}^{n} E[X_i] \\ &= \mu \quad \text{since } E[X_i] \equiv \mu \end{aligned}$$

That is, the expected value of the sample mean is μ, the mean of the distribution. When the distribution mean μ is unknown, the sample mean is often used in statistics to estimate it. ■

Example 2d. ***Boole's inequality.*** Let $A_1, \ldots, A_n$ denote events and define the indicator variables $X_i, i = 1, \ldots, n$ by

$$X_i = \begin{cases} 1 & \text{if } A_i \text{ occurs} \\ 0 & \text{otherwise} \end{cases}$$

Let

$$X = \sum_{i=1}^{n} X_i$$

so X denotes the number of the events A_i that occur. Finally, let

$$Y = \begin{cases} 1 & \text{if } X \geq 1 \\ 0 & \text{otherwise} \end{cases}$$

so Y is equal to 1 if at least one of the A_i occurs and is 0 otherwise. Now, it is immediate that

$$X \geq Y$$

so

$$E[X] \geq E[Y]$$

But since

$$E[X] = \sum_{i=1}^{n} E[X_i] = \sum_{i=1}^{n} P(A_i)$$

and

$$E[Y] = P\{\text{at least one of the } A_i \text{ occur}\} = P\left(\bigcup_{i=1}^{n} A_i\right)$$

we obtain Boole's inequality, namely that

$$P\left(\bigcup_{i=1}^{n} A_i\right) \le \sum_{i=1}^{n} P(A_i)$$ ■

The next three examples show how Equation (2.2) can be used to calculate the expected value of binomial, negative binomial, and hypergeometric random variables. These derivations should be compared with those presented in Chapter 4.

Example 2e. ***Expectation of a binomial random variable.*** Let X be a binomial random variable with parameters n and p. Recalling that such a random variable represents the number of successes in n independent trials when each trial has probability p of being a success, we have that

$$X = X_1 + X_2 + \cdots + X_n$$

where

$$X_i = \begin{cases} 1 & \text{if the } i\text{th trial is a success} \\ 0 & \text{if the } i\text{th trial is a failure} \end{cases}$$

Hence, X_i is a Bernoulli random variable having expectation $E[X_i] = 1(p) + 0(1 - p)$. Thus

$$E[X] = E[X_1] + E[X_2] + \cdots + E[X_n] = np$$ ■

Example 2f. ***Mean of a negative binomial random variable.*** If independent trials, having a constant probability p of being successes, are performed, determine the expected number of trials required to amass a total of r successes.

Solution If X denotes the number of trials needed to amass a total of r successes, then X is a negative binomial random variable. It can be represented by

$$X = X_1 + X_2 + \cdots + X_r$$

where X_1 is the number of trials required to obtain the first success, X_2 the number of additional trials until the second success is obtained, X_3 the number of additional trials until the third success is obtained, and so on. That is, X_i represents the number of additional trials required, after the $(i - 1)$st success, until a total of i successes are amassed. A little thought reveals that each of the random variables X_i is a geometric random variable with parameter p. Hence, from the results of Example 9b of Chapter 4, $E[X_i] = 1/p, i = 1, 2, \ldots, r$; and thus

$$E[X] = E[X_1] + \cdots + E[X_r] = \frac{r}{p}$$ ■

Example 2g. ***Mean of a hypergeometric random variable.*** If n balls are randomly selected from an urn containing N balls of which m are white, find the expected number of white balls selected.

Solution Let X denote the number of white balls selected, and represent X as

$$X = X_1 + \cdots + X_m$$

where

$$X_i = \begin{cases} 1 & \text{if the } i\text{th white ball is selected} \\ 0 & \text{otherwise} \end{cases}$$

Now,

$$\begin{aligned} E[X_i] &= P\{X_i = 1\} \\ &= P\{i\text{th white ball is selected}\} \\ &= \frac{\binom{1}{1}\binom{N-1}{n-1}}{\binom{N}{n}} \\ &= \frac{n}{N} \end{aligned}$$

Hence

$$E[X] = E[X_1] + \cdots + E[X_m] = \frac{mn}{N}$$

We could also have obtained the preceding result by using the alternative representation

$$X = Y_1 + \cdots + Y_n$$

where

$$Y_i = \begin{cases} 1 & \text{if the } i\text{th ball selected is white} \\ 0 & \text{otherwise} \end{cases}$$

Since the ith ball selected is equally likely to be any of the N balls, it follows that

$$E[Y_i] = \frac{m}{N}$$

so

$$E[X] = E[Y_1] + \cdots + E[Y_n] = \frac{nm}{N}$$ ■

Example 2h. ***Expected number of matches.*** A group of N people throw their hats into the center of a room. The hats are mixed up, and each person randomly selects one. Find the expected number of people that select their own hat.

Solution Letting X denote the number of matches, we can most easily compute $E[X]$ by writing

$$X = X_1 + X_2 + \cdots + X_N$$

where

$$X_i = \begin{cases} 1 & \text{if the } i\text{th person selects his own hat} \\ 0 & \text{otherwise} \end{cases}$$

Since, for each i, the ith person is equally likely to select any of the N hats,

$$E[X_i] = P\{X_i = 1\} = \frac{1}{N}$$

Thus

$$E[X] = E[X_1] + \cdots + E[X_N] = \left(\frac{1}{N}\right)N = 1$$

Hence, on the average, exactly one person selects his own hat. ∎

Example 2i. The following problem was posed and solved in the eighteenth century by Daniel Bernoulli. Suppose that a jar contains $2N$ cards, two of them marked 1, two marked 2, two marked 3, and so on. Draw out m cards at random. What is the expected number of pairs that still remain in the jar? (Interestingly enough, Bernoulli proposed the above as a possible probabilistic model for determining the number of marriages that remain intact when there is a total of m deaths among the N married couples.)

Solution Define for $i = 1, 2, \ldots, N$,

$$X_i = \begin{cases} 1 & \text{if the } i\text{th pair remains in the jar} \\ 0 & \text{otherwise} \end{cases}$$

Now,

$$\begin{aligned} E[X_i] &= P\{X_i = 1\} \\ &= \frac{\binom{2N-2}{m}}{\binom{2N}{m}} \\ &= \frac{\dfrac{(2N-2)!}{m!\,(2N-2-m)!}}{\dfrac{(2N)!}{m!\,(2N-m)!}} \\ &= \frac{(2N-m)(2N-m-1)}{(2N)(2N-1)} \end{aligned}$$

Hence the desired result is

$$E[X_1 + X_2 + \cdots + X_N] = E[X_1] + \cdots + E[X_N]$$
$$= \frac{(2N - m)(2N - m - 1)}{2(2N - 1)}$$ ■

Example 2j. ***Coupon-collecting problems.*** Suppose that there are N different types of coupons and each time one obtains a coupon it is equally likely to be any one of the N types.

(a) Find the expected number of different types of coupons that are contained in a set of n coupons.

(b) Find the expected number of coupons one need amass before obtaining a complete set of at least one of each type.

Solution (a) Let X denote the number of different types of coupons in the set of n coupons. We compute $E[X]$ by using the representation

$$X = X_1 + \cdots + X_N$$

where

$$X_i = \begin{cases} 1 & \text{if at least one type } i \text{ coupon is contained in the set of } n \\ 0 & \text{otherwise} \end{cases}$$

Now,

$$\begin{aligned} E[X_i] &= P\{X_i = 1\} \\ &= 1 - P\{\text{no type } i \text{ coupons are contained in the set of } n\} \\ &= 1 - \left(\frac{N-1}{N}\right)^n \end{aligned}$$

Hence

$$E[X] = E[X_1] + \cdots + E[X_N] = N\left[1 - \left(\frac{N-1}{N}\right)^n\right]$$

(b) Let Y denote the number of coupons collected before a complete set is attained. We compute $E[Y]$ by using the same technique as we used in computing the mean of a negative binomial random variable (Example 2f). That is, define $Y_i, i = 0, 1, \ldots, N - 1$ to be the number of additional coupons that need be obtained after i distinct types have been collected in order to obtain another distinct type, and note that

$$Y = Y_0 + Y_1 + \cdots + Y_{N-1}$$

When i distinct types of coupons have already been collected, it follows that a new coupon obtained will be of a distinct type with probability $(N - i)/N$. Therefore,

$$P\{Y_i = k\} = \frac{N-i}{N}\left(\frac{i}{N}\right)^{k-1} \qquad k \geq 1$$

or, in other words, Y_i is a geometric random variable with parameter $(N - i)/N$.

Hence

$$E[Y_i] = \frac{N}{N - i}$$

implying that

$$E[Y] = 1 + \frac{N}{N-1} + \frac{N}{N-2} + \cdots + \frac{N}{1}$$
$$= N\left[1 + \cdots + \frac{1}{N-1} + \frac{1}{N}\right]$$ ■

Example 2k. Ten hunters are waiting for ducks to fly by. When a flock of ducks flies overhead, the hunters fire at the same time, but each chooses his target at random, independently of the others. If each hunter independently hits his target with probability p, compute the expected number of ducks that escape unhurt when a flock of size 10 flies overhead.

Solution Let X_i equal 1 if the ith duck escapes unhurt and 0 otherwise, $i = 1, 2, \ldots, 10$. The expected number of ducks to escape can be expressed as

$$E[X_1 + \cdots + X_{10}] = E[X_1] + \cdots + E[X_{10}]$$

To compute $E[X_i] = P\{X_i = 1\}$, we note that each of the hunters will, independently, hit the ith duck with probability $p/10$, so

$$P\{X_i = 1\} = \left(1 - \frac{p}{10}\right)^{10}$$

Hence

$$E[X] = 10\left(1 - \frac{p}{10}\right)^{10}$$ ■

Example 2l. ***Expected number of runs.*** Suppose that a sequence of n 1's and m 0's is randomly permuted so that each of the $(n + m)!/(n!\, m!)$ possible arrangements is equally likely. Any consecutive string of 1's is said to constitute a run of 1's—for instance, if $n = 6$, $m = 4$, and the ordering is 1, 1, 1, 0, 1, 1, 0, 0, 1, 0, then there are 3 runs of 1's—and we are interested in computing the mean number of such runs. To compute this quantity, let

$$I_i = \begin{cases} 1 & \text{if a run of 1's starts at the } i\text{th position} \\ 0 & \text{otherwise} \end{cases}$$

Therefore, $R(1)$, the number of runs of 1, can be expressed as

$$R(1) = \sum_{i=1}^{n+m} I_i$$

and thus

$$E[R(1)] = \sum_{i=1}^{n+m} E[I_i]$$

Now,

$$E[I_1] = P\{\text{“1” in position 1}\}$$
$$= \frac{n}{n + m}$$

and for $1 < i \le n + m$,

$$E[I_i] = P\{\text{“0” in position } i - 1, \text{“1” in position } i\}$$
$$= \frac{m}{n + m} \frac{n}{n + m - 1}$$

Hence

$$E[R(1)] = \frac{n}{n + m} + (n + m - 1)\frac{nm}{(n + m)(n + m - 1)}$$

Similarly, $E[R(0)]$, the expected number of runs of 0's, is

$$E[R(0)] = \frac{m}{n + m} + \frac{nm}{n + m}$$

and the expected number of runs of either type is

$$E[R(1) + R(0)] = 1 + \frac{2nm}{n + m}$$ ■

Example 2m. Consider a random ordering of $n + m$ items, of which n are special and the other m are ordinary, and suppose that all $(n + m)!$ possible orderings are equally likely. Find $E[X]$, where X is the position of the first special item to appear in the ordering.

Solution Because the first $k - 1$ items are equally likely to be any of the $\binom{n+m}{k-1}$ subsets of $k - 1$ items, it follows that the first $k - 1$ elements in the order will be ordinary with probability $\dfrac{\binom{m}{k-1}}{\binom{n+m}{k-1}}$. Therefore, because X will equal k if the first $k - 1$ elements in the order are ordinary and the kth element is special, we obtain that

$$P\{X = k\} = \frac{\binom{m}{k-1}}{\binom{n+m}{k-1}} \frac{n}{n + m - k + 1}, \qquad k \ge 1 \tag{2.3}$$

However, rather than directly using the preceding probabilities, let us determine $E[X]$ by first expressing X as a sum. To do so, imagine that the ordinary items have names—say, $o_1, \ldots, o_m$. Now, for $i = 1, \ldots, m$, let

$$X_i = \begin{cases} 1, & \text{if } o_i \text{ appears before all } n \text{ special items} \\ 0, & \text{otherwise} \end{cases}$$

Because the position of the first special item in the ordering is one more than the number of ordinary items that precede all the special ones, we have that

$$X = 1 + \sum_{i=1}^{m} X_i$$

Therefore,

$$E[X] = 1 + \sum_{i=1}^{m} E[X_i]$$

However, X_i will equal 1 if item o_i is in front of all n special items; because each of these $n + 1$ items, consisting of the n special items and item o_i, is equally likely to be the first to appear in the ordering, it follows that

$$E[X_i] = P\{X_i = 1\} = \frac{1}{n + 1}$$

Consequently,

$$E[X] = 1 + \frac{m}{n + 1}$$

Whereas the preceding result shows that the expected number of cards of a well-shuffled deck that would need to be turned up until a spade appears is $1 + \frac{39}{14} = 3.786$, and the number of cards that would have to be turned over until an ace appears is $1 + \frac{48}{5} = 10.6$, it has other less straightforward applications.

For a computer science application, suppose that m records, denoted as $r_1, \ldots, r_m$, are to be placed in $n + m$ locations, with at most one record in each location. When m is much smaller than n, an efficient way of allocating the records to locations, so that the records are quickly retrievable when needed, is to use a *hashing function* h that maps record values into locations. Good hashing functions have the property that the successive mapped values of a sequence of records appear to be a sequence of independent random variables that are each equally likely to take on any of the $n + m$ possible location values. Use of the hashing function h on the m records would place record r_1 in location $h(r_1)$; record r_2 would be placed in location $h(r_2)$ provided that location was empty [that is, provided that $h(r_2) \neq h(r_1)$] and would be placed in location $h(r_2) + 1$ if it were occupied. In general, once records $r_1, \ldots, r_k$ were placed, record r_{k+1} would be placed in location $h(r_{k+1}) + j$, where j is the smallest nonnegative integer such that location $h(r_{k+1}) + j$ is empty. In other words, record r_{k+1} is sent to location $h(r_{k+1})$; if that location is filled, it continually moves to the next higher location until it finds an empty location. (All movements are mod $n + m$, so a record that finds location $n + m$ filled would then move to location 1.)

A collision is said to occur each time a record moves to a location that is already filled. Given that the m records have been placed without any collisions occurring [because all the locations $h(r_1), \ldots, h(r_m)$ are distinct], let us find the expected number of collisions caused by placing a new record, under the assumption that the hash value of a record is equally likely to be any of the $n + m$ locations.

To begin, note that because the m records have been placed without any collisions resulting, it follows that they were all initally put into different locations. Because all locations are equally likely, this implies that each of the $\binom{n+m}{m}$ subsets of m locations is equally likely to be the set of locations of these records. Now consider the new record. Its hashing value is equally likely to be any of the $n + m$ values, and the number of collisions that result from its placement is equal to the number of filled locations, starting at its hashing value and continually moving to the right, that need be searched until an unfilled location is found. But by regarding the n unfilled locations as being special and the m filled locations as being ordinary, we see that the number of collisions, call it N, has the same distribution as the number of ordinary items that precede all the special items in a random ordering of n special and m ordinary items. Therefore, $X \equiv N + 1$ has the distribution specified by Equation (2.3), implying that

$$E[N] = E[X] - 1 = \frac{m}{n+1} \qquad \blacksquare$$

Example 2n. ***A random walk in the plane.*** Consider a particle initially located at a given point in the plane and suppose that it undergoes a sequence of steps of fixed length but in a completely random direction. Specifically, suppose that the new position after each step is one unit of distance from the previous position and at an angle of orientation from the previous position that is uniformly distributed over $(0, 2\pi)$ (see Figure 7.1). Compute the expected square of the distance from the origin after n steps.

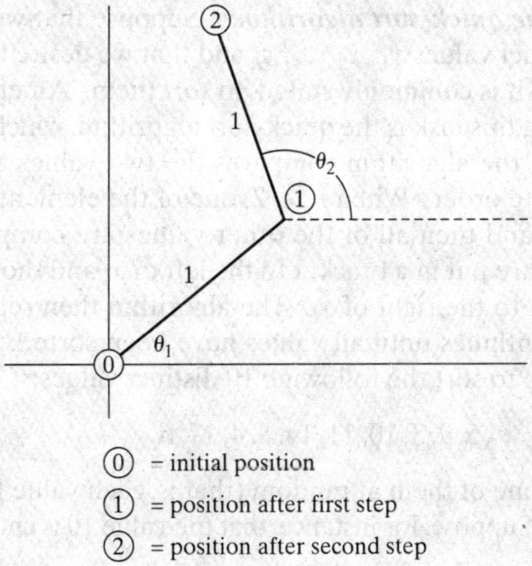

Figure 7.1

Solution Letting (X_i, Y_i) denote the change in position at the ith step, $i = 1, \ldots, n$, in rectangular coordinates, we have that

$$X_i = \cos\theta_i$$

$$Y_i = \sin\theta_i$$

where $\theta_i, i = 1, \ldots, n$ are by assumption, independent, uniform $(0, 2\pi)$ random variables. As the position after n steps has rectangular coordinates $\left(\sum_{i=1}^{n} X_i, \sum_{i=1}^{n} Y_i\right)$, we see that D^2, the square of the distance from the origin, is given by

$$\begin{aligned} D^2 &= \left(\sum_{i=1}^{n} X_i\right)^2 + \left(\sum_{i=1}^{n} Y_i\right)^2 \\ &= \sum_{i=1}^{n} (X_i^2 + Y_i^2) + \sum\sum_{i \neq j} (X_i X_j + Y_i Y_j) \\ &= n + \sum\sum_{i \neq j} (\cos\theta_i \cos\theta_j + \sin\theta_i \sin\theta_j) \end{aligned}$$

where $\cos^2\theta_i + \sin^2\theta_i = 1$. Taking expectations and using the independence of θ_i and θ_j when $i \neq j$ and the fact that

$$2\pi E[\cos\theta_i] = \int_0^{2\pi} \cos u \, du = \sin 2\pi - \sin 0 = 0$$

$$2\pi E[\sin\theta_i] = \int_0^{2\pi} \sin u \, du = \cos 0 - \cos 2\pi = 0$$

we arrive at

$$E[D^2] = n$$ ■

Example 2o. ***Analyzing the quick-sort algorithm.*** Suppose that we are presented with a set of n distinct values $x_1, x_2, \ldots, x_n$ and that we desire to put them in increasing order, or as it is commonly stated, to *sort* them. An efficient procedure for accomplishing this task is the quick-sort algorithm, which is defined as follows. When $n = 2$, the algorithm compares the two values and then puts them in the appropriate order. When $n > 2$, one of the elements is randomly chosen—say it is x_i—and then all of the other values are compared with x_i. Those smaller than x_i are put in a bracket to the left of x_i and those larger than x_i are put in a bracket to the right of x_i. The algorithm then repeats itself on these brackets and continues until all values have been sorted. For instance, suppose that we desire to sort the following 10 distinct values:

$$5, 9, 3, 10, 11, 14, 8, 4, 17, 6$$

We start by choosing one of them at random (that is, each value has probability $\frac{1}{10}$ of being chosen). Suppose, for instance, that the value 10 is chosen. We then compare each of the others to this value, putting in a bracket to the left of 10 all those values smaller than 10 and to the right all those larger. This gives

$$\{5, 9, 3, 8, 4, 6\}, 10, \{11, 14, 17\}$$

We now focus on a bracketed set that contains more than a single value—say the one on the left of the preceding—and randomly choose one of its values—say that 6 is chosen. Comparing each of the values in the bracket with 6 and putting the smaller ones in a new bracket to the left of 6 and the larger ones in a bracket to the right of 6 gives

$$\{5, 3, 4\}, 6, \{9, 8\}, 10, \{11, 14, 17\}$$

If we now consider the leftmost bracket, and randomly choose the value 4 for comparison then the next iteration yields

$$\{3\}, 4, \{5\}, 6, \{9, 8\}, 10, \{11, 14, 17\}$$

This continues until there is no bracketed set that contains more than a single value.

If we let X denote the number of comparisons that it takes the quick-sort algorithm to sort n distinct numbers, then $E[X]$ is a measure of the effectiveness of this algorithm. To compute $E[X]$, we will first express X as a sum of other random variables as follows. To begin, give the following names to the values that are to be sorted: Let 1 stand for the smallest, let 2 stand for the next smallest, and so on. Then, for $1 \le i < j \le n$, let $I(i, j)$ equal 1 if i and j are ever directly compared, and let it equal 0 otherwise. With this definition, it follows that

$$X = \sum_{i=1}^{n-1} \sum_{j=i+1}^{n} I(i, j)$$

implying that

$$\begin{aligned} E[X] &= E\left[\sum_{i=1}^{n-1} \sum_{j=i+1}^{n} I(i, j)\right] \\ &= \sum_{i=1}^{n-1} \sum_{j=i+1}^{n} E[I(i, j)] \\ &= \sum_{i=1}^{n-1} \sum_{j=i+1}^{n} P\{i \text{ and } j \text{ are ever compared}\} \end{aligned}$$

To determine the probability that i and j are ever compared, note that the values $i, i + 1, \ldots, j - 1, j$ will initially be in the same bracket (since all values are initially in the same bracket) and will remain in the same bracket if the number chosen for the first comparison is not between i and j. For instance, if the comparison number is larger than j, then all the values $i, i + 1, \ldots, j - 1, j$ will go in a bracket to the left of the comparison number, and if it is smaller than i, then they will all go in a bracket to the right. Thus all the values i, $i + 1, \ldots, j - 1, j$ will remain in the same bracket until the first time that one of them is chosen as a comparison value. At that point all the other values between i and j will be compared with this comparison value. Now, if this comparison value is neither i nor j, then upon comparison with it, i will go into a left bracket and j into a right bracket, and thus i and j will be in different brackets and so will never be compared. On the other hand, if the comparison value of the set $i, i+1, \ldots, j-1, j$ is either i or j, then there will be a direct comparison between i and j. Now, given that the comparison value is one of the values between i and j, it follows that it is equally likely to be any of these $j-i+1$ values, and thus the probability that it is either i or j is $2/(j-i+1)$. Therefore, we can conclude that

$$P\{i \text{ and } j \text{ are ever compared}\} = \frac{2}{j-i+1}$$

and

$$E[X] = \sum_{i=1}^{n-1} \sum_{j=i+1}^{n} \frac{2}{j-i+1}$$

To obtain a rough approximation of the magnitude of $E[X]$ when n is large, we can approximate the sums by integrals. Now

$$\begin{aligned}
\sum_{j=i+1}^{n} \frac{2}{j-i+1} &\approx \int_{i+1}^{n} \frac{2}{x-i+1}\,dx \\
&= 2\log(x-i+1)\Big|_{i+1}^{n} \\
&= 2\log(n-i+1) - 2\log(2) \\
&\approx 2\log(n-i+1)
\end{aligned}$$

Thus

$$\begin{aligned}
E[X] &\approx \sum_{i=1}^{n-1} 2\log(n-i+1) \\
&\approx 2\int_{1}^{n-1} \log(n-x+1)\,dx \\
&= 2\int_{2}^{n} \log(y)\,dy \\
&= 2(y\log(y) - y)|_2^n \\
&\approx 2n\log(n)
\end{aligned}$$

Thus we see that when n is large, the quicksort algorithm requires, on average, approximately $2n\log(n)$ comparisons to sort n distinct values. ■

Example 2p. ***The probability of a union of events.*** Let $A_1, \ldots A_n$ denote events and define the indicator variables $X_i, i = 1, \ldots, n$, by

$$X_i = \begin{cases} 1 & \text{if } A_i \text{ occurs} \\ 0 & \text{otherwise} \end{cases}$$

Now, note that

$$1 - \prod_{i=1}^{n} (1 - X_i) = \begin{cases} 1 & \text{if } \cup A_i \text{ occurs} \\ 0 & \text{otherwise} \end{cases}$$

Hence

$$E\left[1 - \prod_{i=1}^{n} (1 - X_i)\right] = P\left(\bigcup_{i=1}^{n} A_i\right)$$

Expanding the left side of the preceding yields that

$$P\left(\bigcup_{i=1}^{n} A_i\right) = E\left[\sum_{i=1}^{n} X_i - \sum\sum_{i<j} X_i X_j + \sum\sum\sum_{i<j<k} X_i X_j X_k \right.$$
$$\left. - \cdots + (-1)^{n+1} X_1 \cdots X_n \right] \tag{2.4}$$

However, as

$$X_{i_1} X_{i_2} \cdots X_{i_k} = \begin{cases} 1 & \text{if } A_{i_1} A_{i_2} \cdots A_{i_k} \text{ occurs} \\ 0 & \text{otherwise} \end{cases}$$

we see that

$$E[X_{i_1} \cdots X_{i_k}] = P(A_{i_1} \cdots A_{i_k})$$

and thus (2.4) is just a statement of the well-known formula for the union of events

$$P(\cup A_i) = \sum_i P(A_i) - \sum\sum_{i<j} P(A_i A_j) + \sum\sum\sum_{i<j<k} P(A_i A_j A_k)$$
$$- \cdots + (-1)^{n+1} P(A_1 \cdots A_n)$$ ■

When one is dealing with an infinite collection of random variables X_i, $i \geq 1$, each having a finite expectation, it is not necessarily true that

$$E\left[\sum_{i=1}^{\infty} X_i\right] = \sum_{i=1}^{\infty} E[X_i] \tag{2.5}$$

To determine when (2.5) is valid, we note that $\sum_{i=1}^{\infty} X_i = \lim_{n\to\infty} \sum_{i=1}^{n} X_i$ and thus

$$\begin{aligned} E\left[\sum_{i=1}^{\infty} X_i\right] &= E\left[\lim_{n\to\infty} \sum_{i=1}^{n} X_i\right] \\ &\stackrel{?}{=} \lim_{n\to\infty} E\left[\sum_{i=1}^{n} X_i\right] \\ &= \lim_{n\to\infty} \sum_{i=1}^{n} E[X_i] \\ &= \sum_{i=1}^{\infty} E[X_i] \end{aligned} \tag{2.6}$$

Hence Equation (2.5) is valid whenever we are justified in interchanging the expectation and limit operations in Equation (2.6). Although, in general, this interchange is not justified, it can be shown to be valid in two important special cases:

1. The X_i are all nonnegative random variables (that is, $P\{X_i \geq 0\} = 1$ for all i).

2. $\sum_{i=1}^{\infty} E[|X_i|] < \infty$.

Example 2q. Consider any nonnegative, integer-valued random variable X. If for each $i \geq 1$, we define

$$X_i = \begin{cases} 1 & \text{if } X \geq i \\ 0 & \text{if } X < i \end{cases}$$

then

$$\begin{aligned} \sum_{i=1}^{\infty} X_i &= \sum_{i=1}^{X} X_i + \sum_{i=X+1}^{\infty} X_i \\ &= \sum_{i=1}^{X} 1 + \sum_{i=X+1}^{\infty} 0 \\ &= X \end{aligned}$$

Hence, since the X_i are all nonnegative,

$$\begin{aligned} E[X] &= \sum_{i=1}^{\infty} E(X_i) \\ &= \sum_{i=1}^{\infty} P\{X \geq i\} \end{aligned} \tag{2.7}$$

a useful identity. ■

Example 2r. Suppose that n elements—call them $1, 2, \ldots, n$—must be stored in a computer in the form of an ordered list. Each unit of time a request will be made for one of these elements—i being requested, independently of the past, with probability $P(i)$, $i \geq 1$, $\sum_i P(i) = 1$. Assuming these probabilities are known, what ordering minimizes the average position on the line of the element requested?

Solution Suppose that the elements are numbered so that $P(1) \geq P(2) \geq \cdots \geq P(n)$. To show that $1, 2, \ldots, n$ is the optimal ordering, let X denote the position of the requested element. Now under any ordering—say, $O = i_1, i_2, \ldots, i_n$,

$$\begin{aligned} P_O\{X \geq k\} &= \sum_{j=k}^{n} P(i_j) \\ &\geq \sum_{j=k}^{n} P(j) \\ &= P_{1,2,\ldots,n}\{X \geq k\} \end{aligned}$$

Summing over k and using Equation (2.7) yields

$$E_O[X] \geq E_{1,2,\ldots,n}[X]$$

thus showing that ordering the elements in decreasing order of their request probabilities minimizes the expected position of the element requested. ■

Example 2s. In Example 2m we showed that if X is the position of the first special item in a randomly ordering of n special and m ordinary items, then

$$E[X] = \frac{n + m + 1}{n + 1}$$

Because X is greater than or equal to k if the first $k - 1$ items are ordinary,

$$P\{X \geq k\} = \frac{\binom{m}{k-1}}{\binom{n+m}{k-1}}$$

Therefore, Equation (2.7) yields the combinatorial identity

$$\sum_{k=1}^{n} \frac{\binom{m}{k-1}}{\binom{n+m}{k-1}} = \frac{n + m + 1}{n + 1}$$ ■

*7.2.1 Obtaining Bounds from Expectations via the Probabilistic Method

The probabilistic method is a technique for analyzing the properties of the elements of a set by introducing probabilities on the set, and then studying an element chosen according to those probabilities. This technique was previously seen in Example 4k of Chapter 3, where it was used to show the existence of a set element that satisfied a certain property. In this subsection we show how it can sometimes be used to bound complicated functions.

Let f be a function on the elements of a finite set $\mathcal{S}$, and suppose that we are interested in

$$m = \max_{s \in \mathcal{S}} f(s)$$

A useful lower bound for m can often be obtained by letting S be a random element of $\mathcal{S}$ for which the expected value of $f(S)$ is computable and then noting that $m \geq f(S)$ implies that

$$m \geq E[f(S)]$$

with strict inequality if $f(S)$ is not a constant random variable. That is, $E[f(S)]$ is a lower bound on the maximimum value.

Example 2t. ***The Maximum Number of Hamiltonian Paths in a Tournament.*** A round-robin tournament of $n > 2$ contestants is one in which each of the $\binom{n}{2}$ pair of contestants play each other exactly once. Suppose that the players are numbered $1, 2, 3, \ldots, n$. The permutation $i_1, i_2, \ldots, i_n$ is said to be a *Hamiltonian path* if i_1 beats i_2, i_2 beats $i_3, \ldots$, and i_{n-1} beats i_n. A problem of some interest is to determine the largest possible number of Hamiltonian paths.

As an illustration, suppose that there are 3 players. If one of them wins twice, then there is a single Hamiltonian path (for instance, if 1 wins twice and 2 beats 3 then the only Hamiltonian path is 1, 2, 3); on the other hand, if each of the players wins once, there are 3 Hamiltonian paths (for instance, if 1 beats 2, 2 beats 3, and 3 beats 1, then 1, 2, 3, 2, 3, 1, and 3, 1, 2, are all Hamiltonians). Hence, when $n = 3$, there is a maximum of 3 Hamiltonian paths.

We now show that there is an outcome of the tournament that results in more than $n!/2^{n-1}$ Hamiltonian paths. To begin, let the outcome of the tournament specify the result of each of the $\binom{n}{2}$ games played, and let $\mathcal{S}$ denote the set of all $2^{\binom{n}{2}}$ possible tournament outcomes. Then, with $f(s)$ defined as the number of Hamiltonian paths that result when the outcome is $s \in \mathcal{S}$, we are asked to show that

$$\max_s f(s) \geq \frac{n!}{2^{n-1}}$$

To show this, consider the randomly chosen outcome S that is obtained when the results of the $\binom{n}{2}$ games are independent, with each contestant being equally likely to win each encounter. To determine $E[f(S)]$, the expected number of Hamiltonian paths that result from the outcome S, number the $n!$ permutations, and let for $i = 1, \ldots, n!$,

$$X_i = \begin{cases} 1, & \text{if permutation } i \text{ is a Hamiltonian} \\ 0, & \text{otherwise} \end{cases}$$

Since

$$f(S) = \sum_i X_i$$

it follows that

$$E[f(S)] = \sum_i E[X_i]$$

Because the probability that any specified permutation is a Hamiltonian is, by the assumed independence of game outcomes, $(1/2)^{n-1}$, it follows that

$$E[X_i] = P\{X_i = 1\} = (1/2)^{n-1}$$

Therefore,

$$E[f(S)] = n!(1/2)^{n-1}$$

Because $f(S)$ is not a constant random variable, the preceding equation implies that there is an outcome of the tournament having more than $n!/2^{n-1}$ Hamiltonian paths. ■

Example 2u. A grove of 52 trees is arranged in a circular fashion. If a total of 15 chipmunks live in these trees, show that there is a group of 7 consecutive trees that together house at least 3 chipmunks.

Solution Let the neighborhood of a tree consist of that tree along with the next six trees when moving in the clockwise direction. For any choice of living accommodations of the 15 chipmunks, we want to show that there is a tree that has at least 3 chipmunks living in its neighborhood. To show this, choose a tree at random and let X denote the number of chipmunks that live in its neighborhood. To determine $E[X]$, arbitrarily number the 15 chipmunks, and let, for $i = 1, \ldots, 15$

$$X_i = \begin{cases} 1, & \text{if chipmunk } i \text{ lives in the neighborhood of the randomly chosen tree} \\ 0, & \text{otherwise} \end{cases}$$

Because

$$X = \sum_{i=1}^{15} X_i$$

we obtain that

$$E[X] = \sum_{i=1}^{15} E[X_i]$$

However, because X_i will equal 1 if the randomly chosen tree is any of the 7 trees consisting of the tree in which chipmunk i lives along with its 6 neighboring trees when moving in the countercockwise direction,

$$E[X_i] = P\{X_i = 1\} = \frac{7}{52}$$

Consequently,

$$E[X] = \frac{105}{52} > 2$$

showing that there exists a tree with more than 2 chipmunks living in its neighborhood. ■

*7.2.2 The Maximum-Minimums Identity

We start with an identity relating the maximum of a set of numbers to the minimums of the subsets of these numbers.

Proposition 2.2

For arbitrary numbers $x_i, i = 1, \ldots, n$

$$\max_i x_i = \sum_i x_i - \sum_{i<j} \min(x_i, x_j) + \sum_{i<j<k} \min(x_i, x_j, x_k)$$
$$+ \cdots + (-1)^{n+1} \min(x_1, \ldots, x_n)$$

Proof: We will give a probabilistic proof of the preceding. To begin, assume that all the x_i are in the interval $[0, 1]$. Let U be a uniform $(0, 1)$ random variable, and define the events $A_i, i = 1, \ldots, n$ by $A_i = \{U < x_i\}$. That is, A_i is the event that the uniform random variable is less than x_i. Because at least one of these events A_i will occur if U is less than at least one of the values x_i, we have that

$$\cup_i A_i = \{U < \max_i x_i\}$$

Therefore,

$$P(\cup_i A_i) = P\{U < \max_i x_i\} = \max_i x_i$$

Also,

$$P(A_i) = P\{U < x_i\} = x_i$$

In addition, because all of the events $A_{i_1}, \ldots, A_{i_r}$ will occur if U is less than all the values $x_{i_1}, \ldots, x_{i_r}$, we see that the intersection of these events is

$$A_{i_1} \cdots A_{i_r} = \left\{U < \min_{j=1,\ldots r} x_{i_j}\right\}$$

implying that

$$P(A_{i_1} \cdots A_{i_r}) = P\left\{U < \min_{j=1,\ldots r} x_{i_j}\right\} = \min_{j=1,\ldots r} x_{i_j}$$

Thus the proposition follows from the inclusion-exclusion formula for the probability of the union of events:

$$P(\cup_i A_i) = \sum_i P(A_i) - \sum_{i<j} P(A_i A_j) + \sum_{i<j<k} P(A_i A_j A_k)$$
$$+ \cdots + (-1)^{n+1} P(A_1 \cdots A_n)$$

When the x_i are nonnegative, but not restricted to the unit interval, let c be such that all the x_i are less than c. Then the identity holds for the values $y_i = x_i/c$, and the result follows by multiplying through by c. When the x_i can be negative, let b be such that $x_i + b > 0$ for all i. Therefore, by the preceding,

$$\max_i (x_i + b) = \sum_i (x_i + b) - \sum_{i<j} \min(x_i + b, x_j + b)$$
$$+ \cdots + (-1)^{n+1} \min(x_1 + b, \ldots, x_n + b)$$

Letting

$$M = \sum_i x_i - \sum_{i<j} \min(x_i, x_j) + \cdots + (-1)^{n+1} \min(x_1, \ldots, x_n)$$

we can rewrite the preceding identity as

$$\max_i x_i + b = M + b\left(n - \binom{n}{2} + \cdots + (-1)^{n+1}\binom{n}{n}\right)$$

But

$$0 = (1-1)^n = 1 - n + \binom{n}{2} + \cdots + (-1)^n\binom{n}{n}$$

The preceding two equations show that

$$\max_i x_i = M$$

and the proposition is proven. ✦

It follows from Proposition 2.2 that for any random variables $X_1, \ldots, X_n$

$$\max_i X_i = \sum_i X_i - \sum_{i<j} \min(X_i, X_j) + \cdots + (-1)^{n+1} \min(X_1, \ldots, X_n)$$

Taking expectations of both sides of the preceding equality yields the following relationship between the expected value of the maximum and those of the partial minimums:

$$E\left[\max_i X_i\right] = \sum_i E[X_i] - \sum_{i<j} E[\min(X_i, X_j)]$$
$$+ \cdots + (-1)^{n+1} E[\min(X_1, \ldots, X_n)] \tag{2.8}$$

Example 2v. ***Coupon Collecting with Unequal Probabilities.*** Suppose there are n different types of coupons, and that each time one collects a coupon it is, independent of previous coupons collected, a type i coupon with probability p_i, $\sum_{i=1}^{n} p_i = 1$. Find the expected number of coupons one needs to collect to obtain a complete set of at least one of each type.

Solution If we let X_i denote the number of coupons one needs collect to obtain a type i, then we can express X as

$$X = \max_{i=1,\ldots,n} X_i$$

Because each new coupon obtained is a type i with probability p_i, we see that X_i is a geometric random variable with parameter p_i. Also, because the minimum of X_i and X_j is the number of coupons needed to obtain either a type i or a type j, we see that for $i \neq j$, $\min(X_i, X_j)$ is a geometric random variable with parameter $p_i + p_j$. Similarly, $\min(X_i, X_j, X_k)$, the number needed to obtain any of types i or j or k, is a geometric random variable with parameter $p_i + p_j + p_k$, and so on. Therefore, the identity (2.8) yields that

$$E[X] = \sum_i \frac{1}{p_i} - \sum_{i<j} \frac{1}{p_i + p_j} + \sum_{i<j<k} \frac{1}{p_i + p_j + p_k} + \cdots + (-1)^{n+1} \frac{1}{p_1 + \cdots + p_n}$$

Noting that

$$\int_0^\infty e^{-px}\, dx = \frac{1}{p}$$

and using the identity

$$1 - \prod_{i=1}^{n} (1 - e^{-p_i x}) = \sum_i e^{-p_i x} - \sum_{i<j} e^{-(p_i+p_j)x} + \cdots + (-1)^{n+1} e^{-(p_1+\cdots+p_n)x}$$

shows, upon integrating the preceding identity, that

$$E[X] = \int_0^\infty \left(1 - \prod_{i=1}^{n} (1 - e^{-p_i x})\right) dx$$

a more useful computational form. ■

Example 2w. Suppose that a deck of n cards contains n_i cards of type i for $i = 1, \ldots, r$, $\sum_{i=1}^{r} n_i = n$. The deck is shuffled so that all $n!$ possible orderings of the cards are equally likely, and the cards are then turned over one at a time.

(a) Determine the expected number of cards that need to be turned over until all denominations have appeared at least once.

(b) Determine the expected number of cards that need to be turned over until all the cards of any one of the r types have appeared.

Solution If we let X_i denote the number that need to be turned over before a type i card appears, then X, the number needed to to be turned over to obtain at least one of each type, can be expressed as

$$X = \max_{i=1,\ldots,r} X_i$$

Because n_i of the n cards are type i, it follows from the results of Example 2m that the expected number of cards that need to be turned over to obtain a type i is

$$E[X_i] = 1 + \frac{n - n_i}{n_i + 1} = \frac{n + 1}{n_i + 1}$$

Because for $i \neq j$, $\min(X_i, X_j)$ is the number needed to obtain either a type i or type j card, it again follows from Example 2m that

$$E[\min(X_i, X_j)] = \frac{n+1}{n_i + n_j + 1}$$

As the expected value of the other minimimums are similarly obtained, the identity (2.8) yields that

$$E[X] = \sum_i \frac{n+1}{n_i+1} - \sum_{i<j} \frac{n+1}{n_i+n_j+1} + \cdots + (-1)^{r+1} \frac{n+1}{n_1+\cdots+n_r+1}$$

Noting that

$$\int_0^1 x^m \, dx = \frac{1}{m+1}$$

and using the identity

$$1 - \prod_{i=1}^{r} (1 - x^{n_i}) = \sum_i x^{n_i} - \sum_{i<j} x^{n_i+n_j} + \cdots + (-1)^{r+1} x^{n_1+\cdots+n_r}$$

shows, upon integrating the preceding identity, that

$$E[X] = (n+1) \int_0^1 \left(1 - \prod_{i=1}^{r} (1 - x^{n_i}) \right) dx \tag{2.9}$$

a useful form for computation.

(b) To determine the expected number of cards that need to be turned over before a complete set of at least one of the r types is obtained, note that a complete set would have been exposed when the remaining cards no longer contain at least one of each type. Because the order of the cards when going from the back to the front of the deck is also equally likely to be any of the $n!$ orderings, it follows from part (a) that, when going from the back to the front of the deck, the expected number of cards that need to be turned over to obtain at least one of each type is $E[X]$. Consequently, the expected number of cards that remain in the deck when a complete set of at least one of the r types is obtained is $E[X] - 1$, showing that the expected number of cards it takes to obtain all the cards of some type is $n - E[X] + 1$, where $E[X]$ is given by Equation (2.9). ■

7.3 COVARIANCE, VARIANCE OF SUMS, AND CORRELATIONS

We start with the following proposition, which shows that the expectation of a product of independent random variables is equal to the product of their expectations.

Proposition 3.1

If X and Y are independent, then for any functions h and g,

$$E[g(X)h(Y)] = E[g(X)]E[h(Y)]$$

Proof: Suppose that X and Y are jointly continuous with joint density $f(x, y)$. Then

$$\begin{aligned}
E[g(X)h(Y)] &= \int_{-\infty}^{\infty}\int_{-\infty}^{\infty} g(x)h(y)f(x, y)\,dx\,dy \\
&= \int_{-\infty}^{\infty}\int_{-\infty}^{\infty} g(x)h(y)f_X(x)f_Y(y)\,dx\,dy \\
&= \int_{-\infty}^{\infty} h(y)f_Y(y)\,dy \int_{-\infty}^{\infty} g(x)f_X(x)\,dx \\
&= E[h(Y)]E[g(X)]
\end{aligned}$$

The proof in the discrete case is similar. ✦

Just as the expected value and the variance of a single random variable give us information about this random variable, so does the covariance between two random variables give us information about the relationship between the random variables.

Definition

The covariance between X and Y, denoted by $\mathrm{Cov}(X, Y)$, is defined by

$$\mathrm{Cov}(X, Y) = E[(X - E[X])(Y - E[Y])]$$

Upon expanding the right side of the preceding definition, we see that

$$\begin{aligned}
\mathrm{Cov}(X, Y) &= E[XY - E[X]Y - XE[Y] + E[Y]E[X]] \\
&= E[XY] - E[X]E[Y] - E[X]E[Y] + E[X]E[Y] \\
&= E[XY] - E[X]E[Y]
\end{aligned}$$

Note that if X and Y are independent then, by Proposition 3.1, it follows that $\mathrm{Cov}(X, Y) = 0$. However, the converse is not true. A simple example of two dependent random variables X and Y having zero covariance is obtained by letting X be a random variable such that

$$P\{X = 0\} = P\{X = 1\} = P\{X = -1\} = \tfrac{1}{3}$$

and define

$$Y = \begin{cases} 0 & \text{if } X \neq 0 \\ 1 & \text{if } X = 0 \end{cases}$$

now, $XY = 0$, so $E[XY] = 0$. Also, $E[X] = 0$ and thus

$$\text{Cov}(X, Y) = E[XY] - E[X]E[Y] = 0$$

However, X and Y are clearly not independent.

The following proposition lists some of the properties of covariance.

Proposition 3.2

(i) $\text{Cov}(X, Y) = \text{Cov}(Y, X)$

(ii) $\text{Cov}(X, X) = \text{Var}(X)$

(iii) $\text{Cov}(aX, Y) = a\,\text{Cov}(X, Y)$

(iv) $\text{Cov}\left(\sum_{i=1}^{n} X_i, \sum_{j=1}^{m} Y_j\right) = \sum_{i=1}^{n}\sum_{j=1}^{m} \text{Cov}(X_i, Y_j)$

Proof of Proposition 3.2: Parts (i) and (ii) follow immediately from the definition of covariance, and part (iii) is left as an exercise for the reader. To prove (iv), which states that the covariance operation is additive (as is the operation of taking expectations), let $\mu_i = E[X_i]$ and $\nu_j = E[Y_j]$. Then

$$E\left[\sum_{i=1}^{n} X_i\right] = \sum_{i=1}^{n} \mu_i, \qquad E\left[\sum_{j=1}^{m} Y_j\right] = \sum_{j=1}^{m} \nu_j$$

and

$$\begin{aligned}
\text{Cov}\left(\sum_{i=1}^{n} X_i, \sum_{j=1}^{m} Y_j\right) &= E\left[\left(\sum_{i=1}^{n} X_i - \sum_{i=1}^{n} \mu_i\right)\left(\sum_{j=1}^{m} Y_j - \sum_{j=1}^{m} \nu_j\right)\right] \\
&= E\left[\sum_{i=1}^{n} (X_i - \mu_i) \sum_{j=1}^{m} (Y_j - \nu_j)\right] \\
&= E\left[\sum_{i=1}^{n}\sum_{j=1}^{m} (X_i - \mu_i)(Y_j - \nu_j)\right] \\
&= \sum_{i=1}^{n}\sum_{j=1}^{m} E[(X_i - \mu_i)(Y_j - \nu_j)]
\end{aligned}$$

where the last equality follows because the expected value of a sum of random variables is equal to the sum of the expected values. ✦

It follows from parts (ii) and (iv) of Proposition 3.2, upon taking $Y_j = X_j$, $j = 1, \ldots, n$, that

$$\begin{aligned}\operatorname{Var}\left(\sum_{i=1}^{n} X_i\right) &= \operatorname{Cov}\left(\sum_{i=1}^{n} X_i, \sum_{j=1}^{n} X_j\right) \\ &= \sum_{i=1}^{n}\sum_{j=1}^{n} \operatorname{Cov}(X_i, X_j) \\ &= \sum_{i=1}^{n} \operatorname{Var}(X_i) + \sum\sum_{i \neq j} \operatorname{Cov}(X_i, X_j)\end{aligned}$$

Since each pair of indices $i, j, i \neq j$ appears twice in the double summation, the above is equivalent to the following:

$$\operatorname{Var}\left(\sum_{i=1}^{n} X_i\right) = \sum_{i=1}^{n} \operatorname{Var}(X_i) + 2\sum\sum_{i<j} \operatorname{Cov}(X_i, X_j) \tag{3.1}$$

If $X_1, \ldots, X_n$ are pairwise independent, in that X_i and X_j are independent for $i \neq j$, then Equation (3.1) reduces to

$$\operatorname{Var}\left(\sum_{i=1}^{n} X_i\right) = \sum_{i=1}^{n} \operatorname{Var}(X_i)$$

The following examples illustrate the use of Equation (3.1).

Example 3a. Let $X_1, \ldots, X_n$ be independent and identically distributed random variables having expected value μ and variance σ^2, and as in Example 2c, let $\bar{X} = \sum_{i=1}^{n} X_i/n$ be the sample mean. The quantities $X_i - \bar{X}, i = 1, \ldots, n$, are called *deviations*, as they equal the differences between the individual data and the sample mean. The random variable

$$S^2 = \sum_{i=1}^{n} \frac{(X_i - \bar{X})^2}{n - 1}$$

is called the *sample variance*. Find (a) $\operatorname{Var}(\bar{X})$ and (b) $E[S^2]$.

Solution

$$\begin{aligned}\text{(a) } \operatorname{Var}(\bar{X}) &= \left(\frac{1}{n}\right)^2 \operatorname{Var}\left(\sum_{i=1}^{n} X_i\right) \\ &= \left(\frac{1}{n}\right)^2 \sum_{i=1}^{n} \operatorname{Var}(X_i) \qquad \text{by independence} \\ &= \frac{\sigma^2}{n}\end{aligned}$$

(b) We start with the following algebraic identity:

$$\begin{aligned}(n-1)S^2 &= \sum_{i=1}^{n} (X_i - \mu + \mu - \bar{X})^2 \\ &= \sum_{i=1}^{n} (X_i - \mu)^2 + \sum_{i=1}^{n} (\bar{X} - \mu)^2 - 2(\bar{X} - \mu) \sum_{i=1}^{n} (X_i - \mu) \\ &= \sum_{i=1}^{n} (X_i - \mu)^2 + n(\bar{X} - \mu)^2 - 2(\bar{X} - \mu)n(\bar{X} - \mu) \\ &= \sum_{i=1}^{n} (X_i - \mu)^2 - n(\bar{X} - \mu)^2\end{aligned}$$

Taking expectations of the preceding yields that

$$\begin{aligned}(n-1)E[S^2] &= \sum_{i=1}^{n} E[(X_i - \mu)^2] - nE[(\bar{X} - \mu)^2] \\ &= n\sigma^2 - n\,\mathrm{Var}(\bar{X}) \\ &= (n-1)\sigma^2\end{aligned}$$

where the final equality made use of part (a), and the one preceding it made use of the result, of Example 2c, that $E[\bar{X}] = \mu$. Dividing through by $n - 1$ shows that the expected value of the sample variance is the distribution variance σ^2. ■

Our next example presents an alternative method (to the one used in Chapter 4) for obtaining the variance of a binomial random variable.

Example 3b. ***Variance of a binomial random variable.*** Compute the variance of a binomial random variable X with parameters n and p.

Solution Since such a random variable represents the number of successes in n independent trials when each trial has a common probability p of being a success, we may write

$$X = X_1 + \cdots + X_n$$

where the X_i are independent Bernoulli random variables such that

$$X_i = \begin{cases} 1 & \text{if the } i\text{th trial is a success} \\ 0 & \text{otherwise} \end{cases}$$

Hence, from Equation (3.1) we obtain

$$\mathrm{Var}(X) = \mathrm{Var}(X_1) + \cdots + \mathrm{Var}(X_n)$$

But

$$\begin{aligned}\text{Var}(X_i) &= E[X_i^2] - (E[X_i])^2 \\ &= E[X_i] - (E[X_i])^2 \qquad \text{since } X_i^2 = X_i \\ &= p - p^2\end{aligned}$$

and thus

$$\text{Var}(X) = np(1 - p)$$ ■

Example 3c. ***Variance of the number of matches.*** Compute the variance of X, the number of people that select their own hats in Example 2h.

Solution Using the same representation for X as we do in Example 2h, namely,

$$X = X_1 + \cdots + X_N$$

where

$$X_i = \begin{cases} 1 & \text{if the } i\text{th man selects his own hat} \\ 0 & \text{otherwise} \end{cases}$$

we obtain from Equation (3.1) that

$$\text{Var}(X) = \sum_{i=1}^{N} \text{Var}(X_i) + 2 \sum_{i<j}\sum \text{Cov}(X_i, X_j) \tag{3.2}$$

Since $P\{X_i = 1\} = 1/N$, we see from the preceding example that

$$\text{Var}(X_i) = \frac{1}{N}\left(1 - \frac{1}{N}\right) = \frac{N-1}{N^2}$$

Also,

$$\text{Cov}(X_i, X_j) = E[X_iX_j] - E[X_i]E[X_j]$$

Now,

$$X_iX_j = \begin{cases} 1 & \text{if the } i\text{th and } j\text{th men both select their own hats} \\ 0 & \text{otherwise} \end{cases}$$

and thus

$$\begin{aligned}E[X_iX_j] &= P\{X_i = 1, X_j = 1\} \\ &= P\{X_i = 1\}P\{X_j = 1 | X_i = 1\} \\ &= \frac{1}{N}\frac{1}{N-1}\end{aligned}$$

Hence

$$\text{Cov}(X_i, X_j) = \frac{1}{N(N-1)} - \left(\frac{1}{N}\right)^2 = \frac{1}{N^2(N-1)}$$

and from Equation (3.2),

$$\begin{aligned}\mathrm{Var}(X) &= \frac{N-1}{N} + 2\binom{N}{2}\frac{1}{N^2(N-1)} \\ &= \frac{N-1}{N} + \frac{1}{N} \\ &= 1\end{aligned}$$

Thus both the mean and variance of the number of matches are equal to 1. In a way, this result is not unexpected because, as shown in Example 5m of Chapter 2, when N is large, the probability of i matches is approximately $e^{-1}/i!$. That is, when N is large, the number of matches is approximately distributed as a Poisson random variable with mean 1. Hence, as the mean and variance of a Poisson random variable are equal, the result obtained in this example is not surprising. ■

Example 3d. ***The negative hypergeometric distribution.*** Consider a random ordering of $n + m$ items, of which n are special and m are ordinary, and suppose that all $(n + m)!$ possible orderings are equally likely. If we let X denote the position of the rth special item to appear in the ordering, then X is said to be a *negative hypergeometric random variable* with parameters (r, n, m). (The important special case having $r = 1$, was considered in Example 2m, where its mean was determined.) To understand the rationale behind its name, first note that such a random variable has the same distribution as does the number of balls that need to be withdrawn from an urn that contains n special and m ordinary balls until a total of r special balls have been withdrawn, when each selection is equally likely to be any of the remaining balls in the urn. (To see the connection, imagine that all $n + m$ balls are to be withdrawn and let the ordering be the successive designations of the withdrawn balls.) Consequently, the negative hypergeometric has the same relationship to the hypergeometric as does the negative binomial to the binomial. Namely, a sequence of trials, each of which is either a success (a special ball) or a failure, are observed in sequence. Whereas the binomial and hypergeometric distributions are concerned with the number of successes in a fixed number of trials, the negative binomial and negative hypergeometric distributions are concerned with the number of trials needed to obtain a fixed number of successes. (The difference between the binomial and hypergeometric is that the binomial applies to the ball removing model when withdrawn balls are replaced before the next withdrawal, whereas the hypergeometric applies when the balls are removed without previously withdrawn ones being replaced.)

To derive the probability mass function of the negative hypergeometric random variable X, note that X will equal k if

(a) the first $k - 1$ balls withdrawn consist of $r - 1$ special and $k - r$ ordinary balls, and
(b) the kth ball withdrawn is special.

Therefore,

$$P\{X = k\} = \frac{\binom{n}{r-1}\binom{m}{k-r}}{\binom{n+m}{k-1}} \frac{n - r + 1}{n + m - k + 1}$$

We will not directly utilize the preceding probability mass function to compute the mean and variance of X. Rather, we will follow the approach of Example 2m, and express X as a sum of other random variables. To accomplish this, denote the m ordinary balls as $o_1, o_2, \ldots, o_m$, and define random variables $X_1, \ldots, X_m$ as follows:

$$X_i = \begin{cases} 1, & \text{if } o_i \text{ is removed before a total of } r \text{ special balls have been withdrawn} \\ 0, & \text{otherwise} \end{cases}$$

Because the total number of withdrawals needed to obtain r special balls is equal to r plus the number of ordinary balls withdrawn before this objective is met, we have that

$$X = r + \sum_{i=1}^{m} X_i \tag{3.3}$$

Now consider the $n + 1$ balls consisting of all n special balls along with the ball o_i. Of these $n + 1$ balls, o_i is equally likely to be the first one selected, or the second one selected, ..., or the final one selected. Hence, the probability that it is among the first r of these to be selected is $\frac{r}{n+1}$. Consequently,

$$E[X_i] = P\{X_i = 1\} = \frac{r}{n+1}$$

implying that

$$E[X] = r + \sum_{i=1}^{m} E[X_i] = r\left(1 + \frac{m}{n+1}\right) = \frac{r(n+m+1)}{n+1}$$

To determine the variance of X, we use the representation (3.3) to obtain that

$$\begin{aligned}\text{Var}(X) &= \text{Var}\left(\sum_{i=1}^{m} X_i\right) \\ &= \sum_{i=1}^{m} \text{Var}(X_i) + \sum_{j}\sum_{i \neq j} \text{Cov}(X_i, X_j)\end{aligned}$$

Because X_i is a Bernoulli random variable with parameter $r/(n + 1)$,

$$\text{Var}(X_i) = \frac{r}{n+1}\left(1 - \frac{r}{n+1}\right) = \frac{r(n+1-r)}{(n+1)^2}$$

Also, for $i \neq j$

$$X_iX_j = \begin{cases} 1, & \text{if } o_i \text{ and } o_j \text{ are removed before a total of } r \text{ special balls have been withdrawn} \\ 0, & \text{otherwise} \end{cases}$$

Now, consider the $n + 2$ balls consisting of the n special balls along with o_i and o_j. In order for both o_i and o_j to be removed before a total of r special balls, o_i and o_j must be among the first $r + 1$ of these $n + 2$ balls to be removed. Therefore,

$$E[X_iX_j] = P\{X_iX_j = 1\} = \frac{\binom{2}{2}\binom{n}{r-1}}{\binom{n+2}{r+1}} = \frac{(r+1)r}{(n+2)(n+1)}$$

Consequently, for $i \neq j$,

$$\text{Cov}(X_i, X_j) = \frac{(r+1)r}{(n+2)(n+1)} - \left(\frac{r}{n+1}\right)^2 = \frac{r(n-r+1)}{(n+1)^2(n+2)}$$

Therefore,

$$\begin{aligned}
\text{Var}(X) &= \frac{mr(n+1-r)}{(n+1)^2} + m(m-1)\frac{r(n-r+1)}{(n+1)^2(n+2)} \\
&= \frac{mr(n+1-r)}{(n+1)^2}\left[1 + \frac{m-1}{n+2}\right] \\
&= \frac{mr(n+1-r)(n+m+1)}{(n+1)^2(n+2)}
\end{aligned}$$

■

Example 3e. ***Sampling from a finite population.*** Consider a set of N people each of whom has an opinion about a certain subject that is measured by a real number v, which represents the person's "strength of feeling" about the subject. Let v_i represent the strength of feeling of person $i, i = 1, \ldots N$.

Suppose that these quantities $v_i, i = 1, \ldots, N$ are unknown and to gather information a group of n of the N people is "randomly chosen" in the sense that all of the $\binom{N}{n}$ subsets of size n are equally likely to be chosen. These n people are then questioned and their feelings determined. If S denotes the sum of the n sampled values, determine its mean and variance.

An important application of the preceding is to a forthcoming election in which each person in the population is either for or against a certain candidate or proposition. If we take v_i to equal 1 if person i is in favor and 0 if he or she is against, then $\bar{v} = \sum_{i=1}^{N} v_i/N$ represents the proportion of the population that is in favor. To estimate $\bar{v}$, a random sample of n people is chosen, and these people are polled. The proportion of those polled that are in favor—that is, S/n—is often used as an estimate of $\bar{v}$.

Solution For each person $i, i = 1, \ldots, N$, define an indicator variable I_i to indicate whether or not that person is included in the sample. That is,

$$I_i = \begin{cases} 1 & \text{if person } i \text{ is in the random sample} \\ 0 & \text{otherwise} \end{cases}$$

Now S can be expressed by

$$S = \sum_{i=1}^{N} v_i I_i$$

so

$$E[S] = \sum_{i=1}^{N} v_i E[I_i]$$

$$\begin{aligned} \text{Var}(S) &= \sum_{i=1}^{N} \text{Var}(v_i I_i) + 2 \sum\sum_{i<j} \text{Cov}(v_i I_i, v_j I_j) \\ &= \sum_{i=1}^{N} v_i^2 \text{Var}(I_i) + 2 \sum\sum_{i<j} v_i v_j \text{Cov}(I_i, I_j) \end{aligned}$$

As

$$E[I_i] = \frac{n}{N}$$

$$E[I_i I_j] = \frac{n}{N} \frac{n-1}{N-1}$$

we see that

$$\text{Var}(I_i) = \frac{n}{N}\left(1 - \frac{n}{N}\right)$$

$$\begin{aligned} \text{Cov}(I_i, I_j) &= \frac{n(n-1)}{N(N-1)} - \left(\frac{n}{N}\right)^2 \\ &= \frac{-n(N-n)}{N^2(N-1)} \end{aligned}$$

Hence

$$E[S] = n \sum_{i=1}^{N} \frac{v_i}{N} = n\bar{v}$$

$$\text{Var}(S) = \frac{n}{N}\left(\frac{N-n}{N}\right) \sum_{i=1}^{N} v_i^2 - \frac{2n(N-n)}{N^2(N-1)} \sum\sum_{i<j} v_i v_j$$

The expression for $\text{Var}(S)$ can be simplified somewhat by using the identity $(v_1 + \cdots + v_N)^2 = \sum_{i=1}^{N} v_i^2 + 2 \sum\sum_{i<j} v_i v_j$ to give, after some simplification.

$$\text{Var}(S) = \frac{n(N-n)}{N-1}\left(\frac{\sum_{i=1}^{N} v_i^2}{N} - \bar{v}^2\right)$$

Consider now the special case in which Np of the v's are equal to 1 and the remainder equal to 0. Then in this case S is a hypergeometric random variable and has mean and variance given by

$$E[S] = n\bar{v} = np \qquad \text{since } \bar{v} = \frac{Np}{N} = p$$

$$\text{Var}(S) = \frac{n(N-n)}{N-1}\left(\frac{Np}{N} - p^2\right)$$

$$= \frac{n(N-n)}{N-1}p(1-p)$$

The quantity S/n, equal to the proportion of those sampled that have values equal to 1, is such that

$$E\left[\frac{S}{n}\right] = p$$

$$\text{Var}\left(\frac{S}{n}\right) = \frac{N-n}{n(N-1)}p(1-p)$$ ■

The correlation of two random variables X and Y, denoted by $\rho(X, Y)$, is defined, as long as $\text{Var}(X)\,\text{Var}(Y)$ is positive, by

$$\rho(X, Y) = \frac{\text{Cov}(X, Y)}{\sqrt{\text{Var}(X)\,\text{Var}(Y)}}$$

It can be shown that

$$-1 \le \rho(X, Y) \le 1 \tag{3.4}$$

To prove Equation (3.4), suppose that X and Y have variances given by σ_x^2 and σ_y^2, respectively. Then

$$0 \le \text{Var}\left(\frac{X}{\sigma_x} + \frac{Y}{\sigma_y}\right)$$

$$= \frac{\text{Var}(X)}{\sigma_x^2} + \frac{\text{Var}(Y)}{\sigma_y^2} + \frac{2\,\text{Cov}(X, Y)}{\sigma_x \sigma_y}$$

$$= 2[1 + \rho(X, Y)]$$

implying that

$$-1 \le \rho(X, Y)$$

On the other hand,

$$
\begin{aligned}
0 &\le \mathrm{Var}\left(\frac{X}{\sigma_x} - \frac{Y}{\sigma_y}\right) \\
&= \frac{\mathrm{Var}(X)}{\sigma_x^2} + \frac{\mathrm{Var}\,Y}{(-\sigma_y)^2} - \frac{2\,\mathrm{Cov}(X,Y)}{\sigma_x\sigma_y} \\
&= 2[1 - \rho(X,Y)]
\end{aligned}
$$

implying that

$$\rho(X,Y) \le 1$$

which completes the proof of Equation (3.4).

In fact, since $\mathrm{Var}(Z) = 0$ implies that Z is constant with probability 1 (this intuitive fact will be rigorously proved in Chapter 8), we see from the proof of (3.3) that $\rho(X,Y) = 1$ implies that $Y = a + bX$, where $b = \sigma_y/\sigma_x > 0$ and $\rho(X,Y) = -1$ implies that $Y = a + bX$, where $b = -\sigma_y/\sigma_x < 0$. We leave it as an exercise for the reader to show that the reverse is also true: that if $Y = a + bX$, then $\rho(X,Y)$ is either $+1$ or -1, depending on the sign of b.

The correlation coefficient is a measure of the degree of linearity between X and Y. A value of $\rho(X,Y)$ near $+1$ or -1 indicates a high degree of linearity between X and Y, whereas a value near 0 indicates a lack of such linearity. A positive value of $\rho(X,Y)$ indicates that Y tends to increase when X does, whereas a negative value indicates that Y tends to decrease when X increases. If $\rho(X,Y) = 0$, then X and Y are said to be *uncorrelated*.

Example 3f. Let I_A and I_B be indicator variables for the events A and B. That is,

$$I_A = \begin{cases} 1 & \text{if } A \text{ occurs} \\ 0 & \text{otherwise} \end{cases}$$

$$I_B = \begin{cases} 1 & \text{if } B \text{ occurs} \\ 0 & \text{otherwise} \end{cases}$$

Then

$$
\begin{aligned}
E[I_A] &= P(A) \\
E[I_B] &= P(B) \\
E[I_A I_B] &= P(AB)
\end{aligned}
$$

so

$$
\begin{aligned}
\mathrm{Cov}(I_A, I_B) &= P(AB) - P(A)P(B) \\
&= P(B)[P(A|B) - P(A)]
\end{aligned}
$$

Thus we obtain the quite intuitive result that the indicator variables for A and B are either positively correlated, uncorrelated, or negatively correlated depending on whether $P(A|B)$ is greater than, equal to, or less than $P(A)$. ■

Our next example shows that the sample mean and a deviation from the sample mean are uncorrelated.

Example 3g. Let $X_1, \ldots, X_n$ be independent and identically distributed random variables having variance σ^2. Show that

$$\text{Cov}(X_i - \bar{X}, \bar{X}) = 0$$

Solution

$$\begin{aligned}
\text{Cov}(X_i - \bar{X}, \bar{X}) &= \text{Cov}(X_i, \bar{X}) - \text{Cov}(\bar{X}, \bar{X}) \\
&= \text{Cov}\left(X_i, \frac{1}{n}\sum_{j=1}^{n} X_j\right) - \text{Var}(\bar{X}) \\
&= \frac{1}{n}\sum_{j=1}^{n} \text{Cov}(X_i, X_j) - \frac{\sigma^2}{n} \\
&= \frac{\sigma^2}{n} - \frac{\sigma^2}{n} = 0
\end{aligned}$$

where the next-to-last equality uses the result of Example 3a, and the final equality follows since

$$\text{Cov}(X_i, X_j) = \begin{cases} 0 & \text{if } j \neq i \quad \text{by independence} \\ \sigma^2 & \text{if } j = i \quad \text{since } \text{Var}(X_i) = \sigma^2 \end{cases}$$

Although $\bar{X}$ and the deviation $X_i - \bar{X}$ are uncorrelated, they are not, in general, independent. However, in the special case where the X_i are normal random variables it turns out that not only is $\bar{X}$ independent of a single deviation but it is independent of the entire sequence of deviations $X_j - \bar{X}$, $j = 1, \ldots, n$. This result will be established in Section 7.9, where we will also show that, in this case, the sample mean $\bar{X}$ and the sample variance S^2 are independent, with $(n-1)S^2/\sigma^2$ having a chi-squared distribution with $n - 1$ degrees of freedom. (See Example 3a for the definition of S^2.) ■

Example 3h. Consider m independent trials, each of which results in any of r possible outcomes with probabilities $P_1, P_2, \ldots, P_r$, $\sum_1^r P_i = 1$. If we let N_i, $i = 1, \ldots, r$ denote the number of the m trials that result in outcome i, then N_1, $N_2, \ldots, N_r$ have the multinomial distribution

$$P\{N_1 = n_1, N_2 = n_2, \ldots, N_r = n_r\}$$
$$= \frac{m!}{n_1!\, n_2! \ldots n_r!} P_1^{n_1} P_2^{n_2} \cdots P_r^{n_r} \qquad \sum_{i=1}^{r} n_i = m$$

For $i \neq j$ it seems likely that when N_i is large N_j would tend to be small, and hence it is intuitive that they should be negatively correlated. Let us compute their covariance by using Proposition 3.2(iv) and the representation

$$N_i = \sum_{k=1}^{m} I_i(k) \quad \text{and} \quad N_j = \sum_{k=1}^{m} I_j(k)$$

where

$$I_i(k) = \begin{cases} 1 & \text{if trial } k \text{ results in outcome } i \\ 0 & \text{otherwise} \end{cases}$$

$$I_j(k) = \begin{cases} 1 & \text{if trial } k \text{ results in outcome } j \\ 0 & \text{otherwise} \end{cases}$$

From Proposition 3.2(iv) we have

$$\text{Cov}(N_i, N_j) = \sum_{\ell=1}^{m} \sum_{k=1}^{m} \text{Cov}(I_i(k), I_j(\ell))$$

Now, when $k \neq \ell$,

$$\text{Cov}(I_i(k), I_j(\ell)) = 0$$

since the outcome of trial k is independent of the outcome of trial ℓ. On the other hand,

$$\begin{aligned} \text{Cov}(I_i(\ell), I_j(\ell)) &= E[I_i(\ell)I_j(\ell)] - E[I_i(\ell)]E[I_j(\ell)] \\ &= 0 - P_iP_j = -P_iP_j \end{aligned}$$

where the preceding uses that $I_i(\ell)I_j(\ell) = 0$ since trial ℓ cannot result in both outcome i and outcome j. Hence we obtain that

$$\text{Cov}(N_i, N_j) = -mP_iP_j$$

which is in accord with our intuition that N_i and N_j are negatively correlated.

7.4 CONDITIONAL EXPECTATION

7.4.1 Definitions

Recall that if X and Y are jointly discrete random variables, the conditional probability mass function of X, given that $Y = y$, is defined for all y such that $P\{Y = y\} > 0$, by

$$p_{X|Y}(x|y) = P\{X = x|Y = y\} = \frac{p(x, y)}{p_Y(y)}$$

It is therefore natural to define, in this case, the conditional expectation of X, given that $Y = y$, for all values of y such that $p_Y(y) > 0$ by

$$\begin{aligned} E[X|Y = y] &= \sum_x xP\{X = x|Y = y\} \\ &= \sum_x xp_{X|Y}(x|y) \end{aligned}$$

Example 4a. If X and Y are independent binomial random variables with identical parameters n and p, calculate the conditional expected value of X, given that $X + Y = m$.

Solution Let us first calculate the conditional probability mass function of X, given that $X + Y = m$. For $k \leq \min(n, m)$,

$$\begin{aligned}
P\{X = k|X + Y = m\} &= \frac{P\{X = k, X + Y = m\}}{P\{X + Y = m\}} \\
&= \frac{P\{X = k, Y = m - k\}}{P\{X + Y = m\}} \\
&= \frac{P\{X = k\}P\{Y = m - k\}}{P\{X + Y = m\}} \\
&= \frac{\binom{n}{k}p^k(1 - p)^{n-k}\binom{n}{m - k}p^{m-k}(1 - p)^{n-m+k}}{\binom{2n}{m}p^m(1 - p)^{2n-m}} \\
&= \frac{\binom{n}{k}\binom{n}{m - k}}{\binom{2n}{m}}
\end{aligned}$$

where we have used the fact (see Example 3f of Chapter 6) that $X + Y$ is a binomial random variable with parameters $2n$ and p. Hence the conditional distribution of X, given that $X + Y = m$, is the hypergeometric distribution; thus, from Example 2g, we obtain

$$E[X|X + Y = m] = \frac{m}{2}$$ ■

Similarly, let us recall that if X and Y are jointly continuous, with a joint probability density function $f(x, y)$, the conditional probability density of X, given that $Y = y$, is defined for all values of y such that $f_Y(y) > 0$ by

$$f_{X|Y}(x|y) = \frac{f(x, y)}{f_Y(y)}$$

It is natural, in this case, to define the conditional expectation of X, given that $Y = y$, by

$$E[X|Y = y] = \int_{-\infty}^{\infty} x f_{X|Y}(x|y)\, dx$$

provided that $f_Y(y) > 0$.

Example 4b. Suppose that the joint density of X and Y is given by

$$f(x, y) = \frac{e^{-x/y}e^{-y}}{y} \qquad 0 < x < \infty, 0 < y < \infty$$

Compute $E[X|Y = y]$.

Solution We start by computing the conditional density

$$\begin{aligned} f_{X|Y}(x|y) &= \frac{f(x, y)}{f_Y(y)} \\ &= \frac{f(x, y)}{\int_{-\infty}^{\infty} f(x, y)\,dx} \\ &= \frac{(1/y)e^{-x/y}e^{-y}}{\int_0^{\infty} (1/y)e^{-x/y}e^{-y}\,dx} \\ &= \frac{(1/y)e^{-x/y}}{\int_0^{\infty} (1/y)e^{-x/y}\,dx} \\ &= \frac{1}{y}e^{-x/y} \end{aligned}$$

Hence the conditional distribution of X, given that $Y = y$, is just the exponential distribution with mean y. Thus

$$E[X|Y = y] = \int_0^{\infty} \frac{x}{y}e^{-x/y}\,dx = y$$

■

REMARK. Just as conditional probabilities satisfy all of the properties of ordinary probabilities, so do conditional expectations satisfy the properties of ordinary expectations. For instance, such formulas as

$$E[g(X)|Y = y] = \begin{cases} \sum_x g(x)p_{X|Y}(x|y) & \text{in the discrete case} \\ \int_{-\infty}^{\infty} g(x)f_{X|Y}(x|y)\,dx & \text{in the continuous case} \end{cases}$$

and

$$E\left[\sum_{i=1}^{n} X_i|Y = y\right] = \sum_{i=1}^{n} E[X_i|Y = y]$$

remain valid. As a matter of fact, conditional expectation given $Y = y$ can be thought of as being an ordinary expectation on a reduced sample space consisting only of outcomes for which $Y = y$. ■

7.4.2 Computing Expectations by Conditioning

Let us denote by $E[X|Y]$ that function of the random variable Y whose value at $Y = y$ is $E[X|Y = y]$. Note that $E[X|Y]$ is itself a random variable. An extremely important property of conditional expectation is given by the following proposition.

Proposition 4.1

$$E[X] = E[E[X|Y]] \tag{4.1}$$

If Y is a discrete random variable, then Equation (4.1) states that

$$E[X] = \sum_y E[X|Y = y]P\{Y = y\} \tag{4.1a}$$

whereas if Y is continuous with density $f_Y(y)$, then Equation (4.1) states

$$E[X] = \int_{-\infty}^{\infty} E[X|Y = y]f_Y(y)\,dy \tag{4.1b}$$

We now give a proof of Equation (4.1) in the case where X and Y are both discrete random variables.

Proof of Equation (4.1) When X and Y Are Discrete: We must show that

$$E[X] = \sum_y E[X|Y = y]P\{Y = y\} \tag{4.2}$$

Now, the right-hand side of Equation (4.2) can be written as

$$\begin{aligned}
\sum_y E[X|Y = y]P\{Y = y\} &= \sum_y \sum_x xP\{X = x|Y = y\}P\{Y = y\} \\
&= \sum_y \sum_x x\frac{P\{X = x, Y = y\}}{P\{Y = y\}}P\{Y = y\} \\
&= \sum_y \sum_x xP\{X = x, Y = y\} \\
&= \sum_x x \sum_y P\{X = x, Y = y\} \\
&= \sum_x xP\{X = x\} \\
&= E[X]
\end{aligned}$$

and the result is proved. ✦

One way to understand Equation (4.2) is to interpret it as follows: To calculate $E[X]$, we may take a weighted average of the conditional expected value of X, given that $Y = y$, each of the terms $E[X|Y = y]$ being weighted by the probability of the event on which it is conditioned. (Of what does this remind you?) This is an extremely useful result that often enables us to easily compute expectations by first conditioning on some appropriate random variable. The following examples illustrate its use.

Example 4c. A miner is trapped in a mine containing 3 doors. The first door leads to a tunnel that will take him to safety after 3 hours of travel. The second door leads to a tunnel that will return him to the mine after 5 hours of travel. The third door leads to a tunnel that will return him to the mine after 7 hours. If we assume that the miner is at all times equally likely to choose any one of the doors, what is the expected length of time until he reaches safety?

Solution Let X denote the amount of time (in hours) until the miner reaches safety, and let Y denote the door he initially chooses. Now

$$\begin{aligned} E[X] &= E[X|Y = 1]P\{Y = 1\} + E[X|Y = 2]P\{Y = 2\} \\ &\quad + E[X|Y = 3]P\{Y = 3\} \\ &= \tfrac{1}{3}(E[X|Y = 1] + E[X|Y = 2] + E[X|Y = 3]) \end{aligned}$$

However,

$$\begin{aligned} E[X|Y = 1] &= 3 \\ E[X|Y = 2] &= 5 + E[X] \\ E[X|Y = 3] &= 7 + E[X] \end{aligned} \tag{4.3}$$

To understand why Equation (4.3) is correct, consider, for instance, $E[X|Y = 2]$ and reason as follows: If the miner chooses the second door, he spends 5 hours in the tunnel and then returns to his cell. But once he returns to his cell the problem is as before; thus his expected additional time until safety is just $E[X]$. Hence $E[X|Y = 2] = 5 + E[X]$. The argument behind the other equalities in Equation (4.3) is similar. Hence

$$E[X] = \tfrac{1}{3}(3 + 5 + E[X] + 7 + E[X])$$

or

$$E[X] = 15$$ ■

Example 4d. ***Expectation of a sum of a random number of random variables.*** Suppose that the number of people entering a department store on a given day is a random variable with mean 50. Suppose further that the amounts of money spent by these customers are independent random variables having a common mean of \$8. Assume also that the amount of money spent by a customer is also independent of the total number of customers to enter the store. What is the expected amount of money spent in the store on a given day?

Solution If we let N denote the number of customers that enter the store and X_i the amount spent by the ith such customer, then the total amount of money spent can be expressed as $\sum_{i=1}^{N} X_i$. Now,

$$E\left[\sum_1^N X_i\right] = E\left[E\left[\sum_1^N X_i|N\right]\right]$$

But

$$E\left[\sum_1^N X_i|N = n\right] = E\left[\sum_1^n X_i|N = n\right]$$

$$= E\left[\sum_1^n X_i\right] \quad \text{by the independence of the } X_i \text{ and } N$$

$$= nE[X] \quad \text{where } E[X] = E[X_i]$$

which implies that

$$E\left[\sum_1^N X_i|N\right] = NE[X]$$

and thus

$$E\left[\sum_{i=1}^N X_i\right] = E[NE[X]] = E[N]E[X]$$

Hence, in our example, the expected amount of money spent in the store is 50×8, or \$400. ■

Example 4e. The game of craps is begun by rolling an ordinary pair of dice. If the sum of the dice is 2, 3, or 12, the player loses. If it is 7 or 11, the player wins. If it is any other number i, the player continues to roll the dice until the sum is either 7 or i. If it is 7, the player loses; if it is i, the player wins. Let R denote the number of rolls of the dice in a game of craps. Find

(a) $E[R]$;

(b) $E[R|\text{player wins}]$;

(c) $E[R|\text{player loses}]$.

Solution If we let P_i denote the probability that the sum of the dice is i, then

$$P_i = P_{14-i} = \frac{i-1}{36}, \qquad i = 2, \ldots, 7$$

To compute $E[R]$, condition on S, the initial sum. This gives

$$E[R] = \sum_{i=2}^{12} E[R|S = i]P_i$$

However,

$$E[R|S = i] = \begin{cases} 1, & \text{if } i = 2, 3, 7, 11, 12 \\ 1 + \dfrac{1}{P_i + P_7}, & \text{otherwise} \end{cases}$$

The preceding follows because if the sum is a value i that does not end the game, then the dice will continue to be rolled until the sum is either i or 7, and the number of rolls until this occurs is a geometric random variable with parameter $P_i + P_7$. Therefore,

$$E[R] = 1 + \sum_{i=4}^{6} \frac{P_i}{P_i + P_7} + \sum_{i=8}^{10} \frac{P_i}{P_i + P_7}$$
$$= 1 + 2(3/9 + 4/10 + 5/11) = 3.376$$

To determine $E[R|\text{win}]$, let us start by determining p, the probability that the player wins. Conditioning on S yields

$$p = \sum_{i=2}^{12} P\{\text{win}|S = i\}P_i$$
$$= P_7 + P_{11} + \sum_{i=4}^{6} \frac{P_i}{P_i + P_7}P_i + \sum_{i=8}^{10} \frac{P_i}{P_i + P_7}P_i$$
$$= 0.493$$

where the preceding uses that the probability of obtaining a sum of i before one of 7 is $P_i/(P_i + P_7)$. Now, let us determine the conditional probability mass function of S given that the player wins. Letting $Q_i = P\{S = i|\text{win}\}$, we have

$$Q_2 = Q_3 = Q_{12} = 0, \qquad Q_7 = P_7/p, \qquad Q_{11} = P_{11}/p$$

and for $i = 4, 5, 6, 8, 9, 10$

$$Q_i = \frac{P\{S = i, \text{win}\}}{P\{\text{win}\}}$$
$$= \frac{P_i P\{\text{win}|S = i\}}{p}$$
$$= \frac{P_i^2}{p(P_i + P_7)}$$

Now, conditioning on the initial sum gives that

$$E[R|\text{win}] = \sum_i E[R|\text{win}, S = i]Q_i$$

However, as was noted in Example 2j of Chapter 6, given that the initial sum is i, the number of additional rolls needed and the outcome (whether a win or a loss) are independent. (This is easily seen by first noting that, conditional on an initial sum of i, the outcome is independent of the number of additional dice rolls needed, and then using the symmetry property of independence which states that if event A is independent of event B, then event B is independent of event A.) Therefore, we see that

$$\begin{aligned} E[R|\text{win}] &= \sum_i E[R|S = i]Q_i \\ &= 1 + \sum_{i=4}^{6} \frac{Q_i}{P_i + P_7} + \sum_{i=8}^{10} \frac{Q_i}{P_i + P_7} \\ &= 2.938 \end{aligned}$$

Although we could determine $E[R|\text{player loses}]$ exactly as we did $E[R|\text{player wins}]$, it is easier to use

$$E[R] = E[R|\text{win}]p + E[R|\text{lose}](1 - p)$$

implying that

$$E[R|\text{lose}] = \frac{E[R] - E[R|\text{win}]p}{1 - p} = 3.801 \quad \blacksquare$$

Example 4f. An urn contains a white and b black balls. One ball at a time is randomly withdrawn until the first white ball is drawn. Find the expected number of black balls that are withdrawn.

Solution This problem was previously treated in Example 2m. Here we present a solution using conditioning. Let X denote the number of black balls withdrawn and, to make explicit the dependence on a and b, let $M_{a,b} = E[X]$. We obtain an expression for $M_{a,b}$ by conditioning on the initial ball that is withdrawn. That is, define

$$Y = \begin{cases} 1 & \text{if the first ball selected is white} \\ 0 & \text{if the first ball selected is black} \end{cases}$$

Conditioning on Y yields

$$M_{a,b} = E[X] = E[X|Y = 1]P\{Y = 1\} + E[X|Y = 0]P\{Y = 0\}$$

However,

$$E[X|Y = 1] = 0 \tag{4.4}$$

$$E[X|Y = 0] = 1 + M_{a,b-1} \tag{4.5}$$

To understand Equations (4.4) and (4.5), suppose, for instance, that the first ball withdrawn is black. Then, after the first withdrawal, the situation is exactly the same as if we had started with a white balls and $b - 1$ black balls, which establishes Equation (4.5).

Since $P\{Y = 0\} = b/(a + b)$, we see that

$$M_{a,b} = \frac{b}{a + b}[1 + M_{a,b-1}]$$

Now, $M_{a,0}$ is clearly equal to 0, and we obtain

$$M_{a,1} = \frac{1}{a+1}[1 + M_{a,0}] = \frac{1}{a+1}$$

$$M_{a,2} = \frac{2}{a+2}[1 + M_{a,1}] = \frac{2}{a+2}\left[1 + \frac{1}{a+1}\right] = \frac{2}{a+1}$$

$$M_{a,3} = \frac{3}{a+3}[1 + M_{a,2}] = \frac{3}{a+3}\left[1 + \frac{2}{a+1}\right] = \frac{3}{a+1}$$

By using induction, one can easily verify that

$$M_{a,b} = \frac{b}{a+1}$$ ■

It is also possible to obtain the variance of a random variable by conditioning. We illustrate this by the following example.

Example 4g. ***Variance of the geometric distribution.*** Independent trials each resulting in a success with probability p are successively performed. Let N be the time of the first success. Find $\text{Var}(N)$.

Solution Let $Y = 1$ if the first trial results in a success and $Y = 0$ otherwise. Now,

$$\text{Var}(N) = E[N^2] - (E[N])^2$$

To calculate $E[N^2]$, we condition on Y as follows:

$$E[N^2] = E[E[N^2|Y]]$$

However,

$$E[N^2|Y = 1] = 1$$
$$E[N^2|Y = 0] = E[(1 + N)^2]$$

These two equations follow because, if the first trial results in a success, then clearly $N = 1$; thus $N^2 = 1$. On the other hand, if the first trial results in a failure, then the total number of trials necessary for the first success will have the same distribution as one (the first trial that results in failure) plus the necessary number of additional trials. Since the latter quantity has the same distribution as N, we obtain that $E[N^2|Y = 0] = E[(1 + N)^2]$. Hence we see that

$$\begin{aligned} E[N^2] &= E[N^2|Y = 1]P\{Y = 1\} + E[N^2|Y = 0]P\{Y = 0\} \\ &= p + (1 - p)E[(1 + N)^2] \\ &= 1 + (1 - p)E[2N + N^2] \end{aligned}$$

However, as was shown in Example 9b of Chapter 4, $E[N] = 1/p$; therefore,

$$E[N^2] = 1 + \frac{2(1-p)}{p} + (1-p)E[N^2]$$

or

$$E[N^2] = \frac{2-p}{p^2}$$

Therefore,

$$\begin{aligned}\text{Var}(N) &= E[N^2] - (E[N])^2 \\ &= \frac{2-p}{p^2} - \left(\frac{1}{p}\right)^2 \\ &= \frac{1-p}{p^2}\end{aligned}$$

■

In our next example we determine the expected number of uniform (0, 1) random variables that need to be added for their sum to exceed 1. The answer, surprisingly, is e.

Example 4h. Let $U_1, U_2, \ldots$ be a sequence of independent uniform (0, 1) random variables. Find $E[N]$ when

$$N = \min\left\{n: \sum_{i=1}^{n} U_i > 1\right\}$$

Solution We will solve the above by obtaining a more general result. For $x \in [0, 1]$, let

$$N(x) = \min\left\{n: \sum_{i=1}^{n} U_i > x\right\}$$

and set

$$m(x) = E[N(x)]$$

That is, $N(x)$ is the number of uniform (0, 1) random variables we need add until their sum exceeds x, and $m(x)$ is its expected value. We will now derive an equation for $m(x)$ by conditioning on U_1. This gives, from Equation (4.1b),

$$m(x) = \int_0^1 E[N(x)|U_1 = y]\,dy \tag{4.6}$$

Now,

$$E[N(x)|U_1 = y] = \begin{cases} 1 & \text{if } y > x \\ 1 + m(x-y) & \text{if } y \le x \end{cases} \tag{4.7}$$

The preceding formula is obvious when $y > x$. It is true when $y \le x$ since if the first uniform value is y, then at that point the remaining number of uniforms needed is the same as if we were just starting and were going to add uniforms until their sum exceeded $x - y$. Substituting (4.7) into (4.6) gives that

$$m(x) = 1 + \int_0^x m(x - y)\,dy$$

$$= 1 + \int_0^x m(u)\,du \qquad \text{by letting } u = x - y$$

Differentiating the preceding equation yields that

$$m'(x) = m(x)$$

or, equivalently,

$$\frac{m'(x)}{m(x)} = 1$$

Integrating this gives

$$\log[m(x)] = x + c$$

or

$$m(x) = ke^x$$

Since $m(0) = 1$ we see that $k = 1$, so we obtain that

$$m(x) = e^x$$

Therefore, $m(1)$, the expected number of uniform (0, 1) random variables that need to be added until their sum exceeds 1 is equal to e. ■

7.4.3 Computing Probabilities by Conditioning

Not only can we obtain expectations by first conditioning on an appropriate random variable, but we may also use this approach to compute probabilities. To see this, let E denote an arbitrary event and define the indicator random variable X by

$$X = \begin{cases} 1 & \text{if } E \text{ occurs} \\ 0 & \text{if } E \text{ does not occur} \end{cases}$$

It follows from the definition of X that

$$E[X] = P(E)$$

$$E[X|Y = y] = P(E|Y = y) \qquad \text{for any random variable } Y$$

Therefore, from Equations (4.1a) and (4.1b) we obtain

$$P(E) = \sum_y P(E|Y = y)P(Y = y) \qquad \text{if } Y \text{ is discrete}$$

$$= \int_{-\infty}^{\infty} P(E|Y = y)f_Y(y)\,dy \qquad \text{if } Y \text{ is continuous} \tag{4.8}$$

Note that if Y is a discrete random variable taking on one of the values $y_1, \ldots, y_n$, then, by defining the events $F_i, i = 1, \ldots, n$ by $F_i = \{Y = y_i\}$, Equation (4.8) reduces to the familiar equation

$$P(E) = \sum_{i=1}^{n} P(E|F_i)P(F_i)$$

where $F_1, \ldots, F_n$ are mutually exclusive events whose union is the sample space.

Example 4i. ***The best prize problem.*** Suppose that we are to be presented with n distinct prizes in sequence. After being presented with a prize we must immediately decide whether to accept it or to reject it and consider the next prize. The only information we are given when deciding whether to accept a prize is the relative rank of that prize compared to ones already seen. That is, for instance, when the fifth prize is presented, we learn how it compares with the four prizes already seen. Suppose that once a prize is rejected it is lost, and that our objective is to maximize the probability of obtaining the best prize. Assuming that all $n!$ orderings of the prizes are equally likely, how well can we do?

Solution Rather surprisingly, we can do quite well. To see this, fix a value k, $0 \le k < n$, and consider the strategy that rejects the first k prizes and then accepts the first one that is better than all of those first k. Let $P_k(\text{best})$ denote the probability that the best prize is selected when this strategy is employed. To compute this probability, condition on X, the position of the best prize. This gives

$$\begin{aligned} P_k(\text{best}) &= \sum_{i=1}^{n} P_k(\text{best}|X = i)P(X = i) \\ &= \frac{1}{n}\sum_{i=1}^{n} P_k(\text{best}|X = i) \end{aligned}$$

Now, if the overall best prize is among the first k, then no prize is ever selected under the strategy considered. That is,

$$P_k(\text{best}|X = i) = 0 \quad \text{if } i \le k$$

On the other hand, if the best prize is in position i, where $i > k$, then the best prize will be selected if the best of the first $i - 1$ prizes is among the first k (for then none of the prizes in positions $k + 1, k + 2, \ldots, i - 1$ would be selected). But conditional on the best prize being in position i, it is easy to verify that all possible orderings of the other prizes remain equally likely, which implies that each of the first $i - 1$ prizes is equally likely to be the best of that batch. Hence we see that

$$\begin{aligned} P_k(\text{best}|X = i) &= P\{\text{best of first } i - 1 \text{ is among the first } k|X = i\} \\ &= \frac{k}{i-1} \quad \text{if } i > k \end{aligned}$$

From the preceding, we obtain that

$$\begin{aligned}
P_k(\text{best}) &= \frac{k}{n} \sum_{i=k+1}^{n} \frac{1}{i-1} \\
&\approx \frac{k}{n} \int_{k+1}^{n} \frac{1}{x-1}\,dx \\
&= \frac{k}{n} \log\left(\frac{n-1}{k}\right) \\
&\approx \frac{k}{n} \log\left(\frac{n}{k}\right)
\end{aligned}$$

Now, if we consider the function

$$g(x) = \frac{x}{n} \log\left(\frac{n}{x}\right)$$

then

$$g'(x) = \frac{1}{n} \log\left(\frac{n}{x}\right) - \frac{1}{n}$$

so

$$g'(x) = 0 \Rightarrow \log\left(\frac{n}{x}\right) = 1 \Rightarrow x = \frac{n}{e}$$

Thus, since $P_k(\text{best}) \approx g(k)$, we see that the best strategy of the type considered is to let the first n/e prizes go by and then accept the first one to appear that is better than all of those. In addition, since $g(n/e) = 1/e$, the probability that this strategy selects the best prize is approximately $1/e \approx .36788$.

REMARK. Most people are quite surprised by the size of the probability of obtaining the best prize, thinking that this probability would be close to 0 when n is large. However, even without going through the calculations, a little thought reveals that the probability of obtaining the best prize can be made reasonably large. For consider the strategy of letting half of the prizes go by and then selecting the first one to appear that is better than all of those. The probability that a prize is actually selected is the probability that the overall best is among the second half, and this is $\frac{1}{2}$. In addition, given that a prize is selected, at the time of selection that prize would have been the best of more than $n/2$ prizes to have appeared and would thus have probability of at least $\frac{1}{2}$ of being the overall best. Hence, the strategy of letting the first half of all prizes go by and then accepting the first one that is better than all of those prizes has a probability greater than $\frac{1}{4}$ of obtaining the best prize. ■

Example 4j. Let U be a uniform random variable on $(0, 1)$, and suppose that the conditional distribution of X, given that $U = p$, is binomial with parameters n and p. Find the probability mass function of X.

Solution Conditioning on the value of U gives

$$\begin{aligned} P\{X = i\} &= \int_0^1 P\{X = i|U = p\}f_U(p)\,dp \\ &= \int_0^1 P\{X = i|U = p\}\,dp \\ &= \frac{n!}{i!\,(n-i)!}\int_0^1 p^i(1-p)^{n-i}\,dp \end{aligned}$$

Now it can be shown (a probabilistic proof is given in Section 6.6) that

$$\int_0^1 p^i(1-p)^{n-i}\,dp = \frac{i!(n-i)!}{(n+1)!}$$

Hence we obtain that

$$P\{X = i\} = \frac{1}{n+1} \qquad i = 0, \ldots, n$$

That is, we obtain the surprising result that if a coin whose probability of coming up heads is uniformly distributed over $(0, 1)$ is flipped n times, then the number of heads occurring is equally likely to be any the values $0, \ldots, n$.

Because the preceding conditional distribution has such a nice form, it is worth trying to find another argument to enhance our intuition as to why such a result is true. To do so, let $U, U_1, \ldots, U_n$ be $n + 1$ independent uniform $(0, 1)$ random variables, and let X denote the number of the random variables $U_1, \ldots, U_n$ that are smaller than U. Since all the random variables $U, U_1, \ldots, U_n$ have the same distribution, it follows that U is equally likely to be the smallest, or second smallest, or largest of them; so X is equally likely to be any of the values $0, 1, \ldots, n$. However, given that $U = p$, the number of the U_i that are less than U is a binomial random variable with parameters n and p, thus establishing our previous result. ■

Example 4k. Suppose that X and Y are independent continuous random variables having densities f_X and f_Y, respectively. Compute $P\{X < Y\}$.

Solution Conditioning on the value of Y yields

$$\begin{aligned} P\{X < Y\} &= \int_{-\infty}^{\infty} P\{X < Y|Y = y\}f_Y(y)\,dy \\ &= \int_{-\infty}^{\infty} P\{X < y|Y = y\}f_Y(y)\,dy \\ &= \int_{-\infty}^{\infty} P\{X < y\}f_Y(y)\,dy \qquad \text{by independence} \\ &= \int_{-\infty}^{\infty} F_X(y)f_Y(y)\,dy \end{aligned}$$

where

$$F_X(y) = \int_{-\infty}^{y} f_X(x)\,dx \qquad \blacksquare$$

Example 4l. Suppose that X and Y are independent continuous random variables. Find the distribution of $X + Y$.

Solution By conditioning on the value of Y, we obtain

$$\begin{aligned} P\{X + Y < a\} &= \int_{-\infty}^{\infty} P\{X + Y < a|Y = y\}f_Y(y)\,dy \\ &= \int_{-\infty}^{\infty} P\{X + y < a|Y = y\}f_Y(y)\,dy \\ &= \int_{-\infty}^{\infty} P\{X < a - y\}f_Y(y)\,dy \\ &= \int_{-\infty}^{\infty} F_X(a - y)f_Y(y)\,dy \qquad \blacksquare \end{aligned}$$

7.4.4 Conditional Variance

Just as we have defined the conditional expectation of X given the value of Y, we can also define the conditional variance of X given that $Y = y$, which is defined as follows:

$$\text{Var}(X|Y) \equiv E[(X - E[X|Y])^2|Y]$$

That is, $\text{Var}(X|Y)$ is equal to the (conditional) expected square of the difference between X and its (conditional) mean when the value of Y is given. Or, in other words, $\text{Var}(X|Y)$ is exactly analogous to the usual definition of variance, but now all expectations are conditional on the fact that Y is known.

There is a very useful relationship between $\text{Var}(X)$, the unconditional variance of X, and $\text{Var}(X|Y)$, the conditional variance of X given Y, that can often be applied to compute $\text{Var}(X)$. To obtain this relationship, note first that by the same reasoning that yields $\text{Var}(X) = E[X^2] - (E[X])^2$ we have that

$$\text{Var}(X|Y) = E[X^2|Y] - (E[X|Y])^2$$

so

$$\begin{aligned} E[\text{Var}(X|Y)] &= E[E[X^2|Y]] - E[(E[X|Y])^2] \\ &= E[X^2] - E[(E[X|Y])^2] \end{aligned} \tag{4.9}$$

Also, as $E[E[X|Y]] = E[X]$, we have that

$$\text{Var}(E[X|Y]) = E[(E[X|Y])^2] - (E[X])^2 \tag{4.10}$$

Hence, by adding Equations (4.9) and (4.10), we arrive at the following proposition.

Proposition 4.2 The conditional variance formula

$$\text{Var}(X) = E[\text{Var}(X|Y)] + \text{Var}(E[X|Y])$$

Example 4m. Suppose that by any time t the number of people that have arrived at a train depot is a Poisson random variable with mean λt. If the initial train arrives at the depot at a time (independent of when the passengers arrive) that is uniformly distributed over $(0, T)$, what is the mean and variance of the number of passengers that enter the train?

Solution Let, for each $t \geq 0$, $N(t)$ denote the number of arrivals by t, and let Y denote the time at which the train arrives. The random variable of interest is then $N(Y)$. Conditioning on Y gives

$$\begin{aligned}
E[N(Y)|Y = t] &= E[N(t)|Y = t] \\
&= E[N(t)] \qquad \text{by the independence of } Y \text{ and } N(t) \\
&= \lambda t \qquad \text{since } N(t) \text{ is Poisson with mean } \lambda t
\end{aligned}$$

Hence

$$E[N(Y)|Y] = \lambda Y$$

so taking expectations gives

$$E[N(Y)] = \lambda E[Y] = \frac{\lambda T}{2}$$

To obtain $\text{Var}(N(Y))$, we use the conditional variance formula:

$$\begin{aligned}
\text{Var}(N(Y)|Y = t) &= \text{Var}(N(t)|Y = t) \\
&= \text{Var}(N(t)) \qquad \text{by independence} \\
&= \lambda t
\end{aligned}$$

so

$$\text{Var}(N(Y)|Y) = \lambda Y$$

$$E[N(Y)|Y] = \lambda Y$$

Hence, from the conditional variance formula,

$$\begin{aligned}
\text{Var}(N(Y)) &= E[\lambda Y] + \text{Var}(\lambda Y) \\
&= \lambda \frac{T}{2} + \lambda^2 \frac{T^2}{12}
\end{aligned}$$

where the preceding uses that $\text{Var}(Y) = T^2/12$. ■

Example 4n. ***Variance of a sum of a random number of random variables.*** Let X_1, $X_2, \ldots$ be a sequence of independent and identically distributed random variables and let N be a nonnegative integer-valued random variable that is independent of the sequence X_i, $i \geq 1$. To compute $\text{Var}\left(\sum_{i=1}^{N} X_i\right)$, we condition on N:

$$E\left[\sum_{i=1}^{N} X_i | N\right] = NE[X]$$

$$\text{Var}\left(\sum_{i=1}^{N} X_i | N\right) = N\,\text{Var}(X)$$

The preceding result follows, since given N, $\sum_{i=1}^{N} X_i$ is just the sum of a fixed number of independent random variables, so its expectation and variance is just the sum of the individual means and variances. Hence, from the conditional variance formula,

$$\text{Var}\left(\sum_{i=1}^{N} X_i\right) = E[N]\,\text{Var}(X) + (E[X])^2\,\text{Var}(N)$$ ■

7.5 CONDITIONAL EXPECTATION AND PREDICTION

Sometimes a situation arises where the value of a random variable X is observed and then, based on the observed value, an attempt is made to predict the value of a second random variable Y. Let $g(X)$ denote the predictor; that is, if X is observed to equal x, then $g(x)$ is our prediction for the value of Y. Clearly, we would like to choose g so that $g(X)$ tends to be close to Y. One possible criterion for closeness is to choose g so as to minimize $E[(Y - g(X))^2]$. We now show that under this criterion, the best possible predictor of Y is $g(X) = E[Y|X]$.

Proposition 5.1

$$E[(Y - g(X))^2] \geq E[(Y - E[Y|X])^2]$$

Proof:

$$\begin{aligned} E[(Y - g(X))^2|X] &= E[(Y - E[Y|X] + E[Y|X] - g(X))^2|X] \\ &= E[(Y - E[Y|X])^2|X] \\ &\quad + E[(E[Y|X] - g(X))^2|X] \\ &\quad + 2E[(Y - E[Y|X])(E[Y|X] - g(X))|X] \end{aligned} \tag{5.1}$$

However, given $X, E[Y|X] - g(X)$, being a function of X, can be treated as a constant. Thus

$$\begin{aligned} E[(Y - E[Y|X])(E[Y|X] - g(X))|X] \\ = (E[Y|X] - g(X))E[Y - E[Y|X]|X] \\ = (E[Y|X] - g(X))(E[Y|X] - E[Y|X]) \\ = 0 \end{aligned} \tag{5.2}$$

Thus, from Equations (5.1) and (5.2), we obtain

$$E[(Y - g(X))^2|X] \geq E[(Y - E[Y|X])^2|X]$$

and the result follows by taking expectations of both sides of the preceding expression. ✦

REMARK. A second, more intuitive although less rigorous argument verifying Proposition 5.1 is as follows. It is straightforward to verify that $E[(Y - c)^2]$ is minimized at $c = E[Y]$ (see Theoretical Exercise 1). Thus if we want to predict the value of Y when there are no data available to use, the best possible prediction, in the sense of minimizing the mean square error, is to predict that Y will equal its mean. On the other hand, if the value of the random variable X is observed to be x, then the prediction problem remains exactly as in the previous (no data) case with the exception that all probabilities and expectations are now conditional on the event that $X = x$. Hence it follows that the best prediction in this situation is to predict that Y will equal its conditional expected value given that $X = x$, thus establishing Proposition 5.1. ■

Example 5a. Suppose that the son of a man of height x (in inches) attains a height that is normally distributed with mean $x + 1$ and variance 4. What is the best prediction of the height at full growth of the son of a man who is 6 feet tall?

Solution Formally, this model can be written as

$$Y = X + 1 + e$$

where e is a normal random variable, independent of X, having mean 0 and variance 4. The X and Y, of course, represent the heights of the man and his son, respectively. The best prediction $E[Y|X = 72]$ is thus equal to

$$\begin{aligned} E[Y|X = 72] &= E[X + 1 + e|X = 72] \\ &= 73 + E[e|X = 72] \\ &= 73 + E(e) \qquad \text{by independence} \\ &= 73 \end{aligned}$$

■

Example 5b. Suppose that if a signal value s is sent from location A, then the signal value received at location B is normally distributed with parameters $(s, 1)$. If S, the value of the signal sent at A, is normally distributed with parameters (μ, σ^2), what is the best estimate of the signal sent if R, the value received at B, is equal to r?

Solution Let us start by computing the conditional density of S given R as follows:

$$f_{S|R}(s|r) = \frac{f_{S,R}(s,r)}{f_R(r)}$$
$$= \frac{f_S(s)f_{R|S}(r|s)}{f_R(r)}$$
$$= Ke^{-(s-\mu)^2/2\sigma^2}e^{-(r-s)^2/2}$$

where K does not depend on s. Now,

$$\frac{(s-\mu)^2}{2\sigma^2} + \frac{(r-s)^2}{2} = s^2\left(\frac{1}{2\sigma^2} + \frac{1}{2}\right) - \left(\frac{\mu}{\sigma^2} + r\right)s + C_1$$
$$= \frac{1+\sigma^2}{2\sigma^2}\left[s^2 - 2\left(\frac{\mu + r\sigma^2}{1+\sigma^2}\right)s\right] + C_1$$
$$= \frac{1+\sigma^2}{2\sigma^2}\left(s - \frac{(\mu + r\sigma^2)}{1+\sigma^2}\right)^2 + C_2$$

where C_1 and C_2 do not depend on s. Hence

$$f_{S|R}(s|r) = C\exp\left\{\frac{-\left[s - \dfrac{(\mu + r\sigma^2)}{1+\sigma^2}\right]^2}{2\left(\dfrac{\sigma^2}{1+\sigma^2}\right)}\right\}$$

where C does not depend on s. Hence we may conclude that the conditional distribution of S, the signal sent, given that r is received, is normal with mean and variance now given by

$$E[S|R=r] = \frac{\mu + r\sigma^2}{1+\sigma^2}$$

$$\text{Var}(S|R=r) = \frac{\sigma^2}{1+\sigma^2}$$

Hence, from Proposition 5.1, given that the value received is r, the best estimate, in the sense of minimizing the mean square error, for the signal sent is

$$E[S|R=r] = \frac{1}{1+\sigma^2}\mu + \frac{\sigma^2}{1+\sigma^2}r$$

Writing the conditional mean as we did previously is informative, for it shows that it equals a weighted average of μ, the a priori expected value of the signal and r, the value received. The relative weights given to μ and r are in the same proportion to each other as 1 (the conditional variance of the received signal when s is sent) is to σ^2 (the variance of the signal to be sent). ■

Example 5c. In digital signal processing, raw continuous analog data X must be quantized, or discretized, in order to obtain a digital representation. In order to quantize the raw data X, an increasing set of numbers $a_i, i = 0, \pm 1, \pm 2, \ldots$, such that $\lim_{i \to +\infty} a_i = \infty$, $\lim_{i \to -\infty} a_i = -\infty$, is fixed and the raw data are then quantized according to the interval $(a_i, a_{i+1}]$ in which X lies. Let us denote by y_i the discretized value when $X \in (a_i, a_{i+1}]$, and let Y denote the observed discretized value—that is,

$$Y = y_i \qquad \text{if } a_i < X \leq a_{i+1}$$

The distribution of Y is given by

$$P\{Y = y_i\} = F_X(a_{i+1}) - F_X(a_i)$$

Suppose now that we want to choose the values $y_i, i = 0, \pm 1, \pm 2, \ldots$ so as to minimize $E[(X - Y)^2]$, the expected mean square difference between the raw data and their quantized version.

(a) Find the optimal values $y_i, i = 0, \pm 1, \ldots$.
For the optimal quantizer Y show that
(b) $E[Y] = E[X]$, so the mean square error quantizer preserves the input mean;
(c) $\text{Var}(Y) = \text{Var}(X) - E[(X - Y)^2]$.

Solution (a) For any quantizer Y, upon conditioning on the value of Y we obtain

$$E[(X - Y)^2] = \sum_i E[(X - y_i)^2 | a_i < X \leq a_{i+1}] P\{a_i < X \leq a_{i+1}\}$$

Now, if we let

$$I = i \qquad \text{if } a_i < X \leq a_{i+1}$$

then

$$E[(X - y_i)^2 | a_i < X \leq a_{i+1}] = E[(X - y_i)^2 | I = i]$$

and by Proposition 5.1 this quantity is minimized when

$$\begin{aligned} y_i &= E[X | I = i] \\ &= E[X | a_i < X \leq a_{i+1}] \\ &= \int_{a_i}^{a_{i+1}} \frac{x f_X(x)\, dx}{F_X(a_{i+1}) - F_X(a_i)} \end{aligned}$$

Now, since the optimal quantizer is given by $Y = E[X|I]$, it follows that

(b) $E[Y] = E[X]$

$$\begin{aligned} \text{(c)} \quad \text{Var}(X) &= E[\text{Var}(X|I)] + \text{Var}(E[X|I]) \\ &= E[E[(X - Y)^2 | I]] + \text{Var}(Y) \\ &= E[(X - Y)^2] + \text{Var}(Y) \end{aligned}$$ ∎

It sometimes happens that the joint probability distribution of X and Y is not completely known; or if it is known, it is such that the calculation of $E[Y|X = x]$ is mathematically intractable. If, however, the means and variances of X and Y and the correlation of X and Y are known, then we can at least determine the best *linear* predictor of Y with respect to X.

To obtain the best linear predictor of Y with respect to X, we need to choose a and b so as to minimize $E[(Y - (a + bX))^2]$. Now,

$$\begin{aligned} E[(Y - (a + bX))^2] &= E[Y^2 - 2aY - 2bXY + a^2 + 2abX + b^2X^2] \\ &= E[Y^2] - 2aE[Y] - 2bE[XY] + a^2 \\ &\quad + 2abE[X] + b^2E[X^2] \end{aligned}$$

Taking partial derivatives, we obtain

$$\begin{aligned} \frac{\partial}{\partial a}E[(Y - a - bX)^2] &= -2E[Y] + 2a + 2bE[X] \\ \frac{\partial}{\partial b}E[(Y - a - bX)^2] &= -2E[XY] + 2aE[X] + 2bE[X^2] \end{aligned} \tag{5.3}$$

Equating Equations (5.3) to 0 and solving for a and b yields the solutions

$$\begin{aligned} b &= \frac{E[XY] - E[X]E[Y]}{E[X^2] - (E[X])^2} = \frac{\text{Cov}(X, Y)}{\sigma_x^2} = \rho\frac{\sigma_y}{\sigma_x} \\ a &= E[Y] - bE[X] = E[Y] - \frac{\rho\sigma_y E[X]}{\sigma_x} \end{aligned} \tag{5.4}$$

where $\rho = \text{Correlation}(X, Y)$, $\sigma_y^2 = \text{Var}(Y)$, and $\sigma_x^2 = \text{Var}(X)$. It is easy to verify that the values of a and b from Equation (5.4) minimize $E[(Y - a - bX)^2]$, and thus the best (in the sense of mean square error) linear predictor Y with respect to X is

$$\mu_y + \frac{\rho\sigma_y}{\sigma_x}(X - \mu_x)$$

where $\mu_y = E[Y]$ and $\mu_x = E[X]$.

The mean square error of this predictor is given by

$$\begin{aligned} &E\left[\left(Y - \mu_y - \rho\frac{\sigma_y}{\sigma_x}(X - \mu_x)\right)^2\right] \\ &\quad = E[(Y - \mu_y)^2] + \rho^2\frac{\sigma_y^2}{\sigma_x^2}E[(X - \mu_x)^2] - 2\rho\frac{\sigma_y}{\sigma_x}E[(Y - \mu_y)(X - \mu_x)] \\ &\quad = \sigma_y^2 + \rho^2\sigma_y^2 - 2\rho^2\sigma_y^2 \\ &\quad = \sigma_y^2(1 - \rho^2) \end{aligned} \tag{5.5}$$

We note from Equation (5.5) that if ρ is near $+1$ or -1, then the mean square error of the best linear predictor is near zero.

Example 5d. An example in which the conditional expectation of Y given X is linear in X, and hence the best linear predictor of Y with respect to X is the best overall predictor, is when X and Y have a bivariate normal distribution. In this case their joint density is given by

$$f(x, y) = \frac{1}{2\pi\sigma_x\sigma_y\sqrt{1-\rho^2}} \exp\left\{-\frac{1}{2(1-\rho^2)}\left[\left(\frac{x-\mu_x}{\sigma_x}\right)^2 - \frac{2\rho(x-\mu_x)(y-\mu_y)}{\sigma_x\sigma_y} + \left(\frac{y-\mu_y}{\sigma_y}\right)^2\right]\right\}$$

We leave it for the reader to verify that the conditional density of Y, given $X = x$, is given by

$$f_{Y|X}(y|x) = \frac{1}{\sqrt{2\pi}\,\sigma_y\sqrt{1-\rho^2}} \cdot \exp\left\{-\frac{1}{2\sigma_y^2(1-\rho^2)}\left(y - \mu_y - \frac{\rho\sigma_y}{\sigma_x}(x-\mu_x)\right)^2\right\}$$

Hence the conditional distribution of Y, given $X = x$, is the normal distribution with mean

$$E[Y|X = x] = \mu_y + \rho\frac{\sigma_y}{\sigma_x}(x - \mu_x)$$

and variance $\sigma_y^2(1 - \rho^2)$. ■

7.6 MOMENT GENERATING FUNCTIONS

The moment generating function $M(t)$ of the random variable X is defined for all real values of t by

$$M(t) = E[e^{tX}] = \begin{cases} \sum_x e^{tx}p(x) & \text{if } X \text{ is discrete with mass function } p(x) \\ \int_{-\infty}^{\infty} e^{tx}f(x)\,dx & \text{if } X \text{ is continuous with density } f(x) \end{cases}$$

We call $M(t)$ the moment generating function because all of the moments of X can be obtained by successively differentiating $M(t)$ and then evaluating the result at $t = 0$. For example,

$$\begin{aligned} M'(t) &= \frac{d}{dt}E[e^{tX}] \\ &= E\left[\frac{d}{dt}(e^{tX})\right] \qquad (6.1) \\ &= E[Xe^{tX}] \end{aligned}$$

where we have assumed that the interchange of the differentiation and expectation operators is legitimate. That is, we have assumed that

$$\frac{d}{dt}\left[\sum_x e^{tx}p(x)\right] = \sum_x \frac{d}{dt}[e^{tx}p(x)]$$

in the discrete case, and

$$\frac{d}{dt}\left[\int e^{tx}f(x)\,dx\right] = \int \frac{d}{dt}[e^{tx}f(x)]\,dx$$

in the continuous case. This assumption can almost always be justified and, indeed, is valid for all of the distributions considered in this book. Hence, from Equation (6.1) we obtain, by evaluating at $t = 0$, that

$$M'(0) = E[X]$$

Similarly,

$$\begin{aligned} M''(t) &= \frac{d}{dt}M'(t) \\ &= \frac{d}{dt}E[Xe^{tX}] \\ &= E\left[\frac{d}{dt}(Xe^{tX})\right] \\ &= E[X^2e^{tX}] \end{aligned}$$

and thus

$$M''(0) = E[X^2]$$

In general, the nth derivative of $M(t)$ is given by

$$M^n(t) = E[X^n e^{tX}] \qquad n \geq 1$$

implying that

$$M^n(0) = E[X^n] \qquad n \geq 1$$

We now compute $M(t)$ for some common distributions.

Example 6a. ***Binomial distribution with parameters n and p.*** If X is a binomial random variable with parameters n and p, then

$$\begin{aligned} M(t) &= E[e^{tX}] \\ &= \sum_{k=0}^{n} e^{tk}\binom{n}{k}p^k(1-p)^{n-k} \\ &= \sum_{k=0}^{n} \binom{n}{k}(pe^t)^k(1-p)^{n-k} \\ &= (pe^t + 1 - p)^n \end{aligned}$$

where the last equality follows from the binomial theorem. Differentiation yields

$$M'(t) = n(pe^t + 1 - p)^{n-1} pe^t$$

and thus

$$E[X] = M'(0) = np$$

which checks with the result obtained in Example 1c. Differentiating a second time yields

$$M''(t) = n(n - 1)(pe^t + 1 - p)^{n-2}(pe^t)^2 + n(pe^t + 1 - p)^{n-1} pe^t$$

so

$$E[X^2] = M''(0) = n(n - 1)p^2 + np$$

The variance of X is given by

$$\begin{aligned} \text{Var}(X) &= E[X^2] - (E[X])^2 \\ &= n(n - 1)p^2 + np - n^2p^2 \\ &= np(1 - p) \end{aligned}$$

verifying the result of Example 3b. ■

Example 6b. ***Poisson distribution with mean λ.*** If X is a Poisson random variable with parameter λ, then

$$\begin{aligned} M(t) &= E[e^{tX}] \\ &= \sum_{n=0}^{\infty} \frac{e^{tn}e^{-\lambda}\lambda^n}{n!} \\ &= e^{-\lambda} \sum_{n=0}^{\infty} \frac{(\lambda e^t)^n}{n!} \\ &= e^{-\lambda}e^{\lambda e^t} \\ &= \exp\{\lambda(e^t - 1)\} \end{aligned}$$

Differentiation yields

$$\begin{aligned} M'(t) &= \lambda e^t \exp\{\lambda(e^t - 1)\} \\ M''(t) &= (\lambda e^t)^2 \exp\{\lambda(e^t - 1)\} + \lambda e^t \exp\{\lambda(e^t - 1)\} \end{aligned}$$

and thus

$$\begin{aligned} E[X] &= M'(0) = \lambda \\ E[X^2] &= M''(0) = \lambda^2 + \lambda \\ \text{Var}(X) &= E[X^2] - (E[X])^2 \\ &= \lambda \end{aligned}$$

Hence both the mean and the variance of the Poisson random variable equal λ. ■

Example 6c. ***Exponential distribution with parameter λ***

$$\begin{aligned} M(t) &= E[e^{tX}] \\ &= \int_0^\infty e^{tx}\lambda e^{-\lambda x}\,dx \\ &= \lambda\int_0^\infty e^{-(\lambda-t)x}\,dx \\ &= \frac{\lambda}{\lambda - t} \qquad \text{for } t < \lambda \end{aligned}$$

We note from this derivation that for the exponential distribution, $M(t)$ is only defined for values of t less than λ. Differentiation of $M(t)$ yields

$$M'(t) = \frac{\lambda}{(\lambda - t)^2} \qquad M''(t) = \frac{2\lambda}{(\lambda - t)^3}$$

Hence

$$E[X] = M'(0) = \frac{1}{\lambda} \qquad E[X^2] = M''(0) = \frac{2}{\lambda^2}$$

The variance of X is given by

$$\begin{aligned} \text{Var}(X) &= E[X^2] - (E[X])^2 \\ &= \frac{1}{\lambda^2} \end{aligned}$$

■

Example 6d. ***Normal distribution.*** We first compute the moment generating function of a unit normal random variable with parameters 0 and 1. Letting Z be such a random variable, we have

$$\begin{aligned} M_Z(t) &= E[e^{tZ}] \\ &= \frac{1}{\sqrt{2\pi}}\int_{-\infty}^{\infty} e^{tx}e^{-x^2/2}\,dx \\ &= \frac{1}{\sqrt{2\pi}}\int_{-\infty}^{\infty} \exp\left\{-\frac{(x^2 - 2tx)}{2}\right\}dx \\ &= \frac{1}{\sqrt{2\pi}}\int_{-\infty}^{\infty} \exp\left\{-\frac{(x - t)^2}{2} + \frac{t^2}{2}\right\}dx \\ &= e^{t^2/2}\frac{1}{\sqrt{2\pi}}\int_{-\infty}^{\infty} e^{-(x-t)^2/2}\,dx \\ &= e^{t^2/2} \end{aligned}$$

Hence the moment generating function of the unit normal random variable Z is given by $M_Z(t) = e^{t^2/2}$. To obtain the moment generating function of an arbitrary normal random variable, we recall (see Section 5.4) that $X = \mu + \sigma Z$ will have a normal distribution with parameters μ and σ^2 whenever Z is a unit normal random variable. Hence the moment generating function of such a random variable is given by

$$\begin{aligned}
M_X(t) &= E[e^{tX}] \\
&= E[e^{t(\mu+\sigma Z)}] \\
&= E[e^{t\mu}e^{t\sigma Z}] \\
&= e^{t\mu}E[e^{t\sigma Z}] \\
&= e^{t\mu}M_Z(t\sigma) \\
&= e^{t\mu}e^{(t\sigma)^2/2} \\
&= \exp\left\{\frac{\sigma^2t^2}{2} + \mu t\right\}
\end{aligned}$$

By differentiating, we obtain

$$M'_X(t) = (\mu + t\sigma^2)\exp\left\{\frac{\sigma^2t^2}{2} + \mu t\right\}$$

$$M''_X(t) = (\mu + t\sigma^2)^2\exp\left\{\frac{\sigma^2t^2}{2} + \mu t\right\} + \sigma^2\exp\left\{\frac{\sigma^2t^2}{2} + \mu t\right\}$$

and thus

$$\begin{aligned}
E[X] &= M'(0) = \mu \\
E[X^2] &= M''(0) = \mu^2 + \sigma^2
\end{aligned}$$

implying that

$$\begin{aligned}
\text{Var}(X) &= E[X^2] - E([X])^2 \\
&= \sigma^2
\end{aligned}$$

■

Tables 7.1 and 7.2 give the moment generating function for some common discrete and continuous distributions.

An important property of moment generating functions is that the moment generating function of the sum of independent random variables equals the product of the individual moment generating functions. To prove this, suppose that X and Y are independent and have moment generating functions $M_X(t)$ and $M_Y(t)$, respectively. Then $M_{X+Y}(t)$, the moment generating function of $X + Y$, is given by

$$\begin{aligned}
M_{X+Y}(t) &= E[e^{t(X+Y)}] \\
&= E[e^{tX}e^{tY}] \\
&= E[e^{tX}]E[e^{tY}] \\
&= M_X(t)M_Y(t)
\end{aligned}$$

TABLE 7.1 DISCRETE PROBABILITY DISTRIBUTION

	Probability mass function, $p(x)$	Moment generating function, $M(t)$	Mean	Variance
Binomial with parameters n, p; $0 \leq p \leq 1$	$\binom{n}{x}p^x(1-p)^{n-x}$ $x = 0, 1, \ldots, n$	$(pe^t + 1 - p)^n$	np	$np(1-p)$
Poisson with parameter $\lambda > 0$	$e^{-\lambda}\frac{\lambda^x}{x!}$ $x = 0, 1, 2, \ldots$	$\exp\{\lambda(e^t - 1)\}$	λ	λ
Geometric with parameter $0 \leq p \leq 1$	$p(1-p)^{x-1}$ $x = 1, 2, \ldots$	$\frac{pe^t}{1-(1-p)e^t}$	$\frac{1}{p}$	$\frac{1-p}{p^2}$
Negative binomial with parameters r, p; $0 \leq p \leq 1$	$\binom{n-1}{r-1}p^r(1-p)^{n-r}$ $n = r, r+1, \ldots$	$\left[\frac{pe^t}{1-(1-p)e^t}\right]^r$	$\frac{r}{p}$	$\frac{r(1-p)}{p^2}$

where the next-to-last equality follows from Proposition 3.1, since X and Y are independent.

Another important result is that the moment generating function uniquely determines the distribution. That is, if $M_X(t)$ exists and is finite in some region about $t = 0$, then the distribution of X is uniquely determined. For instance, if $M_X(t) = \left(\frac{1}{2}\right)^{10}(e^t + 1)^{10}$, then it follows from Table 7.1 that X is a binomial random variable with parameters 10 and $\frac{1}{2}$.

Example 6e. Suppose that the moment generating function of a random variable X is given by $M(t) = e^{3(e^t-1)}$. What is $P\{X = 0\}$?

Solution We see from Table 7.1 that $M(t) = e^{3(e^t-1)}$ is the moment generating function of a Poisson random variable with mean 3. Hence, by the one-to-one correspondence between moment generating functions and distribution functions, it follows that X must be a Poisson random variable with mean 3. Thus $P\{X = 0\} = e^{-3}$. ■

Example 6f. ***Sums of independent binomial random variables.*** If X and Y are independent binomial random variables with parameters (n, p) and (m, p), respectively, what is the distribution of $X + Y$?

Solution The moment generating function of $X + Y$ is given by

$$\begin{aligned} M_{X+Y}(t) = M_X(t)M_Y(t) &= (pe^t + 1 - p)^n(pe^t + 1 - p)^m \\ &= (pe^t + 1 - p)^{m+n} \end{aligned}$$

TABLE 7.2 CONTINUOUS PROBABILITY DISTRIBUTION

	Probability mass function, $f(x)$	Moment generating function, $M(t)$	Mean	Variance
Uniform over (a, b)	$f(x) = \begin{cases} \frac{1}{b-a} & a < x < b \\ 0 & \text{otherwise} \end{cases}$	$\frac{e^{tb} - e^{ta}}{t(b-a)}$	$\frac{a+b}{2}$	$\frac{(b-a)^2}{12}$
Exponential with parameter $\lambda > 0$	$f(x) = \begin{cases} \lambda e^{-\lambda x} & x \geq 0 \\ 0 & x < 0 \end{cases}$	$\frac{\lambda}{\lambda - t}$	$\frac{1}{\lambda}$	$\frac{1}{\lambda^2}$
Gamma with parameters $(s, \lambda), \lambda > 0$	$f(x) = \begin{cases} \frac{\lambda e^{-\lambda x}(\lambda x)^{s-1}}{\Gamma(s)} & x \geq 0 \\ 0 & x < 0 \end{cases}$	$\left(\frac{\lambda}{\lambda - t}\right)^s$	$\frac{s}{\lambda}$	$\frac{s}{\lambda^2}$
Normal with parameters (μ, σ^2)	$f(x) = \frac{1}{\sqrt{2\pi}\,\sigma} e^{-(x-\mu)^2/2\sigma^2} \quad -\infty < x < \infty$	$\exp\left\{\mu t + \frac{\sigma^2 t^2}{2}\right\}$	μ	σ^2

However, $(pe^t + 1 - p)^{m+n}$ is the moment generating function of a binomial random variable having parameters $m + n$ and p. Thus this must be the distribution of $X + Y$. ■

Example 6g. ***Sums of independent Poisson random variables.*** Calculate the distribution of $X + Y$ when X and Y are independent Poisson random variables with means λ_1 and λ_2, respectively.

Solution

$$\begin{aligned} M_{X+Y}(t) &= M_X(t)M_Y(t) \\ &= \exp\{\lambda_1(e^t - 1)\}\exp\{\lambda_2(e^t - 1)\} \\ &= \exp\{(\lambda_1 + \lambda_2)(e^t - 1)\} \end{aligned}$$

Hence $X + Y$ is Poisson distributed with mean $\lambda_1 + \lambda_2$, verifying the result given in Example 3d of Chapter 6. ■

Example 6h. ***Sums of independent normal random variables.*** Show that if X and Y are independent normal random variables with parameters (μ_1, σ_1^2) and (μ_2, σ_2^2), respectively, then $X + Y$ is normal with mean $\mu_1 + \mu_2$ and variance $\sigma_1^2 + \sigma_2^2$.

Solution

$$\begin{aligned} M_{X+Y}(t) &= M_X(t)M_Y(t) \\ &= \exp\left\{\frac{\sigma_1^2 t^2}{2} + \mu_1 t\right\}\exp\left\{\frac{\sigma_2^2 t^2}{2} + \mu_2 t\right\} \\ &= \exp\left\{\frac{(\sigma_1^2 + \sigma_2^2)t^2}{2} + (\mu_1 + \mu_2)t\right\} \end{aligned}$$

which is the moment generating function of a normal random variable with mean $\mu_1 + \mu_2$ and variance $\sigma_1^2 + \sigma_2^2$. Hence the result follows because the moment generating function uniquely determines the distribution. ■

Example 6i. Compute the moment generating function of a chi-squared random variable with n degrees of freedom.

Solution We can represent such a random variable as

$$Z_1^2 + \cdots + Z_n^2$$

where $Z_1, \ldots, Z_n$ are independent standard normal random variables. Let $M(t)$ be its moment generating function. By the preceding,

$$M(t) = (E[e^{tZ^2}])^n$$

where Z is a standard normal. Now,

$$E[e^{tZ^2}] = \frac{1}{\sqrt{2\pi}} \int_{-\infty}^{\infty} e^{tx^2} e^{-x^2/2}\, dx$$

$$= \frac{1}{\sqrt{2\pi}} \int_{-\infty}^{\infty} e^{-x^2/2\sigma^2}\, dx \qquad \text{where } \sigma^2 = (1-2t)^{-1}$$

$$= \sigma$$

$$= (1-2t)^{-1/2}$$

where the next-to-last equality uses that the normal density with mean 0 and variance σ^2 integrates to 1. Therefore,

$$M(t) = (1-2t)^{-n/2}$$

■

Example 6j. ***Moment generating function of the sum of a random number of random variables.*** Let $X_1, X_2, \ldots$ be a sequence of independent and identically distributed random variables, and let N be a nonnegative, integer-valued random variable that is independent of the sequence $X, i \geq 1$. We want to compute the moment generating function of

$$Y = \sum_{i=1}^{N} X_i$$

(In Example 4d, Y was interpreted as the amount of money spent in a store on a given day when both the amount spent by a customer and the number of such customers are random variables.)

To compute the moment generating function of Y, we first condition on N as follows:

$$E\left[\exp\left\{t \sum_{1}^{N} X_i\right\} \middle| N = n\right] = E\left[\exp\left\{t \sum_{1}^{n} X_i\right\} \middle| N = n\right]$$

$$= E\left[\exp\left\{t \sum_{1}^{n} X_i\right\}\right]$$

$$= [M_X(t)]^n$$

where

$$M_X(t) = E[e^{tX_i}]$$

Hence

$$E[e^{tY} | N] = (M_X(t))^N$$

and thus

$$M_Y(t) = E[(M_X(t))^N]$$

The moments of Y can now be obtained upon differentiation, as follows:

$$M'_Y(t) = E[N(M_X(t))^{N-1} M'_X(t)]$$

so

$$\begin{aligned}E[Y] &= M'_Y(0)\\&= E[N(M_X(0))^{N-1}M'_X(0)]\\&= E[NEX]\\&= E[N]E[X]\end{aligned} \tag{6.2}$$

verifying the result of Example 4d. (In this last set of equalities we have used the fact that $M_X(0) = E[e^{0X}] = 1$.)

Also,

$$M''_Y(t) = E[N(N-1)(M_X(t))^{N-2}(M'_X(t))^2 + N(M_X(t))^{N-1}M''_X(t)]$$

so

$$\begin{aligned}E[Y^2] &= M''_Y(0)\\&= E[N(N-1)(E[X])^2 + NE[X^2]]\\&= (E[X])^2(E[N^2] - E[N]) + E[N]E[X^2]\\&= E[N](E[X^2] - (E[X])^2) + (E[X])^2E[N^2]\\&= E[N]\,\mathrm{Var}(X) + (E[X])^2E[N^2]\end{aligned} \tag{6.3}$$

Hence, from Equations (6.2) and (6.3), we see that

$$\begin{aligned}\mathrm{Var}(Y) &= E[N]\,\mathrm{Var}(X) + (E[X])^2(E[N^2] - (E[N])^2)\\&= E[N]\,\mathrm{Var}(X) + (E[X])^2\,\mathrm{Var}(N)\end{aligned}$$

■

Example 6k. Let Y denote a uniform random variable on $(0, 1)$, and suppose that conditional on $Y = p$, the random variable X has a binomial distribution with parameters n and p. In Example 4j we showed that X is equally likely to take on any of the values $0, 1, \ldots, n$. Establish this result by using moment generating functions.

Solution To compute the moment generating function of X, start by conditioning on the value of Y. Using the formula for the binomial moment generating function gives

$$E[e^{tX}|Y = p] = (pe^t + 1 - p)^n$$

Hence, since Y is uniform on $(0, 1)$, we obtain upon taking expectations,

$$\begin{aligned}E[e^{tX}] &= \int_0^1 (pe^t + 1 - p)^n\,dp\\&= \frac{1}{e^t - 1}\int_1^{e^t} y^n\,dy \qquad \text{(by the substitution } y = pe^t + 1 - p)\\&= \frac{1}{n+1}\frac{e^{t(n+1)} - 1}{e^t - 1}\\&= \frac{1}{n+1}(1 + e^t + e^{2t} + \cdots + e^{nt})\end{aligned}$$

As the preceding is the moment generating function of a random variable that is equally likely to be any of the values $0, 1, \ldots, n$, the result follows from the fact that the moment generating function of a random variable uniquely determines its distribution. ■

7.6.1 Joint Moment Generating Functions

It is also possible to define the joint moment generating function of two or more random variables. This is done as follows. For any n random variables $X_1, \ldots, X_n$, the joint moment generating function, $M(t_1, \ldots, t_n)$, is defined for all real values of $t_1, \ldots, t_n$ by

$$M(t_1, \ldots, t_n) = E[e^{t_1X_1 + \cdots + t_nX_n}]$$

The individual moment generating functions can be obtained from $M(t_1, \ldots, t_n)$ by letting all but one of the t_j be 0. That is,

$$M_{X_i}(t) = E[e^{tX_i}] = M(0, \ldots, 0, t, 0, \ldots, 0)$$

where the t is in the ith place.

It can be proved (although the proof is too advanced for this text) that $M(t_1, \ldots, t_n)$ uniquely determines the joint distribution of $X_1, \ldots, X_n$. This result can then be used to prove that the n random variables $X_1, \ldots, X_n$ are independent if and only if

$$M(t_1, \ldots, t_n) = M_{X_1}(t_1) \cdots M_{X_n}(t_n) \tag{6.4}$$

This follows because, if the n random variables are independent, then

$$\begin{aligned}
M(t_1, \ldots, t_n) &= E[e^{(t_1X_1 + \cdots + t_nX_n)}] \\
&= E[e^{t_1X_1} \cdots e^{t_nX_n}] \\
&= E[e^{t_1X_1}] \cdots E[e^{t_nX_n}] \qquad \text{by independence} \\
&= M_{X_1}(t_1) \cdots M_{X_n}(t_n)
\end{aligned}$$

On the other hand, if Equation (6.4) is satisfied, then the joint moment generating function $M(t_1, \ldots, t_n)$ is the same as the joint moment generating function of n independent random variables, the ith of which has the same distribution as X_i. As the joint moment generating function uniquely determines the joint distribution, this must be the joint distribution; hence the random variables are independent.

Example 61. Let X and Y be independent normal random variables, each with mean μ and variance σ^2. In Example 7a of Chapter 6 we showed that $X + Y$ and $X - Y$ are independent. Let us now establish this result by computing their joint moment generating function.

$$\begin{aligned}
E[e^{t(X+Y)+s(X-Y)}] &= E[e^{(t+s)X+(t-s)Y}] \\
&= E[e^{(t+s)X}]E[e^{(t-s)Y}] \\
&= e^{\mu(t+s)+\sigma^2(t+s)^2/2}e^{\mu(t-s)+\sigma^2(t-s)^2/2} \\
&= e^{2\mu t+\sigma^2t^2}e^{\sigma^2s^2}
\end{aligned}$$

But we recognize the preceding as the joint moment generating function of the sum of a normal random variable with mean 2μ and variance $2\sigma^2$ and an independent normal random variable with mean 0 and variance $2\sigma^2$. As the joint moment generating function uniquely determines the joint distribution, it thus follows that $X + Y$ and $X - Y$ are independent normal random variables. ■

In the next example we use the joint moment generating function to verify a result that was established in Example 2b of Chapter 6.

Example 6m. Suppose that the number of events that occur is a Poisson random variable with mean λ, and that each event is independently counted with probability p. Show that the number of counted events and the number of uncounted events are independent Poisson random variables with respective means λp and $\lambda(1 - p)$.

Solution Let X denote the total number of events, and let X_c denote the number of them that are counted. To compute the joint moment generating function of X_c, the number of events that are counted, and $X - X_c$, the number that are uncounted, start by conditioning on X to obtain

$$\begin{aligned} E[e^{sX_c + t(X - X_c)}|X = n] &= e^{tn}E[e^{(s-t)X_c}|X = n] \\ &= e^{tn}(pe^{s-t} + 1 - p)^n \\ &= (pe^s + (1 - p)e^t)^n \end{aligned}$$

where the preceding equation follows since conditional on $X = n$, X_c is a binomial random variable with parameters n and p. Hence

$$E[e^{sX_c + t(X - X_c)}|X] = (pe^s + (1 - p)e^t)^X$$

Taking expectations of both sides of the preceding yields that

$$E[e^{sX_c + t(X - X_c)}] = E[(pe^s + (1 - p)e^t)^X]$$

Now, since X is Poisson with mean λ, it follows that $E[e^{tX}] = e^{\lambda(e^t - 1)}$. Therefore, for any positive value a we see (by letting $a = e^t$) that $E[a^X] = e^{\lambda(a-1)}$. Thus

$$\begin{aligned} E[e^{sX_c + t(X - X_c)}] &= e^{\lambda(pe^s + (1-p)e^t - 1)} \\ &= e^{\lambda p(e^s - 1)}e^{\lambda(1-p)(e^t - 1)} \end{aligned}$$

As the preceding is the joint moment generating function of independent Poisson random variables with respective means λp and $\lambda(1 - p)$, the result is proven. ■

7.7 ADDITIONAL PROPERTIES OF NORMAL RANDOM VARIABLES

7.7.1 The Multivariate Normal Distribution

Let $Z_1, \ldots, Z_n$ be a set of n independent unit normal random variables. If, for some constants a_{ij}, $1 \le i \le m$, $1 \le j \le n$, and μ_i, $1 \le i \le m$,

$$
\begin{aligned}
X_1 &= a_{11}Z_1 + \cdots + a_{1n}Z_n + \mu_1 \\
X_2 &= a_{21}Z_1 + \cdots + a_{2n}Z_n + \mu_2 \\
&\vdots \\
X_i &= a_{i1}Z_1 + \cdots + a_{in}Z_n + \mu_i \\
&\vdots \\
X_m &= a_{m1}Z_1 + \cdots + a_{mn}Z_n + \mu_m
\end{aligned}
$$

then the random variables $X_1, \ldots, X_m$ are said to have a multivariate normal distribution.

It follows from the fact that the sum of independent normal random variables is itself a normal random variable that each X_i is a normal random variable with mean and variance given by

$$E[X_i] = \mu_i$$

$$\operatorname{Var}(X_i) = \sum_{j=1}^{n} a_{ij}^2$$

Let us now consider

$$M(t_1, \ldots, t_m) = E[\exp\{t_1X_1 + \cdots + t_mX_m\}]$$

the joint moment generating function of $X_1, \ldots, X_m$. The first thing to note is that since $\sum_{i=1}^{m} t_iX_i$ is itself a linear combination of the independent normal random variables $Z_1, \ldots, Z_n$, it is also normally distributed. Its mean and variance are

$$E\left[\sum_{i=1}^{m} t_iX_i\right] = \sum_{i=1}^{m} t_i\mu_i$$

and

$$
\begin{aligned}
\operatorname{Var}\left(\sum_{i=1}^{m} t_iX_i\right) &= \operatorname{Cov}\left(\sum_{i=1}^{m} t_iX_i, \sum_{j=1}^{m} t_jX_j\right) \\
&= \sum_{i=1}^{m}\sum_{j=1}^{m} t_it_j \operatorname{Cov}(X_i, X_j)
\end{aligned}
$$

Now if Y is a normal random variable with mean μ and variance σ^2, then

$$E[e^Y] = M_Y(t)|_{t=1} = e^{\mu+\sigma^2/2}$$

Thus we see that

$$M(t_1, \ldots, t_m) = \exp\left\{\sum_{i=1}^{m} t_i\mu_i + \frac{1}{2}\sum_{i=1}^{m}\sum_{j=1}^{m} t_i t_j \operatorname{Cov}(X_i, X_j)\right\}$$

which shows that the joint distribution of $X_1, \ldots, X_m$ is completely determined from a knowledge of the values of $E[X_i]$ and $\operatorname{Cov}(X_i, X_j)$, $i, j = 1, \ldots, m$.

7.7.2 The Joint Distribution of the Sample Mean and Sample Variance

Let $X_1, \ldots, X_n$ be independent normal random variables, each with mean μ and variance σ^2. Let $\bar{X} = \sum_{i=1}^{n} X_i/n$ denote their sample mean. Since the sum of independent normal random variables is also a normal random variable, it follows that $\bar{X}$ is a normal random variable with (from Examples 2c and 3a) expected value μ and variance σ^2/n.

Now, recall from Example 3f that

$$\operatorname{Cov}(\bar{X}, X_i - \bar{X}) = 0, \qquad i = 1, \ldots, n \tag{7.1}$$

Also, note that since $\bar{X}, X_1 - \bar{X}, X_2 - \bar{X}, \ldots, X_n - \bar{X}$ are all linear combinations of the independent standard normals $(X_i - \mu)/\sigma$, $i = 1, \ldots, n$, it follows that $\bar{X}$, $X_i - \bar{X}, i = 1, \ldots, n$ has a joint distribution that is multivariate normal. If we let Y be a normal random variable with mean μ and variance σ^2/n that is independent of the $X_i, i = 1, \ldots, n$, then $Y, X_i - \bar{X}, i = 1, \ldots, n$ also has a multivariate normal distribution and indeed, because of (7.1), has the same expected values and covariances as the random variables $\bar{X}$, $X_i - \bar{X}$, $i = 1, \ldots, n$. But since a multivariate normal distribution is determined completely by its expected values and covariances, we can conclude that $Y, X_i - \bar{X}, i = 1, \ldots, n$ and $\bar{X}, X_i - \bar{X}, i = 1, \ldots, n$ have the same joint distribution, thus showing that $\bar{X}$ is independent of the sequence of deviations $X_i - \bar{X}, i = 1, \ldots, n$.

Since $\bar{X}$ is independent of the sequence of deviations $X_i - \bar{X}, i = 1, \ldots, n$, it follows that it is also independent of the sample variance $S^2 \equiv \sum_{i=1}^{n} (X_i - \bar{X})^2/(n - 1)$.

Since we already know that $\bar{X}$ is normal with mean μ and variance σ^2/n, it remains only to determine the distribution of S^2. To accomplish this, recall from Example 3a the algebraic identity

$$\begin{aligned}(n-1)S^2 &= \sum_{i=1}^{n} (X_i - \bar{X})^2 \\ &= \sum_{i=1}^{n} (X_i - \mu)^2 - n(\bar{X} - \mu)^2\end{aligned}$$

Upon dividing the preceding equation by σ^2, we obtain that

$$\frac{(n-1)S^2}{\sigma^2} + \left(\frac{\bar{X} - \mu}{\sigma/\sqrt{n}}\right)^2 = \sum_{i=1}^{n} \left(\frac{X_i - \mu}{\sigma}\right)^2 \tag{7.2}$$

Now,

$$\sum_{i=1}^{n}\left(\frac{X_i - \mu}{\sigma}\right)^2$$

is the sum of the squares of n independent standard normal random variables, and so is a chi-squared random variable with n degrees of freedom. Hence, from Example 6i, its moment generating function is $(1 - 2t)^{-n/2}$. Also,

$$\left(\frac{\bar{X} - \mu}{\sigma/\sqrt{n}}\right)^2$$

is the square of a standard normal and so is a chi-squared random variable with 1 degree of freedom, and so has moment generating function $(1 - 2t)^{-1/2}$. Now, we have seen previously that the two random variables on the left side of Equation (7.2) are independent. Hence, as the moment generating function of the sum of independent random variables is equal to the product of their individual moment generating functions, we see that

$$E[e^{t(n-1)S^2/\sigma^2}](1 - 2t)^{-1/2} = (1 - 2t)^{-n/2}$$

or

$$E[e^{t(n-1)S^2/\sigma^2}] = (1 - 2t)^{-(n-1)/2}$$

But as $(1 - 2t)^{-(n-1)/2}$ is the moment generating function of a chi-squared random variable with $n - 1$ degrees of freedom, we can conclude, since the moment generating function uniquely determines the distribution of the random variable, that this is the distribution of $(n - 1)S^2/\sigma^2$.

Summing up, we have shown the following.

Proposition 7.1

If $X_1, \ldots, X_n$ are independent and identically distributed normal random variables with mean μ and variance σ^2, then the sample mean $\bar{X}$ and the sample variance S^2 are independent. $\bar{X}$ is a normal random variable with mean μ and variance σ^2/n; $(n - 1)S^2/\sigma^2$ is a chi-squared random variable with $n - 1$ degrees of freedom.

7.8 GENERAL DEFINITION OF EXPECTATION

Up to this point we have defined expectations only for discrete and continuous random variables. However, there also exist random variables that are neither discrete nor continuous, and they too may possess an expectation. As an example of such a random variable, let X be a Bernoulli random variable with parameter $p = \frac{1}{2}$, and let Y be a uniformly distributed random variable over the interval $[0, 1]$. Furthermore, suppose that X and Y are independent and define the new random variable W by

$$W = \begin{cases} X & \text{if } X = 1 \\ Y & \text{if } X \neq 1 \end{cases}$$

Clearly, W is neither a discrete (since its set of possible values $[0, 1]$ is uncountable) nor a continuous (since $P\{W = 1\} = \frac{1}{2}$) random variable.

In order to define the expectation of an arbitrary random variable, we require the notion of a Stieltjes integral. Before defining the Stieltjes integral, let us recall that for any function g, $\int_a^b g(x)\,dx$ is defined by

$$\int_a^b g(x)\,dx = \lim \sum_{i=1}^{n} g(x_i)(x_i - x_{i-1})$$

where the limit is taken over all $a = x_0 < x_1 < x_2 \cdots < x_n = b$ as $n \to \infty$ and $\max_{i=1,\ldots,n} (x_i - x_{i-1}) \to 0$.

For any distribution function F, we define the Stieltjes integral of the nonnegative function g over the interval $[a, b]$ by

$$\int_a^b g(x)\,dF(x) = \lim \sum_{i=1}^{n} g(x_i)[F(x_i) - F(x_{i-1})]$$

where, as before, the limit is taken over all $a = x_0 < x_1 < \cdots < x_n = b$ as $n \to \infty$ and $\max_{i=1,\ldots,n} (x_i - x_{i-1}) \to 0$. Further, we define the Stieltjes integral over the whole real line by

$$\int_{-\infty}^{\infty} g(x)\,dF(x) = \lim_{\substack{a \to -\infty \\ b \to +\infty}} \int_a^b g(x)\,dF(x)$$

Finally, if g is not a nonnegative function, we define g^+ and g^- by

$$g^+(x) = \begin{cases} g(x) & \text{if } g(x) \geq 0 \\ 0 & \text{if } g(x) < 0 \end{cases}$$

$$g^-(x) = \begin{cases} 0 & \text{if } g(x) \geq 0 \\ -g(x) & \text{if } g(x) < 0 \end{cases}$$

As $g(x) = g^+(x) - g^-(x)$ and g^+ and g^- are both nonnegative functions, it is natural to define

$$\int_{-\infty}^{\infty} g(x)\,dF(x) = \int_{-\infty}^{\infty} g^+(x)\,dF(x) - \int_{-\infty}^{\infty} g^-(x)\,dF(x)$$

and we say that $\int_{-\infty}^{\infty} g(x)\,dF(x)$ exists as long as $\int_{-\infty}^{\infty} g^+(x)\,dF(x)$ and $\int_{-\infty}^{\infty} g^-(x)\,dF(x)$ are not both equal to $+\infty$.

If X is an arbitrary random variable having cumulative distribution F, we define the expected value of X by

$$E[X] = \int_{-\infty}^{\infty} x\,dF(x) \tag{8.1}$$

It can be shown that if X is a discrete random variable with mass function $p(x)$, then

$$\int_{-\infty}^{\infty} x\,dF(x) = \sum_{x:p(x)>0} xp(x)$$

whereas if X is a continuous random variable with density function $f(x)$, then

$$\int_{-\infty}^{\infty} x\,dF(x) = \int_{-\infty}^{\infty} xf(x)\,dx$$

The reader should note that Equation (8.1) yields an intuitive definition of $E[X]$; for consider the approximating sum

$$\sum_{i=1}^{n} x_i[F(x_i) - F(x_{i-1})]$$

of $E[X]$. As $F(x_i) - F(x_{i-1})$ is just the probability that X will be in the interval $(x_{i-1}, x_i]$, the approximating sum multiplies the approximate value of X when it is in the interval $(x_{i-1}, x_i]$ by the probability that it will be in that interval and then sums over all the intervals. Clearly, as these intervals get smaller and smaller in length, we obtain the "expected value" of X.

Stieltjes integrals are mainly of theoretical interest because they yield a compact way of defining and dealing with the properties of expectation. For instance, use of Stieltjes integrals avoids the necessity of having to give separate statements and proofs of theorems for the continuous and the discrete cases. However, their properties are very much the same as those of ordinary integrals, and all of the proofs presented in this chapter can easily be translated into proofs in the general case.

SUMMARY

If X and Y have a joint probability mass function $p(x, y)$, then

$$E[g(X,Y)] = \sum_{y}\sum_{x} g(x,y)p(x,y)$$

whereas if they have a joint density function $f(x, y)$, then

$$E[g(X,Y)] = \int_{-\infty}^{\infty}\int_{-\infty}^{\infty} g(x,y)f(x,y)\,dx\,dy$$

A consequence of the preceding is that

$$E[X + Y] = E[X] + E[Y]$$

which generalizes to

$$E\left[\sum_{i=1}^{n} X_i\right] = \sum_{i=1}^{n} E[X_i]$$

The *covariance* between random variables X and Y is given by

$$\text{Cov}(X,Y) = E[(X - E[X])(Y - E[Y])] = E[XY] - E[X]E[Y]$$

A useful identity is that

$$\text{Cov}\left(\sum_{i=1}^{n} X_i, \sum_{j=1}^{m} Y_j\right) = \sum_{i=1}^{n}\sum_{j=1}^{m} \text{Cov}(X_i, Y_j)$$

When $n = m$, and $Y_i = X_i, i = 1, \ldots, n$, the preceding gives that

$$\text{Var}\left(\sum_{i=1}^{n} X_i\right) = \sum_{i=1}^{n} \text{Var}(X_i) + 2\sum\sum_{i<j} \text{Cov}(X_i, Y_j)$$

The correlation between X and Y, denoted by $\rho(X, Y)$, is defined by

$$\rho(X, Y) = \frac{\text{Cov}(X, Y)}{\sqrt{\text{Var}(X)\,\text{Var}(Y)}}$$

If X and Y are jointly discrete random variables, then the conditional expected value of X given that $Y = y$ is defined by

$$E[X|Y = y] = \sum_x xP\{X = x|Y = y\}$$

If they are jointly continuous random variables, then

$$E[X|Y = y] = \int_{-\infty}^{\infty} xf_{X|Y}(x|y)$$

where

$$f_{X|Y}(x|y) = \frac{f(x, y)}{f_Y(y)}$$

is the conditional probability density of X given that $Y = y$. Conditional expectations, which are similar to ordinary expectations except that all probabilities are now computed conditional on the event that $Y = y$, satisfy all the properties of ordinary expectations.

Let $E[X|Y]$ denote that function of Y whose value at $Y = y$ is $E[X|Y = y]$. A very useful identity is that

$$E[X] = E[E[X|Y]]$$

In the case of discrete random variables, this reduces to the identity

$$E[X] = \sum_y E[X|Y = y]P\{Y = y\}$$

and, in the continuous case, to

$$E[X] = \int_{-\infty}^{\infty} E[X|Y = y]f_Y(y)$$

The preceding equations can often be applied to obtain $E[X]$ by first "conditioning" on the value of some other random variable Y. In addition, since for any event A, $P(A) = E[I_A]$, where I_A is 1 if A occurs and 0 otherwise, we can also use them to compute probabilities.

The conditional variance of X given that $Y = y$ is defined by

$$\text{Var}(X|Y = y) = E[(X - E[X|Y = y])^2|Y = y]$$

Let $\text{Var}(X|Y)$ be that function of Y whose value at $Y = y$ is $\text{Var}(X|Y = y)$. The following is known as the *conditional variance formula:*

$$\text{Var}(X) = E[\text{Var}(X|Y)] + \text{Var}(E[X|Y])$$

Suppose that the random variable X is to be observed and, based on its value, one must then predict the value of the random variable Y. In such a situation, it turns out that, among all predictors, $E[Y|X]$ has the smallest expectation of the square of the difference between it and Y.

The *moment generating function* of the random variable X is defined by

$$M(t) = E[e^{tX}]$$

The moments of X can be obtained by successively differentiating $M(t)$ and then evaluating the resulting quantity at $t = 0$. Specifically, we have that

$$E[X^n] = \frac{d^n}{dt^n}M(t)\bigg|_{t=0} \qquad n = 1, 2, \ldots$$

Two useful results concerning moment generating functions are, first, that the moment generating function uniquely determines the distribution function of the random variable, and second, that the moment generating function of the sum of independent random variables is equal to the product of their moment generating functions. These results lead to simple proofs that the sum of independent normal (Poisson) [gamma] random variables remains a normal (Poisson) [gamma] random variable.

If $X_1, \ldots, X_m$ are all linear combinations of a finite set of independent standard normal random variables, then they are said to have a *multivariate normal distribution.* Their joint distribution is specified by the values of $E[X_i]$, $\text{Cov}(X_i, X_j)$, $i, j = 1, \ldots, m$.

If $X_1, \ldots, X_n$ are independent and identically distributed normal random variables, then their *sample mean*

$$\bar{X} = \sum_{i=1}^{n} \frac{X_i}{n}$$

and their *sample variance*

$$S^2 = \sum_{i=1}^{n} \frac{(X_i - \bar{X})^2}{n - 1}$$

are independent. The sample mean $\bar{X}$ is a normal random variable with mean μ and variance σ^2/n; the random variable $(n - 1)S^2/\sigma^2$ is a chi-squared random variable with $n - 1$ degrees of freedom.

PROBLEMS

1. A player throws a fair die and simultaneously flips a fair coin. If the coin lands heads, then she wins twice, and if tails, then one-half of the value that appears on the die. Determine her expected winnings.

2. The game of Clue involves 6 suspects, 6 weapons, and 9 rooms. One of each is randomly chosen and the object of the game is to guess the chosen three.
(a) How many solutions are possible?
In one version of the game, after the selection is made each of the players is then randomly given three of the remaining cards. Let S, W, and R be, respectively, the numbers of suspects, weapons, and rooms in the set of three cards given to a specified player. Also, let X denote the number of solutions that are possible after that player observes his or her three cards.
(b) Express X in terms of S, W, and R. **(c)** Find $E[X]$.

3. If X and Y are independent uniform $(0, 1)$ random variables, show that

$$E[|X - Y|^{\alpha}] = \frac{2}{(\alpha + 1)(\alpha + 2)} \quad \text{for} \quad \alpha > 0$$

4. Let X and Y be independent random variables, both being equally likely to be any of the values $1, 2, \ldots, m$. Show that

$$E[|X - Y|] = \frac{(m - 1)(m + 1)}{3m}$$

5. The county hospital is located at the center of a square whose sides are 3 miles wide. If an accident occurs within this square, then the hospital sends out an ambulance. The road network is rectangular, so the travel distance from the hospital, whose coordinates are $(0, 0)$, to the point (x, y) is $|x| + |y|$. If an accident occurs at a point that is uniformly distributed in the square, find the expected travel distance of the ambulance.

6. A fair die is rolled 10 times. Calculate the expected sum of the 10 rolls.

7. Suppose that A and B each randomly, and independently, choose 3 of 10 objects. Find the expected number of objects
(a) chosen by both A and B; **(b)** not chosen by either A or B;
(c) chosen by exactly one of A and B.

8. N people arrive separately to a professional dinner. Upon arrival, each person looks to see if he or she has any friends among those present. That person then either sits at the table of a friend or at an unoccupied table if none of those present is a friend. Assuming that each of the $\binom{N}{2}$ pairs of people are, independently, friends with probability p, find the expected number of occupied tables.

HINT: Let X_i equal 1 or 0 depending on whether the ith arrival sits at a previously unoccupied table.

9. A total of n balls, numbered 1 through n, are put into n urns, also numbered 1 through n in such a way that ball i is equally likely to go into any of the urns $1, 2, \ldots, i$. Find
(a) the expected number of urns that are empty;
(b) the probability that none of the urns is empty.

10. Consider 3 trials, each having the same probability of success. Let X denote the total number of successes in these trials. If $E[X] = 1.8$, what is
(a) the largest possible value of $P\{X = 3\}$;
(b) the smallest possible value of $P\{X = 3\}$?
In both cases construct a probability scenario that results in $P\{X = 3\}$ having the stated value.

HINT: For part (b) you might start by letting U be a uniform random variable on $(0,1)$ and then defining the trials in terms of the value of U.

11. Consider n independent flips of a coin having probability p of landing heads. Say that a changeover occurs whenever an outcome differs from the one preceding it. For instance, if $n = 5$ and the outcome is $H\ H\ T\ H\ T$, then there is a total of 3 changeovers. Find the expected number of changeovers.

HINT: Express the number of changeovers as the sum of $n - 1$ Bernoulli random variables.

12. A group of n men and n women are lined up at random.
(a) Find the expected number of men who have a woman next to them.
(b) Repeat part (a), but now assuming that the group is randomly seated at a round table.

13. A set of 1000 cards, numbered 1 through 1000, are distributed among 1000 people. Compute the expected number of cards that are given to people whose age matches the number on the card.

14. An urn has m black balls. At each stage a black ball is removed and a new ball, that is black with probability p and white with probability $1 - p$, is put in its place. Find the expected number of stages needed until there are no more black balls in the urn.

NOTE: The preceding has possible applications to understanding the AIDS disease. Part of the body's immune system consists of a certain class of cells, known as T-cells. There are 2 types of T-cells, called CD4 and CD8. Now while the total number of T-cells of AIDS sufferers is (at least in the early stages of the disease) the same as that of healthy individuals, it has recently been discovered that the mix of CD4 and CD8 T-cells is different. Roughly 60 percent of the T-cells of a healthy person are of the CD4 type. whereas for AIDS sufferers the percentage of the T-cells that are of CD4 type appears to decrease continually. A recent model proposes that the HIV virus (the virus that causes AIDS) attacks CD4 cells, and that the body's mechanism for replacing killed T-cells does not differentiate between whether the killed T-cell was CD4 or CD8. Instead, it just produces a new T-cell that is CD4 with probability .6 and CD8 with probability .4. However, while this would seem to be a very efficient way of replacing killed T-cells when each one killed is equally likely to be any of the body's T-cells (and thus has probability .6 of being CD4), it has dangerous consequences when facing a virus that targets only the CD4 T-cells.

15. In Example 2h, say that i and j, $i \neq j$, form a matched pair if i chooses the hat belonging to j and j chooses the hat belonging to i. Find the expected number of matched pairs.

16. Let Z be a standard normal random variable, and for a fixed x, set

$$X = \begin{cases} Z & \text{if } Z > x \\ 0 & \text{otherwise} \end{cases}$$

Show that $E[X] = \dfrac{1}{\sqrt{2\pi}} e^{-x^2/2}$.

17. A deck of n cards, numbered 1 through n, is thoroughly shuffled so that all possible $n!$ orderings can be assumed to be equally likely. Suppose you are to make n guesses sequentially, where the ith one is a guess of the card in position i. Let N denote the number of correct guesses.
(a) If you are not given any information about your earlier guesses show that, for any strategy, $E[N] = 1$.

(b) Suppose that after each guess you are shown the card that was in the position in question. What do you think is the best strategy? Show that under this strategy

$$E[N] = \frac{1}{n} + \frac{1}{n-1} + \cdots + 1$$

$$\approx \int_1^n \frac{1}{x}\,dx = \log n$$

(c) Suppose that you are told after each guess whether you are right or wrong. In this case it can be shown that the strategy that maximizes $E[N]$ is one which keeps on guessing the same card until you are told you are correct and then changes to a new card. For this strategy show that

$$E[N] = 1 + \frac{1}{2!} + \frac{1}{3!} + \cdots + \frac{1}{n!}$$

$$\approx e - 1$$

HINT: For all parts, express N as the sum of indicator (that is, Bernoulli) random variables.

18. Cards from an ordinary deck of 52 playing cards are turned face up one at a time. If the first card is an ace, or the second a deuce, or the third a three, or ..., or the thirteenth a king, or the fourteenth an ace, and so on, we say that a match occurs. Note that we do not require that the $(13n + 1)$th card be any particular ace for a match to occur but only that it be an ace. Compute the expected number of matches that occur.

19. A certain region is inhabited by r distinct types of a certain kind of insect species, and each insect caught will, independently of the types of the previous catches, be of type i with probability

$$P_i, i = 1, \ldots, r \qquad \sum_1^r P_i = 1$$

(a) Compute the mean number of insects that are caught before the first type 1 catch.
(b) Compute the mean number of types of insects that are caught before the first type 1 catch.

20. An urn contains n balls—the ith having weight $W(i)$, $i = 1, \ldots, n$. The balls are removed without replacement one at a time according to the following rule: At each selection, the probability that a given ball in the urn is chosen is equal to its weight divided by the sum of the weights remaining in the urn. For instance, if at some time $i_1, \ldots, i_r$ is the set of balls remaining in the urn, then the next selection will be i_j with probability $W(i_j)\Big/\sum_{k=1}^{r} W(i_k)$, $j = 1, \ldots, r$. Compute the expected number of balls that are withdrawn before ball number 1.

21. For a group of 100 people, compute
(a) the expected number of days of the year that are birthdays of exactly 3 people:
(b) the expected number of distinct birthdays.

22. How many times would you expect to roll a fair die before all 6 sides appeared at least once?

23. Urn 1 contains 5 white and 6 black balls, while urn 2 contains 8 white and 10 black balls. Two balls are randomly selected from urn 1 and are then put in urn 2. If 3 balls are then randomly selected from urn 2, compute the expected number of white balls in the trio.

HINT: Let $X_i = 1$ if the ith white ball initially in urn 1 is one of the three selected, and let $X_i = 0$ otherwise. Similarly, let $Y_i = 1$ if the ith white ball from urn 2 is one of the three selected, and let $Y_i = 0$ otherwise. The number of white balls in the trio can now be written as $\sum_1^5 X_i + \sum_1^8 Y_i$.

24. A bottle initially contains m large pills and n small pills. Each day a patient randomly chooses one of the pills. If a small pill is chosen, then that pill is eaten. If a large pill is chosen, then the pill is broken in two; one part is returned to the bottle (and is now considered a small pill) and the other part is then eaten.

(a) Let X denote the number of small pills in the bottle after the last large pill has been chosen and its smaller half returned. Find $E[X]$.

HINT: Define $n + m$ indicator variables, one for each of the small pills initially present and one for each of the m small pills created when a large one is split in two. Now use the argument of Example 2m.

(b) Let Y denote the day on which the last large pill is chosen. Find $E[Y]$.

HINT: What is the relationship between X and Y?

25. Let $X_1, X_2, \ldots$ be a sequence of independent and identically distributed continuous random variables. Let $N \geq 2$ be such that

$$X_1 \geq X_2 \geq \cdots \geq X_{N-1} < X_N$$

That is, N is the point at which the sequence stops decreasing. Show that $E[N] = e$.

HINT: First find $P\{N \geq n\}$.

26. If $X_1, X_2, \ldots, X_n$ are independent and identically distributed random variables having uniform distributions over $(0, 1)$, find

(a) $E[\max(X_1, \ldots, X_n)]$; **(b)** $E[\min(X_1, \ldots, X_n)]$.

***27.** If 101 items are distributed among 10 boxes, then at least one of the boxes must contain more than 10 items. Use the probabilistic method to prove this result.

***28.** The k of r out of n circular reliability system, $k \leq r \leq n$, consists of n components that are arranged in a circular fashion. Each component is either functional or failed, and the system functions if there is no block of r consecutive components of which at least k are failed. Show that there is no way to arrange 47 components, 8 of which are failed, to make a functional 3 of 12 out of 47 circular system.

***29.** There are 4 different types of coupons, the first 2 of which compose one group and the second 2 another group. Each new coupon obtained is type i with probability p_i, where $p_1 = p_2 = 1/8$, $p_3 = p_4 = 3/8$. Find the expected number of coupons that one must obtain to have at least one of

(a) all 4 types; **(b)** all the types of the first group;

(c) all the types of the second group; **(d)** all the types of either group.

30. If X and Y are independent and identically distributed with mean μ and variance σ^2, find

$$E[(X - Y)^2]$$

31. In Problem 6, calculate the variance of the sum of the rolls.

32. In Problem 9, compute the variance of the number of empty urns.

33. If $E[X] = 1$ and $\text{Var}(X) = 5$ find

(a) $E[(2 + X)^2]$; **(b)** $\text{Var}(4 + 3X)$.

34. If 10 married couples are randomly seated at a round table, compute (a) the expected number and (b) the variance of the number of wives who are seated next to their husbands.

35. Cards from an ordinary deck are turned face up one at a time. Compute the expected number of cards that need be turned face up in order to obtain
(a) 2 aces; **(b)** 5 spades; **(c)** all 13 hearts.

36. Let X be the number of 1's and Y the number of 2's that occur in n rolls of a fair die. Compute $\text{Cov}(X, Y)$.

37. A die is rolled twice. Let X equal the sum of the outcomes, and let Y equal the first outcome minus the second. Compute $\text{Cov}(X, Y)$.

38. The random variables X and Y have a joint density function given by

$$f(x, y) = \begin{cases} 2e^{-2x}/x & 0 \le x < \infty, 0 \le y \le x \\ 0 & \text{otherwise} \end{cases}$$

Compute $\text{Cov}(X, Y)$.

39. Let $X_1, \ldots$ be independent with common mean μ and common variance σ^2, and set $Y_n = X_n + X_{n+1} + X_{n+2}$. For $j \ge 0$, find $\text{Cov}(Y_n, Y_{n+j})$.

40. The joint density function of X and Y is given by

$$f(x, y) = \frac{1}{y} e^{-(y+x/y)}, \qquad x > 0, y > 0$$

Find $E[X]$, $E[Y]$, and show that $\text{Cov}(X, Y) = 1$.

41. A pond contains 100 fish, of which 30 are carp. If 20 fish are caught, what are the mean and variance of the number of carp among these 20? What assumptions are you making?

42. A group of 20 people—consisting of 10 men and 10 women—are randomly arranged into 10 pairs of 2 each. Compute the expectation and variance of the number of pairs that consist of a man and a woman. Now suppose the 20 people consisted of 10 married couples. Compute the mean and variance of the number of married couples that are paired together.

43. Let $X_1, X_2, \ldots, X_n$ be independent random variables having an unknown continuous distribution function F, and let $Y_1, Y_2, \ldots, Y_m$ be independent random variables having an unknown continuous distribution function G. Now order those $n + m$ variables and let

$$I_i = \begin{cases} 1 & \text{if the } i\text{th smallest of the } n + m \text{ variables is from the } X \text{ sample} \\ 0 & \text{otherwise} \end{cases}$$

The random variable $R = \sum_{i=1}^{n+m} iI_i$ is the sum of the ranks of the X sample and is the basis of a standard statistical procedure (called the Wilcoxon sum of ranks test) for testing whether F and G are identical distributions. This test accepts the hypothesis that $F = G$ when R is neither too large nor too small. Assuming that the hypothesis of equality is in fact correct, compute the mean and variance of R.

HINT: Use the results of Example 3e.

44. There are two distinct methods for manufacturing certain goods, the quality of goods produced by method i being a continuous random variable having distribution F_i, $i = 1, 2$. Suppose that n goods are produced by method 1 and m by method 2. Rank the $n + m$ goods according to quality and let

$$X_j = \begin{cases} 1 & \text{if the } j\text{th best was produced from method 1} \\ 2 & \text{otherwise} \end{cases}$$

For the vector $X_1, X_2, \ldots, X_{n+m}$, which consists of n 1's and m 2's, let R denote the number of runs of 1. For instance, if $n = 5$, $m = 2$, and $X = 1, 2, 1, 1, 1, 1, 2$, then $R = 2$. If $F_1 = F_2$ (that is, if the two methods produce identically distributed goods), what are the mean and variance of R?

45. If X_1, X_2, X_3, X_4 are (pairwise) uncorrelated random variables each having mean 0 and variance 1, compute the correlations of
(a) $X_1 + X_2$ and $X_2 + X_3$; **(b)** $X_1 + X_2$ and $X_3 + X_4$.

46. Consider the following dice game, as played at a certain gambling casino: Players 1 and 2 roll in turn a pair of dice. The bank then rolls the dice to determine the outcome according to the following: player $i, i = 1, 2$, wins if his roll is strictly greater than the bank's. Let for $i = 1, 2$,

$$I_i = \begin{cases} 1 & \text{if } i \text{ wins} \\ 0 & \text{otherwise} \end{cases}$$

and show that I_1 and I_2 are positively correlated. Explain why this result was to be expected.

47. Consider a graph having n vertices labeled $1, 2, \ldots, n$, and suppose that between each of the $\binom{n}{2}$ pairs of distinct verticcs an edge is, independently, present with probability p. The degree of vertex i, designated as D_i, is the number of edges that have vertex i as one of its vertices.
(a) What is the distribution of D_i?
(b) Find $\rho(D_i, D_j)$, the correlation between D_i and D_j.

48. A fair die is successively rolled. Let X and Y denote, respectively, the number of rolls necessary to obtain a 6 and a 5. Find
(a) $E[X]$; **(b)** $E[X|Y = 1]$; **(c)** $E[X|Y = 5]$.

49. An urn contains 4 white and 6 black balls. Two successive random samples of sizes 3 and 5, respectively, are drawn from the urn without replacement. Let X and Y denote the number of white balls in the two samples, and compute $E[X|Y = i]$, for $i = 1, 2, 3, 4$.

50. The joint density of X and Y is given by

$$f(x, y) = \frac{e^{-x/y}e^{-y}}{y} \qquad 0 < x < \infty, 0 < y < \infty$$

Compute $E[X^2|Y = y]$.

51. The joint density of X and Y is given by

$$f(x, y) = \frac{e^{-y}}{y} \qquad 0 < x < y, 0 < y < \infty$$

Compute $E[X^3|Y = y]$.

52. A population is made up of r disjoint subgroups. Let p_i denote the proportion of the population that is in subgroup $i, i = 1, \ldots, r$. If the average weight of the members of subgroup i is $w_i, i = 1, \ldots, r$, what is the average weight of the members of the population?

53. A prisoner is trapped in a cell containing 3 doors. The first door leads to a tunnel that returns him to his cell after 2 days travel. The second leads to a tunnel that returns him to his cell after 4 days travel. The third door leads to freedom after 1 day of travel. If it is assumed that the prisoner will always select doors 1, 2, and 3 with respective probabilities .5, .3, and .2. what is the expected number of days until the prisoner reaches freedom?

54. Consider the following dice game. A pair of dice are rolled. If the sum is 7, then the game ends and you win 0. If thc sum is not 7, then you have the option of either stopping the game and receiving an amount equal to that sum or starting over again. For each value of $i, i = 2, \ldots, 12$, find your expected return if you employ the strategy of stopping the first time that a value at least as large as i appears. What value of i leads to the largest expected return?

HINT: Let X_i denote the return when you use the critical value i. To compute $E[X_i]$, condition on the initial sum.

55. Ten hunters are waiting for ducks to fly by. When a flock of ducks flies overhead, the hunters fire at the same time, but each chooses his target at random, independently of the others. If each hunter independently hits his target with probability .6, compute the expected number of ducks that are hit. Assume that the number of ducks in a flock is a Poisson random variable with mean 6.

56. The number of people who enter an elevator on the ground floor is a Poisson random variable with mean 10. If there are N floors above the ground floor and if each person is equally likely to get off at any one of these N floors, independently of where the others get off, compute the expected number of stops that the elevator will make before discharging all of its passengers.

57. Suppose that the expected number of accidents per week at an industrial plant is 5. Suppose also that the numbers of workers injured in each accident are independent random variables with a common mean of 2.5. If the number of workers injured in each accident is independent of the number of accidents that occur, compute the expected number of workers injured in a week.

58. A coin having probability p of coming up heads is continually flipped until both heads and tails have appeared. Find
(a) the expected number of flips;
(b) the probability that the last flip lands heads.

59. A person continually flips a coin until a run of 3 consecutive heads appears. Assuming that each flip independently lands heads with probability p, determine the expected number of flips required.

HINT: Let T denote the first flip that lands on tails and let it be 0 if all flips land on heads, and then condition on T.

60. There are $n + 1$ participants in a game. Each person, independently, is a winner with probability p. The winners share a total prize of 1 unit. (For instance, if 4 people win, then each of them receives $\frac{1}{4}$, whereas if there are no winners, then none of the participants receive anything.) Let A denote a specified one of the players, and let X denote the amount that is received by A.
(a) Compute the expected total prize shared by the players.
(b) Argue that $E[X] = \dfrac{1 - (1 - p)^{n+1}}{n + 1}$.
(c) Compute $E[X]$ by conditioning on whether A is a winner, and conclude that

$$E[(1 + B)^{-1}] = \frac{1 - (1 - p)^{n+1}}{(n + 1)p}$$

when B is a binomial random variable with parameters n and p.

61. Each of $m + 2$ players pays 1 unit to a kitty in order to play the following game. A fair coin is to be flipped successively n times, where n is an odd number, and the successive outcomes noted. Each player writes down, before the flips, a prediction of the outcomes. For instance, if $n = 3$, then a player might write down (H, H, T), which means that he or she predicts that the first flip will land heads, the second heads, and the third tails. After the coins are flipped, the players count their total number of correct predictions. Thus, if the actual outcomes are all heads, then the player who wrote (H, H, T) would have 2 correct predictions. The total kitty of $m + 2$ is then evenly split up among those players having the largest number of correct predictions.

Since each of the coin flips is equally likely to land on either heads or tails, m of the players have decided to make their predictions in a totally random fashion. Specifically, they will each flip one of their own fair coins n times and then use the result as their prediction. However, the final 2 of the players have formed a syndicate and will use the following strategy. One of them will make predictions in the same random fashion as the other m players, but the other one will then predict exactly the opposite of the first. That is, when the randomizing member of the syndicate predicts an H, the other member predicts a T. For instance, if the randomizing member of the syndicate predicts (H, H, T), then the other one predicts (T, T, H).

(a) Argue that exactly one of the syndicate members will have more than $n/2$ correct predictions. (Remember, n is odd.)

(b) Let X denote the number of the m nonsyndicate players that have more than $n/2$ correct predictions. What is the distribution of X?

(c) With X as defined in part (b), argue that

$$E[\text{payoff to the syndicate}] = (m + 2)E\left[\frac{1}{X + 1}\right]$$

(d) Use part (c) of Problem 60 to conclude that

$$E[\text{payoff to the syndicate}] = \frac{2(m + 2)}{m + 1}\left[1 - \left(\frac{1}{2}\right)^{m+1}\right]$$

and explicitly compute this when $m = 1, 2,$ and 3. As it can be shown that

$$\frac{2(m + 2)}{m + 1}\left[1 - \left(\frac{1}{2}\right)^{m+1}\right] > 2$$

it follows that the syndicate's strategy always gives it a positive expected profit.

62. Let $U_1, U_2, \ldots$ be a sequence of independent uniform $(0, 1)$ random variables. In Example 4h we showed that for $0 \le x \le 1$, $E[N(x)] = e^x$, where

$$N(x) = \min\left\{n: \sum_{i=1}^{n} U_i > x\right\}$$

This problem gives another approach to establishing this result.

(a) Show by induction on n that for $0 < x \le 1$ and all $n \ge 0$,

$$P\{N(x) \ge n + 1\} = \frac{x^n}{n!}$$

HINT: First condition on U_1 and then use the induction hypothesis.

(b) Use part (a) to conclude that

$$E[N(x)] = e^x$$

63. An urn contains 30 balls, of which 10 are red and 8 are blue. From this urn, 12 balls are randomly withdrawn. Let X denote the number of red, and Y the number of blue balls that are withdrawn. Find $\text{Cov}(X, Y)$

(a) by defining appropriate indicator (that is, Bernoulli) random variables

$$X_i, Y_j \text{ such that } X = \sum_{i=1}^{10} X_i, Y = \sum_{j=1}^{8} Y_j$$

(b) by conditioning (on either X or Y) to determine $E[XY]$.

64. Type i light bulbs function for a random amount of time having mean μ_i and standard deviation $\sigma_i, i = 1, 2$. A light bulb randomly chosen from a bin of bulbs is a type 1 bulb with probability p, and a type 2 bulb with probability $1 - p$. Let X denote the lifetime of this bulb. Find
(a) $E[X]$; **(b)** $\text{Var}(X)$.

65. The number of winter storms in a good year is a Poisson random variable with mean 3, whereas the number in a bad year is a Poisson random variable with mean 5. If next year will be a good year with probability .4 or a bad year with probability .6, find the expected value and variance of the number of storms that will occur.

66. In Example 4c compute the variance of the length of time until the miner reaches safety.

67. Consider a gambler who at each gamble either wins or loses her bet with probabilities p and $1 - p$. When $p > \frac{1}{2}$, a popular gambling system, known as the Kelley strategy, is to always bet the fraction $2p - 1$ of your current fortune. Compute the expected fortune after n gambles of a gambler who starts with x units and employs the Kelley strategy.

68. The number of accidents that a person has in a given year is a Poisson random variable with mean λ. However, suppose that the value of λ changes from person to person, being equal to 2 for 60 percent of the population and 3 for the other 40 percent. If a person is chosen at random, what is the probability that he will have (a) 0 accidents and (b) exactly 3 accidents in a year? What is the conditional probability that he will have 3 accidents in a given year, given that he had no accidents the preceding year?

69. Repeat Problem 68 when the proportion of the population having a value of λ less than x is equal to $1 - e^{-x}$.

70. Consider an urn containing a large number of coins and suppose that each of the coins has some probability p of turning up heads when it is flipped. However, this value of p varies from coin to coin. Suppose that the composition of the urn is such that if a coin is selected at random from the urn, then its p-value can be regarded as being the value of a random variable that is uniformly distributed over $[0, 1]$. If a coin is selected at random from the urn and flipped twice, compute the probability that
(a) the first flip is a head; **(b)** both flips are heads.

71. In Problem 70, suppose that the coin is tossed n times. Let X denote the number of heads that occur. Show that

$$P\{X = i\} = \frac{1}{n + 1} \qquad i = 0, 1, \ldots, n$$

HINT: Make use of the fact that

$$\int_0^1 x^{a-1}(1 - x)^{b-1}\, dx = \frac{(a - 1)!\,(b - 1)!}{(a + b - 1)!}$$

when a and b are positive integers.

72. Suppose that in Problem 70 we continue to flip the coin until a head appears. Let N denote the number of flips needed. Find
(a) $P\{N \geq i\}, i \geq 0$; **(b)** $P\{N = i\}$; **(c)** $E[N]$.

73. In Example 5b let S denote the signal sent and R the signal received.
(a) Compute $E[R]$. **(b)** Compute $\text{Var}(R)$.
(c) Is R normally distributed? **(d)** Compute $\text{Cov}(R, S)$.

74. In Example 5c, suppose that X is uniformly distributed over $(0, 1)$. If the discretized regions are determined by $a_0 = 0$, $a_1 = \frac{1}{2}$, $a_2 = 1$, determine the optimal quantizer Y and compute $E[(X - Y)^2]$.

75. The moment generating function of X is given by $M_X(t) = \exp\{2e^t - 2\}$ and that of Y by $M_Y(t) = \left(\frac{3}{4}e^t + \frac{1}{4}\right)^{10}$. If X and Y are independent, what are
(a) $P\{X + Y = 2\}$; **(b)** $P\{XY = 0\}$; **(c)** $E[XY]$?

76. Let X be the value of the first die and Y the sum of the values when two dice are rolled. Compute the joint moment generating function of X and Y.

77. The joint density of X and Y is given by

$$f(x, y) = \frac{1}{\sqrt{2\pi}} e^{-y} e^{-(x-y)^2/2} \qquad 0 < y < \infty, -\infty < x < \infty$$

(a) Compute the joint moment generating function of X and Y.
(b) Compute the individual moment generating functions.

78. Two envelopes, each containing a check, are placed in front of you. You are to choose one of the envelopes, open it, and see the amount of the check. At this point you can either accept that amount or you can exchange it for the check in the unopened envelope. What should you do? Is it possible to devise a strategy that does better than just accepting the first envelope?

Let A and B, $A < B$, denote the (unknown) amounts of the checks. and note that the strategy that randomly selects an envelope and always accepts its check has an expected return of $(A + B)/2$. Consider the following strategy: Let $F(\cdot)$ be any strictly increasing (that is, continuous) distribution function. Randomly choose an envelope and open it. If the discovered check has value x then accept it with probability $F(x)$, and with probability $1 - F(x)$ exchange it.

(a) Show that if you employ the latter strategy, then your expected return is greater than $(A + B)/2$.

HINT: Condition on whether the first envelope has value A or B.

Now consider the strategy that fixes a value x, and then accepts the first check if its value is greater than x and exchanges it otherwise.

(b) Show that for any x, the expected return under the x-strategy is always at least $(A + B)/2$, and that it is strictly larger than $(A + B)/2$ if x lies between A and B.

(c) Let X be a continuous random variable on the whole line, and consider the following strategy: Generate the value of X, and if $X = x$ then employ the x-strategy of part (b). Show that the expected return under this strategy is greater than $(A + B)/2$.

THEORETICAL EXERCISES

1. Show that $E[(X - a)^2]$ is minimized at $a = E[X]$.

2. Suppose that X is a continuous random variable with density function f. Show that $E[|X - a|]$ is minimized when a is equal to the median of F.

HINT: Write

$$E[|X - a|] = \int |x - a| f(x)\, dx$$

Now break up the integral into the regions where $x < a$ and where $x > a$ and differentiate.

3. Prove Proposition 2.1 when
(a) X and Y have a joint probability mass function;
(b) X and Y have a joint probability density function and $g(x, y) \geq 0$ for all x, y.

4. Let X be a random variable having finite expectation μ and variance σ^2, and let $g(\cdot)$ be a twice differentiable function. Show that

$$E[g(X)] \approx g(\mu) + \frac{g''(\mu)}{2}\sigma^2$$

HINT: Expand $g(\cdot)$ in a Taylor series about μ. Use the first three terms and ignore the remainder.

5. Let $A_1, A_2, \ldots, A_n$ be arbitrary events, and define $C_k = \{\text{at least } k \text{ of the } A_i \text{ occur}\}$. Show that

$$\sum_{k=1}^{n} P(C_k) = \sum_{k=1}^{n} P(A_k)$$

HINT: Let X denote the number of the A_i that occur. Show that both sides of the preceding are equal to $E[X]$.

6. In the text we noted that

$$E\left[\sum_{i=1}^{\infty} X_i\right] = \sum_{i=1}^{\infty} E[X_i]$$

when the X_i are all nonnegative random variables. Since an integral is a limit of sums, one might expect that

$$E\left[\int_0^{\infty} X(t)\,dt\right] = \int_0^{\infty} E[X(t)]\,dt$$

whenever $X(t), 0 \leq t < \infty$, are all nonnegative random variables; and this result is indeed true. Use it to give another proof of the result that, for a nonnegative random variable X,

$$E[X] = \int_0^{\infty} P\{X > t\}\,dt$$

HINT: Define, for each nonnegative t, the random variable $X(t)$ by

$$X(t) = \begin{cases} 1 & \text{if } t < X \\ 0 & \text{if } t \geq X \end{cases}$$

Now relate $\int_0^{\infty} X(t)\,dt$ to X.

7. We say that X is *stochastically larger* than Y, written $X \geq_{\text{st}} Y$, if for all t.

$$P\{X > t\} \geq P\{Y > t\}$$

Show that if $X \geq_{\text{st}} Y$, then $E[X] \geq E[Y]$ when
(a) X and Y are nonnegative random variables;
(b) X and Y are arbitrary random variables.

HINT: Write X as

$$X = X^+ - X^-$$

where

$$X^+ = \begin{cases} X & \text{if } X \geq 0 \\ 0 & \text{if } X < 0 \end{cases}, \quad X^- = \begin{cases} 0 & \text{if } X \geq 0 \\ -X & \text{if } X < 0 \end{cases}$$

Similarly, represent Y as $Y^+ - Y^-$. Then make use of part (a).

8. Show that X is stochastically larger than Y if and only if

$$E[f(X)] \geq E[f(Y)]$$

for all increasing functions f.

HINT: If $X \geq_{st} Y$, show that $E[f(X)] \geq E[f(Y)]$ by showing that $f(X) \geq_{st} f(Y)$ and then using Theoretical Exercise 7. To show that $E[f(X)] \geq E[f(Y)]$ for all increasing functions f implies that $P\{X > t\} \geq P\{Y > t\}$, define an appropriate increasing function f.

9. A coin having probability p of landing heads is flipped n times. Compute the expected number of runs of heads of size 1, of size 2, of size k, $1 \leq k \leq n$.

10. Let $X_1, X_2, \ldots, X_n$ be independent and identically distributed positive random variables. Find, for $k \leq n$,

$$E\left[\frac{\sum_{i=1}^{k} X_i}{\sum_{i=1}^{n} X_i}\right]$$

11. Consider n independent trials each resulting in any one of r possible outcomes with probabilities $P_1, P_2, \ldots, P_r$. Let X denote the number of outcomes that never occur in any of the trials. Find $E[X]$ and show that among all probability vectors $P_1, \ldots, P_r$, $E[X]$ is minimized when $P_i = 1/r$, $i = 1, \ldots, r$.

12. Independent trials are performed. If the ith such trial results in a success with probability P_i, compute (a) the expected number, and (b) the variance, of the number of successes that occur in the first n trials. Does independence make a difference in part (a)? In part (b)?

13. Let $X_1, \ldots, X_n$ be independent and identically distributed continuous random variables. We say that a record value occurs at time j, $j \leq n$, if $X_j \geq X_i$ for all $1 \leq i \leq j$. Show that

(a) $E[\text{number of record values}] = \sum_{j=1}^{n} 1/j$;

(b) $\text{Var}(\text{number of record values}) = \sum_{j=1}^{n} (j-1)/j^2$.

14. For Example 2j show that the variance of the number of coupons needed to amass a full set is equal to

$$\sum_{i=1}^{N-1} \frac{iN}{(N-i)^2}$$

When N is large, this can be shown to be approximately equal (in the sense that their ratio approaches 1 as $N \to \infty$) to $N^2\pi^2/6$.

15. Consider n independent trials, the ith of which results in a success with probability P_i.

(a) Compute the expected number of successes in the n trials—call it μ.

(b) For fixed value of μ, what choice of $P_1, \ldots, P_n$ maximizes the variance of the number of successes?

(c) What choice minimizes the variance?

***16.** Suppose that each of the elements of $S = \{1, 2, \ldots, n\}$ is to be colored either red or blue. If $A_1, \ldots, A_r$ are subsets of S, show that there is a way of doing the coloring so that at most $\sum_{i=1}^{r} (1/2)^{|A_i|-1}$ of these subsets have all their elements the same color (where $|A|$ denotes the number of elements in the set A).

17. Suppose that X_1 and X_2 are independent random variables having a common mean μ. Suppose also that $\text{Var}(X_1) = \sigma_1^2$ and $\text{Var}(X_2) = \sigma_2^2$. The value of μ is unknown and it is proposed to estimate μ by a weighted average of X_1 and X_2. That is, $\lambda X_1 + (1 - \lambda)X_2$ will be used as an estimate of μ, for some appropriate value of λ. Which value of λ yields the estimate having the lowest possible variance? Explain why it is desirable to use this value of λ.

18. In Example 3h we showed that the covariance of the multinomial random variables N_i and N_j is equal to $-mP_iP_j$ by expressing N_i and N_j as the sum of indicator variables. This result could also have been obtained by using the formula

$$\text{Var}(N_i + N_j) = \text{Var}(N_i) + \text{Var}(N_j) + 2\,\text{Cov}(N_i, N_j)$$

(a) What is the distribution of $N_i + N_j$?

(b) Use the preceding identity to show that $\text{Cov}(N_i, N_j) = -mP_iP_j$.

19. If X and Y are identically distributed, not necessarily independent, show that

$$\text{Cov}(X + Y, X - Y) = 0$$

20. *The Conditional Covariance Formula.* The conditional covariance of X and Y, given Z, is defined by

$$\text{Cov}(X, Y|Z) \equiv E[(X - E[X|Z])(Y - E[Y|Z])|Z]$$

(a) Show that

$$\text{Cov}(X, Y|Z) = E[XY|Z] - E[X|Z]E[Y|Z]$$

(b) Prove the conditional covariance formula

$$\text{Cov}(X, Y) = E[\text{Cov}(X, Y|Z)] + \text{Cov}(E[X|Z], E[Y|Z])$$

(c) Set $X = Y$ in part (b) and obtain the conditional variance formula.

21. Let $X_{(i)}, i = 1, \ldots, n$, denote the order statistics from a set of n uniform $(0, 1)$ random variables and note that the density function of $X_{(i)}$ is given by

$$f(x) = \frac{n!}{(i-1)!\,(n-i)!} x^{i-1}(1-x)^{n-i} \qquad 0 < x < 1$$

(a) Compute $\text{Var}(X_{(i)}), i = 1, \ldots, n$.

(b) Which value of i minimizes and which value maximizes $\text{Var}(X_{(i)})$?

22. If $Y = a + bX$, show that

$$\rho(X, Y) = \begin{cases} +1 & \text{if } b > 0 \\ -1 & \text{if } b < 0 \end{cases}$$

23. If Z is a standard normal random variable and if Y is defined by $Y = a + bZ + cZ^2$, show that

$$\rho(Y, Z) = \frac{b}{\sqrt{b^2 + 2c^2}}$$

24. Prove the Cauchy–Schwarz inequality, namely, that

$$(E[XY])^2 \le E[X^2]E[Y^2]$$

HINT: Unless $Y = -tX$ for some constant, in which case this inequality holds with equality, if follows that for all t,

$$0 < E[(tX + Y)^2] = E[X^2]t^2 + 2E[XY]t + E[Y^2]$$

Hence the roots of the quadratic equation

$$E[X^2]t^2 + 2E[XY]t + E[Y^2] = 0$$

must be imaginary, which implies that the discriminant of this quadratic equation must be negative.

25. Show that if X and Y are independent, then

$$E[X|Y = y] = E[X] \quad \text{for all } y$$

(a) in the discrete case; **(b)** in the continuous case.

26. Prove that $E[g(X)Y|X] = g(X)E[Y|X]$.

27. Prove that if $E[Y|X = x] = E[Y]$ for all x, then X and Y are uncorrelated; and give a counterexample to show that the converse is not true.

HINT: Prove and use the fact that $E[XY] = E[XE[Y|X]]$.

28. Show $\text{Cov}(X, E[Y|X]) = \text{Cov}(X, Y)$.

29. Let $X_1, \ldots, X_n$ be independent and identically distributed random variables. Find

$$E[X_1|X_1 + \cdots + X_n = x]$$

30. Consider Example 3h, which is concerned with the multinomial distribution. Use conditional expectation to compute $E[N_iN_j]$ and then use this to verify the formula for $\text{Cov}(N_i, N_j)$ given in Example 3h.

31. An urn initially contains b black and w white balls. At each stage we add r black balls and then withdraw, at random, r from the $b + w + r$. Show that

$$E[\text{number of white balls after stage } t] = \left(\frac{b + w}{b + w + r}\right)^t w$$

32. Suppose that conditional on $Y = y$, the random variables X_1 and X_2 are independent with mean y. Show that

$$\text{Cov}(X_1, X_2) = \text{Var}(Y)$$

33. A coin, which lands on heads with probability p, is continually flipped. Compute the expected number of flips that are made until a string of r heads in a row is obtained.

HINT: Condition on the time of the first occurrence of tails, to obtain the equation

$$E[X] = (1 - p)\sum_{i=1}^{r} p^{i-1}(i + E[X]) + (1 - p)\sum_{i=r+1}^{\infty} p^{i-1}r$$

Simplify and solve for $E[X]$.

34. For another approach to Theoretical Exercise 33, let T_r denote the number of flips required to obtain a run of r consecutive heads.

(a) Determine $E[T_r|T_{r-1}]$.
(b) Determine $E[T_r]$ in terms of $E[T_{r-1}]$.
(c) What is $E[T_1]$?
(d) What is $E[T_r]$?

35. (a) Prove that

$$E[X] = E[X|X < a]P\{X < a\} + E[X|X \geq a]P\{X \geq a\}$$

HINT: Define an appropriate random variable and then compute $E[X]$ by conditioning on it.

(b) Use part (a) to prove Markov's inequality, which states that if $P\{X \geq 0\} = 1$, then for $a > 0$,

$$P\{X \geq a\} \leq \frac{E[X]}{a}$$

36. One ball at a time is randomly selected from an urn containing a white and b black balls until all of the remaining balls are of the same color. Let $M_{a,b}$ denote the expected number of balls left in the urn when the experiment ends. Compute a recursive formula for $M_{a,b}$ and solve when $a = 3, b = 5$.

37. An urn contains a white and b black balls. After a ball is drawn, it is returned to the urn if it is white; but if it is black, it is replaced by a white ball from another urn. Let M_n denote the expected number of white balls in the urn after the foregoing operation has been repeated n times.

(a) Derive the recursive equation

$$M_{n+1} = \left(1 - \frac{1}{a+b}\right)M_n + 1$$

(b) Use part (a) to prove that

$$M_n = a + b - b\left(1 - \frac{1}{a+b}\right)^n$$

(c) What is the probability that the $(n+1)$st ball drawn is white?

38. The best linear predictor of Y with respect to X_1 and X_2 is equal to $a + bX_1 + cX_2$, where a, b, and c are chosen to minimize

$$E[(Y - (a + bX_1 + cX_2))^2]$$

Determine a, b, and c.

39. The best quadratic predictor of Y with respect to X is $a + bX + cX^2$, where a, b, and c are chosen to minimize $E[(Y - (a + bX + cX^2))^2]$. Determine a, b, and c.

40. X and Y are jointly normally distributed with joint density function given by

$$f(x, y) = \frac{1}{2\pi\sigma_x\sigma_y\sqrt{1-\rho^2}} \times \exp\left\{-\frac{1}{2(1-\rho^2)}\left[\left(\frac{x-\mu_x}{\sigma_x}\right)^2 + \left(\frac{y-\mu_y}{\sigma_y}\right)^2 - 2\rho\frac{(x-\mu_x)(y-\mu_y)}{\sigma_x\sigma_y}\right]\right\}$$

(a) Show that the conditional distribution of Y, given $X = x$, is normal with mean $\mu_y + \rho\frac{\sigma_y}{\sigma_x}(x - \mu_x)$ and variance $\sigma_y^2(1 - \rho^2)$.

(b) Show that $\text{Corr}(X, Y) = \rho$.

(c) Argue that X and Y are independent if and only if $\rho = 0$.

41. Let X be a normal random variable with parameters $\mu = 0$ and $\sigma^2 = 1$ and let I, independent of X, be such that $P\{I = 1\} = \frac{1}{2} = P\{I = 0\}$. Now define Y by

$$Y = \begin{cases} X & \text{if } I = 1 \\ -X & \text{if } I = 0 \end{cases}$$

In words, Y is equally likely to equal either X or $-X$.

(a) Are X and Y independent?

(b) Are I and Y independent?

(c) Show that Y is normal with mean 0 and variance 1.

(d) Show that $\text{Cov}(X, Y) = 0$.

42. It follows from Proposition 5.1 and the fact that the best linear predictor of Y with respect to X is $\mu_y + \rho\frac{\sigma_y}{\sigma_x}(X - \mu_x)$ that if

$$E[Y|X] = a + bX$$

then

$$a = \mu_y - \rho\frac{\sigma_y}{\sigma_x}\mu_x \qquad b = \rho\frac{\sigma_y}{\sigma_x}$$

(Why?) Verify this directly.

43. For random variables X and Z show that

$$E[(X - Y)^2] = E[X^2] - E[Y^2]$$

where

$$Y = E[X|Z]$$

44. Consider a population consisting of individuals able to produce offspring of the same kind. Suppose that each individual will, by the end of its lifetime, have produced j new offspring with probability P_j, $j \geq 0$, independently of the number produced by any other individual. The number of individuals initially present, denoted by X_0, is called the size of the zeroth generation. All offspring of the zeroth generation constitute the first generation, and their number is denoted by X_1. In general, let X_n denote the size of the nth generation. Let $\mu = \sum_{j=0}^{\infty} jP_j$ and $\sigma^2 = \sum_{j=0}^{\infty} (j - \mu)^2 P_j$ denote, respectively, the mean and the variance of the number of offspring produced by a single individual. Suppose that $X_0 = 1$—that is, initially there is a single individual in the population.

(a) Show that

$$E[X_n] = \mu E[X_{n-1}]$$

(b) Use part (a) to conclude that

$$E[X_n] = \mu^n$$

(c) Show that

$$\text{Var}(X_n) = \sigma^2\mu^{n-1} + \mu^2\,\text{Var}(X_{n-1})$$

(d) Use part (c) to conclude that

$$\text{Var}(X_n) = \begin{cases} \sigma^2\mu^{n-1}\left(\dfrac{\mu^n - 1}{\mu - 1}\right) & \text{if } \mu \neq 1 \\ n\sigma^2 & \text{if } \mu = 1 \end{cases}$$

The model just described is known as a *branching process*, and an important question for a population that evolves along such lines is the probability that the population will eventually die out. Let π denote this probability when the population starts with a single individual. That is,

$$\pi = P\{\text{population eventually dies out} \mid X_0 = 1\}$$

(e) Argue that π satisfies

$$\pi = \sum_{j=0}^{\infty} P_j \pi^j$$

HINT: Condition on the number of offspring of the initial member of the population.

45. Verify the formula for the moment generating function of a uniform random variable that is given in Table 7.2. Also, differentiate to verify the formulas for the mean and variance.

46. For a standard normal random variable Z, let $\mu_n = E[Z^n]$. Show that

$$\mu_n = \begin{cases} 0 & \text{when } n \text{ is odd} \\ \dfrac{(2j)!}{2^j j!} & \text{when } n = 2j \end{cases}$$

HINT: Start by expanding the moment generating function of Z into a Taylor series about 0 to obtain

$$E[e^{tZ}] = e^{t^2/2} = \sum_{j=0}^{\infty} \frac{(t^2/2)^j}{j!}$$

47. Let X be a normal random variable with mean μ and variance σ^2. Use the results of Theoretical Exercise 46 to show that

$$E[X^n] = \sum_{j=0}^{[n/2]} \frac{\binom{n}{2j}\mu^{n-2j}\sigma^{2j}(2j)!}{2^j j!}$$

In the preceding equation, $[n/2]$ is the largest integer less than or equal to $n/2$. Check your answer by letting $n = 1$ and $n = 2$.

48. If $Y = aX + b$, where a and b are constants, express the moment generating function of Y in terms of the moment generating function of X.

49. The positive random variable X is said to be a *lognormal* random variable with parameters μ and σ^2 if $\log(X)$ is a normal random variable with mean μ and variance σ^2. Use the normal moment generating function to find the mean and variance of a lognormal random variable.

50. Let X have moment generating function $M(t)$, and define $\Psi(t) = \log M(t)$. Show that

$$\Psi''(t)|_{t=0} = \text{Var}(X)$$

51. Use Table 7.2 to determine the distribution of $\sum_{i=1}^{n} X_i$ when $X_1, \ldots, X_n$ are independent and identically distributed exponential random variables, each having mean $1/\lambda$.

52. Show how to compute $\text{Cov}(X, Y)$ from the joint moment generating function of X and Y.

53. Suppose that $X_1, \ldots, X_n$ have a multivariate normal distribution. Show that $X_1, \ldots, X_n$ are independent random variables if and only if

$$\text{Cov}(X_i, X_j) = 0 \qquad \text{when } i \neq j$$

54. If Z is a standard normal random variable, what is $\text{Cov}(Z, Z^2)$?

SELF-TEST PROBLEMS AND EXERCISES

1. Consider a list of m names, where the same name may appear more than once on the list. Let $n(i)$ denote the number of times that the name in position i appears on the list, $i = 1, \ldots, m$, and let d denote the number of distinct names on the list.
(a) Express d in terms of the variables $m, n(i), i = 1, \ldots, m$.
Let U be a uniform $(0, 1)$ random variable, and let $X = [mU] + 1$.
(b) What is the probability mass function of X?
(c) Argue that $E[m/n(X)] = d$.

2. An urn has n white and m black balls that are removed one at a time in a randomly chosen order. Find the expected number of instances in which a white ball is immediately followed by a black one.

3. Twenty individuals, consisting of 10 married couples, are to be seated at five different tables, with four people at each table.
(a) If the seating is done "at random," what is the expected number of married couples that are seated at the same table?
(b) If two men and two women are randomly chosen to be seated at each table, what is the expected number of married couples that are seated at the same table?

4. If a die is to be rolled until all sides have appeared at least once, find the expected number of times that outcome 1 appears.

5. A deck of $2n$ cards consists of n red and n black cards. These cards are shuffled and then turned over one at a time. Suppose that each time a red card is turned over we win 1 unit if more red cards than black cards have been turned over by that time. (For instance, if $n = 2$ and the result is r b r b, then we would win a total of 2 units.) Find the expected amount that we win.

6. Let $A_1, A_2, \ldots, A_n$ be events, and let N denote the number of them that occur. Also, let $I = 1$ if all of these events occur, and let it be 0 otherwise. Prove Bonferroni's inequality, namely that

$$P(A_1 \cdots A_n) \geq \sum_{i=1}^{n} P(A_i) - (n - 1)$$

HINT: Argue first that $N \leq n - 1 + I$.

7. Let X be the smallest value obtained when k numbers are randomly chosen from the set $1, \ldots, n$. Find $E[X]$ by relating X to the random variable of Example 2m.

8. An arriving plane carries r families. A total of n_j of these families have checked in a total of j pieces of luggage, $\sum_j n_j = r$. Suppose that when the plane lands, the $N = \sum_j jn_j$ pieces of luggage come out of the plane in a random order. As soon as a family collects all of its luggage, it immediately departs the airport. If the Sanchez family checked in j pieces of luggage, find the expected number of families that depart after they do.

***9.** Nineteen items on the rim of a circle of radius 1 are to be chosen. Show that for any choice of these points there will be an arc of (arc) length 1 that contains at least 4 of them.

10. Let X be a Poisson random variable with mean λ. Show that if λ is not too small, then

$$\text{Var}(\sqrt{X}) \approx .25$$

HINT: Use the result of Theoretical Exercise 4 to approximate $E[\sqrt{X}]$.

11. Suppose in Self-Test Problem 3 that the 20 people are to be seated at seven tables, three of which have 4 seats and four of which have 2 seats. If the people are randomly seated, find the expected value of the number of married couples that are seated at the same table.

12. Individuals 1 through n, $n > 1$, are to be recruited into a firm in the following manner. Individual 1 starts the firm and recruits individual 2. Individuals 1 and 2 will then compete to recruit individual 3. Once individual 3 is recruited, individuals 1, 2, and 3 will compete to recruit individual 4, and so on. Suppose that when individuals $1, 2, \ldots, i$ compete to recruit individual $i + 1$, each of them is equally likely to be the successful recruiter.

(a) Find the expected number of the individuals $1, \ldots, n$ who did not recruit anyone else.

(b) Derive an expression for the variance of the number of individuals who did not recruit anyone else and evaluate it for $n = 5$.

13. The nine players on a basketball team consist of 2 centers, 3 forwards, and 4 backcourt players. If the players are paired up at random into three groups of size 3 each, find (a) the expected value and (b) the variance of the number of triplets consisting of one of each type of player.

14. A deck of 52 cards is shuffled and a bridge hand of 13 cards is dealt out. Let X and Y denote, respectively, the number of aces and the number of spades in the dealt hand.

(a) Show that X and Y are uncorrelated.

(b) Are they independent?

15. Each coin in a bin has a value attached to it. Each time that a coin with value p is flipped it lands on heads with probability p. When a coin is randomly chosen from the bin, its value is uniformly distributed on $(0, 1)$. Suppose that after the coin is chosen but before it is flipped, you must predict whether it will land heads or tails. You will win 1 if you are correct and will lose 1 otherwise.

(a) What is your expected gain if you are not told the value of the coin?

(b) Suppose now that you are allowed to inspect the coin before it is flipped, with the result of your inspection being that you learn the value of the coin. As a function of p, the value of the coin, what prediction should you make?

(c) Under the conditions of part (b), what is your expected gain?

16. In Self-Test Problem 1 we showed how to use the value of a uniform $(0, 1)$ random variable (commonly called a *random number*) to obtain the value of a random variable whose mean is equal to the expected number of distinct names on a list. However, its use required that one chooses a random position and then determine the number of times that the name in that position appears on the list. Another approach, which can be more efficient when there is a large amount of name replication, is as follows. As before, start by choosing the random variable X as in Problem 1. Now identify the name in position X, and then go through the list starting at the beginning until that name appears. Let I equal 0 if you encounter that name before getting to position X, and let I equal 1 if your first encounter with the name is at position X. Show that $E[mI] = d$.

HINT: Compute $E[I]$ by using conditional expectation.

17. A total of m items are to be sequentially distributed among n cells, with each item independently being put in cell j with probability $p_j, j = 1, \ldots, n$. Find the expected number of collisions that occur, where a collision occurs whenever an item is put into a nonempty cell.

18. Let X be the length of the initial run in a random ordering of n ones and m zeroes. That is, if the first k values are the same (either all ones or all zeroes) then $X \geq k$. Find $E[X]$.

19. There are n items in a box labeled H and m in a box labeled T. A coin that comes up heads with probability p and tails with probability $1 - p$ is flipped. Each time it comes up heads an item is removed from the H box, and each time it comes up tails an item is removed from the T box. (If a box is empty and its outcome occurs then no items are removed.) Find the expected number of coin flips needed for both boxes to become empty.

HINT: Condition on the number of heads in the first $n + m$ flips.

20. Let X be a nonnegative random variable having distribution function F. With $\bar{F}(x) = 1 - F(x)$, show that

$$E[X^n] = \int_0^\infty x^{n-1}\bar{F}(x)\,dx$$

HINT: Start with the identity

$$X^n = n\int_0^X x^{n-1}\,dx$$
$$= n\int_0^\infty x^{n-1} I_X(x)\,dx$$

where

$$I_X(x) = \begin{cases} 1, & \text{if } x < X \\ 0, & \text{otherwise} \end{cases}$$

CHAPTER 8

Limit Theorems

8.1 INTRODUCTION

The most important theoretical results in probability theory are limit theorems. Of these, the most important are those that are classified either under the heading *laws of large numbers* or under the heading *central limit theorems*. Usually, theorems are considered to be laws of large numbers if they are concerned with stating conditions under which the average of a sequence of random variables converges (in some sense) to the expected average. On the other hand, central limit theorems are concerned with determining conditions under which the sum of a large number of random variables has a probability distribution that is approximately normal.

8.2 CHEBYSHEV'S INEQUALITY AND THE WEAK LAW OF LARGE NUMBERS

We start this section by proving a result known as Markov's inequality.

Proposition 2.1 Markov's inequality

If X is a random variable that takes only nonnegative values, then for any value $a > 0$,

$$P\{X \geq a\} \leq \frac{E[X]}{a}$$

Proof: For $a > 0$, let

$$I = \begin{cases} 1 & \text{if } X \geq a \\ 0 & \text{otherwise} \end{cases}$$

and note that since $X \geq 0$,

$$I \leq \frac{X}{a}$$

Taking expectations of the preceding yields that

$$E[I] \leq \frac{E[X]}{a}$$

which, since $E[I] = P\{X \geq a\}$, proves the result. ✦

As a corollary, we obtain Proposition 2.2.

Proposition 2.2 *Chebyshev's inequality*

If X is a random variable with finite mean μ and variance σ^2, then for any value $k > 0$,

$$P\{|X - \mu| \geq k\} \leq \frac{\sigma^2}{k^2}$$

Proof: Since $(X - \mu)^2$ is a nonnegative random variable, we can apply Markov's inequality (with $a = k^2$) to obtain

$$P\{(X - \mu)^2 \geq k^2\} \leq \frac{E[(X - \mu)^2]}{k^2} \tag{2.1}$$

But since $(X - \mu)^2 \geq k^2$ if and only if $|X - \mu| \geq k$, Equation (2.1) is equivalent to

$$P\{|X - \mu| \geq k\} \leq \frac{E[(X - \mu)^2]}{k^2} = \frac{\sigma^2}{k^2}$$

and the proof is complete. ✦

The importance of Markov's and Chebyshev's inequalities is that they enable us to derive bounds on probabilities when only the mean, or both the mean and the variance, of the probability distribution are known. Of course, if the actual distribution were known, then the desired probabilities could be exactly computed and we would not need to resort to bounds.

Example 2a. Suppose that it is known that the number of items produced in a factory during a week is a random variable with mean 50.

(a) What can be said about the probability that this week's production will exceed 75?

(b) If the variance of a week's production is known to equal 25, then what can be said about the probability that this week's production will be between 40 and 60?

Solution Let X be the number of items that will be produced in a week

(a) by Markov's inequality

$$P\{X > 75\} \leq \frac{E[X]}{75} = \frac{50}{75} = \frac{2}{3}$$

(b) by Chebyshev's inequality

$$P\{|X - 50| \geq 10\} \leq \frac{\sigma^2}{10^2} = \frac{1}{4}$$

Hence

$$P\{|X - 50| < 10\} \geq 1 - \tfrac{1}{4} = \tfrac{3}{4}$$

so the probability that this week's production will be between 40 and 60 is at least .75. ■

As Chebyshev's inequality is valid for all distributions of the random variable X, we cannot expect the bound on the probability to be very close to the actual probability in most cases. For instance, consider Example 2b.

Example 2b. If X is uniformly distributed over the interval $(0, 10)$, then, as $E[X] = 5$, $\text{Var}(X) = \frac{25}{3}$, it follows from Chebyshev's inequality that

$$P\{|X - 5| > 4\} \leq \frac{25}{3(16)} \approx .52$$

whereas the exact result is

$$P\{|X - 5| > 4\} = .20$$

Thus, although Chebyshev's inequality is correct, the upper bound that it provides is not particularly close to the actual probability.

Similarly, if X is a normal random variable with mean μ and variance σ^2, Chebyshev's inequality states that

$$P\{|X - \mu| > 2\sigma\} \leq \tfrac{1}{4}$$

whereas the actual probability is given by

$$P\{|X - \mu| > 2\sigma\} = P\left\{\left|\frac{X - \mu}{\sigma}\right| > 2\right\} = 2[1 - \Phi(2)] \approx .0456$$ ■

Chebyshev's inequality is often used as a theoretical tool in proving results. This is illustrated first by Proposition 2.3 and then, most importantly, by the weak law of large numbers.

Proposition 2.3

If $\text{Var}(X) = 0$, then

$$P\{X = E[X]\} = 1$$

In other words, the only random variables having variances equal to 0 are those that are constant with probability 1.

Proof: By Chebyshev's inequality we have, for any $n \geq 1$

$$P\left\{|X - \mu| > \frac{1}{n}\right\} = 0$$

Letting $n \to \infty$ and using the continuity property of probability yields

$$0 = \lim_{n\to\infty} P\left\{|X - \mu| > \frac{1}{n}\right\} = P\left\{\lim_{n\to\infty}\left\{|X - \mu| > \frac{1}{n}\right\}\right\}$$
$$= P\{X \neq \mu\}$$

and the result is established. ✦

Theorem 2.1 The weak law of large numbers

Let $X_1, X_2, \ldots$ be a sequence of independent and identically distributed random variables, each having finite mean $E[X_i] = \mu$. Then, for any $\varepsilon > 0$,

$$P\left\{\left|\frac{X_1 + \cdots + X_n}{n} - \mu\right| \geq \varepsilon\right\} \to 0 \quad \text{as} \quad n \to \infty$$

Proof: We shall prove the result only under the additional assumption that the random variables have a finite variance σ^2. Now, as

$$E\left[\frac{X_1 + \cdots + X_n}{n}\right] = \mu \quad \text{and} \quad \text{Var}\left(\frac{X_1 + \cdots + X_n}{n}\right) = \frac{\sigma^2}{n}$$

it follows from Chebyshev's inequality that

$$P\left\{\left|\frac{X_1 + \cdots + X_n}{n} - \mu\right| \geq \varepsilon\right\} \leq \frac{\sigma^2}{n\varepsilon^2}$$

and the result is proved. ✦

The weak law of large numbers was originally proved by James Bernoulli for the special case where the X_i are 0—1 (that is, Bernoulli) random variables. His statement and proof of this theorem were presented in his book *Ars Conjectandi*, which was published in 1713, 8 years after his death, by his nephew Nicholas Bernoulli. It should be noted that as Chebyshev's inequality was not known in his time, Bernoulli had to resort to a quite ingenious proof to establish the result. The general form of the weak law of large numbers presented in Theorem 2.1 was proved by the Russian mathematician Khintchine.

8.3 THE CENTRAL LIMIT THEOREM

The central limit theorem is one of the most remarkable results in probability theory. Loosely put, it states that the sum of a large number of independent random variables has a distribution that is approximately normal. Hence it not only provides a simple method for computing approximate probabilities for sums of independent random variables, but it also helps explain the remarkable fact that the empirical frequencies of so many natural populations exhibit bell-shaped (that is, normal) curves.

In its simplest form the central limit theorem is as follows.

Theorem 3.1 The central limit theorem

Let $X_1, X_2, \ldots$ be a sequence of independent and identically distributed random variables each having mean μ and variance σ^2. Then the distribution of

$$\frac{X_1 + \cdots + X_n - n\mu}{\sigma\sqrt{n}}$$

tends to the standard normal as $n \to \infty$. That is, for $-\infty < a < \infty$,

$$P\left\{\frac{X_1 + \cdots + X_n - n\mu}{\sigma\sqrt{n}} \le a\right\} \to \frac{1}{\sqrt{2\pi}} \int_{-\infty}^{a} e^{-x^2/2}\,dx \qquad \text{as} \qquad n \to \infty$$

The key to the proof of the central limit theorem is the following lemma, which we state without proof.

Lemma 3.1

Let $Z_1, Z_2, \ldots$ be a sequence of random variables having distribution functions F_{Z_n} and moment generating functions $M_{Z_n}, n \ge 1$; and let Z be a random variable having distribution function F_Z and moment generating function M_Z. If $M_{Z_n}(t) \to M_Z(t)$ for all t, then $F_{Z_n}(t) \to F_Z(t)$ for all t at which $F_Z(t)$ is continuous.

If we let Z be a standard normal random variable, then, as $M_Z(t) = e^{t^2/2}$, it follows from Lemma 3.1 that if $M_{Z_n}(t) \to e^{t^2/2}$ as $n \to \infty$, then $F_{Z_n}(t) \to \Phi(t)$ as $n \to \infty$. We are now ready to prove the central limit theorem.

Proof of the Central Limit Theorem: Let us assume at first that $\mu = 0$ and $\sigma^2 = 1$. We shall prove the theorem under the assumption that the moment generating function of the X_i, $M(t)$, exists and is finite. Now the moment generating function of $X_i/\sqrt{n}$ is given by

$$E\left[\exp\left\{\frac{tX_i}{\sqrt{n}}\right\}\right] = M\left(\frac{t}{\sqrt{n}}\right)$$

and thus the moment generating function of $\sum_{i=1}^{n} X_i/\sqrt{n}$ is given by $\left[M\left(\frac{t}{\sqrt{n}}\right)\right]^n$. Let

$$L(t) = \log M(t)$$

and note that

$$\begin{aligned}
L(0) &= 0 \\
L'(0) &= \frac{M'(0)}{M(0)} \\
&= \mu \\
&= 0 \\
L''(0) &= \frac{M(0)M''(0) - [M'(0)]^2}{[M(0)]^2} \\
&= E[X^2] \\
&= 1
\end{aligned}$$

Now, to prove the theorem, we must show that $[M(t/\sqrt{n})]^n \to e^{t^2/2}$ as $n \to \infty$, or, equivalently, that $nL(t/\sqrt{n}) \to t^2/2$ as $n \to \infty$. To show this, note that

$$\begin{aligned}
\lim_{n\to\infty} \frac{L(t/\sqrt{n})}{n^{-1}} &= \lim_{n\to\infty} \frac{-L'(t/\sqrt{n})n^{-3/2}t}{-2n^{-2}} \qquad \text{by L'Hospital's rule} \\
&= \lim_{n\to\infty} \left[\frac{L'(t/\sqrt{n})t}{2n^{-1/2}}\right] \\
&= \lim_{n\to\infty} \left[\frac{-L''(t/\sqrt{n})n^{-3/2}t^2}{-2n^{-3/2}}\right] \qquad \text{again by L'Hospital's rule} \\
&= \lim_{n\to\infty} \left[L''\left(\frac{t}{\sqrt{n}}\right)\frac{t^2}{2}\right] \\
&= \frac{t^2}{2}
\end{aligned}$$

Thus the central limit theorem is proved when $\mu = 0$ and $\sigma^2 = 1$. The result now follows in the general case by considering the standardized random variables $X_i^* = (X_i - \mu)/\sigma$ and applying the preceding result, since $E[X_i^*] = 0$, $\text{Var}(X_i^*) = 1$. ✦

REMARK. Although Theorem 3.1 only states that for each a,

$$P\left\{\frac{X_1 + \cdots + X_n - n\mu}{\sigma\sqrt{n}} \le a\right\} \to \Phi(a)$$

it can, in fact, be shown that the convergence is uniform in a. [We say that $f_n(a) \to f(a)$ uniformly in a, if for each $\varepsilon > 0$, there exists an N such that $|f_n(a) - f(a)| < \varepsilon$ for all a whenever $n \ge N$.] ■

The first version of the central limit theorem was proved by DeMoivre around 1733 for the special case where the X_i are Bernoulli random variables with $p = \frac{1}{2}$. This was subsequently extended by Laplace to the case of arbitrary p. (Since a binomial random variable may be regarded as the sum of n independent and identically distributed Bernoulli random variables, this justifies the normal approximation to the binomial that was presented in Section 5.4.1.) Laplace also discovered the more general form of the central limit theorem given in Theorem 3.1. His proof, however, was not completely rigorous and, in fact, cannot easily be made rigorous. A truly rigorous proof of the central limit theorem was first presented by the Russian mathematician Liapounoff in the period 1901–1902.

This important theorem is illustrated by the central limit theorem module on the website. This website plots the density function of the sum of n independent and identically distributed random variables that each take on one of the values 0, 1, 2, 3, 4. When using it, one enters the probability mass function and the desired value of n. Figure 8.1 shows the resulting plots for a specified probability mass function when (a) $n = 5$, (b) $n = 10$, (c) $n = 25$, and (d) $n = 100$.

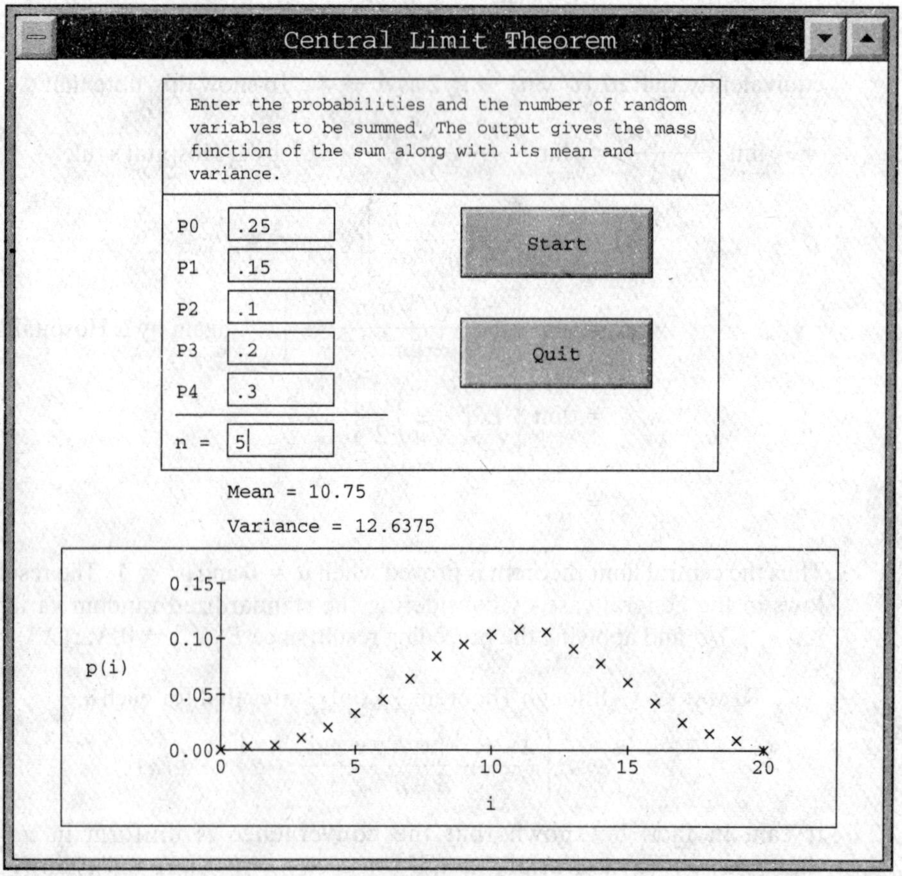

Figure 8.1(a)

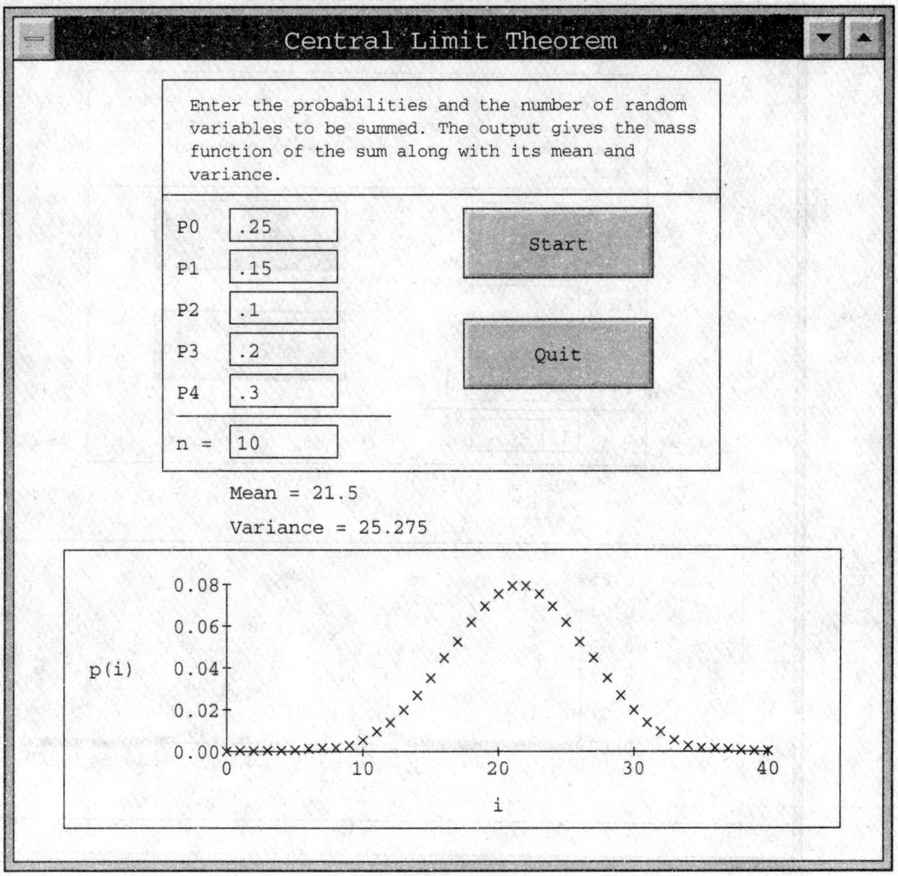

Figure 8.1(b)

Example 3a. An astronomer is interested in measuring, in light years, the distance from his observatory to a distant star. Although the astronomer has a measuring technique, he knows that, because of changing atmospheric conditions and normal error, each time a measurement is made it will not yield the exact distance but merely an estimate. As a result the astronomer plans to make a series of measurements and then use the average value of these measurements as his estimated value of the actual distance. If the astronomer believes that the values of the measurements are independent and identically distributed random variables having a common mean d (the actual distance) and a common variance of 4 (light years), how many measurements need he make to be reasonably sure that his estimated distance is accurate to within $\pm .5$ light year?

Solution Suppose that the astronomer decides to make n observations. If $X_1, X_2, \ldots, X_n$ are the n measurements, then, from the central limit theorem, it follows that

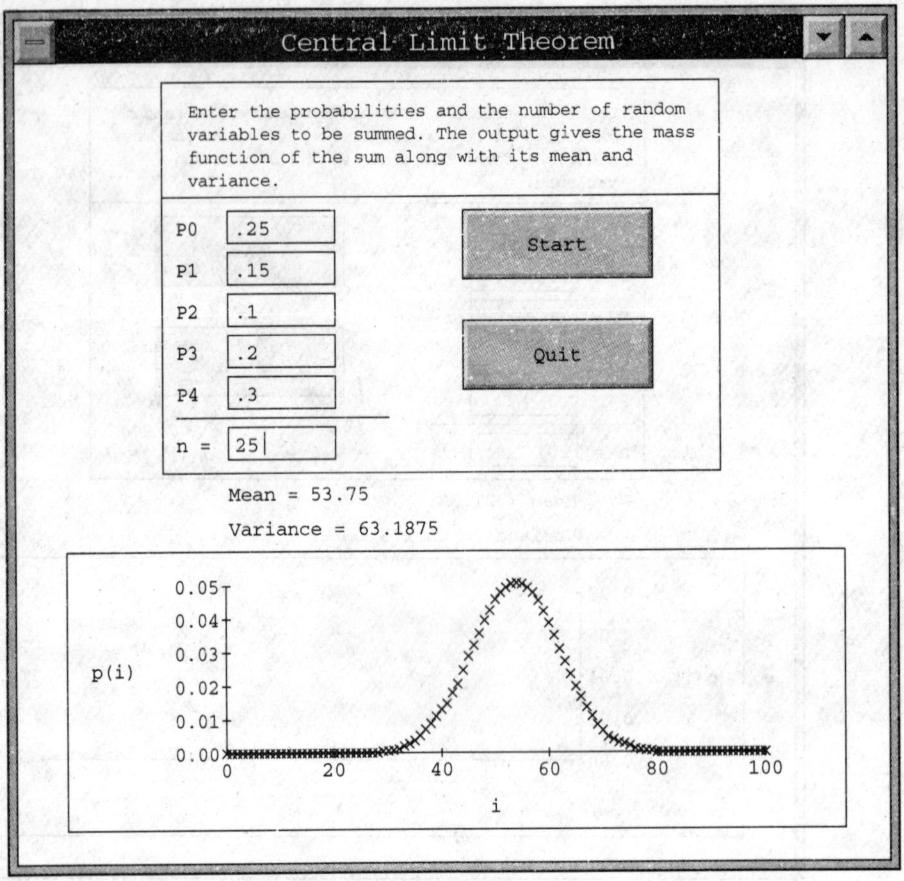

Figure 8.1(c)

$$Z_n = \frac{\sum_{i=1}^{n} X_i - nd}{2\sqrt{n}}$$

has approximately a standard normal distribution. Hence

$$P\left\{-.5 \le \frac{\sum_{i=1}^{n} X_i}{n} - d \le .5\right\} = P\left\{-.5\frac{\sqrt{n}}{2} \le Z_n \le .5\frac{\sqrt{n}}{2}\right\}$$

$$\approx \Phi\left(\frac{\sqrt{n}}{4}\right) - \phi\left(-\frac{\sqrt{n}}{4}\right) = 2\Phi\left(\frac{\sqrt{n}}{4}\right) - 1$$

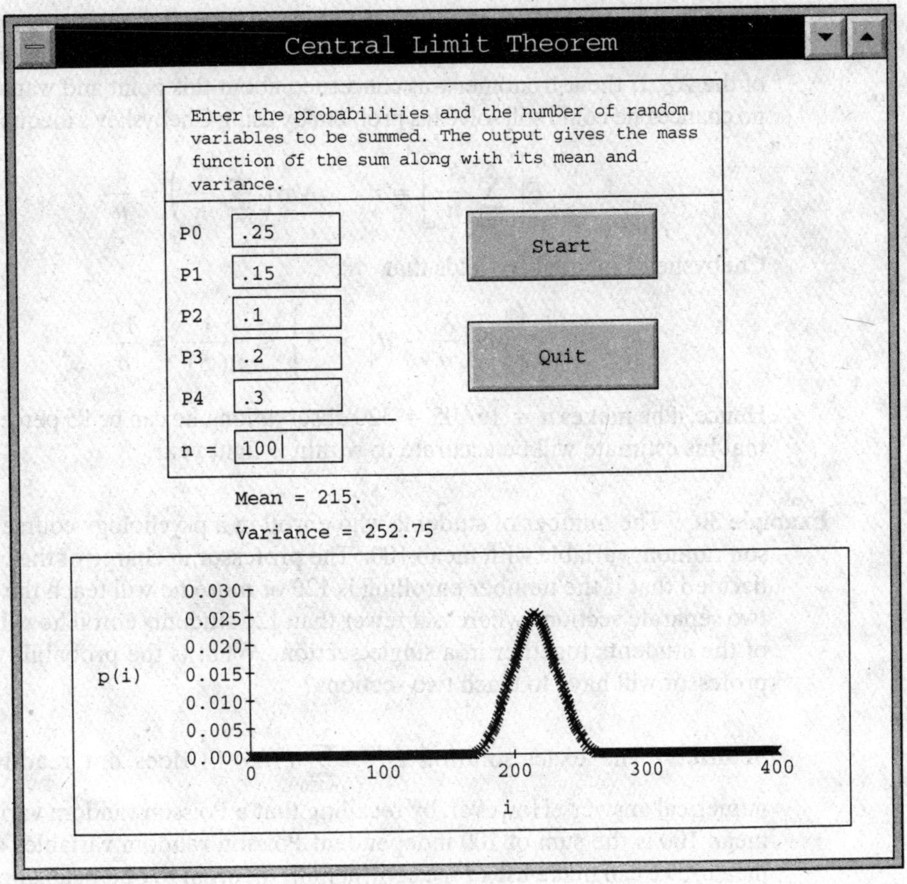

Figure 8.1(d)

Therefore, if the astronomer wanted, for instance, to be 95 percent certain that his estimated value is accurate to within .5 light year, he should make n^* measurements, where n^* is such that

$$2\Phi\left(\frac{\sqrt{n^*}}{4}\right) - 1 = .95 \qquad \text{or} \qquad \Phi\left(\frac{\sqrt{n^*}}{4}\right) = .975$$

and thus from Table 5.1 of Chapter 5,

$$\frac{\sqrt{n^*}}{4} = 1.96 \qquad \text{or} \qquad n^* = (7.84)^2 \approx 61.47$$

As n^* is not integral valued, he should make 62 observations.

It should, however, be noted that the preceding analysis has been done under the assumption that the normal approximation will be a good approximation when

$n = 62$. Although this will usually be the case, in general the question of how large n need be before the approximation is "good" depends on the distribution of the X_i. If the astronomer was concerned about this point and wanted to take no chances, he could still solve his problem by using Chebyshev's inequality. Since

$$E\left[\sum_{i=1}^{n} \frac{X_i}{n}\right] = d \qquad \text{Var}\left(\sum_{i=1}^{n} \frac{X_i}{n}\right) = \frac{4}{n}$$

Chebyshev's inequality yields that

$$P\left\{\left|\sum_{i=1}^{n} \frac{X_i}{n} - d\right| > .5\right\} \le \frac{4}{n(.5)^2} = \frac{16}{n}$$

Hence, if he makes $n = 16/.05 = 320$ observations, he can be 95 percent certain that his estimate will be accurate to within .5 light year. ■

Example 3b. The number of students who enroll in a psychology course is a Poisson random variable with mean 100. The professor in charge of the course has decided that if the number enrolling is 120 or more he will teach the course in two separate sections, whereas if fewer than 120 students enroll he will teach all of the students together in a single section. What is the probability that the professor will have to teach two sections?

Solution The exact solution $e^{-100} \sum_{i=120}^{\infty} (100)^i/i!$ does not readily yield a numerical answer. However, by recalling that a Poisson random variable with mean 100 is the sum of 100 independent Poisson random variables each with mean 1, we can make use of the central limit theorem to obtain an approximate solution. If X denotes the number of students that enroll in the course, we have

$$\begin{aligned} P\{X \ge 120\} &= P\{X \ge 119.5\} \quad \text{(the continuity correction)} \\ &= P\left\{\frac{X - 100}{\sqrt{100}} \ge \frac{119.5 - 100}{\sqrt{100}}\right\} \\ &\approx 1 - \Phi(1.95) \\ &\approx .0256 \end{aligned}$$

where we have used the fact that the variance of a Poisson random variable is equal to its mean. ■

Example 3c. If 10 fair dice are rolled, find the approximate probability that the sum obtained is between 30 and 40, inclusive.

Solution Let X_i denote the value of the ith die, $i = 1, 2, \ldots, 10$. Since $E(X_i) = \frac{7}{2}$, $\text{Var}(X_i) = E[X_i^2] - (E[X_i])^2 = \frac{35}{12}$, the central limit theorem yields

$$P\{29.5 \le X \le 40.5\} = P\left\{\frac{29.5-35}{\sqrt{\frac{350}{12}}} \le \frac{X-35}{\sqrt{\frac{350}{12}}} \le \frac{40.5-35}{\sqrt{\frac{350}{12}}}\right\}$$
$$\approx 2\Phi(1.0184) - 1$$
$$\approx .692$$
■

Example 3d. Let $X_i, i = 1, \ldots, 10$, be independent random variables, each uniformly distributed over $(0, 1)$. Calculate an approximation to $P\left\{\sum_{i=1}^{10} X_i > 6\right\}$.

Solution Since $E[X_i] = \frac{1}{2}$, $\text{Var}(X_i) = \frac{1}{12}$, we have by the central limit theorem

$$P\left\{\sum_{1}^{10} X_i > 6\right\} = P\left\{\frac{\sum_{1}^{10} X_i - 5}{\sqrt{10\left(\frac{1}{12}\right)}} > \frac{6-5}{\sqrt{10\left(\frac{1}{12}\right)}}\right\}$$
$$\approx 1 - \Phi\left(\sqrt{1.2}\right)$$
$$\approx .1367$$

Hence only 14 percent of the time will $\sum_{i=1}^{10} X_i$ be greater than 6. ■

Central limit theorems also exist when the X_i are independent but not necessarily identically distributed random variables. One version, by no means the most general, is as follows.

Theorem 3.2 Central limit theorem for independent random variables

Let $X_1, X_2, \ldots$ be a sequence of independent random variables having respective means and variances $\mu_i = E[X_i], \sigma_i^2 = \text{Var}(X_i)$. If (a) the X_i are uniformly bounded; that is, if for some M, $P\{|X_i| < M\} = 1$ for all i, and (b) $\sum_{i=1}^{\infty} \sigma_i^2 = \infty$, then

$$P\left\{\frac{\sum_{i=1}^{n}(X_i - \mu_i)}{\sqrt{\sum_{i=1}^{n}\sigma_i^2}} \le a\right\} \to \Phi(a) \quad \text{as} \quad n \to \infty$$

Historical Note

Pierre Simon, Marquis de Laplace

The central limit theorem was originally stated and proved by the French mathematician Pierre Simon, the Marquis de Laplace, who came to this theorem from his observations that errors of measurement (which can usually be regarded as being the sum of a large number of tiny forces) tend to be normally distributed. Laplace, who was also a famous astronomer (and indeed was called "the Newton of France"), was one of the great early contributors to both probability and statistics. Laplace was also a popularizer of the uses of probability in everyday life. He strongly believed in its importance, as is indicated by the following quotations of his taken from his published book *Analytical Theory of Probability*. *"We see that the theory of probability is at bottom only common sense reduced to calculation; it makes us appreciate with exactitude what reasonable minds feel by a sort of instinct, often without being able to account for it. ... It is remarkable that this science, which originated in the consideration of games of chance, should become the most important object of human knowledge. ... The most important questions of life are, for the most part, really only problems of probability."*

The application of the central limit theorem to show that measurement errors are approximately normally distributed is regarded as an important contribution to science. Indeed, in the seventeenth and eighteenth centuries the central limit theorem was often called the *law of frequency of errors*. Listen to the words of Francis Galton (taken from his book *Natural Inheritance*, published in 1889): "I know of scarcely anything so apt to impress the imagination as the wonderful form of cosmic order expressed by the 'Law of Frequency of Error.' The Law would have been personified by the Greeks and deified, if they had known of it. It reigns with serenity and in complete self-effacement amidst the wildest confusion. The huger the mob and the greater the apparent anarchy, the more perfect is its sway. It is the supreme law of unreason."

8.4 THE STRONG LAW OF LARGE NUMBERS

The *strong law of large numbers* is probably the best-known result in probability theory. It states that the average of a sequence of independent random variables having a common distribution will, with probability 1, converge to the mean of that distribution.

Theorem 4.1 The strong law of large numbers

Let $X_1, X_2, \ldots$ be a sequence of independent and identically distributed random variables, each having a finite mean $\mu = E[X_i]$. Then, with probability 1,

$$\frac{X_1 + X_2 + \cdots + X_n}{n} \to \mu \quad \text{as} \quad n \to \infty^{\dagger}$$

As an application of the strong law of large numbers, suppose that a sequence of independent trials of some experiment is performed. Let E be a fixed event of the experiment and denote by $P(E)$ the probability that E occurs on any particular trial. Letting

$$X_i = \begin{cases} 1 & \text{if } E \text{ occurs on the } i\text{th trial} \\ 0 & \text{if } E \text{ does not occur on the } i\text{th trial} \end{cases}$$

we have by the strong law of large numbers that with probability 1,

$$\frac{X_1 + \cdots + X_n}{n} \to E[X] = P(E) \tag{4.1}$$

Since $X_1 + \cdots + X_n$ represents the number of times that the event E occurs in the first n trials, we may interpret Equation (4.1) as stating that, with probability 1, the limiting proportion of time that the event E occurs is just $P(E)$.

Although the theorem can be proven without this assumption, our proof of the strong law of large numbers will assume that the random variables X_i have a finite fourth moment. That is, we will suppose that $E[X_i^4] = K < \infty$.

Proof of the Strong Law of Large Numbers: To begin, assume that μ, the mean of the X_i, is equal to 0. Let $S_n = \sum_{i=1}^{n} X_i$ and consider

$$E[S_n^4] = E[(X_1 + \cdots + X_n)(X_1 + \cdots + X_n) \times (X_1 + \cdots + X_n)(X_1 + \cdots + X_n)]$$

Expanding the right side of the preceding will result in terms of the form

$$X_i^4, \quad X_i^3 X_j, \quad X_i^2 X_j^2, \quad X_i^2 X_j X_k, \quad \text{and} \quad X_i X_j X_k X_l$$

where i, j, k, l are all different. As all the X_i have mean 0, it follows by independence that

†That is, the strong law of large numbers states that

$$P\left\{\lim_{n\to\infty} (X_1 + \cdots + X_n)/n = \mu\right\} = 1$$

$$E[X_i^3X_j] = E[X_i^3]E[X_j] = 0$$
$$E[X_i^2X_jX_k] = E[X_i^2]E[X_j]E[X_k] = 0$$
$$E[X_iX_jX_kX_l] = 0$$

Now, for a given pair i and j there will be $\binom{4}{2} = 6$ terms in the expansion that will equal $X_i^2X_j^2$. Hence it follows upon expanding the preceding product and taking expectations term by term that

$$\begin{aligned} E[S_n^4] &= nE[X_i^4] + 6\binom{n}{2}E[X_i^2X_j^2] \\ &= nK + 3n(n-1)E[X_i^2]E[X_j^2] \end{aligned}$$

where we have once again made use of the independence assumption. Now, since

$$0 \le \text{Var}(X_i^2) = E[X_i^4] - (E[X_i^2])^2$$

we see that

$$(E[X_i^2])^2 \le E[X_i^4] = K$$

Therefore, from the preceding we have that

$$E[S_n^4] \le nK + 3n(n-1)K$$

which implies that

$$E\left[\frac{S_n^4}{n^4}\right] \le \frac{K}{n^3} + \frac{3K}{n^2}$$

Therefore, it follows that

$$E\left[\sum_{n=1}^{\infty} \frac{S_n^4}{n^4}\right] = \sum_{n=1}^{\infty} E\left[\frac{S_n^4}{n^4}\right] < \infty$$

But the preceding implies that with probability 1, $\sum_{n=1}^{\infty} S_n^4/n^4 < \infty$. (For if there is a positive probability that the sum is infinite, then its expected value is infinite.) But the convergence of a series implies that its nth term goes to 0; so we can conclude that with probability 1,

$$\lim_{n\to\infty} \frac{S_n^4}{n^4} = 0$$

But if $S_n^4/n^4 = (S_n/n)^4$ goes to 0, then so must S_n/n; so we have proven that with probability 1,

$$\frac{S_n}{n} \to 0 \quad \text{as} \quad n \to \infty$$

When μ, the mean of the X_i, is not equal to 0, we can apply the preceding argument to the random variables $X_i - \mu$ to obtain that with probability 1,

$$\lim_{n\to\infty} \sum_{i=1}^{n} \frac{(X_i - \mu)}{n} = 0$$

or, equivalently,

$$\lim_{n\to\infty} \sum_{i=1}^{n} \frac{X_i}{n} = \mu$$

which proves the result. ✦

The strong law is illustrated by two modules on the website that consider independent and identically distributed random variables which take on one of the values 0, 1, 2, 3, 4. The modules simulate the values of n such random variables; the proportions of time that each outcome occurs, as well as the resulting sample mean $\sum_{i=1}^{n} X_i/n$, are then indicated and plotted. When using these modules, which differ only in the type of graph presented, one enters the probabilities and the desired value of n. Figure 8.2 gives the results of a simulation using a specified probability mass function and (a) $n = 100$, (b) $n = 1000$, and (c) $n = 10{,}000$.

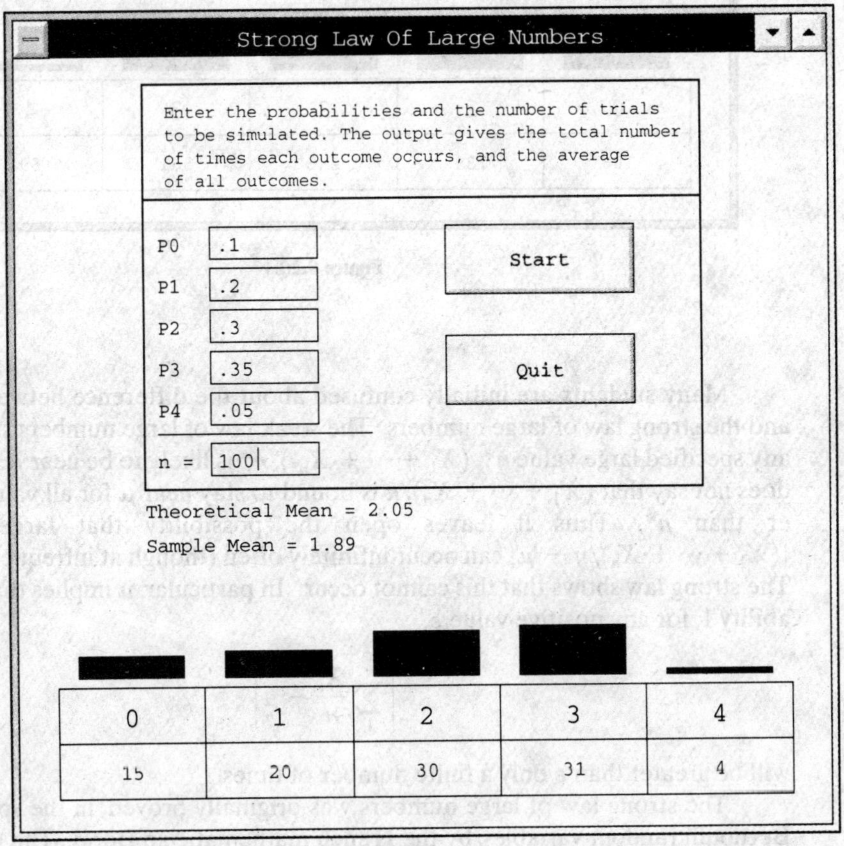

Figure 8.2(a)

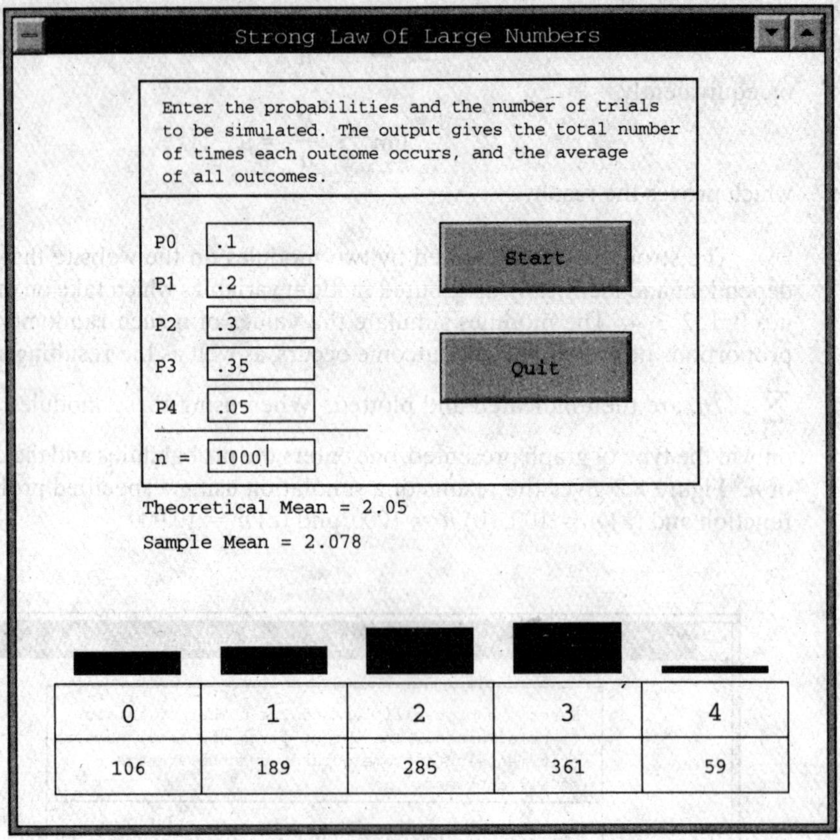

Figure 8.2(b)

Many students are initially confused about the difference between the weak and the strong law of large numbers. The weak law of large numbers states that for any specified large value n^*, $(X_1 + \cdots + X_{n^*})/n^*$ is likely to be near μ. However, it does not say that $(X_1 + \cdots + X_n)/n$ is bound to stay near μ for all values of n larger than n^*. Thus it leaves open the possibility that large values of $|(X_1 + \cdots + X_n)/n - \mu|$ can occur infinitely often (though at infrequent intervals). The strong law shows that this cannot occur. In particular, it implies that with probability 1, for any positive value ε,

$$\left| \sum_1^n \frac{X_i}{n} - \mu \right|$$

will be greater than ε only a finite number of times.

The strong law of large numbers was originally proved, in the special case of Bernoulli random variables, by the French mathematician Borel. The general form

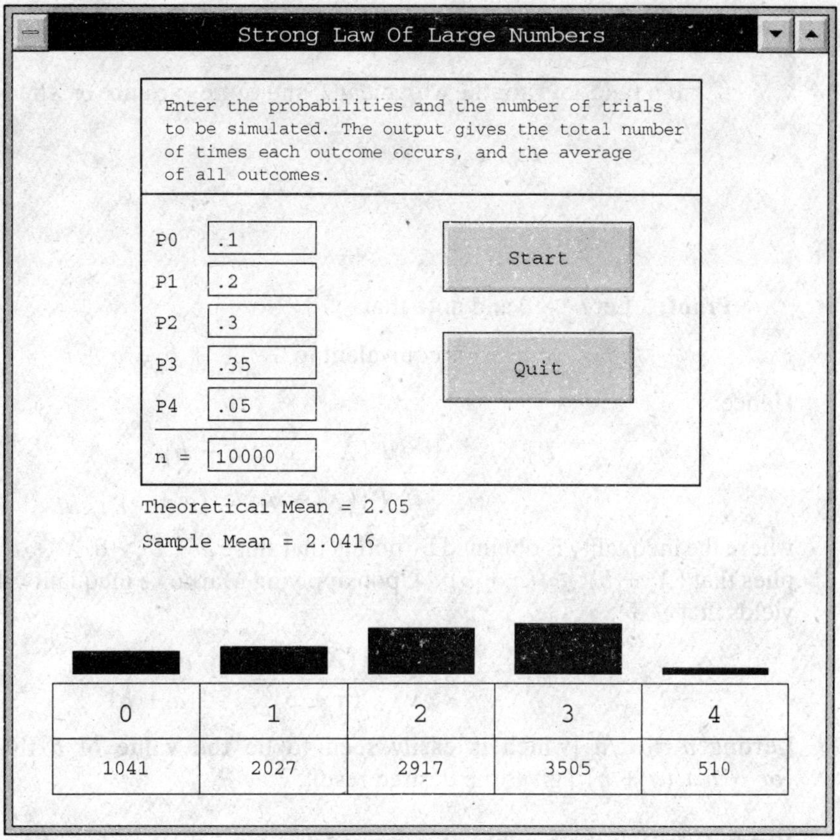

Figure 8.2(c)

of the strong law presented in Theorem 4.1 was proved by the Russian mathematician A. N. Kolmogorov.

8.5 OTHER INEQUALITIES

We are sometimes confronted with situations in which we are interested in obtaining an upper bound for a probability of the form $P\{X - \mu \geq a\}$, where a is some positive value and when only the mean $\mu = E[X]$ and variance $\sigma^2 = \text{Var}(X)$ of the distribution of X are known. Of course, since $X - \mu \geq a > 0$ implies that $|X - \mu| \geq a$, it follows from Chebyshev's inequality that

$$P\{X - \mu \geq a\} \leq P\{|X - \mu| \geq a\} \leq \frac{\sigma^2}{a^2} \quad \text{when} \quad a > 0$$

However, as the following proposition shows, it turns out that we can do better.

Proposition 5.1 One-sided Chebyshev inequality

If X is a random variable with mean 0 and finite variance σ^2, then for any $a > 0$,

$$P\{X \geq a\} \leq \frac{\sigma^2}{\sigma^2 + a^2}$$

Proof: Let $b > 0$ and note that

$$X \geq a \qquad \text{is equivalent to} \qquad X + b \geq a + b$$

Hence

$$\begin{aligned} P\{X \geq a\} &= P\{X + b \geq a + b\} \\ &\leq P\{(X + b)^2 \geq (a + b)^2\} \end{aligned}$$

where the inequality is obtained by noting that since $a + b > 0$, $X + b \geq a + b$ implies that $(X + b)^2 \geq (a + b)^2$. Upon applying Markov's inequality, the preceding yields that

$$P\{X \geq a\} \leq \frac{E[(X + b)^2]}{(a + b)^2} = \frac{\sigma^2 + b^2}{(a + b)^2}$$

Letting $b = \sigma^2/a$ [which is easily seen to be the value of b that minimizes $(\sigma^2 + b^2)/(a + b)^2$] gives the desired result.

Example 5a. If the number of items produced in a factory during a week is a random variable with mean 100 and variance 400, compute an upper bound on the probability that this week's production will be at least 120.

Solution It follows from the one-sided Chebyshev inequality that

$$P\{X \geq 120\} = P\{X - 100 \geq 20\} \leq \frac{400}{400 + (20)^2} = \frac{1}{2}$$

Hence the probability that this week's production will be 120 or more is at most $\frac{1}{2}$.

If we attempted to obtain a bound by applying Markov's inequality, then we would have obtained

$$P\{X \geq 120\} \leq \frac{E(X)}{120} = \frac{5}{6}$$

which is a far weaker bound than the preceding one. ∎

Suppose now that X has mean μ and variance σ^2. As both $X - \mu$ and $\mu - X$ have mean 0 and variance σ^2, we obtain from the one-sided Chebyshev inequality that for $a > 0$,

$$P\{X - \mu \geq a\} \leq \frac{\sigma^2}{\sigma^2 + a^2}$$

and

$$P\{\mu - X \geq a\} \leq \frac{\sigma^2}{\sigma^2 + a^2}$$

Thus we have the following corollary.

Corollary 5.1

If $E[X] = \mu$, $\text{Var}(X) = \sigma^2$, then for $a > 0$,

$$P\{X \geq \mu + a\} \leq \frac{\sigma^2}{\sigma^2 + a^2}$$

$$P\{X \leq \mu - a\} \leq \frac{\sigma^2}{\sigma^2 + a^2}$$

Example 5b. A set of 200 people, consisting of 100 men and 100 women, is randomly divided into 100 pairs of 2 each. Give an upper bound to the probability that at most 30 of these pairs will consist of a man and a woman.

Solution Number the men, arbitrarily, from 1 to 100 and let for $i = 1, 2, \ldots 100$,

$$X_i = \begin{cases} 1 & \text{if man } i \text{ is paired with a woman} \\ 0 & \text{otherwise} \end{cases}$$

Then X, the number of man–woman pairs, can be expressed as

$$X = \sum_{i=1}^{100} X_i$$

As man i is equally likely to be paired with any of the other 199 people, of which 100 are women, we have

$$E[X_i] = P\{X_i = 1\} = \frac{100}{199}$$

Similarly, for $i \neq j$,

$$\begin{aligned} E[X_i X_j] &= P\{X_i = 1, X_j = 1\} \\ &= P\{X_i = 1\}P\{X_j = 1 | X_i - 1\} = \frac{100}{199}\frac{99}{197} \end{aligned}$$

where $P\{X_j = 1|X_i = 1\} = 99/197$ since, given that man i is paired with a woman, man j is equally likely to be paired with any of the remaining 197 people, of which 99 are women. Hence we obtain that

$$\begin{aligned} E[X] &= \sum_{i=1}^{100} E[X_i] \\ &= (100)\frac{100}{199} \\ &\approx 50.25 \\ \text{Var}(X) &= \sum_{i=1}^{100} \text{Var}(X_i) + 2\sum_{i<j}\sum \text{Cov}(X_i, X_j) \\ &= 100\frac{100}{199}\frac{99}{199} + 2\binom{100}{2}\left[\frac{100}{199}\frac{99}{197} - \left(\frac{100}{199}\right)^2\right] \\ &\approx 25.126 \end{aligned}$$

The Chebyshev inequality yields that

$$P\{X \le 30\} \le P\{|X - 50.25| \ge 20.25\} \le \frac{25.126}{(20.25)^2} \approx .061$$

and thus there are fewer than 6 chances in a hundred that fewer than 30 men will be paired with women. However, we can improve on this bound by using the one-sided Chebyshev inequality, which yields that

$$\begin{aligned} P\{X \le 30\} &= P\{X \le 50.25 - 20.25\} \\ &\le \frac{25.126}{25.126 + (20.25)^2} \\ &\approx .058 \end{aligned}$$

■

When the moment generating function of the random variable X is known, we can obtain even more effective bounds on $P\{X \ge a\}$. Let

$$M(t) = E[e^{tX}]$$

be the moment generating function of the random variable X. Then for $t > 0$,

$$\begin{aligned} P\{X \ge a\} &= P\{e^{tX} \ge e^{ta}\} \\ &\le E[e^{tX}]e^{-ta} \qquad \text{by Markov's inequality} \end{aligned}$$

Similarly, for $t < 0$,

$$\begin{aligned} P\{X \le a\} &= P\{e^{tX} \ge e^{ta}\} \\ &\le E[e^{tX}]e^{-ta} \end{aligned}$$

Thus we have the following inequalities, known as *Chernoff bounds*.

Proposition 5.2 Chernoff bounds

$$P\{X \geq a\} \leq e^{-ta}M(t) \qquad \text{for all} \qquad t > 0$$

$$P\{X \leq a\} \leq e^{-ta}M(t) \qquad \text{for all} \qquad t < 0$$

Since the Chernoff bounds hold for all t in either the positive or negative quadrant, we obtain the best bound on $P\{X \geq a\}$ by using the t that minimizes $e^{-ta}M(t)$.

Example 5c. ***Chernoff bounds for the standard normal random variable.*** If Z is a standard normal random variable, then its moment generating function is $M(t) = e^{t^2/2}$, so the Chernoff bound on $P\{Z \geq a\}$ is given by

$$P\{Z \geq a\} \leq e^{-ta}e^{t^2/2} \qquad \text{for all} \qquad t > 0$$

Now the value of t, $t > 0$, that minimizes $e^{t^2/2-ta}$ is the value that minimizes $t^2/2 - ta$, which is $t = a$. Thus for $a > 0$ we see that

$$P\{Z \geq a\} \leq e^{-a^2/2}$$

Similarly, we can show that for $a < 0$,

$$P\{Z \leq a\} \leq e^{-a^2/2}$$

■

Example 5d. ***Chernoff bounds for the Poisson random variable.*** If X is a Poisson random variable with parameter λ, then its moment generating function is $M(t) = e^{\lambda(e^t-1)}$. Hence the Chernoff bound on $P\{X \geq i\}$ is

$$P\{X \geq i\} \leq e^{\lambda(e^t-1)}e^{-it} \qquad t > 0$$

Minimizing the right side of the preceding is equivalent to minimizing $\lambda(e^t - 1) - it$, and calculus shows that the minimal value occurs when $e^t = i/\lambda$. Provided that $i/\lambda > 1$, this minimizing value of t will be positive. Therefore, assuming that $i > \lambda$ and letting $e^t = i/\lambda$ in the Chernoff bound yields that

$$P\{X \geq i\} \leq e^{\lambda(i/\lambda-1)}\left(\frac{\lambda}{i}\right)^i$$

or, equivalently,

$$P\{X \geq i\} \leq \frac{e^{-\lambda}(e\lambda)^i}{i^i}$$

■

Example 5e. Consider a gambler who on every play is equally likely, independent of the past, to either win or lose 1 unit. That is, if X_i is the gambler's winnings on the ith play, then the X_i are independent and

$$P\{X_i = 1\} = P\{X_i = -1\} = \tfrac{1}{2}$$

Let $S_n = \sum_{i=1}^{n} X_i$ denote the gambler's winnings after n plays. We will use the Chernoff bound on $P\{S_n \geq a\}$. To start, note that the moment generating function of X_i is

$$E[e^{tX}] = \frac{e^t + e^{-t}}{2}$$

Now, using the McLaurin expansions of e^t and e^{-t}, we see that

$$\begin{aligned}
e^t + e^{-t} &= 1 + t + \frac{t^2}{2!} + \frac{t^3}{3!} + \cdots + \left(1 - t + \frac{t^2}{2!} - \frac{t^3}{3!} + \cdots\right) \\
&= 2\left\{1 + \frac{t^2}{2!} + \frac{t^4}{4!} + \cdots\right\} \\
&= 2\sum_{n=0}^{\infty} \frac{t^{2n}}{(2n)!} \\
&\leq 2\sum_{n=0}^{\infty} \frac{(t^2/2)^n}{n!} \qquad \text{since } (2n)! \geq n!\,2^n \\
&= 2e^{t^2/2}
\end{aligned}$$

Therefore,

$$E[e^{tX}] \leq e^{t^2/2}$$

Since the moment generating function of the sum of independent random variables is the product of their moment generating functions, we have that

$$\begin{aligned}
E[e^{tS_n}] &= (E[e^{tX}])^n \\
&\leq e^{nt^2/2}
\end{aligned}$$

Using the preceding result along with the Chernoff bound gives that

$$P\{S_n \geq a\} \leq e^{-ta}e^{nt^2/2} \qquad t > 0$$

The value of t that minimizes the right side of the preceding is the value that minimizes $nt^2/2 - ta$, and this value is $t = a/n$. Supposing that $a > 0$ (so that this minimizing t is positive) and letting $t = a/n$ in the preceding inequality yields that

$$P\{S_n \geq a\} \leq e^{-a^2/2n} \qquad a > 0$$

For instance, this inequality yields that

$$P\{S_{10} \geq 6\} \leq e^{-36/20} \approx .1653$$

whereas the exact probability is

$$\begin{aligned}
P\{S_{10} \geq 6\} &= P\{\text{gambler wins at least 8 of the first 10 games}\} \\
&= \frac{\binom{10}{8} + \binom{10}{9} + \binom{10}{10}}{2^{10}} = \frac{56}{1024} \approx .0547
\end{aligned}$$

■

The next inequality is one having to do with expectations rather than probabilities. Before stating it, we need the following definition.

Definition

A twice-differentiable real-valued function $f(x)$ is said to be *convex* if $f''(x) \geq 0$ for all x; similarly, it is said to be *concave* if $f''(x) \leq 0$.

Some examples of convex functions are $f(x) = x^2$, $f(x) = e^{ax}$, $f(x) = -x^{1/n}$ for $x \geq 0$. If $f(x)$ is convex, then $g(x) = -f(x)$ is concave, and vice versa.

Proposition 5.3 Jensen's inequality

If $f(x)$ is a convex function, then

$$E[f(X)] \geq f(E[X])$$

provided that the expectations exist and are finite.

Proof: Expanding $f(x)$ in a Taylor's series expansion about $\mu = E[X]$ yields

$$f(x) = f(\mu) + f'(\mu)(x - \mu) + \frac{f''(\xi)(x - \mu)^2}{2}$$

where ξ is some value between x and μ. Since $f''(\xi) \geq 0$, we obtain

$$f(x) \geq f(\mu) + f'(\mu)(x - \mu)$$

Hence

$$f(X) \geq f(\mu) + f'(\mu)(X - \mu)$$

Taking expectations yields

$$E[f(X)] \geq f(\mu) + f'(\mu)E[X - \mu] = f(\mu)$$

and the inequality is established. ✦

Example 5f. An investor is faced with the following choices: She can either invest all of her money in a risky proposition that would lead to a random return X that has mean m; or she can put the money into a risk-free venture that will lead to a return of m with probability 1. Suppose that her decision will be made on the basis of maximizing the expected value of $u(R)$, where R is her return and u is her utility function. By Jensen's inequality it follows that if u is a concave function, then $E[u(X)] \leq u(m)$, so the risk-free alternative is preferable; whereas if u is convex, then $E[u(X)] \geq u(m)$, so the risk investment alternative would be preferred. ■

8.6 BOUNDING THE ERROR PROBABILITY WHEN APPROXIMATING A SUM OF INDEPENDENT BERNOULLI RANDOM VARIABLES BY A POISSON RANDOM VARIABLE

In this section we establish bounds on how closely a sum of independent Bernoulli random variables is approximated by a Poisson random variable with the same mean. Suppose that we want to approximate the sum of independent Bernoulli random variables with respective means $p_1, p_2, \ldots, p_n$. Starting with a sequence $Y_1, \ldots, Y_n$ of independent Poisson random variables, with Y_i having mean p_i, we will construct a sequence of independent Bernoulli random variables $X_1, \ldots, X_n$ with parameters $p_1, \ldots, p_n$ such that

$$P\{X_i \neq Y_i\} \leq p_i^2 \quad \text{for each } i$$

Letting $X = \sum_{i=1}^{n} X_i$ and $Y = \sum_{i=1}^{n} Y_i$, we will use the preceding fact to conclude that

$$P\{X \neq Y\} \leq \sum_{i=1}^{n} p_i^2$$

Finally, we will show that the preceding inequality implies that for any set of real numbers A,

$$|P\{X \in A\} - P\{Y \in A\}| \leq \sum_{i=1}^{n} p_i^2$$

Since X is the sum of independent Bernoulli random variables and Y is a Poisson random variable, the preceding inequality will yield the desired bound.

To show how the preceding is accomplished, let $Y_i, i = 1, \ldots, n$ be independent Poisson random variables with respective means p_i. Now let $U_1, \ldots, U_n$ be independent random variables that are also independent of the Y_i's and which are such that

$$U_i = \begin{cases} 0 & \text{with probability} \quad (1 - p_i)e^{p_i} \\ 1 & \text{with probability} \quad 1 - (1 - p_i)e^{p_i} \end{cases}$$

The preceding definition implicity makes use of the inequality

$$e^{-p} \geq 1 - p$$

in assuming that $(1 - p_i)e^{p_i} \leq 1$.

Now define the random variables $X_i, i = 1, \ldots, n$, by

$$X_i = \begin{cases} 0 & \text{if } Y_i = U_i = 0 \\ 1 & \text{otherwise} \end{cases}$$

Note that

$$P\{X_i = 0\} = P\{Y_i = 0\}P\{U_i = 0\} = e^{-p_i}(1 - p_i)e^{p_i} = 1 - p_i$$

$$P\{X_i = 1\} = 1 - P\{X_i = 0\} = p_i$$

Now if X_i is equal to 0, then so must Y_i equal 0 (by the definition of X_i). Therefore, we see that

$$
\begin{aligned}
P\{X_i \neq Y_i\} &= P\{X_i = 1, Y_i \neq 1\} \\
&= P\{Y_i = 0, X_i = 1\} + P\{Y_i > 1\} \\
&= P\{Y_i = 0, U_i = 1\} + P\{Y_i > 1\} \\
&= e^{-p_i}[1 - (1 - p_i)e^{p_i}] + 1 - e^{-p_i} - p_i e^{-p_i} \\
&= p_i - p_i e^{-p_i} \\
&\leq p_i^2 \qquad (\text{since } 1 - e^{-p} \leq p)
\end{aligned}
$$

Now let $X = \sum_{i=1}^{n} X_i$ and $Y = \sum_{i=1}^{n} Y_i$ and note that X is the sum of independent Bernoulli random variables and Y is Poisson with the expected value $E[Y] = E[X] = \sum_{i=1}^{n} p_i$. Note also that the inequality $X \neq Y$ implies that $X_i \neq Y_i$ for some i, so

$$
\begin{aligned}
P\{X \neq Y\} &\leq P\{X_i \neq Y_i \text{ for some } i\} \\
&\leq \sum_{i=1}^{n} P\{X_i \neq Y_i\} \qquad (\text{Boole's inequality}) \\
&\leq \sum_{i=1}^{n} p_i^2
\end{aligned}
$$

For any event B, let I_B, the indicator variable for the event B, be defined by

$$
I_B = \begin{cases} 1 & \text{if } B \text{ occurs} \\ 0 & \text{otherwise} \end{cases}
$$

Note that for any set of real numbers A,

$$
I_{\{X \in A\}} - I_{\{Y \in A\}} \leq I_{\{X \neq Y\}}
$$

The preceding follows since, as an indicator variable is either 0 or 1, the left-hand side equals 1 only when $I_{\{X \in A\}} = 1$ and $I_{\{Y \in A\}} = 0$. But this would imply that $X \in A$ and $Y \notin A$, which means that $X \neq Y$, so the right side would also equal 1. Upon taking expectations of the preceding inequality, we obtain that

$$
P\{X \in A\} - P\{Y \in A\} \leq P\{X \neq Y\}
$$

By reversing X and Y, we obtain in the same manner that

$$
P\{Y \in A\} - P\{X \in A\} \leq P\{X \neq Y\}
$$

and thus we can conclude that

$$
|P\{X \in A\} - P\{Y \in A\}| \leq P\{X \neq Y\}
$$

Therefore, we have proven that with $\lambda = \sum_{i=1}^{n} p_i$,

$$
\left| P\left\{ \sum_{i=1}^{n} X_i \in A \right\} - \sum_{i \in A} \frac{e^{-\lambda}\lambda^i}{i!} \right| \leq \sum_{i=1}^{n} p_i^2
$$

REMARK. When all the p_i are equal to p, X is a binomial random variable. Thus the preceding shows that for any set of nonnegative integers A,

$$\left| \sum_{i \in A} \binom{n}{i} p^i (1-p)^{n-i} - \sum_{i \in A} \frac{e^{-np}(np)^i}{i!} \right| \le np^2 \qquad \blacksquare$$

SUMMARY

Two useful probability bounds are provided by the *Markov* and *Chebyshev* inequalities. The Markov inequality is concerned with nonnegative random variables and says that for X of that type

$$P\{X \ge a\} \le \frac{E[X]}{a}$$

for every positive value a. The Chebyshev inequality, which is a simple consequence of the Markov inequality, states that if X has mean μ and variance σ^2, then for every positive k,

$$P\{|X - \mu| \ge k\sigma\} \le \frac{1}{k^2}$$

The two most important theoretical results in probability are the *central limit theorem* and the *strong law of large numbers*. Both are concerned with a sequence of independent and identically distributed random variables. The central limit theorem says that if the random variables have a finite mean μ and a finite variance σ^2, then the distribution of the sum of the first n of them is, for large n, approximately that of a normal random variable with mean $n\mu$ and variance $n\sigma^2$. That is, if $X_i, i \ge 1$, is the sequence, then the central limit theorem states that for every real number a,

$$\lim_{n\to\infty} P\left\{\frac{X_1 + \cdots + X_n - n\mu}{\sigma\sqrt{n}} \le a\right\} = \frac{1}{\sqrt{2\pi}} \int_{-\infty}^{a} e^{-x^2/2}\,dx$$

The *strong law of large numbers* requires only that the random variables in the sequence have a finite mean μ. It states that with probability 1, the average of the first n of them will converge to μ as n goes to infinity. This implies that if A is any specified event of an experiment for which independent replications are performed, then the limiting proportion of experiments whose outcomes are in A will, with probability 1, equal $P(A)$. Therefore, if we accept the interpretation that "with probability 1" means "with certainty," we obtain the theoretical justification for the long-run relative frequency interpretation of probabilities.

PROBLEMS

1. Suppose that X is a random variable with mean and variance both equal to 20. What can be said about $P\{0 < X < 40\}$?
2. From past experience a professor knows that the test score of a student taking her final examination is a random variable with mean 75.
   (a) Give an upper bound for the probability that a student's test score will exceed 85. Suppose, in addition, the professor knows that the variance of a student's test score is equal to 25.
   (b) What can be said about the probability that a student will score between 65 and 85?
   (c) How many students would have to take the examination to ensure, with probability at least .9, that the class average would be within 5 of 75? Do not use the central limit theorem.
3. Use the central limit theorem to solve part (c) of Problem 2.
4. Let $X_1, \ldots, X_{20}$ be independent Poisson random variables with mean 1.
   (a) Use the Markov inequality to obtain a bound on

$$P\left\{\sum_1^{20} X_i > 15\right\}$$

   (b) Use the central limit theorem to approximate $P\left\{\sum_1^{20} X_i > 15\right\}$.
5. Fifty numbers are rounded off to the nearest integer and then summed. If the individual round-off errors are uniformly distributed over $(-.5, .5)$, what is the probability that the resultant sum differs from the exact sum by more than 3?
6. A die is continually rolled until the total sum of all rolls exceeds 300. What is the probability that at least 80 rolls are necessary?
7. One has 100 lightbulbs whose lifetimes are independent exponentials with mean 5 hours. If the bulbs are used one at a time, with a failed bulb being replaced immediately by a new one, what is the probability that there is still a working bulb after 525 hours?
8. In Problem 7 suppose that it takes a random time, uniformly distributed over $(0, .5)$, to replace a failed bulb. What is the probability that all bulbs have failed by time 550?
9. If X is a gamma random variable with parameters $(n, 1)$ how large need n be so that

$$P\left\{\left|\frac{X}{n} - 1\right| > .01\right\} < .01?$$

10. Civil engineers believe that W, the amount of weight (in units of 1000 pounds) that a certain span of a bridge can withstand without structural damage resulting, is normally distributed with mean 400 and standard deviation 40. Suppose that the weight (again, in units of 1000 pounds) of a car is a random variable with mean 3 and standard deviation .3. How many cars would have to be on the bridge span for the probability of structural damage to exceed .1?
11. Many people believe that the daily change of price of a company's stock on the stock market is a random variable with mean 0 and variance σ^2. That is, if Y_n represents the price of the stock on the nth day, then

$$Y_n = Y_{n-1} + X_n \qquad n \geq 1$$

where $X_1, X_2, \ldots$ are independent and identically distributed random variables with mean 0 and variance σ^2. Suppose that the stock's price today is 100. If $\sigma^2 = 1$, what can you say about the probability that the stock's price will exceed 105 after 10 days?

12. We have 100 components that we will put in use in a sequential fashion. That is, component 1 is initially put in use, and upon failure it is replaced by component 2, which is itself replaced upon failure by component 3, and so on. If the lifetime of component i is exponentially distributed with mean $10 + i/10$, $i = 1, \ldots, 100$, estimate the probability that the total life of all components will exceed 1200. Now repeat when the life distribution of component i is uniformly distributed over $(0, 20 + i/5)$, $i = 1, \ldots, 100$.

13. Student scores on exams given by a certain instructor have mean 74 and standard deviation 14. This instructor is about to give two exams, one to a class of size 25 and the other to a class of size 64.

(a) Approximate the probability that the average test score in the class of size 25 exceeds 80.

(b) Repeat part (a) for the class of size 64.

(c) Approximate the probability that the average test score in the larger class exceeds that of the other class by over 2.2 points.

(d) Approximate the probability that the average test score in the smaller class exceeds that of the other class by over 2.2 points.

14. A certain component is critical to the operation of an electrical system and must be replaced immediately upon failure. If the mean lifetime of this type of component is 100 hours and its standard deviation is 30 hours, how many of these components must be in stock so that the probability that the system is in continual operation for the next 2000 hours is at least .95?

15. An insurance company has 10,000 automobile policyholders. The expected yearly claim per policyholder is \$240 with a standard deviation of \$800. Approximate the probability that the total yearly claim exceeds \$2.7 million.

16. Redo Example 5b under the assumption that the number of man–woman pairs is (approximately) normally distributed. Does this seem like a reasonable supposition?

17. Repeat part (a) of Problem 2 when it is known that the variance of a student's test score is equal to 25.

18. A lake contains 4 distinct types of fish. Suppose that each fish caught is equally likely to be any one of these types. Let Y denote the number of fish that need be caught to obtain at least one of each type.

(a) Give an interval (a, b) such that $P\{a \le Y \le b\} \ge .90$.

(b) Using the one-sided Chebyshev inequality, how many fish need we plan on catching so as to be at least 90 percent certain of obtaining at least one of each type?

19. If X is a nonnegative random variable with mean 25, what can be said about

(a) $E[X^3]$; **(b)** $E[\sqrt{X}]$;

(c) $E[\log X]$; **(d)** $E[e^{-X}]$?

20. Let X be a nonnegative random variable. Prove that

$$E[X] \le (E[X^2])^{1/2} \le (E[X^3])^{1/3} \le \cdots$$

21. Would the results of Example 5f change if the investor were allowed to divide her money and invest the fraction α, $0 < \alpha < 1$ in the risky proposition and invest the remainder in the risk-free venture? Her return for such a split investment would be $R = \alpha X + (1 - \alpha)m$.

22. Let X be a Poisson random variable with mean 20.

(a) Use the Markov inequality to obtain an upper bound on

$$p = P\{X \ge 26\}$$

(b) Use the one-sided Chebyshev inequality to obtain an upper bound on p.

(c) Use the Chernoff bound to obtain an upper bound on p.

(d) Approximate p by making use of the central limit theorem.

(e) Determine p by running an appropriate program.

THEORETICAL EXERCISES

1. If X has variance σ^2, then σ, the positive square root of the variance, is called the *standard deviation*. If X has mean μ and standard deviation σ, show that

$$P\{|X - \mu| \geq k\sigma\} \leq \frac{1}{k^2}$$

2. If X has mean μ and standard deviation σ, the ratio $r \equiv |\mu|/\sigma$ is called the *measurement signal-to-noise ratio* of X. The idea is that X can be expressed as $X = \mu + (X - \mu)$, with μ representing the signal and $X - \mu$ the noise. If we define $|(X - \mu)/\mu| \equiv D$ as the relative deviation of X from its signal (or mean) μ, show that for $\alpha > 0$,

$$P\{D \leq \alpha\} \geq 1 - \frac{1}{r^2\alpha^2}$$

3. Compute the measurement signal-to-noise ratio—that is, $|\mu|/\sigma$ where $\mu = E[X]$, $\sigma^2 = \text{Var}(X)$—of the following random variables:
(a) Poisson with mean λ;
(b) binomial with parameters n and p;
(c) geometric with mean $1/p$;
(d) uniform over (a, b);
(e) exponential with mean $1/\lambda$;
(f) normal with parameters μ, σ^2.

4. Let $Z_n, n \geq 1$ be a sequence of random variables and c a constant such that for each $\varepsilon > 0$, $P\{|Z_n - c| > \varepsilon\} \to 0$ as $n \to \infty$. Show that for any bounded continuous function g,

$$E[g(Z_n)] \to g(c) \quad \text{as} \quad n \to \infty$$

5. Let $f(x)$ be a continuous function defined for $0 \leq x \leq 1$. Consider the functions

$$B_n(x) = \sum_{k=0}^{n} f\left(\frac{k}{n}\right)\binom{n}{k}x^k(1 - x)^{n-k}$$

(called *Bernstein polynomials*) and prove that

$$\lim_{n\to\infty} B_n(x) = f(x)$$

HINT: Let $X_1, X_2, \ldots$ be independent Bernoulli random variables with mean x. Show and then use the fact (by making use of the result of Theoretical Exercise 4) that

$$B_n(x) = E\left[f\left(\frac{X_1 + \cdots + X_n}{n}\right)\right]$$

As it can be shown that the convergence of $B_n(x)$ to $f(x)$ is uniform in x, the preceding provides a probabilistic proof to the famous Weierstrass theorem of analysis, which states that any continuous function on a closed interval can be approximated arbitrarily closely by a polynomial.

6. **(a)** Let X be a discrete random variable, whose possible values are $1, 2, \ldots$. If $P\{X = k\}$ is nonincreasing in $k = 1, 2, \ldots$, prove that

$$P\{X = k\} \leq 2\frac{E[X]}{k^2}$$

(b) Let X be a nonnegative continuous random variable having a nonincreasing density function. Show that

$$f(x) \leq \frac{2E[X]}{x^2} \quad \text{for all} \quad x > 0$$

7. Suppose that a fair die is rolled 100 times. Let X_i be the value obtained on the ith roll. Compute an approximation for

$$P\left\{\prod_{1}^{100} X_i \le a^{100}\right\} \qquad 1 < a < 6$$

8. Explain why a gamma random variable with parameters (t, λ) has an approximately normal distribution when t is large.

9. Suppose a fair coin is tossed 1000 times. If the first 100 tosses all result in heads, what proportion of heads would you expect on the final 900 tosses? Comment on the statement that "the strong law of large numbers swamps but does not compensate."

10. If X is a Poisson random variable with mean λ, show that for $i < \lambda$,

$$P\{X \le i\} \le \frac{e^{-\lambda}(e\lambda)^i}{i^i}$$

11. Let X be a binomial random variable with parameters n and p. Show that for $i > np$:

(a) $\underset{t>0}{\text{minimum}}\ e^{-ti}E[e^{tX}]$ occurs when t is such that $e^t = \dfrac{iq}{(n-i)p}$, where $q = 1 - p$.

(b) $P\{X \ge i\} \le \dfrac{n^n}{i^i(n-i)^{n-i}}p^i(1-p)^{n-i}$.

12. The Chernoff bound on a standard normal random variable Z gives that $P\{Z > a\} \le e^{-a^2/2}, a > 0$. Show, by considering the density of Z, that the right side of the inequality can be reduced by a factor 2. That is, show that

$$P\{Z > a\} \le \tfrac{1}{2}e^{-a^2/2} \qquad a > 0$$

13. If $E[X] < 0$ and $\theta \ne 0$ is such that $E[e^{\theta X}] = 1$, show that $\theta > 0$.

SELF-TEST PROBLEMS AND EXERCISES

1. The number of automobiles sold weekly at a certain dealership is a random variable with expected value 16. Give an upper bound to the probability that
(a) next week's sales exceed 18; **(b)** next week's sales exceed 25.

2. Suppose in Problem 1 that the variance of the number of automobiles sold weekly is 9.
(a) Give a lower bound to the probability that next week's sales are between 10 and 22 inclusively.
(b) Give an upper bound to the probability that next week's sales exceed 18.

3. If

$$E[X] = 75 \quad E[Y] = 75 \quad \text{Var}(X) = 10 \quad \text{Var}(Y) = 12 \quad \text{Cov}(X,Y) = -3$$

give an upper bound to
(a) $P\{|X - Y| > 15\}$; **(b)** $P\{X > Y + 15\}$;
(c) $P\{Y > X + 15\}$.

4. Suppose that the number of units produced daily at factory A is a random variable with mean 20 and standard deviation 3 and the number produced at factory B is a random variable with mean 18 and standard deviation 6. Assuming independence, derive an upper bound for the probability that more units are produced today at factory B than at factory A.

5. The amount of time that a certain type of component functions before failing is a random variable with probability density function

$$f(x) = 2x \qquad 0 < x < 1$$

Once the component fails it is immediately replaced by another one of the same type. If we let X_i denote the lifetime of the ith component to be put in use, then $S_n = \sum_{i=1}^{n} X_i$ represents the time of the nth failure. The long-term rate at which failures occur, call it r, is defined by

$$r = \lim_{n\to\infty} \frac{n}{S_n}$$

Assuming that the random variables $X_i, i \geq 1$ are independent, determine r.

6. In Self-Test Problem 5, how many components would one need to have on hand to be approximately 90 percent certain that the stock will last at least 35 days?

7. The servicing of a machine requires two separate steps, with the time needed for the first step being an exponential random variable with mean .2 hour and the time for the second step being an independent exponential random variable with mean .3 hour. If a repair person has 20 machines to service, approximate the probability that all the work can be completed in 8 hours.

8. On each bet, a gambler loses 1 with probability .7, loses 2 with probability .2, or wins 10 with probability .1. Approximate the probability that the gambler will be losing after his first 100 bets.

9. Determine t so that the probability that the repair person in Self-Test Problem 7 finishes the 20 jobs within time t is approximately equal to .95.

10. A tobacco company claims that the amount of nicotine in one of its cigarettes is a random variable with mean 2.2 mg and standard deviation .3 mg. However, the average nicotine content of 100 randomly chosen cigarettes was 3.1 mg. Approximate the probability that the average would have been as high as or higher than 3.1 if the company's claims were true.

11. A clinic is equally likely to have 2, 3, or 4 doctors volunteer for service on a given day. No matter how may volunteer doctors there are on a given day, the numbers of patients seen by these doctors are independent Poisson random variables with mean 30. Let X denote the number of patients seen in the clinic on a given day.

(a) Find $E[X]$. **(b)** Find $\text{Var}(X)$.

(c) Use a table of the standard normal probability distribution to approximate $P\{X > 65\}$.

CHAPTER 9

Additional Topics in Probability

9.1 THE POISSON PROCESS

Before defining a Poisson process, recall that a function f is said to be $o(h)$ if $\lim_{h\to 0} f(h)/h = 0$. That is, f is $o(h)$ if, for small values of h, $f(h)$ is small even in relation to h. Suppose now that "events" are occurring at random time points and let $N(t)$ denote the number of events that occur in the time interval $[0, t]$. The collection of random variables $\{N(t), t \geq 0\}$ is said to be a *Poisson process having rate* λ, $\lambda > 0$ if

(i) $N(0) = 0$.

(ii) The numbers of events that occur in disjoint time intervals are independent.

(iii) The distribution of the number of events that occur in a given interval depends only on the length of that interval and not on its location.

(iv) $P\{N(h) = 1\} = \lambda h + o(h)$.

(v) $P\{N(h) \geq 2\} = o(h)$.

Thus condition (i) states that the process begins at time 0. Condition (ii), the *independent increment* assumption, states, for instance, that the number of events by time t [that is, $N(t)$] is independent of the number of events that occur between t and $t + s$ [that is, $N(t + s) - N(t)$]. Condition (iii), the *stationary increment* assumption, states that the probability distribution of $N(t + s) - N(t)$ is the same for all values of t.

In Chapter 4 we presented an argument, based on the Poisson distribution being a limiting version of the binomial distribution, that the foregoing conditions imply that $N(t)$ has a Poisson distribution with mean λt. We will now obtain this result by a different method.

Lemma 1.1

For a Poisson process with rate λ,

$$P\{N(t) = 0\} = e^{-\lambda t}$$

Proof: Let $P_0(t) = P\{N(t) = 0\}$. We derive a differential equation for $P_0(t)$ in the following manner:

$$\begin{aligned}P_0(t + h) &= P\{N(t + h) = 0\}\\ &= P\{N(t) = 0, N(t + h) - N(t) = 0\}\\ &= P\{N(t) = 0\}P\{N(t + h) - N(t) = 0\}\\ &= P_0(t)[1 - \lambda h + o(h)]\end{aligned}$$

where the final two equations follow from condition (ii) plus the fact that conditions (iv) and (v) imply that $P\{N(h) = 0\} = 1 - \lambda h + o(h)$. Hence

$$\frac{P_0(t + h) - P_0(t)}{h} = -\lambda P_0(t) + \frac{o(h)}{h}$$

Now, letting $h \to 0$, we obtain

$$P_0'(t) = -\lambda P_0(t)$$

or, equivalently,

$$\frac{P_0'(t)}{P_0(t)} = -\lambda$$

which implies, by integration, that

$$\log P_0(t) = -\lambda t + c$$

or

$$P_0(t) = Ke^{-\lambda t}$$

Since $P_0(0) = P\{N(0) = 0\} = 1$, we arrive at

$$P_0(t) = e^{-\lambda t}$$ ✦

For a Poisson process, let us denote by T_1 the time of the first event. Further, for $n > 1$, let T_n denote the elapsed time between the $(n - 1)$st and the nth event. The sequence $\{T_n, n = 1, 2, \dots\}$ is called the *sequence of interarrival times*. For instance, if $T_1 = 5$ and $T_2 = 10$, then the first event of the Poisson process would have occurred at time 5 and the second at time 15.

We shall now determine the distribution of the T_n. To do so, we first note that the event $\{T_1 > t\}$ takes place if and only if no events of the Poisson process occur in the interval $[0, t]$, and thus

$$P\{T_1 > t\} = P\{N(t) = 0\} = e^{-\lambda t}$$

Hence T_1 has an exponential distribution with mean $1/\lambda$. Now,

$$P\{T_2 > t\} = E[P\{T_2 > t|T_1\}]$$

However,

$$\begin{aligned} P\{T_2 > t | T_1 = s\} &= P\{0 \text{ events in } (s, s+t] | T_1 = s\} \\ &= P\{0 \text{ events in } (s, s+t]\} \\ &= e^{-\lambda t} \end{aligned}$$

where the last two equations followed from the assumptions about independent and stationary increments. Therefore, from the preceding we conclude that T_2 is also an exponential random variable with mean $1/\lambda$, and furthermore, that T_2 is independent of T_1. Repeating the same argument yields Proposition 1.1.

Proposition 1.1

$T_1, T_2, \ldots$ are independent exponential random variables each with mean $1/\lambda$.

Another quantity of interest is S_n, the arrival time of the nth event, also called the *waiting time* until the nth event. It is easily seen that

$$S_n = \sum_{i=1}^{n} T_i \qquad n \geq 1$$

and hence from Proposition 1.1 and the results of Section 5.6.1, it follows that S_n has a gamma distribution with parameters n and λ. That is, the probability density of S_n is given by

$$f_{S_n}(x) = \lambda e^{-\lambda x} \frac{(\lambda x)^{n-1}}{(n-1)!} \qquad x \geq 0$$

We are now ready to prove that $N(t)$ is a Poisson random variable with mean λt.

Theorem 1.1

For a Poisson process with rate λ,

$$P\{N(t) = n\} = \frac{e^{-\lambda t}(\lambda t)^n}{n!}$$

Proof: Note that the nth event of the Poisson process will occur before or at time t if and only if the number of events that occur by t is at least n. That is,

$$N(t) \geq n \Leftrightarrow S_n \leq t$$

so

$$P\{N(t) = n\} = P\{N(t) \geq n\} - P\{N(t) \geq n + 1\}$$
$$= P\{S_n \leq t\} - P\{S_{n+1} \leq t\}$$
$$= \int_0^t \lambda e^{-\lambda x} \frac{(\lambda x)^{n-1}}{(n-1)!} dx - \int_0^t \lambda e^{-\lambda x} \frac{(\lambda x)^n}{n!} dx$$

But the integration by parts formula $\int u\,dv = uv - \int v\,du$ yields, with $u = e^{-\lambda x}$, $dv = \lambda[(\lambda x)^{n-1}/(n-1)!]\,dx$,

$$\int_0^t \lambda e^{-\lambda x} \frac{(\lambda x)^{n-1}}{(n-1)!} dx = e^{-\lambda t} \frac{(\lambda t)^n}{n!} + \int_0^t \lambda e^{-\lambda x} \frac{(\lambda x)^n}{n!} dx$$

which completes the proof. ✦

9.2 MARKOV CHAINS

Consider a sequence of random variables $X_0, X_1, \ldots$, and suppose that the set of possible values of these random variables is $\{0, 1, \ldots, M\}$. It will be helpful to interpret X_n as being the state of some system at time n, and, in accordance with this interpretation, we say that the system is in state i at time n if $X_n = i$. The sequence of random variables is said to form a *Markov chain* if each time the system is in state i there is some fixed probability—call it P_{ij}—that it will next be in state j. That is, for all $i_0, \ldots, i_{n-1}, i, j$,

$$P\{X_{n+1} = j | X_n = i, X_{n-1} = i_{n-1}, \ldots, X_1 = i_1, X_0 = i_0\} = P_{ij}$$

The values P_{ij}, $0 \leq i \leq M$, $0 \leq j \leq N$, are called the *transition probabilities* of the Markov chain and they satisfy (why?)

$$P_{ij} \geq 0 \qquad \sum_{j=0}^{M} P_{ij} = 1 \qquad i = 0, 1, \ldots, M$$

It is convenient to arrange the transition probabilities P_{ij} in a square array as follows:

$$\begin{Vmatrix} P_{00} & P_{01} & \cdots & P_{0M} \\ P_{10} & P_{11} & \cdots & P_{1M} \\ \vdots & & & \\ P_{M0} & P_{M1} & \cdots & P_{MM} \end{Vmatrix}$$

Such an array is called a *matrix*.

Knowledge of the transition probability matrix and the distribution of X_0 enables us, in theory, to compute all probabilities of interest. For instance, the joint probability mass function of $X_0, \ldots, X_n$ is given by

$$P\{X_n = i_n, X_{n-1} = i_{n-1}, \ldots, X_1 = i_1, X_0 = i_0\}$$
$$= P\{X_n = i_n | X_{n-1} = i_{n-1}, \ldots, X_0 = i_0\} P\{X_{n-1} = i_{n-1}, \ldots, X_0 = i_0\}$$
$$= P_{i_{n-1}, i_n} P\{X_{n-1} = i_{n-1}, \ldots, X_0 = i_0\}$$

and continual repetition of this argument yields that the preceding is equal to

$$P_{i_{n-1}, i_n} P_{i_{n-2}, i_{n-1}} \cdots P_{i_1, i_2} P_{i_0, i_1} P\{X_0 = i_0\}$$

Example 2a. Suppose that whether or not it rains tomorrow depends on previous weather conditions only through whether or not it is raining today. Suppose further that if it is raining today, then it will rain tomorrow with probability α, and, if it is not raining today, then it will rain tomorrow with probability β.

If we say that the system is in state 0 when it rains and state 1 when it does not, then the preceding system is a two-state Markov chain having transition probability matrix

$$\begin{Vmatrix} \alpha & 1 - \alpha \\ \beta & 1 - \beta \end{Vmatrix}$$

That is, $P_{00} = \alpha = 1 - P_{01}, P_{10} = \beta = 1 - P_{11}$. ■

Example 2b. Consider a gambler who at each play of the game either wins 1 unit with probability p or loses 1 unit with probability $1 - p$. If we suppose that the gambler will quit playing when his fortune hits either 0 or M, then the gambler's sequence of fortunes is a Markov chain having transition probabilities

$$P_{i,i+1} = p = 1 - P_{i,i-1} \qquad i = 1, \ldots, M - 1$$

$$P_{00} = P_{MM} = 1$$ ■

Example 2c. The physicists P. and T. Ehrenfest considered a conceptual model for the movement of molecules in which M molecules are distributed among 2 urns. At each time point one of the molecules is chosen at random and is removed from its urn and placed in the other one. If we let X_n denote the number of molecules in the first urn immediately after the nth exchange, then $\{X_0, X_1, \ldots\}$ is a Markov chain with transition probabilities

$$P_{i,i+1} = \frac{M - i}{M} \qquad 0 \le i \le M$$

$$P_{i,i-1} = \frac{i}{M} \qquad 0 \le i \le M$$

$$P_{ij} = 0 \qquad \text{if } |j - i| > 1$$ ■

Thus, for a Markov chain, P_{ij} represents the probability that a system in state i will enter state j at the next transition. We can also define the two-stage transition probability, $P_{ij}^{(2)}$, that a system presently in state i will be in state j after two additional transitions. That is,

$$P_{ij}^{(2)} = P\{X_{m+2} = j | X_m = i\}$$

The $P_{ij}^{(2)}$ can be computed from the P_{ij} as follows:

$$
\begin{aligned}
P_{ij}^{(2)} &= P\{X_2 = j \mid X_0 = i\} \\
&= \sum_{k=0}^{M} P\{X_2 = j, X_1 = k \mid X_0 = i\} \\
&= \sum_{k=0}^{M} P\{X_2 = j \mid X_1 = k, X_0 = i\} P\{X_1 = k \mid X_0 = i\} \\
&= \sum_{k=0}^{M} P_{kj} P_{ik}
\end{aligned}
$$

In general, we define the n-stage transition probabilities, denoted as $P_{ij}^{(n)}$, by

$$P_{ij}^{(n)} = P\{X_{n+m} = j \mid X_m = i\}$$

Proposition 2.1, known as the Chapman–Kolmogorov equations, shows how the $P_{ij}^{(n)}$ can be computed.

Proposition 2.1 The Chapman–Kolmogorov equations

$$P_{ij}^{(n)} = \sum_{k=0}^{M} P_{ik}^{(r)} P_{kj}^{(n-r)} \qquad \text{for all } 0 < r < n$$

Proof:

$$
\begin{aligned}
P_{ij}^{(n)} &= P\{X_n = j \mid X_0 = i\} \\
&= \sum_{k} P\{X_n = j, X_r = k \mid X_0 = i\} \\
&= \sum_{k} P\{X_n = j \mid X_r = k, X_0 = i\} P\{X_r = k \mid X_0 = i\} \\
&= \sum_{k} P_{kj}^{(n-r)} P_{ik}^{(r)}
\end{aligned}
$$

✦

Example 2d. ***A random walk.*** An example of a Markov chain having a countably infinite state space is the *random walk*, which tracks a particle as it moves along a one-dimensional axis. Suppose that at each point in time the particle will move either one step to the right or one step to the left with respective probabilities p and $1 - p$. That is, suppose the particle's path follows a Markov chain with transition probabilities

$$P_{i,i+1} = p = 1 - P_{i,i-1} \qquad i = 0, \pm 1, \ldots$$

If the particle is at state i, then the probability it will be at state j after n transitions is the probability that $(n - i + j)/2$ of these steps are to the right and $n - [(n - i + j)/2] = (n + i - j)/2$ are to the left. As each step will be to the right, independently of the other steps, with probability p, it follows that the above is just the binomial probability

$$P_{ij}^{n} = \binom{n}{(n-i+j)/2} p^{(n-i+j)/2}(1-p)^{(n+i-j)/2}$$

where $\binom{n}{x}$ is taken to equal 0 when x is not a nonnegative integer less than or equal to n. The preceding can be rewritten as

$$P_{i,i+2k}^{2n} = \binom{2n}{n+k} p^{n+k}(1-p)^{n-k} \qquad k = 0, \pm 1, \ldots, \pm n$$

$$P_{i,i+2k+1}^{2n+1} = \binom{2n+1}{n+k+1} p^{n+k+1}(1-p)^{n-k}$$

$$k = 0, \pm 1, \ldots, \pm\, n, -(n+1) \qquad \blacksquare$$

Although the $P_{ij}^{(n)}$ denote conditional probabilities, we can, by conditioning on the initial state, use them to derive expressions for unconditional probabilities. For instance,

$$\begin{aligned} P\{X_n = j\} &= \sum_i P\{X_n = j | X_0 = i\} P\{X_0 = i\} \\ &= \sum_i P_{ij}^{(n)} P\{X_0 = i\} \end{aligned}$$

For a large number of Markov chains it turns out that $P_{ij}^{(n)}$ converges, as $n \to \infty$, to a value Π_j that depends only on j. That is, for large values of n, the probability of being in state j after n transitions is approximately equal to Π_j no matter what the initial state was. It can be shown that a sufficient condition for a Markov chain to possess this property is that for some $n > 0$,

$$P_{ij}^{(n)} > 0 \qquad \text{for all } i, j = 0, 1, \ldots, M \tag{2.1}$$

Markov chains that satisfy Equation (2.1) are said to be *ergodic*. Since Proposition 2.1 yields

$$P_{ij}^{(n+1)} = \sum_{k=0}^{M} P_{ik}^{(n)} P_{kj}$$

it follows, by letting $n \to \infty$, that for ergodic chains

$$\Pi_j = \sum_{k=0}^{M} \Pi_k P_{kj} \tag{2.2}$$

Furthermore, since $1 = \sum_{j=0}^{M} P_{ij}^{(n)}$, we also obtain, by letting $n \to \infty$,

$$\sum_{j=0}^{M} \Pi_j = 1 \tag{2.3}$$

In fact, it can be shown that the $\Pi_j, 0 \le j \le M$, are the unique nonnegative solutions of Equations (2.2) and (2.3). All this is summed up in Theorem 2.1, which we state without proof.

Theorem 2.1

For an ergodic Markov chain

$$\Pi_j = \lim_{n\to\infty} P_{ij}^{(n)}$$

exists, and the $\Pi_j, 0 \le j \le M$, are the unique nonnegative solutions of

$$\Pi_j = \sum_{k=0}^{M} \Pi_k P_{kj}$$

$$\sum_{j=0}^{M} \Pi_j = 1$$

Example 2e. Consider Example 2a, in which we assume that if it rains today, then it will rain tomorrow with probability α; and, if it does not rain today, then it will rain tomorrow with probability β. From Theorem 2.1 it follows that the limiting probabilities of rain and of no rain, Π_0 and Π_1, are given by

$$\Pi_0 = \alpha\Pi_0 + \beta\Pi_1$$

$$\Pi_1 = (1 - \alpha)\Pi_0 + (1 - \beta)\Pi_1$$

$$\Pi_0 + \Pi_1 = 1$$

which yields

$$\Pi_0 = \frac{\beta}{1 + \beta - \alpha} \qquad \Pi_1 = \frac{1 - \alpha}{1 + \beta - \alpha}$$

For instance, if $\alpha = .6$, $\beta = .3$, then the limiting probability of rain on the nth day is $\Pi_0 = \frac{3}{7}$. ■

The quantity Π_j is also equal to the long-run proportion of time that the Markov chain is in state $j, j = 0, \ldots, M$. To see intuitively why this might be so, let P_j denote the long-run proportion of time the chain is in state j. (It can be proven, using the strong law of large numbers, that for an ergodic chain such long-run proportions exist and are constants.) Now, since the proportion of time the chain is in state k is P_k and since, when in state k, the chain goes to state j with probability P_{kj}, it follows that the proportion of time the Markov chain is entering state j from state k is equal to $P_k P_{kj}$. Summing over all k shows that P_j, the proportion of time the Markov chain is entering state j, satisfies

$$P_j = \sum_k P_k P_{kj}$$

Since clearly it is also true that

$$\sum_j P_j = 1$$

it thus follows, since by Theorem 2.1 the $\Pi_j, j = 0, \ldots, M$ are the unique solution of the preceding, that $P_j = \Pi_j, j = 0, \ldots, M$. The longrun proportion interpretation of Π_j, is generally valid even when the chain is not ergodic.

Example 2f. Suppose in Example 2c that we are interested in the proportion of time there are j molecules in urn 1, $j = 0, \ldots, M$. By Theorem 2.1 these quantities will be the unique solution of

$$\Pi_0 = \Pi_1 \times \frac{1}{M}$$

$$\Pi_j = \Pi_{j-1} \times \frac{M - j + 1}{M} + \Pi_{j+1} \times \frac{j + 1}{M} \qquad j = 1, \ldots, M$$

$$\Pi_M = \Pi_{M-1} \times \frac{1}{M}$$

$$\sum_{j=0}^{M} \Pi_j = 1$$

However, as it is easily checked that

$$\Pi_j = \binom{M}{j}\left(\frac{1}{2}\right)^M \qquad j = 0, \ldots, M$$

satisfy the preceding equations, it follows that these are the long-run proportions of time that the Markov chain is in each of the states. (See Problem 11 for an explanation of how one might have guessed at the foregoing solution.) ■

9.3 SURPRISE, UNCERTAINTY, AND ENTROPY

Consider an event E that can occur when an experiment is performed. How surprised would we be to hear that E does, in fact, occur? It seems reasonable to suppose that the amount of surprise engendered by the information that E has occurred should depend on the probability of E. For instance, if the experiment consists of rolling a pair of dice, then we would not be too surprised to hear that E has occurred when E represents the event that the sum of the dice is even (and thus has probability $\frac{1}{2}$), whereas we would certainly be more surprised to hear that E has occurred when E is the event that the sum of the dice is 12 (and thus has probability $\frac{1}{36}$).

In this section we attempt to quantify the concept of surprise. To begin, let us agree to suppose that the surprise one feels upon learning that an event E has occurred depends only on the probability of E; and let us denote by $S(p)$ the surprise evoked by the occurrence of an event having probability p. We determine the functional form of $S(p)$ by first agreeing on a set of reasonable conditions that

$S(p)$ should satisfy, and then proving that these axioms require that $S(p)$ has a specified form. We assume throughout that $S(p)$ is defined for all $0 < p \leq 1$ but is not defined for events having $p = 0$.

Our first condition is just a statement of the intuitive fact that there is no surprise in hearing that an event sure to occur has indeed occurred.

Axiom 1

$$S(1) = 0$$

Our second condition states that the more unlikely an event is to occur, the greater is the surprise evoked by its occurrence.

Axiom 2

$S(p)$ is a strictly decreasing function of p; that is, if $p < q$, then $S(p) > S(q)$.

The third condition is a mathematical statement of the fact that we would intuitively expect a small change in p to correspond to a small change in $S(p)$.

Axiom 3

$S(p)$ is a continuous function of p.

To motivate the final condition, consider two independent events E and F, having respective probabilities $P(E) = p$ and $P(F) = q$. Since $P(EF) = pq$, the surprise evoked by the information that both E and F have occurred is $S(pq)$. Now, suppose that we are first told that E has occurred and then, afterward, that F has also occurred. As $S(p)$ is the surprise evoked by the occurrence of E, it follows that $S(pq) - S(p)$ represents the additional surprise evoked when we are informed that F has also occurred. However, as F is independent of E, the knowledge that E occurred does not change the probability of F, and hence the additional surprise should just be $S(q)$. This reasoning suggests the final condition.

Axiom 4

$$S(pq) = S(p) + S(q) \qquad 0 < p \leq 1, 0 < q \leq 1$$

We are now ready for Theorem 3.1, which yields the structure of $S(p)$.

Theorem 3.1

If $S(\cdot)$ satisfies Axioms 1 through 4, then

$$S(p) = -C \log_2 p$$

where C is an arbitrary positive integer.

Proof: It follows from Axiom 4 that

$$S(p^2) = S(p) + S(p) = 2S(p)$$

and by induction that

$$S(p^m) = m\,S(p) \tag{3.1}$$

Also, since for any integral n, $S(p) = S(p^{1/n} \cdots p^{1/n}) = n\,S(p^{1/n})$, it follows that

$$S(p^{1/n}) = \frac{1}{n} S(p) \tag{3.2}$$

Thus, from Equations (3.1) and (3.2), we obtain

$$\begin{aligned} S(p^{m/n}) &= m\,S(p^{1/n}) \\ &= \frac{m}{n} S(p) \end{aligned}$$

which is equivalent to

$$S(p^x) = xS(p) \tag{3.3}$$

whenever x is a positive rational number. But this implies by the continuity of S (Axiom 3) that Equation (3.3) is valid for all nonnegative x. (Reason this out.)

Now, for any p, $0 < p \le 1$, let $x = -\log_2 p$. Then $p = \left(\frac{1}{2}\right)^x$ and from Equation (3.3),

$$S(p) = S\left(\left(\tfrac{1}{2}\right)^x\right) = xS\left(\tfrac{1}{2}\right) = -C \log_2 p$$

where $C = S\left(\frac{1}{2}\right) > S(1) = 0$ by Axioms 2 and 1. ✦

It is usual to let C equal 1. In this case the surprise is said to be expressed in units of *bits* (short for *binary digits*).

Consider now a random variable X, which must take on one of the values $x_1, \ldots, x_n$ with respective probabilities $p_1, \ldots, p_n$. As $-\log p_i$ represents the surprise evoked if X takes on the value x_i,† it follows that the expected amount of surprise we shall receive upon learning the value of X is given by

$$H(X) = -\sum_{i=1}^{n} p_i \log p_i$$

† For the remainder of this chapter we write $\log x$ for $\log_2 x$. Also, we use $\ln x$ for $\log_e x$.

The quantity $H(X)$ is known in information theory as the *entropy* of the random variable X. (In case one of the $p_i = 0$, we take $0 \log 0$ to equal 0.) It can be shown (and we leave it as an exercise) that $H(X)$ is maximized when all of the p_i are equal. (Is this intuitive?)

As $H(X)$ represents the average amount of surprise one receives upon learning the value of X, it can also be interpreted as representing the amount of *uncertainty* that exists as to the value of X. In fact, in information theory, $H(X)$ is interpreted as the average amount of *information* received when the value of X is observed. Thus the average surprise evoked by X, the uncertainty of X, or the average amount of information yielded by X, all represent the same concept viewed from three slightly different points of view.

Consider now two random variables, X and Y, which take on respective values $x_1, \ldots, x_n$ and $y_1, \ldots, y_m$ with joint mass function

$$p(x_i, y_j) = P\{X = x_i, Y = y_j\}$$

It follows that the uncertainty as to the value of the random vector (X, Y), denoted by $H(X, Y)$, is given by

$$H(X, Y) = -\sum_i \sum_j p(x_i, y_j) \log p(x_i, y_j)$$

Suppose now that Y is observed to equal y_j. In this situation the amount of uncertainty remaining in X is given by

$$H_{Y=y_j}(X) = -\sum_i p(x_i|y_j) \log p(x_i|y_j)$$

where

$$p(x_i|y_j) = P\{X = x_i | Y = y_j\}$$

Hence the average amount of uncertainty that will remain in X after Y is observed is given by

$$H_Y(X) = \sum_j H_{Y=y_j}(X) p_Y(y_j)$$

where

$$p_Y(y_j) = P\{Y = y_j\}$$

Proposition 3.1 relates $H(X, Y)$ to $H(Y)$ and $H_Y(X)$. It states that the uncertainty as to the value of X and Y is equal to the uncertainty of Y plus the average uncertainty remaining in X when Y is to be observed.

Proposition 3.1

$$H(X, Y) = H(Y) + H_Y(X)$$

Proof: Using the identity $p(x_i, y_j) = p_Y(y_j)p(x_i|y_j)$ yields

$$
\begin{aligned}
H(X,Y) &= -\sum_i \sum_j p(x_i, y_j) \log p(x_i, y_j) \\
&= -\sum_i \sum_j p_Y(y_j)p(x_i|y_j)[\log p_Y(y_j) + \log p(x_i|y_j)] \\
&= -\sum_j p_Y(y_j) \log p_Y(y_j) \sum_i p(x_i|y_j) \\
&\quad -\sum_j p_Y(y_j) \sum_i p(x_i|y_j) \log p(x_i|y_j) \\
&= H(Y) + H_Y(X)
\end{aligned}
$$ ◆

It is a fundamental result in information theory that the amount of uncertainty in a random variable X will, on the average, decrease when a second random variable Y is observed. Before proving this, we need the following lemma, whose proof is left as an exercise.

Lemma 3.1

$$\ln x \le x - 1 \qquad x > 0$$

with equality only at $x = 1$.

Theorem 3.2

$$H_Y(X) \le H(X)$$

with equality if and only if X and Y are independent.

Proof:

$$
\begin{aligned}
H_Y(X) - H(X) &= -\sum_i \sum_j p(x_i|y_j) \log[p(x_i|y_j)]p(y_j) \\
&\quad +\sum_i \sum_j p(x_i, y_j) \log p(x_i) \\
&= \sum_i \sum_j p(x_i, y_j) \log\left[\frac{p(x_i)}{p(x_i|y_j)}\right] \\
&\le \log e \sum_i \sum_j p(x_i, y_j)\left[\frac{p(x_i)}{p(x_i|y_j)} - 1\right] \qquad \text{by Lemma 3.1}
\end{aligned}
$$

$$= \log e \left[\sum_i \sum_j p(x_i)p(y_j) - \sum_i \sum_j p(x_i, y_j) \right]$$
$$= \log e[1 - 1]$$
$$= 0$$ ✦

9.4 CODING THEORY AND ENTROPY

Suppose that the value of a discrete random vector X is to be observed at location A and then transmitted to location B via a communication network that consists of two signals, 0 and 1. In order to do this, it is first necessary to encode each possible value of X in terms of a sequence of 0's and 1's. To avoid any ambiguity, it is usually required that no encoded sequence can be obtained from a shorter encoded sequence by adding more terms to the shorter.

For instance, if X can take on four possible values x_1, x_2, x_3, x_4, then one possible coding would be

$$\begin{aligned} x_1 &\leftrightarrow 00 \\ x_2 &\leftrightarrow 01 \\ x_3 &\leftrightarrow 10 \\ x_4 &\leftrightarrow 11 \end{aligned} \tag{4.1}$$

That is, if $X = x_1$, then the message 00 is sent to location B, whereas if $X = x_2$, then 01 is sent to B, and so on. A second possible coding is

$$\begin{aligned} x_1 &\leftrightarrow 0 \\ x_2 &\leftrightarrow 10 \\ x_3 &\leftrightarrow 110 \\ x_4 &\leftrightarrow 111 \end{aligned} \tag{4.2}$$

However, a coding such as

$$\begin{aligned} x_1 &\leftrightarrow 0 \\ x_2 &\leftrightarrow 1 \\ x_3 &\leftrightarrow 00 \\ x_4 &\leftrightarrow 01 \end{aligned}$$

is not allowed because the coded sequences for x_3 and x_4 are both extensions of the one for x_1.

One of the objectives in devising a code is to minimize the expected number of bits (that is, binary digits) that need be sent from location A to location B. For example, if

$$P\{X = x_1\} = \tfrac{1}{2}$$
$$P\{X = x_2\} = \tfrac{1}{4}$$
$$P\{X = x_3\} = \tfrac{1}{8}$$
$$P\{X = x_4\} = \tfrac{1}{8}$$

then the code given by Equation (4.2) would expect to send $\frac{1}{2}(1) + \frac{1}{4}(2) + \frac{1}{8}(3) + \frac{1}{8}(3) = 1.75$ bits; whereas the code given by Equation (4.1) would expect to send 2 bits. Hence, for the preceding set of probabilities, the encoding in Equation (4.2) is more efficient than that in Equation (4.1).

The preceding raises the following question: For a given random vector X, what is the maximum efficiency achievable by an encoding scheme? The answer is that for any coding, the average number of bits that will be sent is at least as large as the entropy of X. To prove this result, known in information theory as the *noiseless coding theorem*, we shall need Lemma 4.1.

Lemma 4.1

Let X take on the possible values $x_1, \ldots, x_N$. Then, in order for it to be possible to encode the values of X in binary sequences (none of which is an extension of another) of respective lengths $n_1, \ldots, n_N$, it is necessary and sufficient that

$$\sum_{i=1}^{N} \left(\tfrac{1}{2}\right)^{n_i} \le 1$$

Proof: For a fixed set of N positive integers $n_1, \ldots, n_N$, let w_j denote the number of the n_i that are equal to j, $j = 1, \ldots$. For there to be a coding that assigns n_i bits to the value x_i, $i = 1, \ldots, N$, it is clearly necessary that $w_1 \le 2$. Furthermore, as no binary sequence is allowed to be an extension of any other, we must have that $w_2 \le 2^2 - 2w_1$. (This follows because 2^2 is the number of binary sequences of length 2, whereas $2w_1$ is the number of sequences that are extensions of the w_1 binary sequence of length 1.) In general, the same reasoning shows that we must have

$$w_n \le 2^n - w_1 2^{n-1} - w_2 2^{n-2} - \cdots - w_{n-1}2 \tag{4.3}$$

for $n = 1, \ldots$. In fact, a little thought should convince the reader that these conditions are not only necessary but are also sufficient for a code to exist that assigns n_i bits to x_i, $i = 1, \ldots, N$.

Rewriting inequality (4.3) as

$$w_n + w_{n-1}2 + w_{n-2}2^2 + \cdots + w_1 2^{n-1} \le 2^n \qquad n = 1, \ldots$$

and dividing by 2^n yields that the necessary and sufficient conditions are

$$\sum_{j=1}^{n} w_j \left(\tfrac{1}{2}\right)^j \le 1 \qquad \text{for all } n \tag{4.4}$$

However, as $\sum_{j=1}^{n} w_j\left(\frac{1}{2}\right)^j$ is increasing in n, it follows that Equation (4.4) will be true if and only if

$$\sum_{j=1}^{\infty} w_j\left(\frac{1}{2}\right)^j \leq 1$$

The result is now established, since by the definition of w_j as the number of n_i that equal j, it follows that

$$\sum_{j=1}^{\infty} w_j\left(\frac{1}{2}\right)^j = \sum_{i=1}^{N} \left(\frac{1}{2}\right)^{n_i}$$ ◆

We are now ready to prove Theorem 4.1.

Theorem 4.1 The noiseless coding theorem

Let X take on the values $x_1, \ldots, x_N$ with respective probabilities $p(x_1), \ldots, p(x_N)$. Then, for any coding of X that assigns n_i bits to x_i,

$$\sum_{i=1}^{N} n_i p(x_i) \geq H(X) = -\sum_{i=1}^{N} p(x_i) \log p(x_i)$$

Proof: Let $P_i = p(x_i)$, $q_i = 2^{-n_i}\Big/\sum_{j=1}^{N} 2^{-n_j}$, $i = 1, \ldots, N$. Then

$$\begin{aligned}
-\sum_{i=1}^{N} P_i \log\left(\frac{P_i}{q_i}\right) &= -\log e \sum_{i=1}^{N} P_i \ln\left(\frac{P_i}{q_i}\right) \\
&= \log e \sum_{i=1}^{N} P_i \ln\left(\frac{q_i}{P_i}\right) \\
&\leq \log e \sum_{i=1}^{N} P_i\left(\frac{q_i}{P_i} - 1\right) \qquad \text{by Lemma 3.1} \\
&= 0 \qquad \text{since } \sum_{i=1}^{N} P_i = \sum_{i=1}^{N} q_i = 1
\end{aligned}$$

Hence

$$\begin{aligned}
-\sum_{i=1}^{N} P_i \log P_i &\leq -\sum_{i=1}^{N} P_i \log q_i \\
&= \sum_{i=1}^{N} n_i P_i + \log\left(\sum_{j=1}^{N} 2^{-n_j}\right) \\
&\leq \sum_{i=1}^{N} n_i P_i \qquad \text{by Lemma 4.1}
\end{aligned}$$ ◆

Example 4a. Consider a random variable X with probability mass function

$$p(x_1) = \tfrac{1}{2} \qquad p(x_2) = \tfrac{1}{4}. \qquad p(x_3) = p(x_4) = \tfrac{1}{8}$$

Since

$$\begin{aligned} H(X) &= -\left[\tfrac{1}{2}\log\tfrac{1}{2} + \tfrac{1}{4}\log\tfrac{1}{4} + \tfrac{1}{4}\log\tfrac{1}{8}\right] \\ &= \tfrac{1}{2} + \tfrac{2}{4} + \tfrac{3}{4} \\ &= 1.75 \end{aligned}$$

it follows from Theorem 4.1 that there is no more efficient coding scheme than

$$\begin{aligned} x_1 &\leftrightarrow 0 \\ x_2 &\leftrightarrow 10 \\ x_3 &\leftrightarrow 110 \\ x_4 &\leftrightarrow 111 \end{aligned}$$

■

For most random vectors there does not exist a coding for which the average number of bits sent attains the lower bound $H(X)$. However, it is always possible to devise a code such that the average number of bits is within 1 of $H(X)$. To prove this, define n_i to be the integer satisfying

$$-\log p(x_i) \le n_i < -\log p(x_i) + 1$$

Now,

$$\sum_{i=1}^{N} 2^{-n_i} \le \sum_{i=1}^{N} 2^{\log p(x_i)} = \sum_{i=1}^{N} p(x_i) = 1$$

so by Lemma 4.1 we can associate sequences of bits, having lengths n_i to the x_i, $i = 1, \ldots, N$. The average length of such a sequence,

$$L = \sum_{i=1}^{N} n_i p(x_i)$$

satisfies

$$-\sum_{i=1}^{N} p(x_i)\log p(x_i) \le L < -\sum_{i=1}^{N} p(x_i)\log p(x_i) + 1$$

or

$$H(X) \le L < H(X) + 1$$

Example 4b. Suppose that 10 independent tosses of a coin, having probability p of coming up heads, are made at location A and the result is to be transmitted to location B. The outcome of this experiment is a random vector $X = (X_1, \ldots, X_{10})$, where X_i is 1 or 0 according to whether or not the outcome of the

ith toss is heads. By the results of this section it follows that L, the average number of bits transmitted by any code, satisfies

$$H(X) \leq L$$

with

$$L \leq H(X) + 1$$

for at least one code. Now, since the X_i are independent, it follows from Proposition 3.1 and Theorem 3.2 that

$$H(X) = H(X_1, \ldots, X_n) = \sum_{i=1}^{N} H(X_i)$$
$$= -10[p \log p + (1 - p)\log(1 - p)]$$

If $p = \frac{1}{2}$, then $H(X) = 10$, and it follows that we can do no better than just encoding X by its actual value. That is, for example, if the first 5 tosses come up heads and the last 5 tails, then the message 1111100000 is transmitted to location B.

However, if $p \neq \frac{1}{2}$, we can often do better by using a different coding scheme. For instance, if $p = \frac{1}{4}$, then

$$H(X) = -10\left(\tfrac{1}{4}\log\tfrac{1}{4} + \tfrac{3}{4}\log\tfrac{3}{4}\right) = 8.11$$

and thus there is an encoding for which the average length of the encoded message is no greater than 9.11.

One simple coding that is more efficient, in this case, than the identity code is to break up $(X_1, \ldots, X_{10})$ into 5 pairs of 2 random variables each and then code each of the pairs as follows:

$$X_i = 0, X_{i+1} = 0 \leftrightarrow 0$$
$$X_i = 0, X_{i+1} = 1 \leftrightarrow 10$$
$$X_i = 1, X_{i+1} = 0 \leftrightarrow 110$$
$$X_i = 1, X_{i+1} = 1 \leftrightarrow 111$$

for $i = 1, 3, 5, 7, 9$. The total message then transmitted is the successive encodings of the preceding pairs.

For instance, if the outcome $TTTHHTTTTH$ is observed, then the message 010110010 is sent. The average number of bits needed to transmit the message using this code is

$$5\left[1\left(\tfrac{3}{4}\right)^2 + 2\left(\tfrac{1}{4}\right)\left(\tfrac{3}{4}\right) + 3\left(\tfrac{1}{4}\right)\left(\tfrac{3}{4}\right) + 3\left(\tfrac{1}{4}\right)^2\right] = \tfrac{135}{16}$$
$$\approx 8.44$$

■

Up to this point we have assumed that the message sent at location A is received, without error, at location B. However, there are always certain errors that can occur because of random disturbances along the communications channel. Such random disturbances might lead, for example, to the message 00101101, sent at A, being received at B in the form 01101101.

Let us suppose that a bit transmitted at location A will be correctly received at location B with probability p, independently from bit to bit. Such a communications system is called a *binary symmetric channel.* Suppose further that $p = .8$ and we want to transmit a message, consisting of a large number of bits, from A to B. Thus direct transmission of the message will result in an error probability of .20, for each bit, which is quite high. One way to reduce this probability of bit error would be to transmit each bit 3 times and then decode by majority rule. That is, we could use the following scheme:

Encode	Decode
$0 \to 000$	000, 001, 010, 100 $\to 0$
$1 \to 111$	111, 110, 101, 011 $\to 1$

Note that if no more than one error occurs in transmission, then the bit will be correctly decoded. Hence the probability of bit error is reduced to

$$(.2)^3 + 3(.2)^2(.8) = .104$$

a considerable improvement. In fact, it is clear that we can make the probability of bit error as small as we want by repeating the bit many times and then decoding by majority rule. For instance, the scheme

Encode	Decode
$0 \to$ string of 17 0's $1 \to$ string of 17 1's	By majority rule

will reduce the probability of bit error to below .01.

The problem with the type of encoding scheme above is that although it decreases the probability of bit error, it does so at the cost of also decreasing the effective rate of bits sent per signal (see Table 9.1).

TABLE 9.1 REPETITION OF BITS ENCODING SCHEME

Probability of error (per bit)	Rate (bits transmitted per signal)
.20	1
.10	$.33\left(=\frac{1}{3}\right)$
.01	$.06\left(=\frac{1}{17}\right)$

In fact, at this point it may appear inevitable to the reader that decreasing the probability of bit error to 0 always results in also decreasing the effective rate at which bits are transmitted per signal to 0. However, it is a remarkable result of information theory, known as the *noisy coding theorem* and due to Claude Shannon, that this is not the case. We now state this result as Theorem 4.2.

Theorem 4.2 The noisy coding theorem

There is a number C such that for any value R which is less than C, and any $\varepsilon > 0$, there exists a coding–decoding scheme that transmits at an average rate of R bits sent per signal and with an error (per bit) probability of less than ε. The largest such value of C, call it C^*,[†] is called the channel capacity, and for the binary symmetric channel,

$$C^* = 1 + p \log p + (1 - p) \log(1 - p)$$

SUMMARY

The *Poisson process* having rate λ is a collection of random variables $\{N(t), t \geq 0\}$ that relate to an underlying process of randomly occurring events. For instance, $N(t)$ represents the number of events that occur between times 0 and t. The defining features of the Poisson process are as follows:

(i) The number of events that occur in disjoint time intervals are independent.
(ii) The distribution of the number of events that occur in an interval depends only on the length of the interval.
(iii) Events occur one at a time.
(iv) Events occur at rate λ.

It can be shown that $N(t)$ is a Poisson random variable with mean λt. In addition, if $T_i, i \geq 1$, are the times between the successive events, then they are independent exponential random variables with rate λ.

A sequence of random variables $X_n, n \geq 0$, that each take on one of the values $0, \ldots, M$, is said to be a *Markov chain* with transition probabilities $P_{i,j}$, if for all $n, i_0, \ldots, i_n, i, j$,

$$P\{X_{n+1} = j | X_n = i, X_{n-1} = i_{n-1}, \ldots, X_0 = i_0\} = P_{i,j}$$

If we interpret X_n as the state of some process at time n, then a Markov chain is a sequence of successive states of a process that has the property that whenever it enters state i, then, independent of all past states, the next state is j with probability $P_{i,j}$, for all states i and j. For many Markov chains the probability of being in state j at time n converges to a limiting value that does not depend on the initial state. If we let

[†] For an entropy interpretation of C^*, see Problem 18.

$\Pi_j, j = 0, \ldots, M$, denote these limiting probabilities, then they are the unique solution of the equations:

$$\Pi_j = \sum_{i=0}^{M} \Pi_i P_{i,j} \qquad j = 0, \ldots, M$$

$$\sum_{j=1}^{M} \Pi_j = 1$$

Moreover, Π_j is equal to the long-run proportion of time that the chain is in state j.

Let X be a random variable that takes on one of n possible values according to the set of probabilities $\{p_1, \ldots, p_n\}$. The quantity

$$H(X) = -\sum_{i=1}^{n} p_i \log_2(p_i)$$

is called the *entropy* of X. It can be interpreted as representing the average amount of uncertainty that exists regarding the value of X, or the average information received when X is observed. The entropy has important implications for binary codings of X.

THEORETICAL EXERCISES AND PROBLEMS

1. Customers arrive at a bank at a Poisson rate λ. Suppose that two customers arrived during the first hour. What is the probability that
(a) both arrived during the first 20 minutes;
(b) at least one arrived during the first 20 minutes?

2. Cars cross a certain point in the highway in accordance with a Poisson process with rate $\lambda = 3$ per minute. If Al blindly runs across the highway, then what is the probability that he will be uninjured if the amount of time that it takes him to cross the road is s seconds? (Assume that if he is on the highway when a car passes by, then he will be injured.) Do this exercise for $s = 2, 5, 10, 20$.

3. Suppose that in Problem 2, Al is agile enough to escape from a single car but if he encounters two or more cars while attempting to cross the road, then he is injured. What is the probability that he will be unhurt if it takes him s seconds to cross? Do this exercise for $s = 5, 10, 20, 30$.

4. Suppose that 3 white and 3 black balls are distributed in two urns in such a way that each contains 3 balls. We say that the system is in state i if the first urn contains i white balls, $i = 0, 1, 2, 3$. At each stage 1 ball is drawn from each urn and the ball drawn from the first urn is placed in the second, and conversely with the ball from the second urn. Let X_n denote the state of the system after the nth stage, and compute the transition probabilities of the Markov chain $\{X_n, n \geq 0\}$.

5. Consider Example 2a. If there is a 50–50 chance of rain today, compute the probability that it will rain 3 days from now if $\alpha = .7, \beta = .3$.

6. Compute the limiting probabilities for the model of Problem 4.

7. A transition probability matrix is said to be doubly stochastic if

$$\sum_{i=0}^{M} P_{ij} = 1$$

for all states $j = 0, 1, \ldots, M$. If such a Markov chain is ergodic, show that $\Pi_j = 1/(M+1), j = 0, 1, \ldots, M$.

8. On any given day Buffy is either cheerful (c), so-so (s), or gloomy (g). If he is cheerful today then he will be c, s, or g tomorrow with respective probabilities .7, .2, .1. If he is so-so today then he will be c, s, or g tomorrow with respective probabilities .4, .3, .3. If he is gloomy today then Buffy will be c, s, or g tomorrow with probabilities .2, .4, .4. What proportion of time is Buffy cheerful?

9. Suppose that whether or not it rains tomorrow depends on past weather conditions only through the last 2 days. Specifically, suppose that if it has rained yesterday and today then it will rain tomorrow with probability .8; if it rained yesterday but not today then it will rain tomorrow with probability .3; if it rained today but not yesterday then it will rain tomorrow with probability .4; and if it has not rained either yesterday or today then it will rain tomorrow with probability .2. What proportion of days does it rain?

10. A certain person goes for a run each morning. When he leaves his house for his run he is equally likely to go out either the front or the back door; and similarly when he returns he is equally likely to go to either the front or back door. The runner owns 5 pairs of running shoes which he takes off after the run at whichever door he happens to be. If there are no shoes at the door from which he leaves to go running he runs barefooted. We are interested in determining the proportion of time that he runs barefooted.

(a) Set this up as a Markov chain. Give the states and the transition probabilities.

(b) Determine the proportion of days that he runs barefooted.

11. This problem refers to Example 2f.

(a) Verify that the proposed value of Π_j satisfies the necessary equations.

(b) For any given molecule what do you think is the (limiting) probability that it is in urn 1?

(c) Do you think that the events that molecule j is in urn 1 at a very large time, $j \geq 1$, would be (in the limit) independent?

(d) Explain why the limiting probabilities are as given.

12. Determine the entropy of the sum obtained when a pair of fair dice is rolled.

13. If X can take on any of n possible values with respective probabilities $P_1, \ldots, P_n$, prove that $H(X)$ is maximized when $P_i = 1/n, i = 1, \ldots, n$. What is $H(X)$ equal to in this case?

14. A pair of fair dice is rolled. Let

$$X = \begin{cases} 1 & \text{if the sum of the dice is 6} \\ 0 & \text{otherwise} \end{cases}$$

and let Y equal the value of the first die. Compute (a) $H(Y)$, (b) $H_Y(X)$, and (c) $H(X, Y)$.

15. A coin having probability $p = \frac{2}{3}$ of coming up heads is flipped 6 times. Compute the entropy of the outcome of this experiment.

16. A random variable can take on any of n possible values $x_1, \ldots, x_n$ with respective probabilities $p(x_i), i = 1, \ldots, n$. We shall attempt to determine the value of X by asking a series of questions, each of which can be answered by "yes" or "no." For instance, we may ask "Is $X = x_1$?" or "Is X equal to either x_1 or x_2 or x_3?", and so on. What can you say about the average number of such questions that you will need to ask to determine the value of X?

17. For any discrete random variable X and function f, show that

$$H(f(X)) \leq H(X)$$

18. In transmitting a bit from location A to location B, if we let X denote the value of the bit sent at location A and Y the value received at location B, then $H(X) - H_Y(X)$ is called the rate of transmission of information from A to B. The maximal rate of transmission, as a function of $P\{X = 1\} = 1 - P\{X = 0\}$, is called the channel capacity. Show that for a binary symmetric channel with $P\{Y = 1|X = 1\} = P\{Y = 0|X = 0\} = p$, the channel capacity is attained by the rate of transmission of information when $P\{X = 1\} = \frac{1}{2}$ and its value is $1 + p \log p + (1 - p) \log(1 - p)$.

SELF-TEST PROBLEMS AND EXERCISES

1. Events occur according to a Poisson process with rate $\lambda = 3$ per hour.
   (a) What is the probability that no events occur between times 8 and 10 in the morning?
   (b) What is the expected value of the number of events that occur between times 8 and 10 in the morning?
   (c) What is the expected occurrence time of the fifth event after 2 P.M.?
2. Customers arrive at a certain retail establishment according to a Poisson process with rate λ per hour. Suppose that two customers arrive during the first hour. Find the probability that
   (a) both arrived in the first 20 minutes;
   (b) at least one arrived in the first 30 minutes.
3. Four out of every five trucks on the road are followed by a car, while one out of every six cars is followed by a truck. What proportion of vehicles on the road are trucks?
4. A certain town's weather is classified each day as being rainy, sunny, or overcast but dry. If it is rainy one day, then it is equally likely to be either sunny or overcast the following day. If it is not rainy, then there is one chance in three that the weather will persist in whatever state it is in for another day, and if it does change, then it is equally likely to become either of the other two states. In the long run, what proportion of days are sunny? What proportion are rainy?
5. Let X be a random variable that takes on 5 possible values with respective probabilities .35, .2, .2, .2, .05. Also, let Y be a random variable that takes on 5 possible values with respective probabilities .05, .35, .1, .15, .35.
   (a) Show that $H(X) > H(Y)$.
   (b) Using the result of Problem 13, give an intuitive explanation for the preceding inequality.

REFERENCES

Sections 1 and 2

KEMENY, J., L. SNELL, and A. KNAPP. *Denumerable Markov Chains.* New York: D. Van Nostrand Company, 1966.

PARZEN, E. *Stochastic Processes.* San Francisco: Holden-Day, Inc., 1962.

ROSS, S. M. *Introduction to Probability Models.* 7th ed. San Diego, Calif.: Academic Press, Inc., 2000.

ROSS, S. M. *Stochastic Processes.* 2nd ed. New York: John Wiley & Sons, Inc., 1996.

Sections 3 and 4

ABRAMSON, N. *Information Theory and Coding.* New York: McGraw-Hill Book Company, 1963.

MCELIECE, R. *Theory of Information and Coding.* Reading, Mass.: Addison-Wesley Publishing Co., Inc., 1977.

PETERSON, W., and E. WELDON. *Error Correcting Codes*, 2nd ed. Cambridge, Mass.: The MIT Press, 1972.

CHAPTER 10

Simulation

10.1 INTRODUCTION

How can we determine the probability of our winning a game of solitaire? (By solitaire we mean any one of the standard solitaire games played with an ordinary deck of 52 playing cards and with some fixed playing strategy.) One possible approach is to start with the reasonable hypothesis that all (52)! possible arrangements of the deck of cards are equally likely to occur and then attempt to determine how many of these lead to a win. Unfortunately, there does not appear to be any systematic method for determining the number of arrangements that lead to a win and, as (52)! is a rather large number and the only way to determine whether or not a particular arrangement leads to a win seems to be by playing the game out, it can be seen that this approach will not work.

In fact, it might appear that the determination of the win probability for solitaire is mathematically intractable. However, all is not lost, for probability falls not only within the realm of mathematics, but also within the realm of applied science; and, as in all applied sciences, experimentation is a valuable technique. For our solitaire example, experimentation takes the form of playing a large number of such games, or, better yet, programming a computer to do so. After playing, say n games, if we let

$$X_i = \begin{cases} 1 & \text{if the } i\text{th game results in a win} \\ 0 & \text{otherwise} \end{cases}$$

then $X_i, i = 1, \ldots, n$ will be independent Bernoulli random variables for which

$$E[X_i] = P\{\text{win at solitaire}\}$$

Hence, by the strong law of large numbers, we know that

$$\sum_{i=1}^{n} \frac{X_i}{n} = \frac{\text{number of games won}}{\text{number of games played}}$$

will, with probability 1, converge to $P\{\text{win at solitaire}\}$. That is, by playing a large number of games we can use the proportion of won games as an estimate of the probability of winning. This method of empirically determining probabilities by means of experimentation is known as *simulation*.

In order to use a computer to initiate a simulation study, we must be able to generate the value of a uniform (0, 1) random variable; such variates are called random numbers. To generate such numbers, most computers have a built-in subroutine, called a random number generator, whose output is a sequence of pseudo random numbers. This is a sequence of numbers that is, for all practical purposes, indistinguishable from a sample from the uniform (0, 1) distribution. Most random number generators start with an initial value X_0, called the seed, and then recursively compute values by specifying positive integers a, c, and m, and then letting

$$X_{n+1} = (aX_n + c) \text{ modulo } m \qquad n \geq 0$$

where the foregoing means that $aX_n + c$ is divided by m and the remainder is taken as the value of X_{n+1}. Thus each X_n is either $0, 1, \ldots, m - 1$ and the quantity X_n/m is taken as an approximation to a uniform (0, 1) random variable. It can be shown that, subject to suitable choices for a, c, and m, the foregoing gives rise to a sequence of numbers that look as if they were generated from independent uniform (0, 1) random variables.

As our starting point in simulation, we shall suppose that we can simulate from the uniform (0, 1) distribution and we shall use the term *random numbers* to mean independent random variables from this distribution.

In the solitaire example we would need to program a computer to play out the game starting with a given ordering of the cards. However, since the initial ordering is supposed to be equally likely to be any of the (52)! possible permutations, it is also necessary to be able to generate a random permutation. Using only random numbers, the following algorithm shows how this can be accomplished. The algorithm begins by randomly choosing one of the elements and then putting it in position n; it then randomly chooses among the remaining elements and puts the choice in position $n - 1$; and so on. It efficiently makes a random choice among the remaining elements by keeping these elements in an ordered list and then randomly choosing a position on that list.

Example 1a. ***Generating a random permutation.*** Suppose we are interested in generating a permutation of the integers $1, 2, \ldots, n$ that is such that all $n!$ possible orderings are equally likely. Starting with any initial permutation we will accomplish this after $n - 1$ steps where at each step we will interchange the positions of two of the numbers of the permutation. Throughout, we will keep track of the permutation by letting $X(i)$, $i = 1, \ldots, n$ denote the number currently in position i. The algorithm operates as follows:

1. Consider any arbitrary permutation and let $X(i)$ denote the element in position $i, i = 1 \ldots, n$. [For instance, we could take $X(i) = i, i = 1, \ldots, n$.]
2. Generate a random variable N_n that is equally likely to equal any of the values $1, 2, \ldots, n$.
3. Interchange the values of $X(N_n)$ and $X(n)$. The value of $X(n)$ will now remain fixed. [For instance, suppose that $n = 4$ and initially $X(i) = i$, $i = 1, 2, 3, 4$. If $N_4 = 3$, then the new permutation is $X(1) = 1, X(2) = 2$, $X(3) = 4, X(4) = 3$, and element 3 will remain in position 4 throughout.]

4. Generate a random variable N_{n-1} that is equally likely to be either $1, 2, \ldots, n-1$.
5. Interchange the values of $X(N_{n-1})$ and $X(n-1)$. [If $N_3 = 1$, then the new permutation is $X(1) = 4, X(2) = 2, X(3) = 1, X(4) = 3$.]
6. Generate N_{n-2}, which is equally likely to be either $1, 2, \ldots, n-2$.
7. Interchange the values of $X(N_{n-2})$ and $X(2)$. [If $N_2 = 1$, then the new permutation is $X(1) = 2, X(2) = 4, X(3) = 1, X(4) = 3$ and this is the final permutation.]
8. Generate N_{n-3}, and so on. The algorithm continues until N_2 is generated and after the next interchange the resulting permutation is the final one.

To implement this algorithm, it is necessary to be able to generate a random variable that is equally likely to be any of the values $1, 2, \ldots, k$. To accomplish this, let U denote a random number—that is, U is uniformly distributed on $(0, 1)$, and note that kU is uniform on $(0, k)$. Hence

$$P\{i - 1 < kU < i\} = \frac{1}{k} \qquad i = 1, \ldots, k$$

so if we take $N_k = [kU] + 1$, where $[x]$ is the integer part of x (that is, it is the largest integer less than or equal to x), then N_k will have the desired distribution.

The algorithm can now be succinctly written as follows:

Step 1. Let $X(1), \ldots, X(n)$ be any permutation of $1, 2, \ldots, n$. [For instance, we can set $X(i) = i, i = 1, \ldots, n$.]
Step 2. Let $I = n$.
Step 3. Generate a random number U and set $N = [IU] + 1$.
Step 4. Interchange the values of $X(N)$ and $X(I)$.
Step 5. Reduce the value of I by 1 and if $I > 1$ go to step 3.
Step 6. $X(1), \ldots, X(n)$ is the desired random generated permutation.

The foregoing algorithm for generating a random permutation is extremely useful. For instance, suppose that a statistician is developing an experiment to compare the effects of m different treatments on a set of n subjects. He decides to split the subjects into m different groups of respective sizes $n_1, n_2, \ldots, n_m$ where $\sum_{i=1}^{m} n_i = n$ with the members of the ith group to receive treatment i. To eliminate any bias in the assignment of subjects to treatments (for instance, it would cloud the meaning of the experimental results if it turned out that all the "best" subjects had been put in the same group), it is imperative that the assignment of a subject to a given group be done "at random." How is this to be accomplished?†

† When $m = 2$, another technique for randomly dividing the subjects was presented in Example 2g of Chapter 6. The preceding procedure is faster but requires more space than the one of Example 2g.

A simple and efficient procedure is to arbitrarily number the subjects 1 through n and then generate a random permutation $X(1), \ldots, X(n)$ of $1, 2, \ldots, n$. Now assign subjects $X(1), X(2), \ldots, X(n_1)$ to be in group 1, $X(n_1 + 1), \ldots, X(n_1 + n_2)$ to be in group 2, and in general group j is to consist of subjects numbered $X(n_1 + n_2 + \cdots + n_{j-1} + k)$, $k = 1, \ldots, n_j$. ■

10.2 GENERAL TECHNIQUES FOR SIMULATING CONTINUOUS RANDOM VARIABLES

In this section we present two general methods for using random numbers to simulate continuous random variables.

10.2.1 The Inverse Transformation Method

A general method for simulating a random variable having a continuous distribution—called the *inverse transformation method*—is based on the following proposition.

Proposition 2.1

Let U be a uniform $(0, 1)$ random variable. For any continuous distribution function F, if we define the random variable Y by

$$Y = F^{-1}(U)$$

then the random variable Y has distribution function F. [$F^{-1}(x)$ is defined to equal that value y for which $F(y) = x$.]

Proof:

$$\begin{aligned} F_Y(a) &= P\{Y \le a\} \\ &= P\{F^{-1}(U) \le a\} \end{aligned} \tag{2.1}$$

Now, since $F(x)$ is a monotone function, it follows that $F^{-1}(U) \le a$ if and only if $U \le F(a)$. Hence, from Equation (2.1), we see that

$$\begin{aligned} F_Y(a) &= P\{U \le F(a)\} \\ &= F(a) \end{aligned}$$

✦

It follows from Proposition 2.1 that we can simulate a random variable X having a continuous distribution function F by generating a random number U and then setting $X = F^{-1}(U)$.

Example 2a. ***Simulating an exponential random variable.*** If $F(x) = 1 - e^{-x}$, then $F^{-1}(u)$ is that value of x such that

$$1 - e^{-x} = u$$

or

$$x = -\log(1 - u)$$

Hence, if U is a uniform $(0, 1)$ variable, then

$$F^{-1}(U) = -\log(1 - U)$$

is exponentially distributed with mean 1. Since $1 - U$ is also uniformly distributed on $(0, 1)$, it follows that $-\log U$ is exponential with mean 1. Since cX is exponential with mean c when X is exponential with mean 1, it follows that $-c \log U$ is exponential with mean c. ■

The results of Example 2a can also be utilized to simulate a gamma random variable.

Example 2b. ***Simulating a gamma (n, λ) random variable.*** To simulate from a gamma distribution with parameters (n, λ), when n is an integer, we use the fact that the sum of n independent exponential random variables each having rate λ has this distribution. Hence, if $U_1, \ldots, U_n$ are independent uniform $(0, 1)$ random variables,

$$X = -\sum_{i=1}^{n} \frac{1}{\lambda} \log U_i = -\frac{1}{\lambda} \log\left(\prod_{i=1}^{n} U_i\right)$$

has the desired distribution. ■

10.2.2 The Rejection Method

Suppose that we have a method for simulating a random variable having density function $g(x)$. We can use this as the basis for simulating from the continuous distribution having density $f(x)$ by simulating Y from g and then accepting this simulated value with a probability proportional to $f(Y)/g(Y)$.

Specifically, let c be a constant such that

$$\frac{f(y)}{g(y)} \leq c \qquad \text{for all } y$$

We then have the following technique for simulating a random variable having density f.

Rejection Method

Step 1. Simulate Y having density g and simulate a random number U.
Step 2. If $U \leq f(Y)/cg(Y)$, set $X = Y$. Otherwise return to step 1.

The rejection method is expressed pictorially in Figure 10.1. We now prove that it works.

Proposition 2.2

The random variable X generated by the rejection method has density function f.

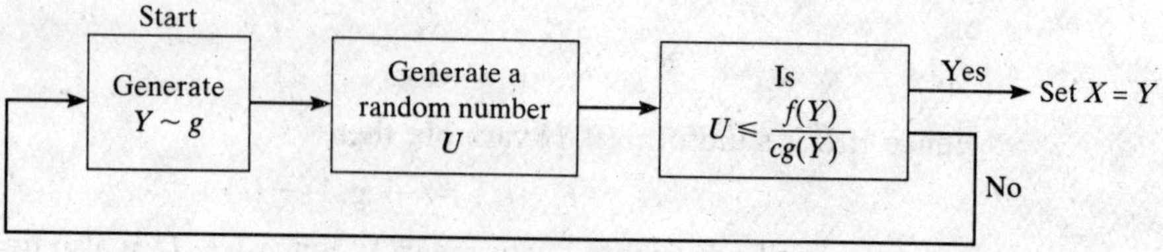

Figure 10.1 Rejection method for simulating a random variable X having density function f.

Proof: Let X be the value obtained and let N denote the number of necessary iterations. Then

$$\begin{aligned} P\{X \le x\} &= P\{Y_N \le x\} \\ &= P\left\{Y \le x \middle| U \le \frac{f(Y)}{cg(Y)}\right\} \\ &= \frac{P\left\{Y \le x, U \le \frac{f(Y)}{cg(Y)}\right\}}{K} \end{aligned}$$

where $K = P\{U \le f(Y)/cg(Y)\}$. Now the joint density function of Y and U is, by independence,

$$f(y,u) = g(y) \qquad 0 < u < 1$$

so, using the foregoing, we have

$$\begin{aligned} P\{X \le x\} &= \frac{1}{K} \iint\limits_{\substack{y \le x \\ 0 \le u \le f(y)/cg(y)}} g(y)\, du\, dy \\ &= \frac{1}{K} \int_{-\infty}^{x} \int_{0}^{f(y)/cg(y)} du\, g(y)\, dy \\ &= \frac{1}{cK} \int_{-\infty}^{x} f(y)\, dy \end{aligned} \tag{2.2}$$

Letting x approach ∞ and using the fact that f is a density gives

$$1 = \frac{1}{cK} \int_{-\infty}^{\infty} f(y)\, dy = \frac{1}{cK}$$

Hence from Equation (2.2) we obtain that

$$P\{X \le x\} = \int_{-\infty}^{x} f(y)\, dy$$

which completes the proof. ✦

REMARKS. (a) It should be noted that the way in which we "accept the value Y with probability $f(Y)/cg(Y)$" is by generating a random number U and then accepting Y if $U \le f(Y)/cg(Y)$.

(b) Since each iteration will, independently, result in an accepted value with probability $P\{U \le f(Y)/cg(Y)\} = K = 1/c$, it follows that the number of iterations has a geometric distribution with mean c. ■

Example 2c. ***Simulating a normal random variable.*** To simulate a unit normal random variable Z (that is, one with mean 0 and variance 1), note first that the absolute value of Z has probability density function

$$f(x) = \frac{2}{\sqrt{2\pi}} e^{-x^2/2} \qquad 0 < x < \infty \tag{2.3}$$

We will start by simulating from the preceding density function by using the rejection method with g being the exponential density function with mean 1—that is,

$$g(x) = e^{-x} \qquad 0 < x < \infty$$

Now, note that

$$\begin{aligned}
\frac{f(x)}{g(x)} &= \sqrt{\frac{2}{\pi}} \exp\left\{\frac{-(x^2 - 2x)}{2}\right\} \\
&= \sqrt{\frac{2}{\pi}} \exp\left\{\frac{-(x^2 - 2x + 1)}{2} + \frac{1}{2}\right\} \\
&= \sqrt{\frac{2e}{\pi}} \exp\left\{\frac{-(x-1)^2}{2}\right\} \qquad (2.4) \\
&\le \sqrt{\frac{2e}{\pi}}
\end{aligned}$$

Hence we can take $c = \sqrt{2e/\pi}$; so, from Equation (2.4),

$$\frac{f(x)}{cg(x)} = \exp\left\{\frac{-(x-1)^2}{2}\right\}$$

Therefore, using the rejection method we can simulate the absolute value of a unit normal random variable as follows:

(a) Generate independent random variables Y and U, Y being exponential with rate 1 and U being uniform on $(0, 1)$.
(b) If $U \le \exp\{-(Y-1)^2/2\}$ set $X = Y$. Otherwise, return to (a).

Once we have simulated a random variable X having density function as in Equation (2.3) we can then generate a unit normal random variable Z by letting Z be equally likely to be either X or $-X$.

In step (b), the value Y is accepted if $U \le \exp\{-(Y-1)^2/2\}$, which is equivalent to $-\log U \ge (Y-1)^2/2$. However, in Example 2a it was shown that $-\log U$ is exponential with rate 1, so steps (a) and (b) are equivalent to

(a′) Generate independent exponentials with rate 1, Y_1 and Y_2.
(b′) If $Y_2 \ge (Y_1 - 1)^2/2$, set $X = Y_1$. Otherwise, return to (a).

Suppose now that the foregoing results in Y_1 being accepted— and so we know that Y_2 is larger than $(Y_1 - 1)^2/2$. By how much does the one exceed the other? To answer this recall that Y_2 is exponential with rate 1, and so, given that it exceeds some value, the amount by which Y_2 exceeds $(Y_1 - 1)^2/2$ [that is, its "additional life" beyond the time $(Y_1 - 1)^2/2$] is (by the memoryless property) also exponentially distributed with rate 1. That is, when we accept step (b′), we obtain not only X (the absolute value of a unit normal) but by computing $Y_2 - (Y_1 - 1)^2/2$ we can also generate an exponential random variable (independent of X) having rate 1.

Hence, summing up, we have the following algorithm that generates an exponential with rate 1 and an independent unit normal random variable.

Step 1. Generate Y_1, an exponential random variable with rate 1.
Step 2. Generate Y_2, an exponential random variable with rate 1.
Step 3. If $Y_2 - (Y_1 - 1)^2/2 > 0$ set $Y = Y_2 - (Y_1 - 1)^2/2$ and go to step 4. Otherwise, go to step 1.
Step 4. Generate a random number U and set

$$Z = \begin{cases} Y_1 & \text{if } U \le \frac{1}{2} \\ -Y_1 & \text{if } U > \frac{1}{2} \end{cases}$$

The random variables Z and Y generated by the foregoing are independent, with Z being normal with mean 0 and variance 1 and Y being exponential with rate 1. (If we want the normal random variable to have mean μ and variance σ^2, just take $\mu + \sigma Z$.)

REMARKS. (a) Since $c = \sqrt{2e/\pi} \approx 1.32$, the foregoing requires a geometrically distributed number of iterations of step 2 with mean 1.32.

(b) If we want to generate a sequence of unit normal random variables, then we can use the exponential random variable Y obtained in step 3 as the initial exponential needed in step 1 for the next normal to be generated. Hence, on the average, we can simulate a unit normal by generating $1.64(= 2 \times 1.32 - 1)$ exponentials and computing 1.32 squares. ■

Example 2d. ***Simulating normal random variables—the polar method.*** It was shown in Example 7b of Chapter 6 that if X and Y are independent unit normal random variables, then their polar coordinates $R = \sqrt{X^2 + Y^2}$, $\Theta = \tan^{-1}(Y/X)$ are independent, with R^2 being exponentially distributed with mean 2 and Θ being uniformly distributed on $(0, 2\pi)$. Hence, if U_1 and U_2 are random numbers then (using the result of Example 2a) we can set

$$R = (-2\log U_1)^{1/2}$$

$$\Theta = 2\pi U_2$$

which yields that

$$\begin{aligned} X &= R\cos\Theta = (-2\log U_1)^{1/2}\cos(2\pi U_2) \\ Y &= R\sin\Theta = (-2\log U_1)^{1/2}\sin(2\pi U_2) \end{aligned} \tag{2.5}$$

are independent unit normals. ■

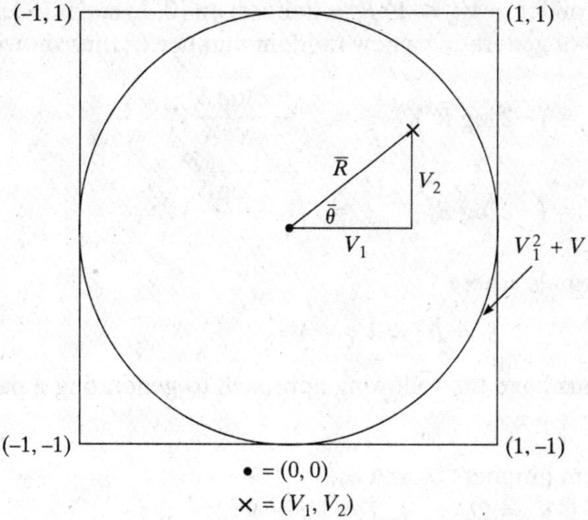

Figure 10.2

The preceding approach to generating unit normal random variables is called the Box–Muller approach. Its efficiency suffers somewhat from its need to compute the sine and cosine values. There is, however, a way to get around this potentially time-consuming difficulty. To begin, note that if U is uniform on $(0, 1)$ then $2U$ is uniform on $(0, 2)$ and so $2U - 1$ is uniform on $(-1, 1)$. Thus, if we generate random numbers U_1 and U_2 and set

$$V_1 = 2U_1 - 1$$

$$V_2 = 2U_2 - 1$$

then (V_1, V_2) is uniformly distributed in the square of area 4 centered at $(0, 0)$ (see Figure 10.2).

Suppose now that we continually generate such pairs (V_1, V_2) until we obtain one that is contained in the disk of radius 1 centered at $(0, 0)$—that is, until (V_1, V_2) is such that $V_1^2 + V_2^2 \le 1$. It now follows that such a pair (V_1, V_2) is uniformly distributed in the disk. If we let $\bar{R}, \bar{\Theta}$ denote the polar coordinates of this pair, then it is easy to verify that $\bar{R}$ and $\bar{\Theta}$ are independent, with $\bar{R}^2$ being uniformly distributed on $(0, 1)$ and $\bar{\Theta}$ uniformly distributed on $(0, 2\pi)$ (see Problem 13).

Since

$$\sin\bar{\Theta} = \frac{V_2}{\bar{R}} = \frac{V_2}{\sqrt{V_1^2 + V_2^2}}$$

$$\cos\bar{\Theta} = \frac{V_1}{\bar{R}} = \frac{V_1}{\sqrt{V_1^2 + V_2^2}}$$

it follows from Equation (2.5) that we can generate independent unit normals X and Y by generating another random number U and setting

$$X = (-2\log U)^{1/2} V_1/\bar{R}$$

$$Y = (-2\log U)^{1/2} V_2/\bar{R}$$

In fact, since (conditional on $V_1^2 + V_2^2 \le 1$) $\bar{R}^2$ is uniform on $(0, 1)$ and is independent of $\bar{\theta}$ we can use it instead of generating a new random number U; thus showing that

$$X = (-2\log \bar{R}^2)^{1/2}\frac{V_1}{\bar{R}} = \sqrt{\frac{-2\log S}{S}}V_1$$

$$Y = (-2\log \bar{R}^2)^{1/2}\frac{V_2}{\bar{R}} = \sqrt{\frac{-2\log S}{S}}V_2$$

are independent unit normals, where

$$S = \bar{R}^2 = V_1^2 + V_2^2$$

Summing up, we thus have the following approach to generating a pair of independent unit normals:

Step 1. Generate random numbers U_1 and U_2.
Step 2. Set $V_1 = 2U_1 - 1$, $V_2 = 2U_2 - 1$, $S = V_1^2 + V_2^2$.
Step 3. If $S > 1$, return to step 1.
Step 4. Return the independent unit normals

$$X = \sqrt{\frac{-2\log S}{S}}V_1, Y = \sqrt{\frac{-2\log S}{S}}V_2$$

The preceding is called the polar method. Since the probability that a random point in the square will fall within the circle is equal to $\pi/4$ (the area of the circle divided by the area of the square), it follows that, on average, the polar method will require $4/\pi \approx 1.273$ iterations of step 1. Hence it will, on average, require 2.546 random numbers, 1 logarithm, 1 square root, 1 division, and 4.546 multiplications to generate 2 independent unit normals.

Example 2e. ***Simulating a chi-squared random variable.*** The chi-squared distribution with n degrees of freedom is the distribution of $\chi_n^2 = Z_1^2 + \cdots + Z_n^2$, where $Z_i, i = 1, \ldots, n$ are independent unit normals. Now it was shown in Section 6.3 of Chapter 6 that $Z_1^2 + Z_2^2$ has an exponential distribution with rate $\frac{1}{2}$. Hence, when n is even, say $n = 2k$, χ_{2k}^2 has a gamma distribution with parameters $\left(k, \frac{1}{2}\right)$. Hence, $-2\log(\prod_{i=1}^k U_i)$ has a chi-squared distribution with $2k$ degrees of freedom. We can simulate a chi-squared random variable with $2k + 1$ degrees of freedom by first simulating a unit normal random variable Z and then adding Z^2 to the foregoing. That is,

$$\chi_{2k+1}^2 = Z^2 - 2\log\left(\prod_{i=1}^k U_i\right)$$

where $Z, U_1, \ldots, U_n$ are independent with Z being a unit normal and the others being uniform $(0, 1)$ random variables.

10.3 SIMULATING FROM DISCRETE DISTRIBUTIONS

All of the general methods for simulating random variables from continuous distributions have analogs in the discrete case. For instance, if we want to simulate a random variable Z having probability mass function

$$P\{X = x_j\} = P_j, \qquad j = 0, 1, \ldots, \qquad \sum_j P_j = 1$$

we can use the following discrete time analog of the inverse transform technique.

To simulate X for which $P\{X = x_j\} = P_j$, let U be uniformly distributed over $(0, 1)$, and set

$$X = \begin{cases} x_1 & \text{if } U \le P_1 \\ x_2 & \text{if } P_1 < U \le P_1 + P_2 \\ \vdots & \\ x_j & \text{if } \sum_1^{j-1} P_i < U \le \sum_i^j P_i \\ \vdots & \end{cases}$$

Since

$$P\{X = x_j\} = P\left\{\sum_1^{j-1} P_i < U \le \sum_1^j P_i\right\} = P_j$$

we see that X has the desired distribution.

Example 3a. ***The geometric distribution.*** Suppose that independent trials each of which results in a "success" with probability p, $0 < p < 1$, are continually performed until a success occurs. Letting X denote the number of necessary trials, then

$$P\{X = i\} = (1 - p)^{i-1}p \qquad i \ge 1$$

which is seen by noting that $X = i$ if the first $i - 1$ trials are all failures and the ith is a success. The random variable X is said to be a geometric random variable with parameter p. Since

$$\begin{aligned} \sum_{i=1}^{j-1} P\{X = i\} &= 1 - P\{X > j - 1\} \\ &= 1 - P\{\text{first } j - 1 \text{ are all failures}\} \\ &= 1 - (1 - p)^{j-1} \qquad j \ge 1 \end{aligned}$$

we can simulate such a random variable by generating a random number U and then setting X equal to that value j for which

$$1 - (1 - p)^{j-1} < U \le 1 - (1 - p)^j$$

or, equivalently, for which

$$(1 - p)^j \le 1 - U < (1 - p)^{j-1}$$

Since $1 - U$ has the same distribution as U, we can thus define X by

$$\begin{aligned} X &= \min\{j : (1 - p)^j \le U\} \\ &= \min\{j : j \log(1 - p) \le \log U\} \\ &= \min\left\{j : j \ge \frac{\log U}{\log(1 - p)}\right\} \end{aligned}$$

where the inequality changed sign since $\log(1 - p)$ is negative [since $\log(1 - p) < \log 1 = 0$]. Using the notation $[x]$ for the integer part of x (that is, $[x]$ is the largest integer less than or equal to x), we can write

$$X = 1 + \left[\frac{\log U}{\log(1 - p)}\right]$$

■

As in the continuous case, special simulating techniques have been developed for the more common discrete distributions. We now present two of these.

Example 3b. ***Simulating a binomial random variable.*** A binomial (n, p) random variable can be easily simulated by recalling that it can be expressed as the sum of n independent Bernoulli random variables. That is, if $U_1, \ldots, U_n$ are independent uniform $(0, 1)$ variables, then letting

$$X_i = \begin{cases} 1 & \text{if } U_i < p \\ 0 & \text{otherwise} \end{cases}$$

it follows that $X \equiv \sum_{i=1}^{n} X_i$ is a binomial random variable with parameters n and p. ■

Example 3c. ***Simulating a Poisson random variable.*** To simulate a Poisson random variable with mean λ, generate independent uniform $(0, 1)$ random variables $U_1, U_2, \ldots$ stopping at

$$N = \min\left\{n : \prod_{i=1}^{n} U_i < e^{-\lambda}\right\}$$

The random variable $X \equiv N - 1$ has the desired distribution. That is, if we continue generating random numbers until their product falls below $e^{-\lambda}$, then the number required, minus 1, is Poisson with mean λ.

That $X \equiv N - 1$ is indeed a Poisson random variable having mean λ can perhaps be most easily seen by noting that

$$X + 1 = \min\left\{n : \prod_{i=1}^{n} U_i < e^{-\lambda}\right\}$$

is equivalent to

$$X = \max\left\{n : \prod_{i=1}^{n} U_i \geq e^{-\lambda}\right\} \quad \text{where} \prod_{i=1}^{0} U_i \equiv 1$$

or, taking logarithms, to

$$X = \max\left\{n : \sum_{i=1}^{n} \log U_i \geq -\lambda\right\}$$

or

$$X = \max\left\{n : \sum_{i=1}^{n} -\log U_i \leq \lambda\right\}$$

However, $-\log U_i$ is exponential with rate 1 and so X can be thought of as being the maximum number of exponentials having rate 1 that can be summed and still be less than λ. But by recalling that the times between successive events of a Poisson process having rate 1 are independent exponentials with rate 1, it follows that X is equal to the number of events by time λ of a Poisson process having rate 1, thus X has a Poisson distribution with mean λ.

10.4 VARIANCE REDUCTION TECHNIQUES

Let $X_1, \ldots, X_n$ have a given joint distribution and suppose that we are interested in computing

$$\theta \equiv E[g(X_1, \ldots, X_n)]$$

where g is some specified function. It sometimes turns out that it is extremely difficult to analytically compute the foregoing, and when such is the case we can attempt to use simulation to estimate θ. This is done as follows: Generate $X_1^{(1)}, \ldots, X_n^{(1)}$ having the same joint distribution as $X_1, \ldots, X_n$ and set

$$Y_1 = g(X_1^{(1)}, \ldots, X_n^{(1)})$$

Now simulate a second set of random variables (independent of the first set) $X_1^{(2)}, \ldots, X_n^{(2)}$ having the distribution of $X_1, \ldots, X_n$ and set

$$Y_2 = g(X_1^{(2)}, \ldots, X_n^{(2)})$$

Continue this until you have generated k (some predetermined number) sets and so have also computed $Y_1, Y_2, \ldots, Y_k$. Now, $Y_1, \ldots, Y_k$ are independent and identically distributed random variables each having the same distribution of $g(X_1, \ldots, X_n)$. Thus, if we let $\bar{Y}$ denote the average of these k random variables—that is,

$$\bar{Y} = \sum_{i=1}^{k} \frac{Y_i}{k}$$

then

$$E[\bar{Y}] = \theta$$

$$E[(\bar{Y} - \theta)^2] = \text{Var}(\bar{Y})$$

Hence, we can use $\bar{Y}$ as an estimate of θ. Since the expected square of the difference between $\bar{Y}$ and θ is equal to the variance of $\bar{Y}$ we would like this quantity to be as small as possible. [In the preceding situation, $\text{Var}(\bar{Y}) = \text{Var}(Y_i)/k$, which is usually not known in advance but must be estimated from the generated values $Y_1, \ldots, Y_n$]. We now present three general techniques for reducing the variance of our estimator.

10.4.1 Use of Antithetic Variables

In the foregoing situation, suppose that we have generated Y_1 and Y_2, which are identically distributed random variables having mean θ. Now

$$\text{Var}\left(\frac{Y_1 + Y_2}{2}\right) = \frac{1}{4}[\text{Var}(Y_1) + \text{Var}(Y_2) + 2\,\text{Cov}(Y_1, Y_2)]$$

$$= \frac{\text{Var}(Y_1)}{2} + \frac{\text{Cov}(Y_1, Y_2)}{2}$$

Hence it would be advantageous (in the sense that the variance would be reduced) if Y_1 and Y_2 rather than being independent were negatively correlated. To see how we could arrange this, let us suppose that the random variables $X_1, \ldots, X_n$ are independent and, in addition, that each is simulated via the inverse transform technique. That is, X_i is simulated from $F_i^{-1}(U_i)$ where U_i is a random number and F_i is the distribution of X_i. Hence, Y_1 can be expressed as

$$Y_1 = g(F_1^{-1}(U_1), \ldots, F_n^{-1}(U_n))$$

Now, since $1 - U$ is also uniform over $(0, 1)$ whenever U is a random number (and is negatively correlated with U), it follows that Y_2 defined by

$$Y_2 = g(F_1^{-1}(1 - U_1), \ldots, F_n^{-1}(1 - U_n))$$

will have the same distribution as Y_1. Hence, if Y_1 and Y_2 were negatively correlated, then generating Y_2 by this means would lead to a smaller variance than if it were generated by a new set of random numbers. (In addition, there is a computational savings because rather than having to generate n additional random numbers, we need only subtract each of the previous n from 1.) Although we cannot, in general, be certain that Y_1 and Y_2 will be negatively correlated, this often turns out to be the case and indeed it can be proven that it will be so whenever g is a monotonic function.

10.4.2 Variance Reduction by Conditioning

Let us start by recalling the conditional variance formula (see Section 7.4.4)

$$\text{Var}(Y) = E[\text{Var}(Y|Z)] + \text{Var}(E[Y|Z])$$

Now suppose that we are interested in estimating $E[g(X_1, \ldots, X_n)]$ by simulating $\mathbf{X} = (X_1, \ldots, X_n)$ and then computing $Y = g(\mathbf{X})$. Now, if for some random variable Z we can compute $E[Y|Z]$ then, as $\text{Var}(Y|Z) \geq 0$, it follows from the preceding conditional variance formula that

$$\text{Var}(E[Y|Z]) \leq \text{Var}(Y)$$

implying, since $E[E[Y|Z]] = E[Y]$, that $E[Y|Z]$ is a better estimator of $E[Y]$ than is Y.

Example 4a. ***Estimation of π.*** Let U_1 and U_2 be random numbers and set $V_i = 2U_i - 1, i = 1, 2$. As noted in Example 2d, (V_1, V_2) will be uniformly distributed in the square of area 4 centered at $(0, 0)$. The probability that this point will fall within the inscribed circle of radius 1 centered at $(0, 0)$ (see Figure 10.2) is equal to $\pi/4$ (the ratio of the area of the circle to that of the square). Hence, upon simulating a large number n of such pairs and setting

$$I_j = \begin{cases} 1 & \text{if the } j\text{th pair falls within the circle} \\ 0 & \text{otherwise} \end{cases}$$

it follows that $I_j, j = 1, \ldots, n$ will be independent and identically distributed random variables having $E[I_j] = \pi/4$. Thus, by the strong law of large numbers

$$\frac{I_1 + \cdots + I_n}{n} \longrightarrow \frac{\pi}{4} \quad \text{as} \quad n \to \infty$$

Therefore, it follows that by simulating a large number of pairs (V_1, V_2) and multiplying the proportion of them that fall within the circle by 4, we can accurately approximate π.

The preceding estimator can, however, be improved upon by using conditional expectation. If we let I be the indicator variable for the pair (V_1, V_2) then, rather than using the observed value of I, it is better to condition on V_1 and so utilize

$$\begin{aligned} E[I|V_1] &= P\{V_1^2 + V_2^2 \le 1|V_1\} \\ &= P\{V_2^2 \le 1 - V_1^2|V_1\} \end{aligned}$$

Now

$$\begin{aligned} P\{V_2^2 \le 1 - V_1^2|V_1 = \nu\} &= P\{V_2^2 \le 1 - \nu^2\} \\ &= P\{-\sqrt{1 - \nu^2} \le V_2 \le \sqrt{1 - \nu^2}\} \\ &= \sqrt{1 - \nu^2} \end{aligned}$$

so

$$E[I|V_1] = \sqrt{1 - V_1^2}$$

Thus, an improvement on using the average value of I to estimate $\pi/4$ is to use the average value of $\sqrt{1 - V_1^2}$. Indeed, since

$$E\left[\sqrt{1 - V_1^2}\right] = \int_{-1}^{1} \tfrac{1}{2}\sqrt{1 - \nu^2}\, d\nu = \int_0^1 \sqrt{1 - u^2}\, du = E\left[\sqrt{1 - U^2}\right]$$

where U is uniform over $(0, 1)$, we can generate n random numbers U and use the average value of $\sqrt{1 - U^2}$ as our estimate of $\pi/4$. (Problem 14 shows that this estimator has the same variance as the average of the n values $\sqrt{1 - V^2}$.)

The preceding estimator of π can be improved even further by noting that the function $g(u) = \sqrt{1 - u^2}$, $0 \le u \le 1$ is a monotone decreasing function of u and so the method of antithetic variables will reduce the variance of the estimator of $E\left[\sqrt{1 - U^2}\right]$. That is, rather than generating n random numbers and using the average value of $\sqrt{1 - U^2}$ as an estimator of $\pi/4$, an improved estimator would be obtained by generating only $n/2$ random numbers U and then using one-half the average of $\sqrt{1 - U^2} + \sqrt{1 - (1 - U)^2}$ as the estimator of $\pi/4$.

The following table gives the estimates of π resulting from simulations, using $n = 10{,}000$, based on the three estimators.

Method	Estimate of π
Proportion of the random points that fall in the circle	3.1612
Average value of $\sqrt{1 - U^2}$	3.128448
Average value of $\sqrt{1 - U^2} + \sqrt{1 - (1 - U)^2}$	3.139578

A further simulation using the final approach and $n = 64{,}000$ yielded the estimate 3.143288. ■

10.4.3 Control Variates

Again suppose that we want to use simulation to estimate $E[g(\mathbf{X})]$, where $\mathbf{X} = (X_1, \ldots, X_n)$. But now suppose that for some function f the expected value of $f(\mathbf{X})$ is known—say $E[f(\mathbf{X})] = \mu$. Then for any constant a we can also use

$$W = g(\mathbf{X}) + a[f(\mathbf{X}) - \mu]$$

as an estimator of $E[g(\mathbf{X})]$. Now

$$\text{Var}(W) = \text{Var}[g(\mathbf{X})] + a^2 \text{Var}[f(\mathbf{X})] + 2a\,\text{Cov}[g(\mathbf{X}), f(\mathbf{X})] \tag{4.1}$$

Simple calculus shows that the foregoing is minimized when

$$a = \frac{-\text{Cov}[f(\mathbf{X}), g(\mathbf{X})]}{\text{Var}[f(\mathbf{X})]} \tag{4.2}$$

and for this value of a

$$\text{Var}(W) = \text{Var}[g(\mathbf{X})] - \frac{[\text{Cov}[f(\mathbf{X}), g(\mathbf{X})]^2}{\text{Var}[f(\mathbf{X})]} \tag{4.3}$$

Unfortunately, neither $\text{Var}[f(\mathbf{X})]$ nor $\text{Cov}[f(\mathbf{X})], g(\mathbf{X})]$ is usually known, so we cannot usually obtain the foregoing reduction in variance. One approach in practice is to use the simulated data to estimate these quantities. This usually yields almost all of the theoretically possible reduction in variance.

SUMMARY

Let F be a continuous distribution function and U a uniform $(0, 1)$ random variable. The random variable $F^{-1}(U)$ has distribution function F, where $F^{-1}(u)$ is that value x such that $F(x) = u$. Applying this result, we can use the values of uniform $(0, 1)$ random variables, called *random numbers*, to generate the values of other random variables. This is called the *inverse transform* method.

Another technique for generating random variables is based on the *rejection* method. Suppose that we have an efficient procedure for generating a random variable from the density function g, and that we desire to generate a random variable having density function f. The rejection method for accomplishing this starts by determining a constant c such that

$$\max \frac{f(x)}{g(x)} \le c$$

It then proceeds as follows:

1. Generate Y having density g.
2. Generate a random number U.
3. If $U \le f(Y)/cg(Y)$, set $X = Y$ and stop.
4. Return to step 1.

The number of passes through step 1 is a geometric random variable with mean c.

Standard normal random variables can be efficiently simulated by using the rejection method (with g being exponential with mean 1) or by using a technique known as the polar algorithm.

To estimate a quantity θ, one often generates the values of a partial sequence of random variables whose expected value is θ. The efficiency of this approach is increased when these random variables have a small variance. Three techniques that can often be used to specify random variables with mean θ and relatively small variances are

1. the use of antithetic variables,
2. the use of conditional expectations,
3. and the use of control variates.

PROBLEMS

1. The following algorithm will generate a random permutation of the elements $1, 2, \ldots, n$. It is somewhat faster than the one presented in Example 1a but is such that no position is fixed until the algorithm ends. In this algorithm, $P(i)$ can be interpreted as the element in position i.

Step 1. Set $k = 1$.

Step 2. Set $P(1) = 1$.

Step 3. If $k = n$, stop. Otherwise, let $k = k + 1$.

Step 4. Generate a random number U and let

$$P(k) = P([kU] + 1)$$
$$P([kU] + 1) = k$$

Go to step 3.

(a) Explain in words what the algorithm is doing.
(b) Show that at iteration k—that is, when the value of $P(k)$ is initially set—that $P(1)$, $P(2), \ldots, P(k)$ is a random permutation of $1, 2, \ldots, k$.

HINT: Use induction and argue that

$$P_k\{i_1, i_2, \ldots, i_{j-1}, k, i_j, \ldots, i_{k-2}, i\}$$
$$= P_{k-1}\{i_1, i_2, \ldots, i_{j-1}, i, i_j, \ldots, i_{k-2}\}\frac{1}{k}$$
$$= \frac{1}{k!} \quad \text{by the induction hypothesis}$$

2. Develop a technique for simulating a random variable having density function

$$f(x) = \begin{cases} e^{2x} & -\infty < x < 0 \\ e^{-2x} & 0 < x < \infty \end{cases}$$

3. Give a technique for simulating a random variable having the probability density function

$$f(x) = \begin{cases} \frac{1}{2}(x - 2) & 2 \le x \le 3 \\ \frac{1}{2}\left(2 - \frac{x}{3}\right) & 3 < x \le 6 \\ 0 & \text{otherwise} \end{cases}$$

4. Present a method to simulate a random variable having distribution function

$$F(x) = \begin{cases} 0 & x \le -3 \\ \frac{1}{2} + \frac{x}{6} & -3 < x < 0 \\ \frac{1}{2} + \frac{x^2}{32} & 0 < x \le 4 \\ 1 & x > 4 \end{cases}$$

5. Use the inverse transformation method to present an approach for generating a random variable from the Weibull distribution

$$F(t) = 1 - e^{-at^{\beta}} \qquad t \ge 0$$

6. Give a method for simulating a random variable having failure rate function

(a) $\lambda(t) = c$; **(b)** $\lambda(t) = ct$;
(c) $\lambda(t) = ct^2$; **(d)** $\lambda(t) = ct^3$.

7. In the following, F is the distribution function

$$F(x) = x^n \qquad 0 < x < 1$$

(a) Give a method for simulating a random variable having distribution F that uses only a single random number.

(b) Let $U_1, \ldots, U_n$ be independent random numbers. Show that

$$P\{\max(U_1, \ldots, U_n) \le x\} = x^n$$

(c) Use part (b) to give a second method of simulating a random variable having distribution F.

8. Suppose it is relatively easy to simulate from F_i for each $i = 1, \ldots, n$. How can we simulate from

(a) $F(x) = \prod_{i=1}^{n} F_i(x)$; **(b)** $F(x) = 1 - \prod_{i=1}^{n} [1 - F_i(x)]$.

9. Suppose we have a method to simulate random variables from the distributions F_1 and F_2. Explain how to simulate from the distribution

$$F(x) = pF_1(x) + (1 - p)F_2(x) \qquad 0 < p < 1$$

Give a method for simulating from

$$F(x) = \begin{cases} \frac{1}{3}(1 - e^{-3x}) + \frac{2}{3}x & 0 < x \le 1 \\ \frac{1}{3}(1 - e^{-3x}) + \frac{2}{3} & x > 1 \end{cases}$$

10. In Example 2c we simulated the absolute value of a unit normal by using the rejection procedure on exponential random variables with rate 1. This raises the question of whether we could obtain a more efficient algorithm by using a different exponential density—that is, we could use the density $g(x) = \lambda e^{-\lambda x}$. Show that the mean number of iterations needed in the rejection scheme is minimized when $\lambda = 1$.

11. Use the rejection method with $g(x) = 1, 0 < x < 1$, to determine an algorithm for simulating a random variable having density function

$$f(x) = \begin{cases} 60x^3(1 - x)^2 & 0 < x < 1 \\ 0 & \text{otherwise} \end{cases}$$

12. Explain how you could use random numbers to approximate $\int_0^1 k(x)\,dx$, where $k(x)$ is an arbitrary function.

HINT: If U is uniform on $(0, 1)$, what is $E[k(U)]$?

13. Let (X, Y) be uniformly distributed in the circle of radius 1 centered at the origin. Its joint density is thus

$$f(x, y) = \frac{1}{\pi} \qquad 0 \le x^2 + y^2 \le 1$$

Let $R = (X^2 + Y^2)^{1/2}$ and $\theta = \tan^{-1}(Y/X)$ denote its polar coordinates. Show that R and θ are independent with R^2 being uniform on $(0, 1)$ and θ being uniform on $(0, 2\pi)$.

14. In Example 4a we have shown that

$$E[(1 - V^2)^{1/2}] = E[(1 - U^2)^{1/2}] = \frac{\pi}{4}$$

when V is uniform $(-1, 1)$ and U is uniform $(0, 1)$. Show that

$$\text{Var}[(1 - V^2)^{1/2}] = \text{Var}[(1 - U^2)^{1/2}]$$

and find their common value.

15. (a) Verify that the minimum of (4.1) occurs when a is as given by (4.2).
(b) Verify that the minimum of (4.1) is given by (4.3).

16. Let X be a random variable on $(0, 1)$ whose density is $f(x)$. Show that we can estimate $\int_0^1 g(x)\,dx$ by simulating X and then taking $g(X)/f(X)$ as our estimate. This method, called importance sampling, tries to choose f similar in shape to g so that $g(X)/f(X)$ has a small variance.

SELF-TEST PROBLEMS AND EXERCISES

1. The random variable X has probability density function

$$f(x) = Ce^x \qquad 0 < x < 1$$

(a) Find the value of the constant C.
(b) Give a method for simulating such a random variable.

2. Give an approach for simulating a random variable having probability density function

$$f(x) = 30(x^2 - 2x^3 + x^4) \qquad 0 < x < 1$$

3. Give an efficient algorithm to simulate the value of a random variable with probability mass function

$$p_1 = .15 \qquad p_2 = .2 \qquad p_3 = .35 \qquad p_4 = .30$$

4. If X is a normal random variable with mean μ and variance σ^2, define a random variable Y that has the same distribution as X and is negatively correlated with it.

5. Let X and Y be independent exponential random variables with mean 1.
(a) Explain how we could use simulation to estimate $E[e^{XY}]$.
(b) Show how to improve the estimation approach in part (a) by using a control variate.

REFERENCE

ROSS, S. M. *Simulation.* San Diego, Calif.: Academic Press, Inc., 1997.

APPENDIX A

Answers to Selected Problems

CHAPTER 1

1. 67,600,000; 19,656,000 **2.** 1296 **4.** 24; 4 **5.** 144; 18 **6.** 2401 **7.** 720; 72; 144; 72 **8.** 120; 1260; 34,650 **9.** 27,720 **10.** 40,320; 10,080; 1152; 2880; 384 **11.** 720; 72; 144 **12.** 24,300,000; 17,100,720 **13.** 190 **14.** 2,598,960 **16.** 42; 94 **17.** 604,800 **18.** 600 **19.** 896; 1000; 910 **20.** 36; 26 **21.** 35 **22.** 18 **23.** 48 **25.** $52!/(13!)^4$ **27.** 27,720 **28.** 65,536; 2520 **29.** 12,600; 945 **30.** 564,480 **31.** 165; 35 **32.** 1287; 14,112 **33.** 220; 572

CHAPTER 2

9. 74 **10.** .4; .1 **11.** 70; 2 **12.** .5; .32; 149/198 **13.** 20,000; 12,000; 11,000; 68,000; 10,000 **14.** 1.057 **15.** .0020; .4226; .0475; .0211; .00024 **17.** 9.10947×10^{-6} **18.** .048 **19.** 5/18 **20.** .9052 **22.** $(n + 1)/2^n$ **23.** 5/12 **25.** .4 **26.** .492929 **27.** .58333 **28.** .0888; .2477; .1243; .2099 **30.** 1/18; 1/6; 1/2 **31.** 2/9; 1/9 **33.** 70/323 **35.** .8363 **36.** .0045; .0588 **37.** .0833; .5 **38.** 4 **39.** .48 **40.** 1/64; 21/64; 36/64; 6/64 **41.** .5177 **44.** .3; .2; .1 **46.** 5 **48.** 1.0604×10^{-3} **49.** .4329 **50.** 2.6084×10^{-6} **52.** .09145; .4268 **53.** 12/35 **54.** .0511 **55.** .2198; .0343

CHAPTER 3

1. 1/3 **2.** 1/6; 1/5; 1/4; 1/3; 1/2; 1 **3.** .339 **5.** 6/91 **6.** 1/2 **7.** 2/3 **8.** 1/2 **9.** 7/11 **10.** .22 **11.** .4848 **12.** .9835 **13.** .0792; .264 **14.** .331; .383; .286; 48.62 **15.** 44.29; 41.18 **16.** .4; 1/26 **17.** .496; 3/14; 9/62 **18.** .504; .3629 **20.** 35/768; 210/768 **21.** 4/9; 1/2 **22.** 1/3; 1/2 **24.** 20/21; 40/41 **26.** 3/128; 29/1536 **27.** .0893 **28.** 7/12; 3/5 **29.** 5/11 **30.** 27/31 **31.** 3/4 **32.** 1/2 **33.** 1/3; 1/5; 1 **34.** 12/37 **35.** 46/185 **36.** 3/13; 5/13; 5/52; 15/52 **37.** 43/459 **38.** 34.48 **39.** 4/9 **41.** 1/11 **44.** 2/3 **45.** 19/268 **46.** 17.5; 38/165; 17/33 **47.** .65; 56/65; 8/65; 1/65; 14/35; 12/35; 9/35 **48.** .11; 16/89; 12/27; 3/5; 9/25 **51.** 9 **53.** (c) 2/3 **56.** 2/3; 1/3; 3/4 **57.** 1/6; 3/20 **61.** 9/13; 1/2 **65.** 9; 9; 18; 110; 4; 4; 8; 120 all over 128 **66.** 1/9; 1/18 **67.** 38/64; 13/64; 13/64 **69.** 1/16; 1/32; 5/16; 1/4; 31/32 **70.** 9/19 **71.** 3/4, 7/12 **73.** $p^2/(1 - 2p + 2p^2)$ **74.** .5550 **76.** .9530 **78.** .5; .6; .8 **79.** 9/19; 6/19; 4/19; 7/15; 53/165; 7/33 **84.** 97/142; 15/26; 33/102

CHAPTER 4

1. $p(4) = 6/91$; $p(2) = 8/91$; $p(1) = 32/91$; $p(0) = 1/91$; $p(-1) = 16/91$; $p(-2) = 28/91$ **4.** 1/2; 5/18; 5/36; 5/84; 5/252; 1/252; 0; 0; 0; 0 **5.** $n - 2i$; $i = 0, \dots, n$ **6.** $p(3) = p(-3) = 1/8$; $p(1) = p(-1) = 3/8$ **12.** $p(4) = 1/16$; $p(3) = 1/8$; $p(2) = 1/16$; $p(0) = 1/2$; $p(-i) = p(i)$; $p(0) = 1$ **13.** $p(0) = .28$; $p(500) = .27$, $p(1000) = .315$; $p(1500) = .09$; $p(2000) = .045$ **14.** $p(0) = 1/2$; $p(1) = 1/6$; $p(2) = 1/12$; $p(3) = 1/20$; $p(4) = 1/5$ **17.** 1/4; 1/6; 1/12; 1/2 **19.** 1/2; 1/10; 1/5; 1/10; 1/10 **20.** .5918; no; −.108 **21.** 39.28; 37 **24.** $p = 11/18$, maximum = 23/72 **26.** 11/2; 17/5 **27.** $A(p + 1/10)$ **28.** 3/5 **31.** p^* **32.** $11 - 10(.9)^{10}$ **33.** 3 **35.** −.067; 1.089 **37.** 82.2; 84.5 **39.** 3/8 **40.** 11/243 **42.** $p \geq 1/2$ **45.** 3 **50.** 1/10; 1/10 **51.** e^{-2}; $1 - 1.2e^{-2}$ **53.** $1 - e^{-.6}$; $1 - e^{-219.18}$ **56.** 253 **57.** .5768; .6070 **59.** .3935; .3033; .0902 **60.** .8886 **61.** .4082 **63.** .0821; .2424 **65.** .3935; .2293; .3935 **66.** .1500; .1012 **68.** 5.8125 **69.** 32/243; 4864/6561; 160/729; 160/729 **73.** $18(17)^{n-1}/(35)^n$ **76.** 3/10; 5/6; 75/138 **77.** .3439

CHAPTER 5

2. $3.5e^{-5/2}$ **3.** no; no **4.** 1/2 **5.** $1 - (.01)^{1/5}$ **6.** $4, 0, \infty$ **7.** 3/5; 6/5 **8.** 2 **10.** 2/3; 2/3 **11.** 2/5 **13.** 2/3; 1/3 **15.** .7977; .6827; .3695; .9522; .1587 **16.** $(.9938)^{10}$ **18.** 22.66 **19.** 14.56 **20.** .9994; .75; .977 **22.** 9.5; .0019 **23.** .9258; .1762 **26.** .0606; .0525 **28.** .8363 **29.** .9993 **32.** e^{-1}; $e^{-1/2}$ **34.** e^{-1}; 1/3 **38.** 3/5 **40.** $1/y$

CHAPTER 6

2. (a) 14/39; 10/39; 10/39; 5/39 (b) 84; 70; 70; 70; 40; 40; 40; 15 all divided by 429 **3.** 15/26; 5/26; 5/26; 1/26 **4.** 25/169; 40/169; 40/169; 64/169 **7.** $p(i, j) = p^2(1 - p)^{i+j}$ **8.** $c = 1/8$; $E[X] = 0$ **9.** $(12x^2 + 6x)/7$; 15/56; .8625; 5/7; 8/7 **10.** 1/2; $1 - e^{-a}$ **11.** .1458 **12.** $39.3e^{-5}$ **13.** 1/6; 1/2 **15.** $\pi/4$ **16.** $n(1/2)^{n-1}$ **17.** 1/3 **18.** 7/9 **19.** 1/2 **21.** 2/5; 2/5 **22.** no; 1/3 **23.** 1/2; 2/3; 1/20; 1/18 **25.** $e^{-1}/i!$ **30.** e^{-2}; $1 - 3e^{-2}$ **32.** .0326 **33.** .3372; .2061 **34.** .0829; .3766 **35.** 1/3; 2/3; 5/12; 7/12 **36.** 5/13; 8/13 **37.** 1/6; 5/6; 1/4; 3/4 **42.** $(y + 1)^2 xe^{-x(y+1)}$; xe^{-xy}; e^{-x} **43.** $1/2 + 3y/(4x) - y^3/(4x^3)$ **47.** $(1 - 2d/L)^3$ **48.** .79297 **49.** $1 - e^{-5\lambda a}$; $(1 - e^{-\lambda a})^5$ **51.** r/π **52.** r **55.** (a) $u/(\nu + 1)^2$

CHAPTER 7

1. 52.5/12 **2.** 324; 198.8 **5.** 3/2 **6.** 35 **7.** .9; 4.9; 4.2 **8.** $(1 - (1 - p)^N)/p$ **10.** .6; 0 **11.** $2(n - 1)p(1 - p)$ **12.** $(3n^2 - n)/(4n - 2)$, $3n^2/(4n - 2)$ **14.** $m/(1 - p)$ **15.** 1/2 **18.** 4 **21.** .9301; 87.5755 **22.** 14.7 **23.** 147/110 **26.** $n/(n + 1)$; $1/(n + 1)$ **29.** $\frac{437}{35}$; 12; 4; $\frac{123}{35}$ **31.** 175/6 **33.** 14 **34.** 20/19; 360/361 **35.** 21.2; 18.929; 49.214 **36.** $-n/36$ **37.** 0 **38.** 1/8 **41.** 6; 112/33 **42.** 100/19; 16,200/6137; 10/19; 3240/6137 **45.** 1/2; 0 **47.** $1/(n - 1)$ **48.** 6; 7; 5.8192 **49.** 9/5; 6/5; 3/5; 0 **50.** $2y^2$ **51.** $y^3/4$ **53.** 12 **54.** 8 **56.** $N(1 - e^{-10/N})$ **57.** 12.5 **63.** −96/145 **65.** 5.16 **66.** 218 **67.** $x[1 + (2p - 1)^2]^n$ **69.** 1/2; 1/16; 2/81 **70.** 1/2, 1/3 **72.** $1/i$; $[i(i + 1)]^{-1}$; ∞ **73.** μ; $1 + \sigma^2$; yes; σ^2

CHAPTER 8

1. $\geq 19/20$ **2.** 15/17; $\geq 3/4$; ≥ 10 **3.** ≥ 3 **4.** $\leq 4/3$; .8428 **5.** .1416 **6.** .9431 **7.** .3085 **8.** .6932 **9.** 66.564 **10.** 117 **11.** $\geq .057$ **13.** .0162; .0003; .2514; .2514 **14.** $n \geq 23$ **17.** ≤ 2 **22.** .769; .357; .4267; .1093; .112184

CHAPTER 9

1. 1/9; 5/9 **3.** .9735; .9098; .7358; .5578 **10.** (b) 1/6 **14.** 2.585; .5417; 3.1267 **15.** 5.5098

APPENDIX B

Solutions to Self-Test Problems and Exercises

CHAPTER 1

1. **(a)** There are 4! different orderings of the letters C, D, E, F. For each of these orderings, we can obtain an ordering with A and B next to each other by inserting A and B, either in the order A, B or in the order B, A, in any of 5 places. Namely, either before the first letter of the permutation of C, D, E, F, or between the first and second, and so on. Hence, there are $2 \cdot 5 \cdot 4! = 240$ arrangements. Another way of solving this is to imagine that B is glued to the back of A. This yields that there are 5! orderings in which A is immediately before B. As there are also 5! orderings in which B is immediately before A, we again obtain a total of $2 \cdot 5! = 240$ different arrangements.

(b) There are a total of $6! = 720$ possible arrangements, and as there are as many with A before B as with B before A, there are 360 arrangements.

(c) Of the 720 possible arrangements, there are as many that have A before B before C, as have any of the 3! possible orderings of A, B, and C. Hence, there are $720/6 = 120$ possible orderings.

(d) Of the 360 arrangements that have A before B, half will have C before D and half D before C. Hence, there are 180 arrangements having A before B and C before D.

(e) Gluing B to the back of A, and D to the back of C, yields $4! = 24$ different orderings in which B immediately follows A and D immediately follows C. Since the order of A and B and of C and D can be reversed, there are thus $4 \cdot 24 = 96$ different arrangements.

(f) There are 5! orderings in which E is last. Hence, there are $6! - 5! = 600$ orderings in which E is not last.

2. 3! 4! 3! 3! since there are 3! possible orderings of countries and then the countrymen must be ordered.

3. **(a)** $10 \cdot 9 \cdot 8 = 720$

(b) $8 \cdot 7 \cdot 6 + 2 \cdot 3 \cdot 8 \cdot 7 = 672$.

The preceding follows since there are $8 \cdot 7 \cdot 6$ choices not including A or B, and there are $3 \cdot 8 \cdot 7$ choices in which a specified one of A and B, but not the other, serves. The latter following since the serving member of the pair can be assigned to any of the 3 offices, the next position can then be filled by any of the other 8 people, and the final position by any of the remaining 7.

(c) $8 \cdot 7 \cdot 6 + 3 \cdot 2 \cdot 8 = 384$. **(d)** $3 \cdot 9 \cdot 8 = 216$.

(e) $9 \cdot 8 \cdot 7 + 9 \cdot 8 = 576$.

4. **(a)** $\binom{10}{7}$ **(b)** $\binom{5}{3}\binom{5}{4} + \binom{5}{4}\binom{5}{3} + \binom{5}{5}\binom{5}{2}$

5. $\binom{7}{3,2,2} = 210$

6. There are $\binom{7}{3} = 35$ choices of the three places for the letters. For each choice, there are $(26)^3(10)^4$ different license plates. Hence, altogether there are a total of $35 \cdot (26)^3 \cdot (10)^4$ different plates.

7. Any choice of r of the n items is equivalent to a choice of $n - r$, namely, those items not selected.

8. **(a)** $10 \cdot 9 \cdot 9 \cdots 9 = 10 \cdot 9^{n-1}$

(b) $\binom{n}{i}9^{n-i}$, since there are $\binom{n}{i}$ choices of the i places to put the zeroes, and then each of the other $n - i$ positions can be any of the digits $1, \ldots, 9$.

9. **(a)** $\binom{3n}{3}$ **(b)** $3\binom{n}{3}$ **(c)** $\binom{3}{1}\binom{2}{1}\binom{n}{2}\binom{n}{1} = 3n^2(n-1)$

(d) n^3 **(e)** $\binom{3n}{3} = 3\binom{n}{3} + 3n^2(n-1) + n^3$

10. There are $9 \cdot 8 \cdot 7 \cdot 6 \cdot 5$ numbers in which no digit is repeated. There are $\binom{5}{2} \cdot 8 \cdot 7 \cdot 6$ numbers in which only one specified digit appears twice, and so there are $9\binom{5}{2} \cdot 8 \cdot 7 \cdot 6$ numbers in which only a single digit appears twice. There are $7 \cdot \frac{5!}{2!2!}$ numbers in which two specified digits appear twice, so there are $\binom{9}{2}7 \cdot \frac{5!}{2!2!}$ numbers in which two digits appear twice. Thus the answer is

$$9 \cdot 8 \cdot 7 \cdot 6 \cdot 5 + 9\binom{5}{2} \cdot 8 \cdot 7 \cdot 6 + \binom{9}{2}7 \cdot \frac{5!}{2!2!}$$

11. There are $\frac{(2n)!}{2^n}$ ways to partition the contestants into a *first* pair, a *second* pair, and so on. Thus, when the ordering of the pairs is not relevant, there are $\frac{(2n)!}{n!\,2^n}$ different sets of first round pairings. Because, for each first round pairing, there are 2^n possible results as to the n winners, it follows that there are $\frac{(2n)!}{n!}$ different first round results.

12. $\binom{8}{3}\binom{7}{3} + \binom{8}{4}\binom{7}{2} = 3430$

The first term gives the number of committees that have 3 women and 3 men; the second gives the number that have 4 women and 2 men.

13. (number of solutions of $x_1 + \cdots + x_5 = 4$)(number of solutions of $x_1 + \cdots + x_5 = 5$) (number of solutions of $x_1 + \cdots + x_5 = 6$) $= \binom{8}{4}\binom{9}{4}\binom{10}{4}$.

14. Since there are $\binom{j-1}{n-1}$ positive vectors whose sum is j, it follows that there are $\sum_{j=n}^{k}\binom{j-1}{n-1}$ such vectors.

15. Let us first determine the number of different results in which k people pass. Because there are $\binom{n}{k}$ different groups of size k, and $k!$ possible orderings of their scores, it follows that there are $\binom{n}{k}k!$ possible results in which k people pass. Consequently, there are a total of $\sum_{k=0}^{n}\binom{n}{k}k!$ possible results.

CHAPTER 2

1. **(a)** $2 \cdot 3 \cdot 4 = 24$ **(b)** $2 \cdot 3 = 6$
(c) $3 \cdot 4 = 12$ **(d)** $AB = \{(c, \text{pasta}, i), (c, \text{rice}, i), (c, \text{potatoes}, i)\}$
(e) 8 **(f)** $ABC = \{(c, \text{rice}, i)\}$

2. Let A be the event that a suit is purchased, B be the event that a shirt is purchased, and C be the event that a tie is purchased. Then

$$P(A \cup B \cup C) = .22 + .30 + .28 - .11 - .14 - .10 + .06 = .51$$

(a) $1 - .51 = .49$
(b) The probability that two or more items are purchased is

$$P(AB \cup AC \cup BC) = .11 + .14 + .10 - .06 - .06 - .06 + .06 = .23$$

Hence, the probability that exactly 1 item is purchased is $.51 - .23 = .28$

3. By symmetry the fourteenth card is equally likely to be any of the 52 cards, and thus the probability is 4/52. A more formal argument is to count the number of the 52! outcomes for which the fourteenth card is an ace. This yields

$$p = \frac{4 \cdot 51 \cdot 50 \cdots 2 \cdot 1}{(52)!} = \frac{4}{52}$$

Letting A be the event that the first ace occurs on the fourteenth card, we have

$$P(A) = \frac{48 \cdot 47 \cdots 36 \cdot 4}{52 \cdot 51 \cdots 40 \cdot 39} = .0312$$

4. Let D denote the event that the minimum temperature is 70 degrees. Then

$$P(A \cup B) = P(A) + P(B) - P(AB) = .7 - P(AB)$$
$$P(C \cup D) = P(C) + P(D) - P(CD) = .2 + P(D) - P(DC)$$

Subtracting one of the preceding equations from the other yields, upon using the fact that $A \cup B = C \cup D$ and $AB = CD$,

$$0 = .5 - P(D)$$

or, $P(D) = .5$.

5. **(a)** $\dfrac{52 \cdot 48 \cdot 44 \cdot 40}{52 \cdot 51 \cdot 50 \cdot 49} = .6761$ **(b)** $\dfrac{52 \cdot 39 \cdot 26 \cdot 13}{52 \cdot 51 \cdot 50 \cdot 49} = .1055$

6. Let R be the event that both balls are red, and let B be the event that they are both black. Then

$$P(R \cup B) = P(R) + P(B) = \frac{3 \cdot 4}{6 \cdot 10} + \frac{3 \cdot 6}{6 \cdot 10} = 1/2$$

7. **(a)** $\dfrac{1}{\binom{40}{8}} = 1.3 \times 10^{-8}$ **(b)** $\dfrac{\binom{8}{7}\binom{32}{1}}{\binom{40}{8}} = 3.3 \times 10^{-6}$

(c) $\dfrac{\binom{8}{6}\binom{32}{2}}{\binom{40}{8}} + 1.3 \times 10^{-8} + 3.3 \times 10^{-6} = 1.8 \times 10^{-4}$

8. (a) $\dfrac{3 \cdot 4 \cdot 4 \cdot 3}{\binom{14}{4}} = .1439$ **(b)** $\dfrac{\binom{4}{2}\binom{4}{2}}{\binom{14}{4}} = .0360$ **(c)** $\dfrac{\binom{8}{4}}{\binom{14}{4}} = .0699$

9. Let $S = \bigcup_{i=1}^{n} A_i$, and consider the experiment of randomly choosing an element of S. Then $P(A) = N(A)/N(S)$, and the results follow from Propositions 4.3 and 4.4.

10. Since there are $5! = 120$ outcomes in which the position of horse number 1 is specified, it follows that $N(A) = 360$. Similarly, $N(B) = 120$, and $N(AB) = 2 \cdot 4! = 48$. Hence, from Self-Test Problem 9, we obtain that $N(A \cup B) = 432$.

11. One way to solve this problem is to start with the complementary probability that at least one suit does not appear. Let A_i be the event that no cards from suit i appear, $i = 1, 2, 3, 4$. Then

$$P\left(\bigcup_{i=1}^{4} A_i\right) = \sum_i P(A_i) - \sum_j \sum_{i:i<j} P(A_i A_j) + \cdots - P(A_1 A_2 A_3 A_4)$$

$$= 4\frac{\binom{39}{5}}{\binom{52}{5}} - \binom{4}{2}\frac{\binom{26}{5}}{\binom{52}{5}} + \binom{4}{3}\frac{\binom{13}{5}}{\binom{52}{5}}$$

$$= 4\frac{\binom{39}{5}}{\binom{52}{5}} - 6\frac{\binom{26}{5}}{\binom{52}{5}} + 4\frac{\binom{13}{5}}{\binom{52}{5}}$$

The desired probability is then 1 minus the preceding. Another way to solve is to let A be the event that all 4 suits are represented, and then use

$$P(A) = P(n, n, n, n, o) + P(n, n, n, o, n) + P(n, n, o, n, n) + P(n, o, n, n, n)$$

where $P(n, n, n, o, n)$, for instance, is the probability that the first card is from a new suit, the second is from a new suit, the third is from a new suit, the fourth is from an old suit (that is, one that has already appeared) and the fifth is from a new suit. This gives

$$P(A) = \frac{52 \cdot 39 \cdot 26 \cdot 13 \cdot 48 + 52 \cdot 39 \cdot 26 \cdot 36 \cdot 13}{52 \cdot 51 \cdot 50 \cdot 49 \cdot 48}$$

$$+ \frac{52 \cdot 39 \cdot 24 \cdot 26 \cdot 13 + 52 \cdot 12 \cdot 39 \cdot 26 \cdot 13}{52 \cdot 51 \cdot 50 \cdot 49 \cdot 48}$$

$$= \frac{52 \cdot 39 \cdot 26 \cdot 13(48 + 36 + 24 + 12)}{52 \cdot 51 \cdot 50 \cdot 49 \cdot 48}$$

$$= .2637$$

12. There are $(10)!/2^5$ different divisions of the 10 players into a first roommate pair, a second roommate pair, and so on. Hence, there are $(10)!(5!\,2^5)$ divisions into 5 roommate pairs. There are $\binom{6}{2}\binom{4}{2}$ ways of choosing the frontcourt and backcourt players to be in the mixed roommate pairs, and then 2 ways of pairing them up. As there is then 1 way to pair up the remaining two backcourt players, and $4!/(2!\,2^2) = 3$ ways of making two roommate pairs from the remaining four frontcourt players, we see that the desired probability is

$$P\{2 \text{ mixed pairs}\} = \frac{\binom{6}{2}\binom{4}{2}(2)(3)}{(10)!/(5!\,2^5)} = .5714$$

13. Let R denote the event that letter R is repeated; similarly, define the events E and V. Then

$$P\{\text{same letter}\} = P(R) + P(E) + P(V) = \frac{2}{7}\frac{1}{8} + \frac{3}{7}\frac{1}{8} + \frac{1}{7}\frac{1}{8} = \frac{3}{28}$$

14. Let $B_1 = A_1, B_i = A_i\left(\bigcup_{j=1}^{i-1} A_j\right)^c, i > 1$. Then

$$\begin{aligned} P\left(\bigcup_{i=1}^{\infty} A_i\right) &= P\left(\bigcup_{i=1}^{\infty} B_i\right) \\ &= \sum_{i=1}^{\infty} P(B_i) \\ &\leq \sum_{i=1}^{\infty} P(A_i) \end{aligned}$$

where the final equality uses the fact that the B_i are mutually exclusive, and the inequality follows since $B_i \subset A_i$.

15.
$$\begin{aligned} P\left(\bigcap_{i=1}^{\infty} A_i\right) &= 1 - P\left(\left(\bigcap_{i=1}^{\infty} A_i\right)^c\right) \\ &= 1 - P\left(\bigcup_{i=1}^{\infty} A_i^c\right) \\ &\geq 1 - \sum_{i=1}^{\infty} P(A_i^c) \\ &= 1 \end{aligned}$$

16. The number of partitions for which $\{1\}$ is a subset is equal to the number of partitions of the remaining $n - 1$ elements into $k - 1$ nonempty subsets, namely, $T_{k-1}(n - 1)$. Because there are $T_k(n - 1)$ partitions of $\{2, \ldots, n - 1\}$ into k nonempty subsets, and then a choice of k of them in which to place element 1, it follows that there are $kT_k(n - 1)$ partitions for which $\{1\}$ is not a subset. Hence, the result follows.

17. Let R, W, B denote, respectively, the events that there are no red, no white, and no blue balls chosen. Then

$$P(R \cup W \cup B) = P(R) + P(W) + P(B) - P(RW) - P(RB) - P(WB) + P(RWB)$$

$$= \frac{\binom{13}{5}}{\binom{18}{5}} + \frac{\binom{12}{5}}{\binom{18}{5}} + \frac{\binom{11}{5}}{\binom{18}{5}} - \frac{\binom{7}{5}}{\binom{18}{5}} - \frac{\binom{6}{5}}{\binom{18}{5}} - \frac{\binom{5}{5}}{\binom{18}{5}}$$

$$\approx 0.2933$$

Thus, the probability that all colors appear in the chosen subset in approximately $1 - 0.2933 = 0.7067$.

CHAPTER 3

1. **(a)** $P(\text{no aces}) = \binom{35}{13} \Big/ \binom{39}{13}$ **(b)** $1 - P(\text{no aces}) - \dfrac{4\binom{35}{12}}{\binom{39}{13}}$

(c) $P(i\text{ aces}) = \dfrac{\binom{3}{i}\binom{36}{13-i}}{\binom{39}{13}}$

2. Let L_i denote the event that the life of the battery is greater than $10{,}000 \times i$ miles.

(a) $P(L_2|L_1) = P(L_1L_2)/P(L_1) = P(L_2)/P(L_1) = 1/2$

(b) $P(L_3|L_1) = P(L_1L_3)/P(L_1) = P(L_3)/P(L_1) = 1/8$

3. Put 1 white and 0 black balls in urn one, and the remaining 9 white and 10 black balls in urn two.

4. Let T be the event that the transferred ball is white, and let W be the event that a white ball is drawn from urn B. Then

$$P(T|W) = \frac{P(W|T)P(T)}{P(W|T)P(T) + P(W|T^c)P(T^c)} = \frac{(2/7)(2/3)}{(2/7)(2/3) + (1/7)(1/3)} = 4/5$$

5. Let B_i denote the event that ball i is black, and let $R_i = B_i^c$.

$$P(B_1|R_2) = \frac{P(R_2|B_1)P(B_1)}{P(R_2|B_1)P(B_1) + P(R_2|R_1)P(R_1)}$$

$$= \frac{[r/[(b+r+c)][b/(b+r)]}{[r/(b+r+c)][b/(b+r)] + [(r+c)/(b+r+c)][r/(b+r)]}$$

$$= \frac{b}{b+r+c}$$

6. Let B denote the event that both cards are aces.

(a)

$$P\{B|\text{yes to ace of spades}\} = \frac{P\{B, \text{yes to ace of spades}\}}{P\{\text{yes to ace of spades}\}}$$

$$= \frac{\binom{1}{1}\binom{3}{1}}{\binom{52}{2}} \Bigg/ \frac{\binom{1}{1}\binom{51}{1}}{\binom{52}{2}}$$

$$= 3/51$$

(b) Since the second card is equally likely to be any of the remaining 51, of which 3 are aces, we see that the answer in this situation is also 3/51.

(c) Since we can always interchange which card is considered first and which is considered second, the result should be the same as in part (b). A more formal argument is as follows:

$$\begin{aligned} P\{B|\text{second is ace}\} &= \frac{P\{B, \text{second is ace}\}}{P\{\text{second is ace}\}} \\ &= \frac{P(B)}{P(B) + P\{\text{first is not ace, second is ace}\}} \\ &= \frac{(4/52)(3/51)}{(4/52)(3/51) + (48/52)(4/51)} \\ &= 3/51 \end{aligned}$$

(d)

$$\begin{aligned} P\{B|\text{at least one}\} &= \frac{P(B)}{P\{\text{at least one}\}} \\ &= \frac{(4/52)(3/51)}{1 - (48/52)(47/51)} \\ &= 1/33 \end{aligned}$$

7.

$$\frac{P(H|E)}{P(G|E)} = \frac{P(HE)}{P(GE)} = \frac{P(H)P(E|H)}{P(G)P(E|G)}$$

The hypothesis H is 1.5 times as likely.

8. Let A denote the event that the plant is alive and let W be the event that it was watered.

(a)

$$\begin{aligned} P(A) &= P(A|W)P(W) + P(A|W^c)P(W^c) \\ &= (.85)(.9) + (.2)(.1) = .785 \end{aligned}$$

(b)

$$\begin{aligned} P(W^c|A^c) &= \frac{P(A^c|W^c)P(W^c)}{P(A^c)} \\ &= \frac{(.8)(.1)}{.215} = \frac{16}{43} \end{aligned}$$

9. Since the black rat has a brown sibling we can conclude that both its parents have one black and one brown gene.

(a)

$$P(2\text{ black} \mid \text{at least one}) = \frac{P(2)}{P(\text{at least one})} = \frac{1/4}{3/4} = \frac{1}{3}$$

(b) Let F be the event that all 5 offspring are black. Let B_2 be the event that the black rat has 2 black genes, and let B_1 be the event that it has 1 black and 1 brown gene.

$$\begin{aligned} P(B_2|F) &= \frac{P(F|B_2)P(B_2)}{P(F|B_2)P(B_2) + P(F|B_1)P(B_1)} \\ &= \frac{(1)(1/3)}{(1)(1/3) + (1/2)^5(2/3)} = \frac{16}{17} \end{aligned}$$

10. Let F be the event that a current flows from A to B, and let C_i be the event that relay i closes. Then

$$P(F) = P(F|C_1)p_1 + P(F|C_1^c)(1 - p_1)$$

Now,

$$\begin{aligned} P(F|C_1) &= P(C_4 \cup C_2C_5) \\ &= P(C_4) + P(C_2C_5) - P(C_4C_2C_5) \\ &= p_4 + p_2p_5 - p_4p_2p_5 \end{aligned}$$

Also,

$$\begin{aligned} P(F|C_1^c) &= P(C_2C_5 \cup C_2C_3C_4) \\ &= p_2p_5 + p_2p_3p_4 - p_2p_3p_4p_5 \end{aligned}$$

Hence, for part (a) we obtain

$$P(F) = p_1(p_4 + p_2p_5 - p_4p_2p_5) + (1 - p_1)p_2(p_5 + p_3p_4 - p_3p_4p_5)$$

For part (b), let $q_i = 1 - p_i$. Then

$$\begin{aligned} P(C_3|F) &= P(F|C_3)P(C_3)/P(F) \\ &= p_3[1 - P(C_1^cC_2^c \cup C_4^cC_5^c)]/P(F) \\ &= p_3(1 - q_1q_2 - q_4q_5 + q_1q_2q_4q_5)/P(F) \end{aligned}$$

11. Let A be the event that component 1 is working, and let F be the event that the system functions.

(a) $$P(A|F) = \frac{P(AF)}{P(F)} = \frac{P(A)}{P(F)} = \frac{1/2}{1 - (1/2)^2} = \frac{2}{3}$$

where $P(F)$ was computed by noting that it is equal to 1 minus the probability that components 1 and 2 are both failed.

(b) $$P(A|F) = \frac{P(AF)}{P(F)} = \frac{P(F|A)P(A)}{P(F)} = \frac{(3/4)(1/2)}{(1/2)^3 + 3(1/2)^3} = \frac{3}{4}$$

where $P(F)$ was computed by noting that it is equal to the probability that all 3 components work plus the three probabilities relating to exactly two of the components working.

12. If we accept that the outcomes of the successive spins are independent then the conditional probability of the next outcome is unchanged by the result that the previous ten spins landed on black.

13. Condition on outcome of initial tosses:

$$\begin{aligned} P(A\,\text{odd}) &= P_1(1 - P_2)(1 - P_3) + (1 - P_1)P_2P_3 + P_1P_2P_3(A\,\text{odd}) \\ &\quad + (1 - P_1)(1 - P_2)(1 - P_3)P(A\,\text{odd}) \end{aligned}$$

so,

$$P(A\,\text{odd}) = \frac{P_1(1 - P_2)(1 - P_3) + (1 - P_1)P_2P_3}{P_1 + P_2 + P_3 - P_1P_2 - P_1P_3 - P_2P_3}$$

14. Let A and B be the events that the first trial is larger and that the second is larger, respectively. Also, let E be the event that the results of the trials are equal. Then

$$1 = P(A) + P(B) + P(E)$$

But, by symmetry, $P(A) = P(B)$, and thus

$$P(B) = \frac{1 - P(E)}{2} = \frac{1 - \sum_{i=1}^{n} p_i^2}{2}$$

Another way of solving the problem is to note that

$$P(B) = \sum_i \sum_{j>i} P\{\text{first trial results in } i, \text{ second trial results in } j\}$$
$$= \sum_i \sum_{j>i} p_i p_j$$

To see that the two expressions derived for $P(B)$ are equal, note that

$$1 = \sum_{i=1}^{n} p_i \sum_{j=1}^{n} p_j$$
$$= \sum_i \sum_j p_i p_j$$
$$= \sum_i p_i^2 + \sum_i \sum_{j \neq i} p_i p_j$$
$$= \sum_i p_i^2 + 2 \sum_i \sum_{j>i} p_i p_j$$

15. Let $E = \{A \text{ gets more heads than } B\}$; then

$$P(E) = P(E|A \text{ leads after both flip } n)P(A \text{ leads after both flip } n)$$
$$+ P(E|\text{ even after both flip } n)P(\text{ even after both flip } n)$$
$$+ P(E|B \text{ leads after both flip } n)P(B \text{ leads after both flip } n)$$
$$= P(A \text{ leads}) + \frac{1}{2}P(\text{even})$$

Now, by symmetry,

$$P(A \text{ leads}) = P(B \text{ leads})$$
$$= \frac{1 - P(\text{even})}{2}$$

Hence,

$$P(E) = \frac{1}{2}$$

16. **(a)** Not true: In rolling 2 dice let $E = \{\text{sum is } 7\}$,

$$F = \{\text{1st die does not land on 4}\}, G = \{\text{2nd die does not land on 3}\}.$$

Then

$$P(E|F \cup G) = \frac{P\{7, \text{not } (4,3)\}}{P\{\text{not } (4,3)\}} = \frac{5/36}{35/36} = 5/35 \neq P(E)$$

(b)

$$\begin{aligned} P(E(F \cup G)) &= P(EF \cup EG) \\ &= P(EF) + P(EG) && \text{since } EFG = \emptyset \\ &= P(E)[P(F) + P(G)] \\ &= P(E)P(F \cup G) && \text{since } FG = \emptyset \end{aligned}$$

(c)

$$\begin{aligned} P(G|EF) &= \frac{P(EFG)}{P(EF)} \\ &= \frac{P(E)P(FG)}{P(EF)} && \text{since } E \text{ is independent of } FG \\ &= \frac{P(E)P(F)P(G)}{P(E)P(F)} && \text{by independence} \\ &= P(G). \end{aligned}$$

17. (a) necessarily false; if they were mutually exclusive then

$$0 = P(AB) \neq P(A)P(B)$$

(b) necessarily false; if they were independent then

$$P(AB) = P(A)P(B) > 0$$

(c) necessarily false; if they were mutually exclusive then

$$P(A \cup B) = P(A) + P(B) = 1.2$$

(d) possibly true.

18. The probabilities in parts (a), (b), and (c) are .5, $(.8)^3 = .512$, $(.9)^7 \approx .4783$.

19. Let $D_i, i = 1, 2$, denote the event that radio i is defective. Also, let A and B be the events that the radios were produced at factory A and at factory B, respectively. Then

$$\begin{aligned} P(D_2|D_1) &= \frac{P(D_1D_2)}{P(D_1)} \\ &= \frac{P(D_1D_2|A)P(A) + P(D_1D_2|B)P(B)}{P(D_1|A)P(A) + P(D_1|B)P(B)} \\ &= \frac{(.05)^2(1/2) + (.01)^2(1/2)}{(.05)(1/2) + (.01)(1/2)} \\ &= 13/300 \end{aligned}$$

20. We are given that $P(AB) = P(B)$ and must show that this implies that $P(B^cA^c) = P(A^c)$. One way is as follows:

$$\begin{aligned} P(B^cA^c) &= P((A \cup B)^c) \\ &= 1 - P(A \cup B) \\ &= 1 - P(A) - P(B) + P(AB) \\ &= 1 - P(A) \\ &= P(A^c) \end{aligned}$$

21. The result is true for $n = 0$. With A_i denoting the event that there are i red balls in the urn after stage n, assume that

$$P(A_i) = \frac{1}{n+1}, \qquad i = 1, \ldots, n+1$$

Now let $B_j, j = 1, \ldots, n+2$; denote the event that there are j red balls in the urn after stage $n + 1$. Then

$$\begin{aligned} P(B_j) &= \sum_{i=1}^{n+1} P(B_j|A_i)P(A_i) \\ &= \frac{1}{n+1} \sum_{i=1}^{n+1} P(B_j|A_i) \\ &= \frac{1}{n+1}[P(B_j|A_{j-1}) + P(B_j|A_j)] \end{aligned}$$

Because there are $n + 2$ balls in the urn after stage n, we see that $P(B_j|A_{j-1})$ is the probability that a red ball is chosen when $j - 1$ of the $n + 2$ balls in the urn are red, and $P(B_j|A_j)$ is the probability that a red ball is not chosen when j of the $n + 2$ balls in the urn are red. Consequently,

$$P(B_j|A_{j-1}) = \frac{j-1}{n+2}, \qquad P(B_j|A_j) = \frac{n+2-j}{n+2}$$

Substituting these into the equation for $P(B_j)$ gives

$$P(B_j) = \frac{1}{n+1}\left[\frac{j-1}{n+2} + \frac{n+2-j}{n+2}\right] = \frac{1}{n+2}$$

This completes the induction proof.

22. If A_i is the event that player i receives an ace, then

$$P(A_i) = 1 - \frac{\binom{2n-2}{n}}{\binom{2n}{n}} = 1 - \frac{1}{2}\frac{n-1}{2n-1} = \frac{3n-1}{4n-2}$$

By arbitrarily numbering the aces and noting that the player who does not receive ace number one will receive n of the remaining $2n - 1$ cards, we see that

$$P(A_1A_2) = \frac{n}{2n-1}$$

Therefore,

$$P(A_2^c|A_1) = 1 - P(A_2|A_1) = 1 - \frac{P(A_1A_2)}{P(A_1)} = \frac{n-1}{3n-1}$$

We may regard the card division outcome as the result of two trials, where trial $i, i = 1, 2$, is said to be a success if ace number i goes to the first player. Because the locations of the two aces become independent as n goes to infinity, with each one being equally likely to be given to either player, it follows that the trials become independent, with each being a success with probability 1/2. Hence, in the limiting case where $n \to \infty$, the problem becomes one of determining the conditional probability that two heads result, given that at least one does, when two fair coins are flipped. Because $\frac{n-1}{3n-1}$ converges to 1/3, the answer agrees with that of Example 2d.

CHAPTER 4

1. Since the probabilities sum to 1, we must have that $4P\{X = 3\} + .5 = 1$, implying that $P\{X = 0\} = .375$, $P\{X = 3\} = .125$. Hence $E[X] = 1(.3) + 2(.2) + 3(.125) = 1.075$.

2. The relationship implies that $p_i = c^i p_0$, $i = 1, 2$, where $p_i = P\{X = i\}$. As these probabilities sum to 1, we see that

$$p_0(1 + c + c^2) = 1 \Rightarrow p_0 = \frac{1}{1 + c + c^2}$$

Hence,

$$E[X] = p_1 + 2p_2 = \frac{c + 2c^2}{1 + c + c^2}$$

3. Letting X be the number of flips, then the probability mass function of X is

$$p_2 = p^2 + (1 - p)^2, \qquad p_3 = 1 - p_2 = 2p(1 - p)$$

Hence,

$$E[X] = 2p_2 + 3p_3 = 2p_2 + 3(1 - p_2) = 3 - p^2 - (1 - p)^2$$

4. The probability that a randomly chosen family will have i children is n_i/m. Hence,

$$E[X] = \sum_{i=1}^{r} in_i/m$$

Also, since there are in_i children in families having i children, it follows that the probability that a randomly chosen child is from a family with i children is $in_i / \sum_{i=1}^{r} in_i$. Therefore,

$$E[Y] = \frac{\sum_{i=1}^{r} i^2 n_i}{\sum_{i=1}^{r} in_i}$$

Thus, we must show

$$\frac{\sum_{i=1}^{r} i^2 n_i}{\sum_{i=1}^{r} in_i} \geq \frac{\sum_{i=1}^{r} in_i}{\sum_{i=1}^{r} n_i}$$

or, equivalently, that

$$\sum_{j=1}^{r} n_j \sum_{i=1}^{r} i^2 n_i \geq \sum_{i=1}^{r} i n_i \sum_{j=1}^{r} j n_j$$

or, equivalently, that

$$\sum_{i=1}^{r} \sum_{j=1}^{r} i^2 n_i n_j \geq \sum_{i=1}^{r} \sum_{j=1}^{r} i j n_i n_j$$

But, for a fixed pair i, j, the coefficient of $n_i n_j$ in the left side summation of the preceding is $i^2 + j^2$, whereas its coefficient in the right-hand summation is $2ij$. Hence, it suffices to show that

$$i^2 + j^2 \geq 2ij$$

which follows since $(i - j)^2 \geq 0$.

5. Let $p = P\{X = 1\}$. Then, $E[X] = p$, $\text{Var}(X) = p(1 - p)$, and so

$$p = 3p(1 - p)$$

implying that $p = 2/3$. Hence, $P\{X = 0\} = 1/3$.

6. If you wager x on a bet that wins the amount wagered with probability p and loses that amount with probability $1 - p$, then your expected winnings are

$$xp - x(1 - p) = (2p - 1)x$$

which is positive (and increasing in x) if and only if $p > 1/2$. Thus, if $p \leq 1/2$ one maximizes one's expected return by wagering 0, and if $p > 1/2$ one maximizes one's expected return by wagering the maximal possible bet. Thus, if the information is that the .6 coin was chosen then you should bet 10, and if the information is that the .3 coin was chosen then you should bet 0. Hence, your expected payoff is

$$\frac{1}{2}(1.2 - 1)10 + \frac{1}{2}0 - C = 1 - C$$

Since your expected payoff is 0 without the information (because in this case the probability of winning is $\frac{1}{2}(.6) + \frac{1}{2}(.3) < 1/2$) it follows that if the information costs less than 1 then it pays to purchase it.

7. **(a)** If you turn over the red paper and observe the value x then your expected return if you switch to the blue paper is

$$2x(1/2) + x/2(1/2) = 5x/4 > x$$

Thus, it would always be better to switch.

(b) Suppose the philanthropist writes the amount x on the red paper and so the amount on the blue paper is either $2x$ or $x/2$. Note that if $x/2 \geq y$ then the amount on the blue paper will be at least y and will thus be accepted. Hence, in this case, the reward is equally likely to be either $2x$ or $x/2$ and so

$$E[R_y(x)] = 5x/4, \qquad \text{if } x/2 \geq y$$

If $x/2 < y \leq 2x$ then the blue paper will be accepted if its value is $2x$ and rejected if it is $x/2$. Therefore,

$$E[R_y(x)] = 2x(1/2) + x(1/2) = 3x/2, \qquad \text{if } x/2 < y \leq 2x$$

Finally, if $2x < y$ then the blue paper will be rejected. Hence, in this case the reward is x, and so

$$R_y(x) = x, \quad \text{if } 2x < y$$

That is, we have shown that the expected return under the y-policy is, when the amount x is written on the red paper,

$$E[R_y(x)] = \begin{cases} x & \text{if } x < y/2 \\ 3x/2 & \text{if } y/2 \le x < 2y \\ 5x/4 & \text{if } x \ge 2y \end{cases}$$

8. Suppose that n independent trials each of which results in a success with probability p are performed. Then the number of successes will be less than or equal to i if and only if the number of failures is greater than or equal to $n - i$. But since each trial is a failure with probability $1 - p$, it follows that the number of failures is a binomial random variable with parameters n and $1 - p$. Hence,

$$\begin{aligned} P\{\text{Bin}(n, p) \le i\} &= P\{\text{Bin}(n, 1 - p) \ge n - i\} \\ &= 1 - P\{\text{Bin}(n, 1 - p) \le n - i - 1\} \end{aligned}$$

The final equality following since the probability that the number of failures is greater than or equal to $n - i$ is 1 minus the probability that it is less than $n - i$.

9. Since $E[X] = np$, $\text{Var}(X) = np(1 - p)$, we are given that $np = 6$, $np(1 - p) = 2.4$. Thus, $1 - p = .4$, or $p = .6$, $n = 10$. Hence,

$$P\{X = 5\} = \binom{10}{5}(.6)^5(.4)^5$$

10. Let X_i denote the number on the ith ball drawn, $i = 1, \ldots, m$. Then

$$\begin{aligned} P\{X \le k\} &= P\{X_1 \le k, X_2 \le k, \ldots, X_m \le k\} \\ &= P\{X_1 \le k\}P\{X_2 \le k\} \cdots P\{X_m \le k\} \\ &= \left(\frac{k}{n}\right)^m \end{aligned}$$

Therefore,

$$P\{X = k\} = P\{X \le k\} - P\{X \le k - 1\} = \left(\frac{k}{n}\right)^m - \left(\frac{k - 1}{n}\right)^m$$

11. (a) Given that A wins the first game, it will win the series if from then on it wins 2 games before team B wins 3 games. Thus

$$P\{A \text{ wins} | A \text{ wins first}\} = \sum_{i=2}^{4} \binom{4}{i} p^i (1 - p)^{4-i}$$

(b)

$$\begin{aligned} P\{A \text{ wins first} | A \text{ wins}\} &= \frac{P\{A \text{ wins} | A \text{ wins first}\} P\{A \text{ wins first}\}}{P\{A \text{ wins}\}} \\ &= \frac{\sum_{i=2}^{4} \binom{4}{i} p^{i+1}(1 - p)^{4-i}}{\sum_{i=3}^{5} \binom{5}{i} p^i (1 - p)^{5-i}} \end{aligned}$$

12. Condition on whether the team wins this weekend, to obtain the solution:

$$.5 \sum_{i=3}^{4} \binom{4}{i} (.4)^i (.6)^{4-i} + .5 \sum_{i=3}^{4} \binom{4}{i} (.7)^i (.3)^{4-i}$$

13. Assuming that the number of hurricanes can be approximated by a Poisson random variable, we obtain the solution

$$\sum_{i=0}^{3} e^{-5.2}(5.2)^i/i!$$

14.

$$\begin{aligned} E[Y] &= \sum_{i=1}^{\infty} iP\{X = i\}/P\{X > 0\} \\ &= E[X]/P\{X > 0\} \\ &= \frac{\lambda}{1 - e^{-\lambda}} \end{aligned}$$

15. **(a)** $\binom{8}{3}(9/19)^3(10/19)^5(9/19) = \binom{8}{3}(9/19)^4(10/19)^5$

(b) If W is her final winnings and X is the number of bets she makes, then since she would have won 4 bets and lost $X - 4$ bets, it follows that

$$W = 20 - 5(X - 4) = 40 - 5X$$

Hence,

$$E[W] = 40 - 5E[X] = 40 - 5[4/(9/19)] = -20/9$$

16. The probability that a round does not result in an "odd person" is equal to 1/4, the probability that all three coins land on the same side.

(a) $(1/4)^2(3/4) = 3/64$ **(b)** $(1/4)^4 = 1/256$

17. Let $q = 1 - p$.

$$\begin{aligned} E[1/X] &= \sum_{i=1}^{\infty} \frac{1}{i} q^{i-1} p \\ &= \frac{p}{q} \sum_{i=1}^{\infty} q^i/i \\ &= \frac{p}{q} \sum_{i=1}^{\infty} \int_0^q x^{i-1}\, dx \\ &= \frac{p}{q} \int_0^q \sum_{i=1}^{\infty} x^{i-1}\, dx \\ &= \frac{p}{q} \int_0^q \frac{1}{1 - x}\, dx \\ &= \frac{p}{q} \int_p^1 \frac{1}{y}\, dy \\ &= -\frac{p}{q} \log(p) \end{aligned}$$

18. Because $\frac{X-b}{a-b}$ will equal 1 with probability p, or 0 with probability $1-p$, it follows that it is a Bernoulli random variable with parameter p. Because the variance of such a Bernoulli random variable is $p(1-p)$ we have that

$$p(1-p) = \text{Var}\left(\frac{X-b}{a-b}\right) = \frac{1}{(a-b)^2}\text{Var}(X-b) = \frac{1}{(a-b)^2}\text{Var}(X)$$

Hence,

$$\text{Var}(X) = (a-b)^2 p(1-p)$$

19. Let X denote the number of games that you play, and Y the number of games that you lose.

(a) After your fourth game you will continue to play until you lose. Therefore, $X-4$ is a geometric random variable with parameter $1-p$, and so

$$E[X] = E[4 + (X-4)] = 4 + E[X-4] = 4 + \frac{1}{1-p}$$

(b) If we let Z denote the number of losses you have in the first 4 games, then Z is a binomial random variable with parameters 4 and $1-p$. Because $Y = Z + 1$, we have that

$$E[Y] = E[Z+1] = E[Z] + 1 = 4(1-p) + 1$$

20. A total of n white balls will be withdrawn before a total of m black balls if and only if there are at least n white balls in the first $n+m-1$ withdrawals. (Compare with *the problem of the points*, Example 4i of Chapter 3.) With X equal to the number of white balls among the first $n+m-1$ withdrawn balls, then X is a hypergeometric random variable, and

$$P\{X \geq n\} = \sum_{i=n}^{n+m-1} P\{X = i\} = \sum_{i=n}^{n+m-1} \frac{\binom{N}{i}\binom{M}{n+m-1-i}}{\binom{N+M}{n+m-1}}$$

CHAPTER 5

1. Let X be the number of minutes played.

(a) $P\{X > 15\} = 1 - P\{X \leq 15\} = 1 - 5(.025) = .875$

(b) $P\{20 < X < 35\} = 10(.05) + 5(.025) = .625$

(c) $P\{X < 30\} = 10(.025) + 10(.05) = .75$

(d) $P\{X > 36\} = 4(.025) = .1$

2. **(a)** $1 = \int_0^1 cx^n\,dx = c/(n+1) \Rightarrow c = n+1$

(b) $P\{X > x\} = (n+1)\int_x^1 x^n\,dx = x^{n+1}\Big|_x^1 = 1 - x^{n+1}$

3. First, let us find c by using that

$$1 = \int_0^2 cx^4\,dx = 32c/5 \Rightarrow c = 5/32$$

(a) $E[X] = \frac{5}{32}\int_0^2 x^5\,dx = \frac{5}{32}\frac{64}{6} = 5/3$

(b) $E[X^2] = \frac{5}{32}\int_0^2 x^6\,dx = \frac{5}{32}\frac{128}{7} = 20/7 \Rightarrow \text{Var}(X) = 20/7 - (5/3)^2 = 5/63$

4. Since

$$1 = \int_0^1 (ax + bx^2)\,dx = a/2 + b/3$$

$$.6 = \int_0^1 (ax^2 + bx^3)\,dx = a/3 + b/4$$

we obtain that $a = 3.6, b = -2.4$. Hence,

(a) $P\{X < 1/2\} = \int_0^{1/2}(3.6x - 2.4x^2)\,dx = (1.8x^2 - .8x^3)\Big|_0^{1/2} = .35$

(b) $E[X^2] = \int_0^1 (3.6x^3 - 2.4x^4)\,dx = .42 \Rightarrow \text{Var}(X) = .06$

5. For $i = 1, \ldots, n$

$$\begin{aligned} P\{X = i\} &= P\{\text{Int}(nU) = i - 1\} \\ &= P\{i - 1 \le nU < i\} \\ &= P\left\{\frac{i-1}{n} \le U < \frac{i}{n}\right\} \\ &= 1/n \end{aligned}$$

6. If you bid x, $70 \le x \le 140$, then you will win the bid and make a profit of $x - 100$ with probability $(140 - x)/70$, or lose the bid and make a profit of 0 otherwise. Therefore, your expected profit if you bid x is

$$\frac{1}{70}(x - 100)(140 - x) = \frac{1}{70}(240x - x^2 - 14000)$$

Differentiating and setting the preceding equal to 0 gives that

$$240 - 2x = 0$$

Therefore, you should bid 120 thousand dollars. Your expected profit will be 40/7 thousand dollars.

7. **(a)** $P\{U > .1\} = 9/10$

(b) $P\{U > .2|U > .1\} = P\{U > .2\}/P\{U > .1\} = 8/9$

(c) $P\{U > .3|U > .2, U > .1\} = P\{U > .3\}/P\{U > .2\} = 7/8$

(d) $P\{U > .3\} = 7/10$

The answer to part (d) could also have been obtained by multiplying the probabilities in parts (a), (b), and (c).

8. Let X be the test score, and let $Z = (X - 100)/15$. Note that Z is a standard normal random variable.

(a) $P\{X > 125\} = P\{Z > 25/15\} \approx .0478$

(b)

$$\begin{aligned} P\{90 < X < 110\} &= P\{-10/15 < Z < 10/15\} \\ &= P\{Z < 2/3\} - P\{Z < -2/3\} \\ &= P\{Z < 2/3\} - [1 - P\{Z < 2/3\}] \\ &\approx .4950 \end{aligned}$$

9. Let X be the travel time. We want to find x such that

$$P\{X > x\} = .05$$

which is equivalent to

$$P\left\{\frac{X - 40}{7} > \frac{x - 40}{7}\right\} = .05$$

That is, we need to find x such that

$$P\left\{Z > \frac{x - 40}{7}\right\} = .05$$

where Z is a standard normal random variable. But

$$P\{Z > 1.645\} = .05$$

and thus

$$\frac{x - 40}{7} = 1.645 \quad \text{or} \quad x = 51.515$$

Therefore, you should leave no later than 8.485 minutes after 12 P.M.

10. Let X be the tire life in units of one thousand, and let $Z = (X - 34)/4$. Note that Z is a standard normal random variable.

(a) $P\{X > 40\} = P\{Z > 1.5\} \approx .0668$

(b) $P\{30 < X < 35\} = P\{-1 < Z < .25\} = P\{Z < .25\} - P\{Z > 1\} \approx .44$

(c) $P\{X > 40 | X > 30\} = P\{X > 40\}/P\{X > 30\}$

$$= P\{Z > 1.5\}/P\{Z > -1\} \approx .079$$

11. Let X be next year's rainfall and let $Z = (X - 40.2)/8.4$.

(a) $P\{X > 44\} = P\{Z > 3.8/8.4\} \approx P\{Z > .4524\} \approx .3255$

(b) $\binom{7}{3}(.3255)^3(.6745)^4$

12. Let M_i and W_i denote, respectively, the numbers of men and women in the samples that earn, in units of one thousand dollars, at least i per year. Also, let Z be a standard normal random variable.

(a)
$$\begin{aligned} P\{W_{25} \geq 70\} &= P\{W_{25} \geq 69.5\} \\ &= P\left\{\frac{W_{25} - 200(.34)}{\sqrt{200(.34)(.66)}} \geq \frac{69.5 - 200(.34)}{\sqrt{200(.34)(.66)}}\right\} \\ &\approx P\{Z \geq .2239\} \\ &\approx .4114 \end{aligned}$$

(b)
$$\begin{aligned} P\{M_{25} \leq 120\} &= P\{M_{25} \leq 120.5\} \\ &= P\left\{\frac{M_{25} - (200)(.587)}{\sqrt{(200)(.587)(.413)}} \leq \frac{120.5 - (200)(.587)}{\sqrt{(200)(.587)(.413)}}\right\} \\ &\approx P\{Z \leq .4452\} \\ &\approx .6719 \end{aligned}$$

(c)
$$P\{M_{20} \geq 150\} = P\{M_{20} \geq 149.5\}$$
$$= P\left\{\frac{M_{20} - (200)(.745)}{\sqrt{(200)(.745)(.255)}} \geq \frac{149.5 - (200)(.745)}{\sqrt{(200)(.745)(.255)}}\right\}$$
$$\approx P\{Z \geq .0811\}$$
$$\approx .4677$$
$$P\{W_{20} \geq 100\} = P\{W_{20} \geq 99.5\}$$
$$= P\left\{\frac{W_{20} - (200)(.534)}{\sqrt{(200)(.534)(.466)}} \geq \frac{99.5 - (200)(.534)}{\sqrt{(200)(.534)(.466)}}\right\}$$
$$\approx P\{Z \geq -1.0348\}$$
$$= \approx .8496$$

Hence,
$$P\{M_{20} \geq 150\}P\{W_{20} \geq 100\} \approx .3974$$

13. The lack of memory property of the exponential gives the result $e^{-4/5}$.

14. **(a)** $e^{-2^2} = e^{-4}$ **(b)** $F(3) - F(1) = e^{-1} - e^{-9}$

(c) $\lambda(t) = 2te^{-t^2}/e^{-t^2} = 2t$

(d) Let Z be a standard normal random variable. Use the identity $E[X] = \int_0^\infty P\{X > x\}\,dx$ to obtain
$$E[X] = \int_0^\infty e^{-x^2}\,dx$$
$$= 2^{-1/2}\int_0^\infty e^{-y^2/2}\,dy$$
$$= 2^{-1/2}\sqrt{2\pi}\,P\{Z > 0\}$$
$$= \sqrt{\pi}/2$$

(e) Use the result of Theoretical Exercise 5 to obtain
$$E[X^2] = \int_0^\infty 2xe^{-x^2}\,dx = -e^{-x^2}\Big|_0^\infty = 1$$

Hence, $\text{Var}(X) = 1 - \pi/4$.

15. **(a)** $P\{X > 6\} = \exp\left\{-\int_0^6 \lambda(t)\,dt\right\} = e^{-3.45}$

(b)
$$P\{X < 8|X > 6\} = 1 - P\{X > 8|X > 6\}$$
$$= 1 - P\{X > 8\}/P\{X > 6\}$$
$$= 1 - e^{-5.65}/e^{-3.45}$$
$$\approx .8892$$

16. For $x \geq 0$
$$F_{1/X}(x) = P\{1/X \leq x\}$$
$$= P\{X \leq 0\} + P\{X \geq 1/x\}$$
$$= 1/2 + 1 - F_X(1/x)$$

Differentiation yields

$$\begin{aligned} f_{1/X}(x) &= x^{-2}f_X(1/x) \\ &= \frac{1}{x^2\pi(1 + (1/x)^2)} \\ &= f_X(x) \end{aligned}$$

The proof when $x < 0$ is similar.

17. If X denotes the number of the first n bets that you win, then the amount that you will be winning after n bets is

$$35X - (n - X) = 36X - n$$

Thus, we want to determine

$$p = P\{36X - n > 0\} = P\{X > n/36\}$$

when X is a binomial random variable with parameters n and $p = 1/38$.

(a) When $n = 34$,

$$\begin{aligned} p &= P\{X \geq 1\} \\ &= P\{X > .5\} \quad \text{(the continuity correction)} \\ &= P\left\{\frac{X - 34/38}{\sqrt{34(1/38)(37/38)}} > \frac{.5 - 34/38}{\sqrt{34(1/38)(37/38)}}\right\} \\ &= P\left\{\frac{X - 34/38}{\sqrt{34(1/38)(37/38)}} > -.4229\right\} \\ &\approx \Phi(.4229) \\ &\approx .6638 \end{aligned}$$

(Because you will be ahead after 34 bets if you win at least one bet, the exact probability in this case is $1 - (37/38)^{34} = .5961$.)

(b) When $n = 1000$,

$$\begin{aligned} p &= P\{X > 27.5\} \\ &= P\left\{\frac{X - 1000/38}{\sqrt{1000(1/38)(37/38)}} > \frac{27.5 - 1000/38}{\sqrt{1000(1/38)(37/38)}}\right\} \\ &\approx 1 - \Phi(.2339) \\ &\approx .4075 \end{aligned}$$

The exact probability, namely the probability that a binomial $n = 1000$, $p = 1/38$ random variable is greater than 27, is .3961.

(c) When $n = 100{,}000$,

$$\begin{aligned} p &= P\{X > 2777.5\} \\ &= P\left\{\frac{X - 100000/38}{\sqrt{100000(1/38)(37/38)}} > \frac{2777.5 - 100000/38}{\sqrt{100000(1/38)(37/38)}}\right\} \\ &\approx 1 - \Phi(2.883) \\ &\approx .0020 \end{aligned}$$

The exact probability in this case is .0021.

18. If X denotes the lifetime of the battery, then the desired probability, $P\{X > s + t|X > t\}$, can be determined as follows:

$$\begin{aligned}P\{X > s + t|X > t\} &= \frac{P\{X > s + t, X > t\}}{P\{X > t\}} \\ &= \frac{P\{X > s + t\}}{P\{X > t\}} \\ &= \frac{P\{X > s + t|\text{battery is type 1}\}p_1 + P\{X > s + t|\text{battery is type 2}\}p_2}{P\{X > t|\text{battery is type 1}\}p_1 + P\{X > t|\text{battery is type 2}\}p_2} \\ &= \frac{e^{-\lambda_1(s+t)}p_1 + e^{-\lambda_2(s+t)}p_2}{e^{-\lambda_1 t}p_1 + e^{-\lambda_2 t}p_2}\end{aligned}$$

Another approach is to directly condition on the type of battery and then use the lack of memory property of exponential random variables. That is, we could do the following:

$$\begin{aligned}P\{X > s + t|X > t\} &= P\{X > s + t|X > t, \text{type 1}\}P\{\text{type 1}|X > t\} \\ &\quad + P\{X > s + t|X > t, \text{type 2}\}P\{\text{type 2}|X > t\} \\ &= e^{-\lambda_1 s}P\{\text{type 1}|X > t\} + e^{-\lambda_2 s}P\{\text{type 2}|X > t\}\end{aligned}$$

Now use that, for $i = 1, 2$

$$\begin{aligned}P\{\text{type } i|X > t\} &= \frac{P\{\text{type } i, X > t\}}{P\{X > t\}} \\ &= \frac{P\{X > t|\text{type } i\}p_i}{P\{X > t|\text{type 1}\}p_1 + P\{X > t|\text{type 2}\}p_2} \\ &= \frac{e^{-\lambda_i t}p_i}{e^{-\lambda_1 t}p_1 + e^{-\lambda_2 t}p_2}\end{aligned}$$

CHAPTER 6

1. **(a)** $3C + 6C = 1 \Rightarrow C = 1/9$

(b) Let $p(i, j) = P\{X = i, Y = j\}$. Then

$$p(1,1) = 4/9, p(1,0) = 2/9, P(0,1) = 1/9, p(0,0) = 2/9$$

(c) $\dfrac{(12)!}{2^6}(1/9)^6(2/9)^6$ **(d)** $\dfrac{(12)!}{(4!)^3}(1/3)^{12}$

(e) $\sum_{i=8}^{12}\binom{12}{i}(2/3)^i(1/3)^{12-i}$

2. **(a)** With $p_j = P\{XYZ = j\}$, we have that

$$p_6 = p_2 = p_4 = p_{12} = 1/4$$

Hence,

$$E[XYZ] = (6 + 2 + 4 + 12)/4 = 6$$

(b) With $q_j = P\{XY + XZ + YZ = j\}$, we have that

$$q_{11} = q_5 = q_8 = q_{16} = 1/4$$

Hence,

$$E[XY + XZ + YZ] = (11 + 5 + 8 + 16)/4 = 10$$

3. In the following, we will make use of the identity

$$\int_0^\infty e^{-x}x^n\,dx = n!$$

which follows since $e^{-x}x^n/n!$, $x > 0$, is the density function of a gamma random variable with parameters $n + 1$ and λ, and must thus integrate to 1.

(a)
$$1 = C\int_0^\infty e^{-y}\int_{-y}^{y}(y - x)\,dx\,dy$$
$$= C\int_0^\infty e^{-y}2y^2\,dy = 4C$$

Hence, $C = 1/4$.

(b) Since the joint density is nonzero only when $y > x$ and $y > -x$, we have that for $x > 0$,

$$f_X(x) = \frac{1}{4}\int_x^\infty (y - x)e^{-y}\,dy$$
$$= \frac{1}{4}\int_0^\infty ue^{-(x+u)}\,du$$
$$= \frac{1}{4}e^{-x}$$

For $x < 0$,

$$f_X(x) = \frac{1}{4}\int_{-x}^\infty (y - x)e^{-y}\,dy$$
$$= \frac{1}{4}[-ye^{-y} - e^{-y} + xe^{-y}]_{-x}^\infty$$
$$= (-2xe^x + e^x)/4$$

(c) $f_Y(y) = \frac{1}{4}e^{-y}\int_{-y}^{y}(y - x)\,dx = \frac{1}{2}y^2e^{-y}$

(d)
$$E[X] = \frac{1}{4}\left[\int_0^\infty xe^{-x}\,dx + \int_{-\infty}^0 (-2x^2e^x + xe^x)\,dx\right]$$
$$= \frac{1}{4}\left[1 - \int_0^\infty (2y^2e^{-y} + ye^{-y})\,dy\right]$$
$$= \frac{1}{4}[1 - 4 - 1] = -1$$

(e) $E[Y] = \frac{1}{2}\int_0^\infty y^3e^{-y}\,dy = 3$

4. (a) Letting $p_j = P\{XYZ = j\}$, we have that

$$p_1 = 1/8, \quad p_2 = 3/8, \quad p_4 = 3/8, \quad p_8 = 1/8$$

(b) Letting $p_j = P\{XY + XZ + YZ = j\}$, we have

$$p_3 = 1/8, \quad p_5 = 3/8, \quad p_8 = 3/8, \quad p_{12} = 1/8$$

(c) Letting $p_j = P\{X^2 + YZ = j\}$, we have

$$p_2 = 1/8, \quad p_3 = 1/4, \quad p_5 = 1/4, \quad p_6 = 1/4, \quad p_8 = 1/8$$

5. (a)

$$\begin{aligned} 1 &= \int_0^1 \int_1^5 (x/5 + cy)\,dy\,dx \\ &= \int_0^1 (4x/5 + 12c)\,dx \\ &= 12c + 2/5 \end{aligned}$$

Hence, $c = 1/20$.

(b) No, the density does not factor.

(c)

$$\begin{aligned} P\{X + Y > 3\} &= \int_0^1 \int_{3-x}^5 (x/5 + y/20)\,dy\,dx \\ &= \int_0^1 [(2 + x)x/5 + 25/40 - (3 - x)^2/40]\,dx \\ &= 1/5 + 1/15 + 5/8 - 19/120 = 11/15 \end{aligned}$$

6. (a) Yes, the joint density function factors.

(b) $f_X(x) = x \int_0^2 y\,dy = 2x, \quad 0 < x < 1$

(c) $f_Y(y) = y \int_0^1 x\,dx = y/2, \quad 0 < y < 2$

(d)

$$\begin{aligned} P\{X < x, Y < y\} &= P\{X < x\}P\{Y < y\} \\ &= \min(1, x^2)\min(1, y^2/4), \quad x > 0, y > 0 \end{aligned}$$

(e) $E[Y] = \int_0^2 y^2/2\,dy = 4/3$

(f)

$$\begin{aligned} P\{X + Y < 1\} &= \int_0^1 x \int_0^{1-x} y\,dy\,dx \\ &= \frac{1}{2}\int_0^1 x(1 - x)^2\,dx = 1/24 \end{aligned}$$

7. Let T_i denote the time at which a type i shock occurs, $i = 1, 2, 3$. For $s > 0, t > 0$,

$$\begin{aligned} P\{X_1 > s, X_2 > t\} &= P\{T_1 > s, T_2 > t, T_3 > \max(s, t)\} \\ &= P\{T_1 > s\}P\{T_2 > t\}P\{T_3 > \max(s, t)\} \\ &= \exp\{-\lambda_1 s\}\exp\{-\lambda_2 t\}\exp\{-\lambda_3 \max(s, t)\} \\ &= \exp\{-(\lambda_1 s + \lambda_2 t + \lambda_3 \max(s, t))\} \end{aligned}$$

8. (a) No, advertisements on pages having many ads are less likely to be chosen than are ones on pages with few ads.

(b) $\dfrac{1}{m}\dfrac{n(i)}{n}$

(c) $\dfrac{\sum_{i=1}^{m} n(i)}{nm} = \bar{n}/n$, where $\bar{n} = \sum_{i=1}^{m} n(i)/m$

(d) $(1 - \bar{n}/n)^{k-1}\frac{1}{m}\frac{n(i)}{n}\frac{1}{n(i)} = (1 - \bar{n}/n)^{k-1}/(nm)$

(e) $\sum_{k=1}^{\infty}\frac{1}{nm}(1 - \bar{n}/n)^{k-1} = \frac{1}{\bar{n}m}$.

(f) The number of iterations is geometric with mean $n\sqrt{n}$

9. (a) $P\{X = i\} = 1/m, \quad i = 1, \ldots, m.$

(b) *Step 2.* Generate a uniform $(0, 1)$ random variable U. If $U < n(X)/n$, go to step 3. Otherwise return to step 1.

Step 3. Generate a uniform $(0, 1)$ random variable U, and select the element on page X in position $[n(X)U] + 1$.

10. Yes, they are independent. This can be easily seen by considering the equivalent question of whether X_N is independent of N. But this is indeed so, since knowing when the first random variable greater than c occurs does not affect the probability distribution of its value, which is the uniform distribution on $(c, 1)$.

11. Letting p_i denote the probability of obtaining i points on a single throw of the dart, then

$$\begin{aligned} p_{30} &= \pi/36 \\ p_{20} &= 4\pi/36 - p_{30} = \pi/12 \\ p_{10} &= 9\pi/36 - p_{20} - p_{30} = 5\pi/36 \\ p_0 &= 1 - p_{10} - p_{20} - p_{30} = 1 - \pi/4 \end{aligned}$$

(a) $\pi/12$

(b) $\pi/9$

(c) $1 - \pi/4$

(d) $\pi(30/36 + 20/12 + 50/36) = 35\pi/9$

(e) $(\pi/4)^2$

(f) $2(\pi/36)(1 - \pi/4) + 2(\pi/12)(5\pi/36)$

12. Let Z be a standard normal random variable.

(a)
$$P\left\{\sum_{i=1}^{4} X_i > 0\right\} = P\left\{\frac{\sum_{i=1}^{4} X_i - 6}{\sqrt{24}} > \frac{-6}{\sqrt{24}}\right\}$$
$$\approx P\{Z > -1.2247\} \approx .8897$$

(b)
$$P\left\{\sum_{i=1}^{4} X_i > 0 \,\middle|\, \sum_{i=1}^{2} X_i = -5\right\} = P\{X_3 + X_4 > 5\}$$
$$= P\left\{\frac{X_3 + X_4 - 3}{\sqrt{12}} > 2/\sqrt{12}\right\}$$
$$\approx P\{Z > .5774\} \approx .2818$$

(c)
$$p\left\{\sum_{i=1}^{4} X_i > 0 | X_1 = 5\right\} = P\{X_2 + X_3 + X_4 > -5\}$$
$$= P\left\{\frac{X_2 + X_3 + X_4 - 4.5}{\sqrt{18}} > -9.5/\sqrt{18}\right\}$$
$$\approx P\{Z > -2.239\} \approx .9874$$

13. In the following, C does not depend on n.

$$P\{N = n|X = x\} = f_{X|N}(x|n)P\{N = n\}/f_X(x)$$
$$= C\frac{1}{(n-1)!}(\lambda x)^{n-1}(1-p)^{n-1}$$
$$= C(\lambda(1-p)x)^{n-1}/(n-1)!$$

which shows that, conditional on $X = x$, $N - 1$ is a Poisson random variable with mean $\lambda(1-p)x$. That is,

$$P\{N = n|X = x\} = P\{N - 1 = n - 1|X = x\}$$
$$= e^{-\lambda(1-p)x}(\lambda(1-p)x)^{n-1}/(n-1)!, n \geq 1.$$

14. **(a)** The Jacobian of the transformation is

$$J = \begin{vmatrix} 1 & 0 \\ 1 & 1 \end{vmatrix} = 1$$

As the equations $u = x, v = x + y$ imply that $x = u, y = v - u$, we obtain that

$$f_{U,V}(u,v) = f_{X,Y}(u, v-u) = 1, \qquad 0 < u < 1, \qquad 0 < v - u < 1$$

or, equivalently,

$$f_{U,V}(u,v) = 1, \qquad \max(v-1, 0) < u < \min(v, 1)$$

(b) For $0 < v < 1$,

$$f_V(v) = \int_0^v du = v$$

For, $1 \leq v \leq 2$,

$$f_V(v) = \int_{v-1}^1 du = 2 - v$$

15. Let U be a uniform random variable on $(7, 11)$. If you bid $x, 7 \leq x \leq 10$, you will be the high bidder with probability

$$(P\{U < x\})^3 = \left(P\left\{\frac{U-7}{4} < \frac{x-7}{4}\right\}\right)^3 = \left(\frac{x-7}{4}\right)^3$$

Hence, your expected gain, call it $E[G(x)]$, if you bid x is

$$E[G(x)] = \frac{1}{4}(x-7)^3(10-x)$$

Calculus shows this is maximized when $x = 37/4$.

16. Let $i_1, i_2, \ldots, i_n$ be a permutation of $1, 2, \ldots, n$.

$$P\{X_1 = i_1, X_2 = i_2, \ldots, X_n = i_n\} = P\{X_1 = i_1\}P\{X_2 = i_2\}\cdots P\{X_n = i_n\}$$
$$= p_{i_1}p_{i_2}\cdots p_{i_n}$$
$$= p_1p_2\cdots p_n$$

Therefore, the desired probability is $n!\, p_1p_2\cdots p_n$, which reduces to $\frac{n!}{n^n}$ when all $p_i = 1/n$.

17. (a) Because $\sum_{i=1}^{n} X_i = \sum_{i=1}^{n} Y_i$, it follows that $N = 2M$.

(b) Consider the $n - k$ coordinates whose Y-values are equal to 0 and call them the red coordinates. Because the k coordinates whose X-values are equal to 1 are equally likely to be any of the $\binom{n}{k}$ sets of k coordinates, it follows that the number of red coordinates among these k coordinates has the same distribution as the number of red balls chosen when one randomly chooses k of a set of n balls of which $n - k$ are red. Therefore, M is a hypergeometric random variable.

(c)
$$E[N] = E[2M] = 2E[M] = \frac{2k(n-k)}{n}$$

(d) Using the formula for the variance of a hypergeometric given in Equation (9.6) of Chapter 4, we obtain that

$$\text{Var}(N) = 4\,\text{Var}(M) = 4\frac{n-k}{n-1}k(1 - k/n)(k/n)$$

***18. (a)** First note that $S_n - S_k = \sum_{i=k+1}^{n} Z_i$ is a normal random variable with mean 0 and variance $n - k$ that is independent of S_k. Consequently, S_n, given that $S_k = y$, is a normal random variable with mean y and variance $n - k$.

(b) Because the conditional density function of S_k given that $S_n = x$ is a density function whose argument is y, anything that does not depend on y can be regarded as a constant. (For instance, x is regarded as a fixed constant.) In the following, the quantities $C_i, i = 1, 2, 3, 4$ are all constants that do not depend on y.

$$\begin{aligned}
f_{S_k|S_n}(y|x) &= \frac{f_{S_k,S_n}(y,x)}{f_{S_n}(x)} \\
&= C_1 f_{S_n|S_k}(x|y) f_{S_k}(y) \qquad \left(\text{where } C_1 = \frac{1}{f_{S_n}(x)}\right) \\
&= C_1 \frac{1}{\sqrt{2\pi}\sqrt{n-k}} e^{-(x-y)^2/2(n-k)} \frac{1}{\sqrt{2\pi}\sqrt{k}} e^{-y^2/2k} \\
&= C_2 \exp\left\{-\frac{(x-y)^2}{2(n-k)} - \frac{y^2}{2k}\right\} \\
&= C_3 \exp\left\{\frac{2xy}{2(n-k)} - \frac{y^2}{2(n-k)} - \frac{y^2}{2k}\right\} \\
&= C_3 \exp\left\{-\frac{n}{2k(n-k)}\left(y^2 - 2\frac{k}{n}xy\right)\right\} \\
&= C_3 \exp\left\{-\frac{n}{2k(n-k)}\left[\left(y - \frac{k}{n}x\right)^2 - \left(\frac{k}{n}x\right)^2\right]\right\} \\
&= C_4 \exp\left\{-\frac{n}{2k(n-k)}\left(y - \frac{k}{n}x\right)^2\right\}
\end{aligned}$$

But we recognize the preceding as the density function of a normal random variable with mean $\frac{k}{n}x$ and variance $\frac{k(n-k)}{n}$.

CHAPTER 7

1. **(a)** $d = \sum_{i=1}^{m} 1/n(i)$

(b) $P\{X = i\} = P\{[mU] = i - 1\} = P\{i - 1 \le mU < i\} = 1/m, \quad i = 1,\ldots,m$

(c) $E\left[\frac{m}{n(X)}\right] = \sum_{i=1}^{m} \frac{m}{n(i)} P\{X = i\} = \sum_{i=1}^{m} \frac{m}{n(i)} \frac{1}{m} = d$

2. Let I_j equal 1 if the jth ball withdrawn is white and the $(j+1)$st is black, and let it equal 0 otherwise. If X is the number of instances in which a white ball is immediately followed by a black one, then we may express X as

$$X = \sum_{j=1}^{n+m-1} I_j$$

and, thus,

$$\begin{aligned} E[X] &= \sum_{j=1}^{n+m-1} E[I_j] \\ &= \sum_{j=1}^{n+m-1} P\{j^{th} \text{ selection is white, } (j+1)^{st} \text{ is black}\} \\ &= \sum_{j=1}^{n+m-1} P\{j^{th} \text{ selection is white}\}\, P\{j+1)^{st} \text{ is black} \mid j^{th} \text{ is white}\} \\ &= \sum_{j=1}^{n+m-1} \frac{n}{n+m} \frac{m}{n+m-1} \\ &= \frac{nm}{n+m} \end{aligned}$$

The preceding used the fact that each of the $n + m$ balls is equally likely to be the jth one selected, and given that selection is a white ball each of the other $n + m - 1$ balls is equally likely to be the next ball chosen.

3. Arbitrarily number the couples, and then let I_j equal 1 if married couple number j, $j = 1,\ldots,10$, is seated at the same table. Then, if X represents the number of married couples that are seated at the same table, we have that

$$X = \sum_{j=1}^{10} I_j$$

and so

$$E[X] = \sum_{j=1}^{10} E[I_j]$$

(a) To compute $E[I_j]$ in this case, consider wife number j. Since each of the $\binom{19}{3}$ groups of size 3 not including her is equally likely to be the remaining members of her table, it follows that the probability that her husband is at her table is

$$\frac{\binom{1}{1}\binom{18}{2}}{\binom{19}{3}} = \frac{3}{19}$$

Hence, $E[I_j] = 3/19$ and so

$$E[X] = 30/19$$

(b) In this case, since the two men at the table of wife j are equally likely to be any of the 10 men, it follows that the probability that one of them is her husband is 2/10, and so

$$E[I_j] = 2/10, \qquad \text{and} \qquad E[X] = 2$$

4. From Example 2j, we know that the expected number of times that the die need be rolled until all sides have appeared at least once is $6(1 + 1/2 + 1/3 + 1/4 + 1/5 + 1/6) = 14.7$. Now, if we let X_i denote the total number of times that side i appears, then since $\sum_{i=1}^{6} X_i$ is equal to the total number of rolls, we have that

$$14.7 = E\left[\sum_{i=1}^{6} X_i\right] = \sum_{i=1}^{6} E[X_i]$$

But, by symmetry, $E[X_i]$ will be the same for all i, and thus it follows from the preceding that $E[X_1] = 14.7/6 = 2.45$.

5. Let I_j equal 1 if we win 1 when the jth red card to show is turned over and let I_j equal 0 otherwise. (For instance, I_1 will equal 1 if the first card turned over is red.) Hence, if X is our total winnings then

$$E[X] = E\left[\sum_{j=1}^{n} I_j\right] = \sum_{j=1}^{n} E[I_j]$$

Now, I_j will equal 1 if j red cards appear before j black cards and, by symmetry, the probability of this event is equal to 1/2. Therefore, $E[I_j] = 1/2$, and $E[X] = n/2$.

6. To see that $N \le n - 1 + I$, note that if all events occur then both sides of the preceding are equal to n, whereas if they do not all occur then the inequality reduces to $N \le n - 1$, which is clearly true in this case. Taking expectations, yields that

$$E[N] \le n - 1 + E[I]$$

However, if we let I_i equal 1 if A_i occurs and 0 otherwise, then

$$E[N] = E\left[\sum_{i=1}^{n} I_i\right] = \sum_{i=1}^{n} E[I_i] = \sum_{i=1}^{n} P(A_i)$$

As $E[I] = P(A_1 \cdots A_n)$, the result follows.

7. Imagine that the values $1, 2, \ldots, n$ are lined up in their numerical order and that the k values selected are considered special. From Example 2m, the position of the first special value,

equal to the smallest value chosen, has mean $1 + \dfrac{n-k}{k+1} = \dfrac{n+1}{k+1}$.

For a more formal argument note that $X \ge j$ if none of the $j - 1$ smallest values are chosen. Hence,

$$P\{X \ge j\} = \frac{\binom{n-j+1}{k}}{\binom{n}{k}} = \frac{\binom{n-k}{j-1}}{\binom{n}{j-1}}$$

which shows that X has the same distribution as the random variable of Example 2m (with the notational change that the total number of balls is now n and the number of special balls is k).

8. Let X denote the number of families that depart after the Sanchez family. Arbitrarily number all the $N - 1$ non-Sanchez families, and let I_r, $1 \leq r \leq N - 1$, equal 1 if family r departs after the Sanchez family. Then

$$X = \sum_{r=1}^{N-1} I_r$$

Taking expectations gives

$$E[X] = \sum_{r=1}^{N-1} P\{\text{family } r \text{ departs after the Sanchez family}\}$$

Now consider any non-Sanchez family who checked in k pieces of luggage. Because each of the $k + j$ pieces of luggage checked in either by this family or by the Sanchez family is equally likely to be the last of these $k + j$ to appear, the probability that this family departs after the Sanchez family is $\frac{k}{k+j}$. Because, the number of non-Sanchez families who checked in k pieces of luggage is n_k when $k \neq j$, or $n_j - 1$ when $k = j$, we obtain that

$$E[X] = \sum_k \frac{kn_k}{k+j} - \frac{1}{2}$$

9. Let the neighborhood of any point on the rim be the arc starting at that point and extending for a length 1. Now consider a uniformly chosen point on the rim of the circle—that is, the probability that it lies on a specified arc of length x is $\frac{x}{2\pi}$—and let X denote the number of points that lie in its neighborhood. With I_j defined to equal 1 if item number j is in the neighborhood of the random point, and to equal 0 otherwise,

$$X = \sum_{j=1}^{19} I_j$$

Taking expectations gives

$$E[X] = \sum_{j=1}^{19} P\{\text{item } j \text{ lies in the neighborhood of the random point}\}$$

But because item j will lie in its neighborhood if the random point is located on the arc of length 1 going from item j in the counterclockwise direction, it follows that

$$P\{\text{item } j \text{ lies in the neighborhood of the random point}\} = \frac{1}{2\pi}$$

Hence,

$$E[X] = \frac{19}{2\pi} > 3$$

Because $E[X] > 3$ at least one of the possible values of X must exceed 3, proving the result.

10. If $g(x) = x^{1/2}$, then

$$g'(x) = \frac{1}{2}x^{-1/2}, \quad g''(x) = -\frac{1}{4}x^{-3/2}$$

and so the Taylor series expansion of $\sqrt{x}$ about λ gives

$$\sqrt{X} \approx \sqrt{\lambda} + \frac{1}{2}\lambda^{-1/2}(X - \lambda) - \frac{1}{8}\lambda^{-3/2}(X - \lambda)^2$$

Taking expectations yields

$$\begin{aligned}
E\left[\sqrt{X}\right] &\approx \sqrt{\lambda} + \frac{1}{2}\lambda^{-1/2}E[X - \lambda] - \frac{1}{8}\lambda^{-3/2}E[(X - \lambda)^2] \\
&= \sqrt{\lambda} - \frac{1}{8}\lambda^{-3/2}\lambda \\
&= \sqrt{\lambda} - \frac{1}{8}\lambda^{-1/2}
\end{aligned}$$

Hence,

$$\begin{aligned}
\text{Var}\left(\sqrt{X}\right) &= E[X] - \left(E\left[\sqrt{X}\right]\right)^2 \\
&\approx \lambda - \left(\sqrt{\lambda} - \frac{1}{8}\lambda^{-1/2}\right)^2 \\
&= 1/4 - \frac{1}{64\lambda} \\
&\approx 1/4
\end{aligned}$$

11. Number the tables so that tables 1, 2, 3 are the ones with four and tables 4, 5, 6, 7 are the ones with two seats. Also, number the women and let $X_{i,j}$ equal 1 if woman i is seated with her husband at table j. Note that

$$E[X_{i,j}] = \frac{\binom{2}{2}\binom{18}{2}}{\binom{20}{4}} = \frac{3}{95}, \qquad j = 1, 2, 3$$

and

$$E[X_{i,j}] = \frac{1}{\binom{20}{2}} = \frac{1}{190}, \qquad j = 4, 5, 6, 7$$

If X denotes the number of married couples that are seated at the same table, we thus have

$$\begin{aligned}
E[X] &= E\left[\sum_{i=1}^{5}\sum_{j=1}^{7} X_{i,j}\right] \\
&= \sum_{i=1}^{5}\sum_{j=1}^{3} E[X_{i,j}] + \sum_{i=1}^{5}\sum_{j=4}^{7} E[X_{i,j}] \\
&= 15(3/95) + 20(1/190) = 11/19
\end{aligned}$$

12. Let X_i equal 1 if individual i does not recruit anyone and let it equal 0 otherwise. Then

$$\begin{aligned}
E[X_i] &= P\{i \text{ does not recruit any of } i + 1, i + 2, \ldots, n\} \\
&= \frac{i - 1}{i}\,\frac{i}{i + 1}\cdots\frac{n - 2}{n - 1} \\
&= \frac{i - 1}{n - 1}
\end{aligned}$$

Hence,

$$E\left[\sum_{i=1}^{n} X_i\right] = \sum_{i=1}^{n} \frac{i-1}{n-1} = \frac{n}{2}$$

From the preceding we also obtain that

$$\text{Var}(X_i) = \frac{i-1}{n-1}\left(1 - \frac{i-1}{n-1}\right) = \frac{(i-1)(n-i)}{(n-1)^2}$$

Now, for $i < j$,

$$E[X_iX_j] = \frac{i-1}{i} \cdots \frac{j-2}{j-1}\frac{j-2}{j}\frac{j-1}{j+1} \cdots \frac{n-3}{n-1}$$
$$= \frac{(i-1)(j-2)}{(n-2)(n-1)}$$

and, thus,

$$\text{Cov}(X_i, X_j) = \frac{(i-1)(j-2)}{(n-2)(n-1)} - \frac{i-1}{n-1}\frac{j-1}{n-1}$$
$$= \frac{(i-1)(j-n)}{(n-2)(n-1)^2}$$

Therefore,

$$\text{Var}\left(\sum_{i=1}^{n} X_i\right) = \sum_{i=1}^{n} \text{Var}(X_i) + 2\sum_{i=1}^{n-1}\sum_{j=i+1}^{n} \text{Cov}(X_i, X_j)$$
$$= \sum_{i=1}^{n} \frac{(i-1)(n-i)}{(n-1)^2} + 2\sum_{i=1}^{n-1}\sum_{j=i+1}^{n} \frac{(i-1)(j-n)}{(n-2)(n-1)^2}$$
$$= \frac{1}{(n-1)^2}\sum_{i=1}^{n}(i-1)(n-i)$$
$$- \frac{1}{(n-2)(n-1)^2}\sum_{i=1}^{n-1}(i-1)(n-i)(n-i-1)$$

13. Let X_i equal 1 if the ith triple consists of one of each type of player. Then

$$E[X_i] = \frac{\binom{2}{1}\binom{3}{1}\binom{4}{1}}{\binom{9}{3}} = \frac{2}{7}$$

Hence, for part (a) we obtain that

$$E\left[\sum_{i=1}^{3} X_i\right] = 6/7$$

It follows from the preceding that

$$\text{Var}(X_i) = (2/7)(1 - 2/7) = 10/49$$

Also, for $i \neq j$,

$$\begin{aligned} E[X_i X_j] &= P\{X_i = 1, X_j = 1\} \\ &= P\{X_i = 1\}P\{X_j = 1 | X_i = 1\} \\ &= \frac{\binom{2}{1}\binom{3}{1}\binom{4}{1}}{\binom{9}{3}} \frac{\binom{1}{1}\binom{2}{1}\binom{3}{1}}{\binom{6}{3}} \\ &= 6/70 \end{aligned}$$

Hence, for part (b) we obtain

$$\begin{aligned} \text{Var}\left(\sum_{i=1}^{3} X_i\right) &= \sum_{i=1}^{3} \text{Var}(X_i) + 2\sum \sum_{j>1} \text{Cov}(X_i, X_j) \\ &= 30/49 + 2\binom{3}{2}\left(\frac{6}{70} - \frac{4}{49}\right) \\ &= \frac{312}{490} \end{aligned}$$

14. Let X_i equal 1 if the ith card is an ace and let it be 0 otherwise, and let Y_j equal 1 if the jth card is a spade and let it be 0 otherwise, $i, j = 1, \ldots, 13$. Now,

$$\begin{aligned} \text{Cov}(X, Y) &= \text{Cov}\left(\sum_{i=1}^{n} X_i, \sum_{j=1}^{n} Y_j\right) \\ &= \sum_{i=1}^{n} \sum_{j=1}^{n} \text{Cov}(X_i, Y_j) \end{aligned}$$

However, X_i is clearly independent of Y_j because knowing the suit of a particular card clearly gives no information about whether it is an ace and thus cannot affect the probability that another specified card is an ace. More formally, let $A_{i,s}, A_{i,h}, A_{i,d}, A_{i,c}$ be the events, respectively, that card i is a spade, a heart, a diamond, or a club. Then

$$\begin{aligned} P\{Y_j = 1\} = \frac{1}{4}(&P\{Y_j = 1|A_{i,s}\} + P\{Y_j = 1|A_{i,h}\} \\ &+ P\{Y_j = 1|A_{i,d}\} + P\{Y_j = 1|A_{i,c}\}) \end{aligned}$$

But, by symmetry, we have that

$$P\{Y_j = 1|A_{i,s}\} = P\{Y_j = 1|A_{i,h}\} = P\{Y_j = 1|A_{i,d}\} = P\{Y_j = 1|A_{i,c}\}$$

Therefore,

$$P\{Y_j = 1\} = P\{Y_j = 1|A_{i,s}\}$$

As the preceding implies that

$$P\{Y_j = 1\} = P\{Y_j = 1|A_{i,s}^c\}$$

we see that Y_j and X_i are independent. Hence, $\text{Cov}(X_i, Y_j) = 0$, and thus $\text{Cov}(X, Y) = 0$.

The random variables X and Y, although uncorrelated, are not independent. This follows, for instance, from the fact that

$$P\{Y = 13|X = 4\} = 0 \neq P\{Y = 13\}$$

15. **(a)** Your expected gain without any information is 0.
(b) You should predict heads if $p > 1/2$ and tails otherwise.
(c) Conditioning on V, the value of the coin, gives

$$\begin{aligned} E[\text{Gain}] &= \int_0^1 E[\text{Gain}|V = p]\,dp \\ &= \int_0^{1/2} [1(1-p) - 1(p)]\,dp + \int_{1/2}^1 [1(p) - 1(1-p)]\,dp \\ &= 1/2 \end{aligned}$$

16. Given that the name chosen appears in $n(X)$ different positions on the list it follows, since each of these positions is equally likely to be the one chosen, that

$$E[I|n(X)] = P\{I = 1|n(X)\} = 1/n(X)$$

Hence,

$$E[I] = E[1/n(X)]$$

and thus, $E[mI] = E[m/n(X)] = d$.

17. Letting X_i equal 1 if a collision occurs when the ith item is placed, and letting it equal 0 otherwise, the total number of collisions X can be expressed as

$$X = \sum_{i=1}^{m} X_i$$

Therefore,

$$E[X] = \sum_{i=1}^{m} E[X_i]$$

To determine $E[X_i]$, condition on the cell in which it is placed.

$$\begin{aligned} E[X_i] &= \sum_j E[X_i|\text{ placed in cell } j]p_j \\ &= \sum_j P\{i \text{ causes collision} \mid \text{placed in cell } j]p_j \\ &= \sum_j [1 - (1 - p_j)^{i-1}]p_j \\ &= 1 - \sum_j (1 - p_j)^{i-1}p_j \end{aligned}$$

The next to last equality used that, conditional on item i being placed in cell j, item i will cause a collision if any of the preceding $i - 1$ items were put in cell j. Thus,

$$E[X] = m - \sum_{i=1}^{m}\sum_{j=1}^{n} (1 - p_j)^{i-1}p_j$$

Interchanging the order of the summations gives

$$E[X] = m - n + \sum_{j=1}^{n} (1 - p_j)^m$$

Looking at the result shows that the preceding could have been more easily derived by taking expectations of both sides of the identity:

$$\text{number of nonempty cells} = m - X$$

The expected number of nonempty cells is then found by defining an indicator variable for each cell, equal to 1 if that cell is nonempty and to 0 otherwise and then taking the expecation of the sum of these indicator variables.

18. Let L denote the length of the initial run. Conditioning on the first value gives

$$E[L] = E[L|\text{first value is one}]\frac{n}{n+m} + E[L|\text{first value is zero}]\frac{m}{n+m}$$

Now, if the first value is one, then the length of the run will be the position of the first zero when considering the remaining $n + m - 1$ values, of which $n - 1$ are ones and m are zeroes. (For instance, if the initial value of the remaining $n + m - 1$ is zero, then $L = 1$.) As a similar result is true given that the first value is a zero, we obtain from the preceding, upon using the result from Example 2m, that

$$\begin{aligned} E[L] &= \frac{n+m}{m+1}\frac{n}{n+m} + \frac{n+m}{n+1}\frac{m}{n+m} \\ &= \frac{n}{m+1} + \frac{m}{n+1} \end{aligned}$$

19. Let X be the number of flips needed for both boxes to become empty, and let Y denote the number of heads in the first $n + m$ flips. Then

$$\begin{aligned} E[X] &= \sum_{i=0}^{n+m} E[X|Y = i]P\{Y = i\} \\ &= \sum_{i=0}^{n+m} E[X|Y = i]\binom{n+m}{i}p^i(1-p)^{n+m-i} \end{aligned}$$

Now, if the number of heads in the first $n + m$ flips is $i, i \le n$, then the number of additional flips is the number of flips needed to obtain an additional $n - i$ heads. Similarly, if the number of heads in the first $n + m$ flips is $i, i > n$, then, because there would have been a total of $n + m - i < m$ tails, the number of additional flips is the number needed to obtain an additional $i - n$ heads. Because the number of flips needed for j outcomes of a particular type is a negative binomial random variable whose mean is j divided by the probability of that outcome, we obtain that

$$E[X] = \sum_{i=0}^{n} \frac{n-i}{p}\binom{n+m}{i}p^i(1-p)^{n+m-i} + \sum_{i=n+1}^{n+m} \frac{i-n}{1-p}\binom{n+m}{i}p^i(1-p)^{n+m-i}$$

20. Take expectations of both sides of the identity given in the hint to obtain

$$\begin{aligned} E[X^n] &= E\left[n\int_0^\infty x^{n-1}I_X(x)\,dx\right] \\ &= n\int_0^\infty E[x^{n-1}I_X(x)]\,dx \\ &= n\int_0^\infty x^{n-1}E[I_X(x)]\,dx \\ &= n\int_0^\infty x^{n-1}\bar{F}(x)\,dx \end{aligned}$$

Taking the expectation inside the integral sign is justified because all the random variables $I_X(x), 0 < x < \infty$, are nonnegative.

CHAPTER 8

1. Let X denote the number of sales made next week, and note that X is integral. From Markov's inequality we obtain the following.

(a) $P\{X > 18\} = P\{X \geq 19\} \leq \dfrac{E[X]}{19} = 16/19$

(b) $P\{X > 25\} = P\{X \geq 26\} \leq \dfrac{E[X]}{26} = 16/26$

2. (a)
$$\begin{aligned} P\{10 \leq X \leq 22\} &= P\{|X - 16| \leq 6\} \\ &= P\{|X - \mu| \leq 6\} \\ &= 1 - P\{|X - \mu| > 6\} \\ &\geq 1 - 9/36 = 3/4 \end{aligned}$$

(b)
$$P\{X \geq 19\} = P\{X - 16 \geq 3\} \leq \frac{9}{9 + 9} = 1/2$$

In part (a) we used Chebyshev's inequality and in part (b) its one sided version (see Proposition 5.1).

3. First note that $E[X - Y] = 0$, and

$$\text{Var}(X - Y) = \text{Var}(X) + \text{Var}(Y) - 2\,\text{Cov}(X, Y) = 28$$

Using Chebyshev's inequality in part (a) and the one-sided version in parts (b) and (c) gives the following results:

(a)
$$P\{|X - Y| > 15\} \leq 28/225$$

(b)
$$P\{X - Y > 15\} \leq \frac{28}{28 + 225} = 28/253$$

(c)
$$P\{Y - X > 15\} \leq \frac{28}{28 + 225} = 28/253$$

4. If X is the number produced at factory A and Y the number produced at factory B, then

$$E[Y - X] = -2, \qquad \text{Var}(Y - X) = 36 + 9 = 45$$

$$P\{Y - X > 0\} = P\{Y - X \geq 1\} = P\{Y - X + 2 \geq 3\} \leq \frac{45}{45 + 9} = 45/54$$

5. Note first that

$$E[X_i] = \int_0^1 2x^2\,dx = 2/3$$

Now use the strong law of large numbers to obtain

$$\begin{aligned} r &= \lim_{n\to\infty} \frac{n}{S_n} \\ &= \lim_{n\to\infty} \frac{1}{S_n/n} \\ &= \frac{1}{\lim_{n\to\infty} S_n/n} \\ &= 1/(2/3) = 3/2 \end{aligned}$$

6. Since $E[X_i] = 2/3$, and

$$E[X_i^2] = \int_0^1 2x^3\,dx = 1/2$$

we have that $\text{Var}(X_i) = 1/2 - (2/3)^2 = 1/18$. Thus, if there are n components on hand, then

$$\begin{aligned} P\{S_n \geq 35\} &= P\{S_n \geq 34.5\} \quad \text{(the continuity correction)} \\ &= P\left\{\frac{S_n - 2n/3}{\sqrt{n/18}} \geq \frac{34.5 - 2n/3}{\sqrt{n/18}}\right\} \\ &\approx P\left\{Z \geq \frac{34.5 - 2n/3}{\sqrt{n/18}}\right\} \end{aligned}$$

where Z is a standard normal random variable. Since

$$P\{Z > -1.284\} = P\{Z < 1.284\} \approx .90$$

we see that n should be chosen so that

$$(34.5 - 2n/3) \approx -1.284\sqrt{n/18}$$

A numerical computation gives the result $n = 55$.

7. If X is the time to service a machine, then

$$E[X] = .2 + .3 = .5$$

Also, using that the variance of an exponential random variable is equal to the square of its mean gives

$$\text{Var}(X) = (.2)^2 + (.3)^2 = .13$$

Therefore, with X_i being the time to service job i, $i = 1, \ldots, 20$, and Z being a standard normal random variable,

$$\begin{aligned} P\{X_1 + \cdots + X_{20} < 8\} &= P\left\{\frac{X_1 + \cdots + X_{20} - 10}{\sqrt{2.6}} < \frac{8 - 10}{\sqrt{2.6}}\right\} \\ &\approx P\{Z < -1.24035\} \\ &\approx .1074 \end{aligned}$$

8. Note first that if X is the gambler's winnings on a single bet, then

$$E[X] = -.7 - .4 + 1 = -.1,\ E[X^2] = .7 + .8 + 10 = 11.5$$
$$\rightarrow \text{Var}(X) = 11.49$$

Therefore, with Z having a standard normal distribution,

$$\begin{aligned} P\{X_1 + \cdots + X_{100} \leq -.5\} &= P\left\{\frac{X_1 + \cdots + X_{100} + 10}{\sqrt{1149}} \leq \frac{-.5 + 10}{\sqrt{1149}}\right\} \\ &\approx P\{Z \leq .2803\} \\ &\approx .6104 \end{aligned}$$

9. Using the notation of Problem 7 we have

$$P\{X_1 + \cdots + X_{20} < t\} = P\left\{\frac{X_1 + \cdots + X_{20} - 10}{\sqrt{2.6}} < \frac{t - 10}{\sqrt{2.6}}\right\}$$

$$\approx P\left\{Z < \frac{t - 10}{\sqrt{2.6}}\right\}$$

Now, $P\{Z < 1.645\} \approx .95$, and so t should be such that

$$\frac{t - 10}{\sqrt{2.6}} \approx 1.645$$

which yields that $t \approx 12.65$.

10. If the claim were true then, by the central limit theorem, the average nicotine content, call it X, would approximately have a normal distribution with mean 2.2 and standard deviation .03. The probability it would thus be as high as 3.1 is

$$P\{X > 3.1\} = P\left\{\frac{X - 2.2}{\sqrt{.03}} > \frac{3.1 - 2.2}{\sqrt{.03}}\right\}$$

$$\approx P\{Z > 5.196\}$$

$$\approx 0$$

where Z is a standard normal random variable.

11. Let N denote the number of doctors that volunteer. Conditional on the event $N = i$, the number of patients seen is distributed as the sum of i independent Poisson random variables with common mean 30. Because the sum of independent Poisson random variables is also a Poisson random variable, it follows that the conditional distribution of X given that $N = i$ is Poisson with mean $30i$. Therefore,

$$E[X|N] = 30N \qquad \text{Var}(X|N) = 30N$$

As a result,

$$E[X] = E[E[X|N]] = 30E[N] = 90$$

Also, by the conditional variance formula,

$$\text{Var}(X) = E[\text{Var}(X|N)] + \text{Var}(E[X|N]) = 30E[N] + (30)^2\,\text{Var}(N)$$

Because

$$\text{Var}(N) = \frac{1}{3}(2^2 + 3^2 + 4^2) - 9 = 2/3$$

we obtain that $\text{Var}(X) = 690$.

To approximate $P\{X > 65\}$, we would not be justified in assuming that the distribution of X is approximately that of a normal random variable with mean 90 and variance 690. What we do know, however, is that

$$P\{X > 65\} = \sum_{i=2}^{4} P\{X > 65|N = i\}P\{N = i\} = \frac{1}{3}\sum_{i=2}^{4} \bar{P}_i(65)$$

where $\bar{P}_i(65)$ is the probability that a Poisson random variable with mean $30i$ is greater than 65. That is,

$$\bar{P}_i(65) = 1 - \sum_{j=0}^{65} e^{-30i}(30i)^j/j!$$

Because a Poisson random variable with mean $30i$ has the same distribution as does the sum of $30i$ independent Poisson random variables with mean 1, it follows from the central limit theorem that its distribution is approximately normal with mean and variance equal to $30i$. Consequently, with X_i being a Poisson random variable with mean $30i$, and Z being a standard normal random variable, we can approximate $\bar{P}_i(65)$ as follows:

$$\begin{aligned}\bar{P}_i(65) &= P\{X > 65\}\\ &= P\{X \geq 65.5\}\\ &= P\left\{\frac{X - 30i}{\sqrt{30i}} \geq \frac{65.5 - 30i}{\sqrt{30i}}\right\}\\ &\approx P\left\{Z \geq \frac{65.5 - 30i}{\sqrt{30i}}\right\}\end{aligned}$$

Therefore,

$$\begin{aligned}\bar{P}_2(65) &\approx P\{Z \geq .7100\} \approx .2389\\ \bar{P}_3(65) &\approx P\{Z \geq -2.583\} \approx .9951\\ \bar{P}_4(65) &\approx P\{Z \geq -4.975\} \approx 1\end{aligned}$$

leading to the result that

$$P\{X > 65\} \approx .7447$$

If we would have mistakenly assumed that X was approximately normal, we would have obtained the approximate answer .8244. (The exact probability is .7440.)

CHAPTER 9

1. From axiom (iii) it follows that the number of events that occur between times 8 and 10 has the same distribution as the number of events that occur by time 2, and thus is a Poisson random variable with mean 6. Hence, we obtain the following solutions for parts (a) and (b).

(a) $P\{N(10) - N(8) = 0\} = e^{-6}$ **(b)** $E[N(10) - N(8)] = 6$

(c) It follows from axioms (ii) and (iii) that from any time point onward the process of events occurring is a Poisson process with rate λ. Hence, the expected time of the fifth event after 2 P.M. is $2 + E[S_5] = 2 + 5/3$. That is, the expected time of this event is 3:40 P.M.

2. (a)
$$\begin{aligned}&P\{N(1/3) = 2|N(1) = 2\}\\ &= \frac{P\{N(1/3) = 2, N(1) = 2\}}{P\{N(1) = 2\}}\\ &= \frac{P\{N(1/3) = 2, N(1) - N(1/3) = 0\}}{P\{N(1) = 2\}}\\ &= \frac{P\{N(1/3) = 2\}P\{N(1) - N(1/3) = 0\}}{P\{N(1) = 2\}} \quad \text{(by axiom}(ii))\\ &= \frac{P\{N(1/3) = 2\}P\{N(2/3) = 0\}}{P\{N(1) = 2\}} \quad \text{(by axiom}(iii))\\ &= \frac{e^{-\lambda/3}(\lambda/3)^2/2!\, e^{-2\lambda/3}}{e^{-\lambda}\lambda^2/2!}\\ &= 1/9\end{aligned}$$

(b)
$$\begin{aligned}
P\{N(1/2) \geq 1 | N(1) = 2\} &= 1 - P\{N(1/2) = 0 | N(1) = 2\} \\
&= 1 - \frac{P\{N(1/2) = 0, N(1) = 2\}}{P\{N(1) = 2\}} \\
&= 1 - \frac{P\{N(1/2) = 0, N(1) - N(1/2) = 2\}}{P\{N(1) = 2\}} \\
&= 1 - \frac{P\{N(1/2) = 0\}P\{N(1) - N(1/2) = 2\}}{P\{N(1) = 2\}} \\
&= 1 - \frac{P\{N(1/2) = 0\}P\{N(1/2) = 2\}}{P\{N(1) = 2\}} \\
&= 1 - \frac{e^{-\lambda/2}e^{-\lambda/2}(\lambda/2)^2/2!}{e^{-\lambda}\lambda^2/2!} \\
&= 1 - 1/4 = 3/4
\end{aligned}$$

3. Fix a point on the road and let X_n equal 0 if the nth vehicle to pass is a car and let it equal 1 if it is a truck, $n \geq 1$. We now suppose that the sequence $X_n, n \geq 1$, is a Markov chain with transition probabilities

$$P_{0,0} = 5/6, \quad P_{0,1} = 1/6, \quad P_{1,0} = 4/5, \quad P_{1,1} = 1/5$$

Hence, the long run proportion of times is the solution of

$$\begin{aligned}
\pi_0 &= \pi_0(5/6) + \pi_1(4/5) \\
\pi_1 &= \pi_0(1/6) + \pi_1(1/5) \\
\pi_0 + \pi_1 &= 1
\end{aligned}$$

Solving the preceding equations gives that

$$\pi_0 = 24/29 \quad \pi_1 = 5/29$$

Thus, $2400/29 \approx 83$ percent of the vehicles on the road are cars.

4. The successive weather classifications constitute a Markov chain. If the states are 0 for rainy, 1 for sunny, and 2 for overcast, then the transition probability matrix is as follows.

$$\mathbf{P} = \begin{matrix} 0 & 1/2 & 1/2 \\ 1/3 & 1/3 & 1/3 \\ 1/3 & 1/3 & 1/3 \end{matrix}$$

The long run proportions satisfy

$$\begin{aligned}
\pi_0 &= \pi_1(1/3) + \pi_2(1/3) \\
\pi_1 &= \pi_0(1/2) + \pi_1(1/3) + \pi_2(1/3) \\
\pi_2 &= \pi_0(1/2) + \pi_1(1/3) + \pi_2(1/3) \\
1 &= \pi_0 + \pi_1 + \pi_2
\end{aligned}$$

The solution of the preceding is

$$\pi_0 = 1/4, \quad \pi_1 = 3/8, \quad \pi_2 = 3/8$$

Hence, three-eights of the days are sunny and one-fourth are rainy.

5. (a) A direct computation yields that

$$H(X)/H(Y) \approx 1.06$$

(b) Both random variables take on two of their values with the same probabilities .35 and .05. The difference is that if they do not take on either of those values then X, but not Y, is equally likely to take on any of its three remaining possible values. Hence, from Problem 13 we would expect the result of part (a).

CHAPTER 10

1. (a)

$$1 = C\int_0^1 e^x\,dx \Rightarrow C = 1/(e-1)$$

(b)

$$F(x) = C\int_0^x e^y\,dy = \frac{e^x - 1}{e - 1}, \qquad 0 \le x \le 1$$

Hence, if we let $X = F^{-1}(U)$, then

$$U = \frac{e^X - 1}{e - 1}$$

or

$$X = \log(U(e-1) + 1)$$

Thus, we can simulate the random variable X by generating a random number U and then setting $X = \log(U(e-1) + 1)$.

2. Use the acceptance-rejection method with $g(x) = 1, 0 < x < 1$. Calculus shows that the maximum value of $f(x)/g(x)$ occurs at a value of $x, 0 < x < 1$, such that

$$2x - 6x^2 + 4x^3 = 0$$

or, equivalently, when

$$4x^2 - 6x + 2 = (4x-2)(x-1) = 0$$

The maximum thus occurs when $x = 1/2$, and

$$C = \max f(x)/g(x) = 30(1/4 - 2/8 + 1/16) = 15/8$$

Hence, the algorithm is as follows:

Step 1. Generate a random number U_1
Step 2. Generate a random number U_2
Step 3. If $U_2 \le 16(U_1^2 - 2U_1^3 + U_1^4)$ set $X = U_1$; else return to Step 1

3. It is most efficient to check the higher probability values first, as in the following algorithm.

Step 1. Generate a random number U
Step 2. If $U \le .35$, set $X = 3$ and stop
Step 3. If $U \le .65$, set $X = 4$ and stop
Step 4. If $U \le .85$, set $X = 2$ and stop
Step 5. $X = 1$

4. $2\mu - X$

5. (a) Generate $2n$ independent exponential random variables with mean 1, X_i, Y_i, $i = 1, \ldots, n$, and then use the estimator $\sum_{i=1}^{n} e^{X_i Y_i}/n$.

(b) We can use XY as a control variate to obtain an estimator of the type

$$\sum_{i=1}^{n} (e^{X_i Y_i} + cX_i Y_i)/n$$

Another possibility would be to use $XY + X^2Y^2/2$ as the control variate and so obtain an estimator of the type

$$\sum_{i=1}^{n} (e^{X_i Y_i} + c[X_i Y_i + X_i^2 Y_i^2/2 - 1/2])/n$$

The motivation behind the preceding is based on the fact that the first three terms of the MacLaurin series expansion of e^{xy} are $1 + xy + (x^2y^2)/2$.

Index